# A LIFE SCIENTIST'S GUIDE TO PHYSICAL CHEMISTRY

Motivating students to engage with physical chemistry through biological examples, this textbook demonstrates how the tools of physical chemistry can be used to illuminate biological questions. It clearly explains key principles and their relevance to life science students, using only the most straightforward and relevant mathematical tools.

More than 350 exercises are spread throughout the chapters, covering a wide range of biological applications and explaining issues that students often find challenging. These, along with problems at the end of each chapter and end-of-term review questions, encourage active and continuous study. Over 130 worked examples, many deriving directly from life sciences, help students connect principles and theories to their own laboratory studies. Connections between experimental measurements and key theoretical quantities are frequently highlighted and reinforced.

Answers to the exercises are included in the book. Fully worked solutions and answers to the review problems, password-protected for instructors, are available at www.cambridge.org/roussel.

MARC R. ROUSSEL is Professor of Chemistry and Biochemistry at the University of Lethbridge, Canada. His research on the dynamics of biological systems lies at the interface of chemistry, biology and mathematics. He has been teaching physical chemistry for more than 15 years.

# A LIFE SCIENTIST'S GUIDE TO PHYSICAL CHEMISTRY

MARC R. ROUSSEL

*Department of Chemistry and Biochemistry*
*University of Lethbridge*
*Canada*

CAMBRIDGE UNIVERSITY PRESS
Cambridge, New York, Melbourne, Madrid, Cape Town,
Singapore, São Paulo, Delhi, Mexico City

Cambridge University Press
The Edinburgh Building, Cambridge CB2 8RU, UK

Published in the United States of America by Cambridge University Press, New York

www.cambridge.org
Information on this title: www.cambridge.org/9781107006782

First published 2012

Printed in the United Kingdom at the University Press, Cambridge

*A catalog record for this publication is available from the British Library*

*Library of Congress Cataloging in Publication data*
Roussel, Marc R., 1966–
A life scientist's guide to physical chemistry / Marc R. Roussel.
pages   cm.
Includes index.
ISBN 978-1-107-00678-2 (hardback) – ISBN 978-0-521-18696-4 (pbk.)
1. Chemistry, Physical and theoretical – Problems, exercises, etc.
2. Chemistry, Physical and theoretical – Mathematical models.
3. Chemistry, Physical and theoretical – Textbooks.   4. Life sciences – Problems, exercises, etc.
5. Life sciences – Mathematical models.   6. Life sciences – Textbooks.   I. Title.
QD456.R68   2012
541 – dc23      2011049198

ISBN 978-1-107-00678-2 Hardback
ISBN 978-0-521-18696-4 Paperback

Additional resources for this publication at www.cambridge.org/9781107006782

*Dedicated to all the students*
*who have studied and will study*
*physical chemistry*
*in my classes at*
*the University of Lethbridge*

# Contents

# Preface

## To the student

I wrote this book for you.

When I came to the University of Lethbridge in 1995, I started teaching physical chemistry to a mixed class of chemistry and biochemistry students. I have been teaching versions of this course ever since. My first year here, I picked a book that *I* liked. Boy, was that a mistake! First of all, the book contained almost no examples that appealed to the biochemists, who were the majority of the students in the class. Second, it was filled with mathematical derivations, which I found very satisfying, but which sometimes obscured the concepts of physical chemistry for the students.

Having made this mistake, I started looking around for other textbooks. The ones that I liked the best, Barrow's *Physical Chemistry for the Life Sciences* and Morris's *A Biologist's Physical Chemistry*, were out of print at the time. We used a few different textbooks over the years, and some of them were very good books, but my students were never completely happy, and so I wasn't happy. In some cases, the books contained too much math and not enough insight. In others, too many equations were presented without derivation or explanation, which undermined the students' understanding of the material. I therefore set about writing a book for my students, and therefore for the broader community of life science students who need a term of physical chemistry. Given my experience with other textbooks, I had a few criteria in mind:

- I wanted it to be a book that students could read and understand, and not just one they would open up when the professor told them to solve problems 1, 2 and 4 from Chapter 8. As a corollary to that, and adapting a phrase from Canadian history, I thought that we should use *calculus if necessary, but not necessarily calculus*. A lot of the theory of thermodynamics, in particular, is very elegantly phrased in terms of multi-variable calculus, but unless you have a couple of years of university-level calculus behind you, it's sometimes hard to appreciate what this beautiful theory is telling you. We do need some calculus in physical chemistry, but in a first course, we don't need nearly as much as you find in many physical chemistry books.

- I wanted to derive as many of the equations as possible, because I do think there is value in knowing where an equation comes from, even if you can't necessarily reproduce every step of the derivation yourself.
- I wanted the book to be relevant to life science students since I had so many of them in my class. Now I'm one of those people who think that scientists should be broadly interested in science, so you will find lots of examples and problems in the book that have nothing to do with the life sciences. Some of them, particularly those that come from other areas of the chemical sciences, were chosen because they provide simple illustrations of key concepts. Ultimately, it's all about learning the concepts so that you can apply them to your studies of living systems, and when it's easier to learn a concept using an uncomplicated example from, say, organic chemistry, then that's what you're going to get. Others were chosen because I think they're cool, and judging from the classroom discussions we've had over the years, so do my students. But you will also find many, many examples and problems directly inspired by the life sciences.
- I sometimes tell my students that an education in science involves learning a progressively more sophisticated set of lies, until finally the lies are so good that you can't tell them apart from reality. In lower-level courses, we tell students a lot of lies. I usually try to tell my students when I'm lying to them, and I've tried to do the same in this book, on the basis that it's important to know how far you can trust a certain equation or theory.
- Physical chemistry is an experimental science, so I wanted that to be reflected in the book. I try whenever possible to talk about experimental methods. I use as much real data as possible in the examples and problems. I also spend a lot more time and space than is customary discussing the analysis of data, particularly in the kinetics chapters. The ability to critically look at data is, I hope, something you will be able to use beyond this course.
- I think that students should have lots of opportunities to test their knowledge. I also think that these opportunities have to be at least a little structured. I've done a couple of things to help you with this:
  (1) The exercises are not concentrated at the ends of the chapters. Rather, I have exercises spread throughout the book, in short exercise groups, as well as some end-of-chapter problems. I'll come back to that in a minute.
  (2) I give answers to *every single problem* at the back of the book. You shouldn't need to wonder if you solved the problem correctly or not.

Drafts of this book have been used here at the University of Lethbridge for some years, so it's been thoroughly field-tested on students just like you. Hopefully, you will find, as they have, that while physical chemistry is a difficult subject, it is one you *can* master with a little effort. You may even discover that you enjoy it.

### Studying physical chemistry

Most of the students who get in trouble in physical chemistry just don't keep up with the material. It's not a subject where cramming is an effective study strategy, so you really need to set some time aside every week to read the book and to solve some problems. Reading a physical chemistry book is something you do with a pad of paper, a pencil and a calculator, because you usually need to work out some of the steps in the derivations or in the examples in order to make sure you understand them properly. Beyond that, I have placed exercise groups after most sections. The idea is that you can verify your learning *right away* with a handful of questions that require the material in the section you have just read. You should be doing those problems as you go, i.e. you should sit down to solve some problems at least once a week, and maybe more often if you can manage it. Is this a lot of work? You bet! But it will make it easier for you to study for the tests, and it should improve your grade. And, let's face it, it's a lot more fun to be successful in a course than not.

When test time comes around, have a look at the ends of the chapters. I've put a selection of problems there, too, so that you can test your understanding of a larger slice of the material.

Finally, at the end of term, when you're preparing for the final exam, I've created a long set of questions to help you review all the material in the book. You may not cover every page of the book in your course, so you may see some problems there that aren't relevant to you. Hopefully by this point you'll be able to recognize these, or your instructor may help you pick through this selection of problems.

You should, I hope, have all the resources at hand that you need to study physical chemistry. Good luck!

**To the instructor**

This book was designed for a single-term course in physical chemistry for life science students (although I use it in a mixed class of biochemists and chemists). I assume that students have acquired the standard background of first-year science courses: two terms of general chemistry, one term of calculus and one term of physics. Particularly when it comes to mathematics, I try to help the students along when we draw on this background, but you may need to fill in some of the blanks if your students come to this course without some of these prerequisites.

Because this is a book intended to be used in a single-term course, some choices had to be made in terms of the topics covered. Quite apart from the time constraints, I also wanted to keep the production costs down, which implies some restraint in terms of topic coverage. This book focuses particularly on thermodynamics and kinetics, because these are the areas of most direct relevance to life science students. I have included a little bit of quantum mechanics and spectroscopy to support the other two topics, although the book was written so that you can bypass this material. I would, however, suggest that you include Section 3.2 on the Boltzmann distribution if you can. This was a late addition to the book, but after I wrote it, I found that I could explain many topics in the rest of the book much more straightforwardly.

Despite my overall attempt to limit the size of the book, I have tried to maintain some flexibility in the sequences you can take through the book. Accordingly, there is a bit more material in this book than you could cover in one term. In my own class, we cover almost all of Parts Two and Three (thermodynamics and kinetics) in a standard three-credit-hour (3 hours per week times 13 weeks) course.

A little while ago, I decided to change all the notations in this book to those recommended in the IUPAC Green Book (*Quantities, Units and Symbols in Physical Chemistry*). I came to this decision after a long period of using a combination of notations commonly used in other textbooks and a few I had invented myself. There are certainly advantages to using one's own notations, but eventually I decided that the proliferation of non-standard notations by textbook authors was doing more harm than good. I think we've all had the experience of reading books or scientific papers where we had to keep flipping back and forth to figure out what $x$ or $y$ represented. While some of the IUPAC notations are, in my opinion, not particularly elegant, they have the advantage of being standard for the field.

## Contacting me

If you have comments or suggestions about this book, feel free to contact me by email: roussel@uleth.ca. I would be happy to hear from both students and instructors. I would be particularly interested to hear from instructors what sections you included or excluded in your course, and whether there is any material that you really think I should add in a future edition, should I ever be asked to write one.

## Acknowledgments

I would like to thank all of the Chemistry 2710, 2720 and 2740 students who have helped me work through the various versions of this book. Your questions and comments have had a great influence both on this text and on my own thinking about some of the problems raised during the study of physical chemistry.

I also must thank my wife Catharine and son Liam for allowing me the time to complete this project. Your support and understanding mean everything to me, always.

# 1

# Orientation: what is physical chemistry about?

Chemistry is traditionally divided into a small number of subfields, namely organic, inorganic, analytical and physical chemistry. It's fairly easy to say what the first three are about, but it's much harder to define physical chemistry. The problem is that physical chemistry is all of the following simultaneously:

- A discipline in its own right, with its own set of problems and techniques;
- The source of the basic theory that underlies all of the chemical sciences;
- A provider of experimental methods used across the chemical sciences.

Note that "chemical sciences" includes biochemistry and materials science, among other fields that depend on physical chemistry for at least some of their theory and methods. Physical chemistry's large mandate means that it's difficult to put a finger on what it is exactly. It's a bit like chemistry itself that way: every time you come up with a definition, you immediately think of half a dozen things done under that heading that don't fit.

Rather than trying to give a simple, neat definition of physical chemistry, I'm going to tell you about the big theories that make up physical chemistry. Hopefully, this will give you an idea of what physical chemistry is about, even if we can't wrap it up in a neat package as we can with the other subfields of chemistry.

Most physical chemists would tell you that physical chemistry has three major subdivisions: quantum mechanics, thermodynamics and kinetics (Figure 1.1). **Quantum mechanics** is the study of the properties of matter at the atomic level. In quantum mechanics, we talk about the forces that hold atoms and molecules together, and about the interaction of matter with light (spectroscopy), among other things. **Thermodynamics** is the study of matter from the other extreme: in thermodynamics, we don't worry about the microscopic details, we just deal with matter as we normally perceive it in terms of variables like temperature, pressure and volume. Chemical thermodynamics concerns itself mainly with the energetics of reactions, which sometimes allows us to say something about which reactions are possible under given conditions. Finally, **kinetics** is the study of the rate of reactions. It turns out that thermodynamics doesn't tell us anything about how fast a reaction will occur, so we need a separate set of theories to treat this important issue.

Figure 1.1 also shows some of the connections between the three major theoretical pillars of physical chemistry. **Statistical thermodynamics** allows us to calculate thermodynamic

1

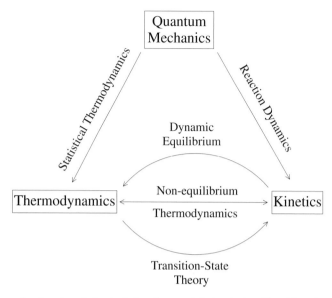

Figure 1.1 The major theories of physical chemistry and their relationships. Note that the figure only shows some of the connections between quantum mechanics, thermodynamics and kinetics.

properties from the quantum properties of matter. **Reaction dynamics** similarly lets us calculate rates of reaction from quantum mechanical principles. The classical theory of **dynamic equilibrium** connects kinetics to equilibrium and thus to a whole body of knowledge in thermodynamics. **Transition-state theory** is a theory of rates of reaction that rests on a foundation of thermodynamic reasoning. **Non-equilibrium thermodynamics** allows us to understand both the energetics and kinetics of reactions in a unified framework. These are just a few of the connections we could put into this diagram.

You will note the emphasis on theories. This is perhaps one of the defining characteristics of physical chemistry: physical chemists like to have a big theoretical umbrella that covers knowledge in the discipline. This is not to say that physical chemists aren't concerned with experiments. Most physical chemists are, in fact, experimentalists. In my experience though, almost all physical chemists ultimately want to connect their measurements to some deeper principles. This is certainly a common attitude among scientists, but perhaps a more intensely felt one among physical chemists than might be the case in other areas of chemistry.

Given the complexity implied by Figure 1.1, how can we proceed to learn physical chemistry? Fortunately, the major theories are coherent entities that can be studied one at a time. Because of the connections between the theories, a knowledge of one will enhance our appreciation of the others, but we can still study kinetics, for example, as a thing in itself. In this book, we will study all three of the major theories, as well as some of the bridges between them. The intention is to provide you with a core of chemical theory that can be applied to a wide variety of problems in the chemical sciences.

## A note on graph axis labels and table headings

We can only graph pure numbers. We could put numbers with units in a table, but to avoid repeating the units, we typically just put them in the table heading, leaving just numbers in the table itself. Throughout this text, you will see graph axis labels and table headings that look like "$\lambda/nm$." The logic behind this notation is as follows: $\lambda$ is a physical observable that has both a value and units. The pure number in the table or graph is what you get by dividing out the units of $\lambda$, in this case nm.

This way of labeling axes and tables may not seem like a huge improvement over just writing "$\lambda$ (nm)." The advantage appears when you have numbers that all share a common multiple of a power of 10 that you want to avoid writing down over and over again. For example, molar absorption coefficients are often a multiple of $10^5 \, L \, mol^{-1}cm^{-1}$. If I'm typing a table of these coefficients, I might not want to repeat '$\times 10^5$' for every entry. I would then label the table heading as "$\varepsilon/10^5 \, L \, mol^{-1}cm^{-1}$," meaning that the number in the table is what you get when you divide $\varepsilon$ by $10^5 \, L \, mol^{-1}cm^{-1}$. For example, if one of the numbers in the table is 1.02, then that means that $\varepsilon/10^5 \, L \, mol^{-1}cm^{-1} = 1.02$, or that $\varepsilon = 1.02 \times 10^5 \, L \, mol^{-1}cm^{-1}$. Once you get used to this way of writing table headings, you will find that it's much clearer than any of the alternatives you routinely run across.

# Part One

Quantum mechanics and spectroscopy

# 2

# A quick tour of quantum mechanical ideas

The objective of this chapter is to go over a few of the basic concepts of quantum mechanics in preparation for a discussion of spectroscopy, which is in many ways the business end of quantum mechanics, at least for chemical scientists. We will also need a few quantum mechanical ideas from time to time in our study of thermodynamics and of kinetics.

Why should we learn quantum mechanics at all? Atoms and molecules are small, and their constituent parts, electrons, protons and neutrons, are even smaller. Early in the twentieth century, we learned that small things don't obey the laws of classical mechanics. A different kind of mechanics, quantum mechanics, is required to understand chemistry on a fundamental level. In fact, we need different mechanical theories to treat extremes of both size and speed. Figure 2.1 summarizes the situation. There isn't a sharp cut-off between the various sectors of this diagram. Also note that some of the theories are more general than others. We could in principle use quantum mechanics or general relativity to predict the trajectories of tennis balls, but it just isn't worth the effort, given that classical mechanics works perfectly well in this range of masses and speeds. On the other hand, classical mechanics doesn't give very good results for things that are either extremely large, or small, or fast.

Like it or not, to discuss phenomena on an atomic scale, we need quantum mechanics. Ordinary (non-relativistic) quantum mechanics is generally adequate, although electrons in heavy atoms sometimes reach relativistic speeds (approaching the speed of light, $c$), requiring relativistic quantum mechanics. We can get away with using classical mechanics to treat large-scale motions of molecules (e.g. motions of domains of proteins). However, many molecular phenomena will remain mysterious to us if we don't arm ourselves with at least a little bit of quantum mechanical theory.

## 2.1 Light

From the seventeenth to the nineteenth century, there were two competing theories on the nature of light. Some evidence (diffraction, refraction etc.) suggested that light was a wave phenomenon. On the other hand, a particle theory was attractive to many workers due to the linear propagation of light rays. Although the wave theory of light was more

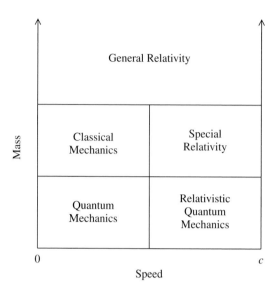

Figure 2.1 Sketch of the domains of validity of different mechanical theories (not to scale). The speed of light ($c$) sets an upper limit on the speeds that can be reached by material objects. Relativistic theories are required for objects whose speeds are close to the speed of light. Classical mechanics is appropriate to objects of moderate size moving at speeds well below $c$. Quantum mechanics is required to treat phenomena on an atomic scale.

broadly successful in this period, there was no clear resolution of the matter until the 1860s.

James Clerk Maxwell's contributions to physics are among the most important and beautiful of the nineteenth century. His crowning achievement was perhaps the unification of the laws of electricity and magnetism into a set of consistent equations which together describe all electrical and magnetic phenomena. As he studied these equations, he made a startling discovery: the equations suggested the possibility of electromagnetic waves. Furthermore, the wave speed, which could be computed from the equations, was extremely close to the best estimate then available of the speed of light. Very soon, everyone became convinced that a final explanation of the nature of light had been discovered: light is an electromagnetic wave, i.e. a traveling wave of oscillating electric and magnetic fields. Figure 2.2 shows a schematic drawing of an electromagnetic wave.

If light is a wave phenomenon, then it obeys the usual laws of wave dynamics. For instance, its frequency ($\nu$) and wavelength ($\lambda$) are related by

$$c = \lambda\nu, \tag{2.1}$$

where $c$ is the wave speed, in this case the speed of light. The SI unit of frequency is the hertz (Hz). One hertz is one cycle per second.

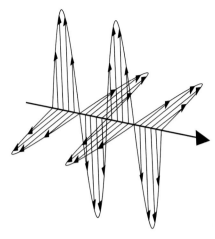

Figure 2.2 Schematic diagram of an electromagnetic wave. The wave is made up of oscillating electric and magnetic fields, represented here by vectors. The vertical vectors (say) represent the electric field at different points along the wave propagation axis, while the horizontal vectors represent the magnetic field. The direction of propagation of the wave is indicated by the large arrow. The wave amplitude is the height of the wave (measured in electric field units). The wavelength is the distance between two successive maxima. The frequency is the number of cycles of the wave observed at a fixed position in space divided by the observation time.

**Example 2.1 Wavelength and frequency** The shortest wavelength of light visible to us is approximately 400 nm. The corresponding frequency is

$$\nu = \frac{c}{\lambda} = \frac{2.997\,924\,58 \times 10^8\,\text{m s}^{-1}}{400 \times 10^{-9}\,\text{m}} = 7.49 \times 10^{14}\,\text{Hz}.$$

Instead of the wavelength or frequency, we sometimes use the **wavenumber** $\tilde{\nu}$ to describe light waves. The wavenumber is just the inverse of the wavelength, so $\tilde{\nu} = \lambda^{-1}$. If the wavelength is the length of one wave, the wavenumber is the number of waves per unit length. Wavenumbers are mostly used in spectroscopy, which we will study in the next chapter, and are usually given in reciprocal centimeters (cm$^{-1}$). This unit is so commonly used that spectroscopists often read values like 1000 cm$^{-1}$ as "one thousand wavenumbers," although this is a bad habit which should be discouraged.

**Example 2.2 Wavelength and wavenumbers** Wavenumbers are most commonly encountered in infrared (IR) spectroscopy. The infrared part of the electromagnetic spectrum ranges from about 750 nm to 1 mm. Let us convert this into a wavenumber range. Let's start with the lower end of the wavelength range. That wavelength, converted to cm, is

$$\frac{(750\,\text{nm})(10^{-9}\,\text{m nm}^{-1})}{10^{-2}\,\text{m cm}^{-1}} = 7.5 \times 10^{-5}\,\text{cm}.$$

This corresponds to a wavenumber of $(7.5 \times 10^{-5} \text{ cm})^{-1} \approx 13\,000 \text{ cm}^{-1}$. If we do the same calculation for the other end of the infrared range, we get $10 \text{ cm}^{-1}$, so the infrared ranges from 10 to $13\,000 \text{ cm}^{-1}$.

Maxwell's electromagnetic theory of light was thought for a few decades to answer all questions about the nature and behavior of light. However, as so often happens in science, an anomaly cropped up. The photoelectric effect, the ejection of electrons from a metal surface when irradiated with light of a sufficiently high frequency, resisted explanation by Maxwell's theory. In a nutshell, the problem was that the energy of a classical electromagnetic wave should be related to its amplitude. Cranking up the intensity should eventually provide enough energy for any desired process, including removing electrons from matter. The frequency shouldn't have anything to do with it.

It was Einstein who provided the resolution of this puzzle in 1905: he postulated that light is made up of particles he called photons. Each photon has an energy related to the frequency of the light by an equation originally proposed by Max Planck to explain blackbody radiation (wherein lies a whole other tale):

$$E = h\nu, \tag{2.2}$$

where $h$ is Planck's constant. This innocent-looking equation revolutionized physics; it links the energy of a *particle* to a *wave* property, the frequency $\nu$. Einstein had thus provided a completely original and unexpected solution to the old debate about the nature of light: light is *both* a particle *and* a wave. Light propagates in space like a wave, but in its interactions with matter, light behaves as if it were made of particles which are absorbed as individual units. This ability of light to behave either like a particle or like a wave, depending on the situation, is called **duality**.

This solves the puzzle of the photoelectric effect: assuming that only one photon is absorbed at a time (an idea known as the law of photochemical equivalence, to which we shall shortly return), then an individual photon either does or does not have enough energy to eject an electron from a metal surface. Since the energy of a photon is proportional to its frequency, it is easy to see that the frequency must be sufficiently high in order to cause a photoelectric effect. In the photon theory, increasing the intensity of a light beam only increases the number of photons delivered by the beam per unit time, and not the energies of the photons.

**Example 2.3 Photon energy** We calculated earlier that the highest frequency of visible light is approximately $7.49 \times 10^{14} \text{ Hz}$ (Example 2.1). The energy of a single photon with this frequency is

$$E = h\nu = (6.626\,068\,8 \times 10^{-34} \text{ J Hz}^{-1})(7.49 \times 10^{14} \text{ Hz}) = 4.96 \times 10^{-19} \text{ J}.$$

The energy of a mole of photons of this frequency is

$$\bar{E} = (4.96 \times 10^{-19} \text{ J})(6.022\,142\,0 \times 10^{23} \text{ mol}^{-1}) = 299 \text{ kJ mol}^{-1}.$$

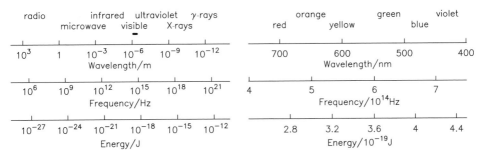

Figure 2.3 Electromagnetic spectrum. The full spectrum is shown on the left, plotted on a logarithmic scale. The visible part of the spectrum (marked by the heavy dash under the word visible) represents only a tiny fraction of the range of wavelengths commonly observed in the natural environment. On the right, we see a blowup of the visible part of the spectrum. The labels ($\gamma$-ray, X-ray etc.) only name an approximate region of the spectrum. The color labels of the visible spectrum are particularly unreliable as there is wide variation in color perception among people.

It turns out that this energy is similar to chemical reaction energies. This observation is of considerable importance in photochemistry and photobiology.

Equations (2.1) and (2.2) can be combined to give a relationship between photon energy and wavelength, or, since the wavenumber is the reciprocal of the wavelength, between energy and wavenumber:

$$E = \frac{hc}{\lambda} = hc\tilde{\nu}. \tag{2.3}$$

While we can only see electromagnetic radiation in a very restricted range, there is neither an upper nor a lower limit to the possible wavelengths of light. Figure 2.3 shows the electromagnetic spectrum. The labels are not to be taken too seriously; there is no exact dividing line between, for instance, $\gamma$-rays and X-rays. However, these labels are convenient identifiers of the spectral region to which a given radiation belongs.

Einstein is of course most famous for his work on relativity. One of the central equations of relativity theory is

$$E^2 = c^2 p^2 + m_0^2 c^4, \tag{2.4}$$

where $E$ is the energy of a particle, $p$ is its momentum and $m_0$ is the rest mass (the mass at zero velocity) of the particle. (In relativity, the mass varies with speed.) If we take the case of a particle at rest ($p = 0$), we recover the most famous version of this equation: $E = m_0 c^2$. Photons represent the opposite extreme. They have a rest mass of zero, so for photons

$$E = cp. \tag{2.5}$$

In other words, a photon has a momentum proportional to its energy. This momentum can be related to the wavelength by

$$E = cp = hc/\lambda;$$
$$\therefore p = h/\lambda. \tag{2.6}$$

Before we proceed to some examples, it is worth recalling that the SI unit of mass is the kilogram, not the gram. Thus, if we consistently work in SI units, the units of momentum obtained from Equation (2.6) will be $\mathrm{kg\,m\,s^{-1}}$.

**Example 2.4 Momentum of a mole of photons** What is the momentum of a mole of 400 nm photons?

$$p = \frac{6.626\,068\,8 \times 10^{-34}\,\mathrm{J\,Hz^{-1}}}{400 \times 10^{-9}\,\mathrm{m}} = 1.66 \times 10^{-27}\,\mathrm{kg\,m\,s^{-1}}$$

per photon or

$$(1.66 \times 10^{-27}\,\mathrm{kg\,m\,s^{-1}})(6.022\,142\,0 \times 10^{23}) = 9.98 \times 10^{-4}\,\mathrm{kg\,m\,s^{-1}}$$

for a mole of photons.

**Example 2.5 Photon pressure and solar sailing** While the momentum of a photon is quite small, the Sun just keeps producing photons so that, away from a planet's gravitational field, this is sufficient to accelerate a spacecraft equipped with a large sail, i.e. a thin sheet of reflective material, to respectable speeds. Solar sailing is made possible by the near-vacuum conditions present in interplanetary space (which minimize frictional losses) and by the microgravity environment (which makes it possible to deploy very large, thin sails). Photon pressure has been used for attitude control on a number of spacecraft. The first spacecraft to actually be propelled by a solar sail is the Japanese craft IKAROS. IKAROS is a 315 kg craft carrying out a variety of science experiments. It has deployed a square solar sail with a 20 m diagonal, corresponding to an area of $200\,\mathrm{m^2}$. As of this writing, IKAROS is near Venus, where the solar flux is[1] $\phi_E = 2563\,\mathrm{J\,m^{-2}s^{-1}}$. This is the amount of electromagnetic radiation from the Sun that would be received on a one-square-meter surface every second near the orbit of Venus. Equation (2.5) allows us to transform this energy flux into a momentum flux:

$$\phi_p = \frac{\phi_E}{c} = \frac{2563\,\mathrm{J\,m^{-2}s^{-1}}}{2.997\,924\,58 \times 10^8\,\mathrm{m\,s^{-1}}} = 8.549 \times 10^{-6}\,(\mathrm{kg\,m\,s^{-1}})\mathrm{m^{-2}s^{-1}}.$$

By multiplying by the surface area of the sail, we get the momentum of the photons passing through that area every second. The maximum push is provided when the sail is perpendicular to the solar flux. In this case, the photons are reflected straight back so the

---

[1] *CRC Handbook of Chemistry and Physics*, 66th edn.; Boca Raton: CRC Press, 1985, p. F-129.

change in momentum of the spacecraft is twice the initial momentum of the photons:

$$\Delta p_{max}/\Delta t = 2(8.549 \times 10^{-6}\,(\text{kg m s}^{-1})\text{m}^{-2}\text{s}^{-1})(200\,\text{m}^2)$$
$$= 3.420 \times 10^{-3}\,\text{kg m s}^{-2}.$$

Since $p = mv$ and $m$ is fixed, $\Delta p = m\Delta v$, so $\Delta p/\Delta t = m\Delta v/\Delta t = ma = F$, where $a$ is the acceleration and $F$ is the force. The spacecraft therefore experiences a maximum force of 3.420 mN. The maximum acceleration of IKAROS's 315 kg is therefore $a = F/m = 1.09 \times 10^{-5}\,\text{m s}^{-2}$. This is really tiny, but remember that this acceleration acts all the time, without the need to find fuel. To put this number into perspective, consider that if this acceleration acted constantly for one year, IKAROS could reach a speed of about $v = at \approx 340\,\text{m s}^{-1}$ due solely to the photon pressure. Provided you have a bit of time, which you do with unmanned probes like IKAROS, this is not a bad way to get around the solar system.

---

### Exercise group 2.1

(1) In an ordinary incandescent light bulb, approximately 35% of the energy used is converted to photons. The rest is lost in the form of heat. Assuming that the average photon has a wavelength of 580 nm (about the middle of the visible range), roughly how many photons are produced per hour by a 100 W light bulb? Express your answer in moles.

(2) When $^{57}$Co decays, it forms an excited $^{57}$Fe nucleus (mass $9.45 \times 10^{-26}$ kg) which then reaches its ground state by emitting a $\gamma$-ray photon of energy $2.31 \times 10^{-15}$ J. When this happens, because of conservation of momentum, the iron nucleus must recoil, i.e. acquire a momentum equal and opposite to that of the emitted particle, much like a canon that recoils after firing. Calculate the momentum of the emitted photon. Then calculate the kinetic energy acquired by the iron nucleus as a result of this process, assuming that the nucleus is initially at rest.

(3) In photochemical reactions, light causes chemical reactions, often by breaking bonds. A carbon–carbon single bond takes (on average) $350\,\text{kJ mol}^{-1}$ to break. Calculate the energy, frequency and wavelength of a photon capable of breaking a carbon–carbon bond. To what spectral region does such a photon belong?

---

## 2.2 Wave properties of matter

Einstein's work implied that light had a dual nature: it is a wave, and it is made up of particles. Whether light shows us its wave or particle properties depends on the experiment performed. In his 1924 Ph.D. thesis, Louis de Broglie completed Einstein's argument: since the theory of relativity implies that matter and energy are interconvertible, then if light (a form of energy) is both a particle and a wave, the same must be true for particles of matter. Indeed, the very same equations must apply. Specifically, the wavelength of a particle of momentum $p$ must be given by Equation (2.6):

$$\lambda = h/p$$

whether the particle is massless (like a photon) or not. De Broglie's hypothesis was very soon confirmed by the observation of a wave-like property of electrons (diffraction) by Davisson and Germer. The de Broglie equation is the foundation of the old quantum theory, a simple theory that can still be very useful, provided we're careful not to ask too much of it.

**Example 2.6 Thermal wavelength of an electron** At 298 K, the average kinetic energy of a particle in a gas is $6 \times 10^{-21}$ J. What is the wavelength of an electron of this energy?
We can calculate the speed from the kinetic energy: $K = \frac{1}{2}mv^2$ so

$$v = \sqrt{\frac{2K}{m}} = \sqrt{\frac{2(6 \times 10^{-21}\,\text{J})}{9.109\,381\,9 \times 10^{-31}\,\text{kg}}} = 1.15 \times 10^5\,\text{m s}^{-1}.$$

$$\therefore p = mv = (9.109\,381\,9 \times 10^{-31}\,\text{kg})(1.15 \times 10^5\,\text{m s}^{-1})$$

$$= 1.05 \times 10^{-25}\,\text{kg m s}^{-1}.$$

$$\therefore \lambda = \frac{h}{p} = \frac{6.626\,068\,8 \times 10^{-34}\,\text{J Hz}^{-1}}{1.05 \times 10^{-25}\,\text{kg m s}^{-1}} = 6\,\text{nm}.$$

This wavelength is called the **thermal wavelength** because it is the wavelength of an average electron in thermal equilibrium near room temperature. On a molecular scale, this is a large distance. For instance, 6 nm would be the approximate length of a chain of 39 carbon atoms. That said, this might be a good wavelength for the operation of an electron microscope, a device in which the wave properties of electrons are used to image small features of (for instance) biological samples on scales that are beyond the resolving power of optical microscopes.

**Example 2.7 Thermal wavelength of a neutron** Let's now calculate the thermal wavelength of a neutron at 298 K.

$$v = \sqrt{\frac{2K}{m}} = \sqrt{\frac{2(6 \times 10^{-21}\,\text{J})}{1.674\,927\,16 \times 10^{-27}\,\text{kg}}} = 2.68 \times 10^3\,\text{m s}^{-1}.$$

$$\therefore p = mv = (1.674\,927\,16 \times 10^{-27}\,\text{kg})(2.68 \times 10^3\,\text{m s}^{-1})$$

$$= 4.48 \times 10^{-24}\,\text{kg m s}^{-1}.$$

$$\therefore \lambda = \frac{h}{p} = \frac{6.626\,068\,8 \times 10^{-34}\,\text{J Hz}^{-1}}{4.48 \times 10^{-24}\,\text{kg m s}^{-1}} = 1.5\,\text{Å}.$$

This is much more closely comparable to atomic dimensions. This observation is the basis of neutron diffraction, a technique that can be used to determine the structures of molecules in solid samples, among other applications.

**Example 2.8 Wavelength of your professor** Just for fun, let's calculate de Broglie wavelength of a 75 kg person walking at a speed of $1 \, \text{m s}^{-1}$.

The momentum of the person is

$$p = mv = 75 \, \text{kg m s}^{-1}.$$

The de Broglie wavelength is therefore

$$\lambda = h/p = 8.8 \times 10^{-36} \, \text{m}.$$

This is tiny. To put this number in perspective, the radii of some of the smallest things we know, neutrons and protons, are around $10^{-15} \, \text{m}$.

These last examples show us why we need quantum mechanics to treat atoms and molecules, but not to treat the motion of ordinary objects: the de Broglie wavelengths of the constituents of atoms and molecules (electrons and nuclei) tend to be large compared to atomic dimensions. The wave properties of these particles therefore cannot be neglected if we hope to describe the behavior of matter on an atomic scale. On the other hand, the de Broglie wavelengths of everyday objects are tiny so that the wave properties of large pieces of matter are negligible. Note that quantum mechanics should apply equally well to large objects as it does to the very small; it's just that the ways in which quantum mechanics differs from classical mechanics become less and less easily observable as objects grow larger so that it's generally not worth the bother once objects exceed a certain size. There is no sharp cut-off, but a gradual transition from quantum to classical behavior. Biopolymers such as proteins and DNA, for instance, are sufficiently large that some of their motions, especially those that involve a large number of atoms, can be described relatively accurately by classical mechanics, but many details of their structure and function can only be understood using quantum mechanics.

---

### Exercise group 2.2

(1) What is the wavelength associated with a neutron whose speed is $8000 \, \text{m s}^{-1}$?
(2) At 10 K, the molecules in a gas have an average kinetic energy of $2 \times 10^{-22} \, \text{J}$. Would quantum mechanical effects be important to understand the motion of helium atoms at this temperature?

---

## 2.3 Probability waves

Sound is a pressure wave, meaning that regions of higher and lower pressure propagate through space. Light, as we have seen, is an electromagnetic wave. But what exactly is it that is propagating through space in a matter wave? The surprising answer to this question was provided by Max Born: the wave associated with a particle is a probability wave. This makes quantum mechanics very different from classical mechanics; in classical mechanics, our knowledge of a system is only limited by our measuring instruments. Quantum mechanics, on the other hand, is intrinsically probabilistic. We will be able to assign probabilities that a particle is in one region or another, but we won't be able to assign precise coordinates to a particle.

We now need a little bit of language from probability theory. In quantum mechanics, we will need to describe how probability varies with position in space. However, we can't talk about the probability that a particle is at some precise coordinate $x$, for two reasons: first, this probability is zero because the true value of the particle position will always differ from $x$ if we look at enough decimal places of the two numbers; second, limitations in the precision of our measuring instruments mean that there will always be some uncertainty in the position. The best we can do is to ask what is the probability that the object is at $x$ plus or minus $\Delta x$. In probability theory, we deal with these problems by introducing a **probability density** $p(x)$ whose value gives us the amount of probability per unit length. More formally, $p(x)\, dx$ is the probability that a particular measurement of the position will fall between $x$ and $x + dx$.

In quantum mechanics, a particle has a **wavefunction** whose square is the probability density for the location of the particle. The absolute value of the wavefunction will therefore be large in regions where the particle is most likely to be found.

The fact that particles have wave properties has interesting consequences. Waves are hard to confine. Think about your neighbor's loud party. You can close all your windows and doors, and still the sound waves propagate through your walls. You have to put a lot of insulation between you and that party before you reach the point where you can't hear the noise anymore. The same thing is true of particles; their wavefunctions can propagate through barriers that, if we were in the domain of classical particles, would be impenetrable. The propagation of the wavefunction through a barrier means that there is a probability that the particle can suddenly appear on the other side, which indeed does happen. This phenomenon is known as **tunneling**. Thus, electrons can sometimes tunnel through insulating materials, creating a tunneling current, i.e. a current that shows up on the other side of an insulator from the side where the electrons were originally circulating. Tunneling is exploited in a number of modern devices, ranging from flash memory to some types of electron microscopes.

## 2.4 Quantization of energy

Using just the principles we have seen so far, we will study a very simple (but often surprisingly useful) model, that of a particle in a one-dimensional box. This will enable us to draw out some key features of quantum mechanics and of its consequences for molecular science.

A box is simply a container with hard walls from which a particle cannot escape. Our box will be considered to be perfect so that no tunneling can occur. A quantum mechanical particle has an associated wave, and waves interfere with themselves when they cross. Our

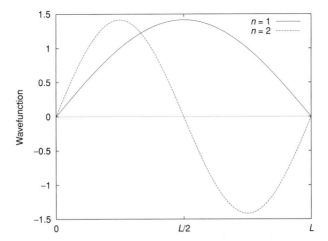

Figure 2.4 Ground state ($n = 1$) and first excited state ($n = 2$) wavefunctions for the particle in a box.

perfect box must reflect the wave at its boundary, which is what we mean when we say that we can't have tunneling. If the wave has an arbitrary wavelength, each pass through the box will have a different phase (different positions for its maxima and minima), and it will destructively interfere with itself, i.e. cancel itself out. The wavelength should therefore be such that the wave, on reflection from one end of the box, has the same shape as the wave before reflection. A little thought reveals that there must be an integer (whole) number of half-waves in the box, i.e.

$$L = n\lambda/2$$

or

$$\lambda = 2L/n,$$

where $n$ is an integer. This relation is called a **quantization condition**. A quantization condition requires a property of a particle (in this case the wavelength) to be related to the integers by an algebraic relationship. The integer $n$ is called a **quantum number**. The first two wavefunctions are shown in Figure 2.4.

The wavelength of the particle is related to the size of its momentum by

$$p = h/\lambda = \frac{nh}{2L},$$

so the momentum is also quantized. The speed can be computed from the momentum:

$$v = p/m = \frac{nh}{2mL}.$$

Finally, the (kinetic) energy is

$$E_n = \frac{1}{2}mv^2 = \frac{n^2h^2}{8mL^2}. \tag{2.7}$$

The energy, too, is quantized.

The sign of $n$ has no physical meaning in this problem since it is just the number of half-waves that fit in the box. You will note in fact that the energy doesn't depend on the sign of the quantum number. We therefore can restrict our attention to the positive integers. Could $n$ be zero? If it were, we would have no half-waves in the box, i.e. the wavefunction would be zero everywhere. However, this would mean that the probability of finding the particle in the box would be zero, or to put it another way, that there was no particle in the box at all. If there is a particle in the box, $n$ therefore cannot be zero. Consequently, the smallest possible value of $n$ is 1.

If $n$ must be at least 1, then the energy of a particle in a box must be at least $E_1 = h^2/8mL^2$. De Broglie's wave–particle duality and the probabilistic interpretation of the wavefunction have forced us to a remarkable conclusion: not only can the energy only take on certain discrete values, but there is a minimum value, called the **zero-point energy**, below which the energy cannot fall. In short, it is not possible for a quantum-mechanical particle to be completely stationary.

A system in its lowest energy level is said to be in its **ground state**. The next lowest energy level is the first **excited state**. The energy level after that is the second excited state and so on.

**Example 2.9 Particle-in-a-box treatment of a chemical bond** A chemical bond is a region of space between two nuclei in which electrons are more likely to be found than not. Treating a bond as a box, let us calculate the minimum speed of an electron in a $\sigma$ bond. A carbon–carbon bond is approximately $1.5$ Å long. The minimum speed is

$$v = \frac{h}{2mL} = \frac{6.626\,0688 \times 10^{-34}\,\mathrm{J\,Hz^{-1}}}{2(9.109\,3819 \times 10^{-31}\,\mathrm{kg})(1.5 \times 10^{-10}\,\mathrm{m})} = 2.4 \times 10^6\,\mathrm{m\,s^{-1}}.$$

This is of course an extremely crude model for a bond, but it does give us an idea of the size of this minimum speed for electrons in typical molecular settings. These speeds are large, but not relativistic, which makes the theory much simpler than it would have been otherwise.

---

### Exercise group 2.3

(1) The professor-in-a-box problem: a 75 kg quantum mechanics professor has been locked into a short, narrow hallway of length 5 m. To pass the time, he decides to calculate the minimum kinetic energy that quantum mechanics requires him to have, assuming that he can treat himself as a particle in a one-dimensional box. What value does he get? Is this a large or a small kinetic energy?

(2) Suppose that a proton is placed in a narrow box of length 1 cm at 25 °C. The average kinetic energy of particles held at 298 K in one dimension is approximately $2 \times 10^{-21}$ J. Compute the quantum number $n$ of a proton in a box at this kinetic energy. Then calculate the energy necessary for this particle to make a transition from your computed value of $n$ to $n + 1$. Do you expect quantum mechanical effects to be important in this system? *Hint*: To calculate the transition energy, it will be convenient to derive a formula first, and only then to plug in the numbers.

## 2.5 A first look at spectroscopy

**Spectroscopy** is the study of the interaction of light with matter. In the next chapter, we will study spectroscopy in greater depth, but here, we will discuss just one type of interaction, namely absorption. Absorption of photons by matter obeys a couple of rules:

(1) Under normal circumstances, one photon is absorbed by one molecule, a rule known as the Stark–Einstein law, or the **law of photochemical equivalence**. Two-photon processes are extremely unlikely at normal illumination levels, but are possible at the very high light intensities that can be reached using some lasers. We will focus here on the normal case in which the law of photochemical equivalence holds.
(2) The energy of the photon is completely converted into molecular energy, so this energy must correspond to the spacing between two energy levels of the molecule.

It is the latter rule that makes spectroscopy an invaluable tool: it gives us a direct probe of the energy levels of a molecule, which in turn are connected to its structure in ways that we will explore in more detail in the next chapter. Here, we will look at the absorption spectroscopy of a particle in a box as a simple model for the spectroscopy of molecules with conjugated $\pi$ bonds.

We will see in the next chapter that the energy of a molecule can be separated into various contributions. One of these is the electronic energy. You will have discussed atomic orbitals in your introductory chemistry course. Molecules also have orbitals. In principle, these **molecular orbitals** extend over the entire molecule, although some orbitals may be strongly localized to a particular part of the molecule. In the case of conjugated $\pi$ bonds (whose Lewis structures appear as alternating single and double bonds "interchanged" by resonance), resonance creates long orbitals through which the electrons can move, i.e. a "box." You can probably think of objections to treating conjugated $\pi$ bonds as a one-dimensional box, but this does turn out to be a reasonable model.

To discuss molecular orbitals, we need an additional principle which you will have seen in your introductory chemistry class, namely the **Pauli exclusion principle**, one version of which states that no two electrons can occupy the same orbital. We are now ready to consider the spectroscopy of molecules with conjugated $\pi$ bonds.

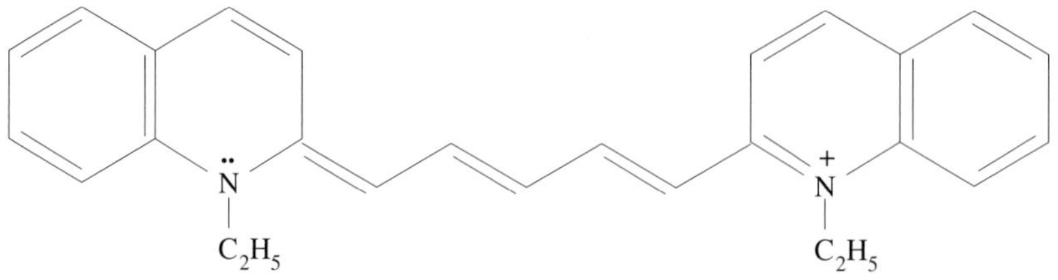

Figure 2.5 The 1,1′-diethyl-2,2′-dicarbocyanine ion. Cyanine dyes such as this one absorb strongly in the visible range and thus are brightly colored compounds. Accordingly, they have traditionally been used as dyes. This particular dye absorbs strongly at the red end of the spectrum, giving it a pleasant blue-green color.

**Example 2.10 Spectroscopy of molecules with conjugated $\pi$ bonds** The 1,1′-diethyl-2,2′-dicarbocyanine ion (Figure 2.5) absorbs light at 708 nm. This dye has 10 $\pi$ electrons in the conjugated $\pi$ chain that extends from one nitrogen atom to the other (the lone pair of electrons on the nitrogen on the left plus the four pairs of $\pi$ electrons in the double bonds), so these electrons fill five molecular orbitals in the molecule's ground state, in accord with the Pauli exclusion principle. Absorbing a photon with a wavelength of 708 nm causes one electron to be transferred from the highest occupied molecular orbital (HOMO) to the lowest unoccupied molecular orbital (LUMO). Treating the $\pi$ orbitals as a one-dimensional box, this would be an $n = 5$ to $n = 6$ transition. The energy of the photon is

$$E_{photon} = \frac{hc}{\lambda} = \frac{(6.626\,068\,8 \times 10^{-34}\,\text{J Hz}^{-1})(2.997\,924\,58 \times 10^{8}\,\text{m s}^{-1})}{708 \times 10^{-9}\,\text{m}}$$
$$= 2.81 \times 10^{-19}\,\text{J}.$$

The difference in energy between the HOMO and LUMO is

$$\Delta E = E_6 - E_5 = \frac{6^2 h^2}{8mL^2} - \frac{5^2 h^2}{8mL^2} = \frac{11h^2}{8mL^2}.$$

Since our box contains electrons, $m$ is the mass of an electron. If we equate the energy of the photon to this energy difference, the only unknown is therefore the length of the box:

$$L^2 = \frac{11h^2}{8mE_{photon}} = \frac{11(6.626\,068\,8 \times 10^{-34}\,\text{J Hz}^{-1})^2}{8(9.109\,381\,9 \times 10^{-31}\,\text{kg})(2.81 \times 10^{-19}\,\text{J})}$$
$$= 2.36 \times 10^{-18}\,\text{m}^2.$$
$$\therefore L = 1.54 \times 10^{-9}\,\text{m}.$$

There are eight bonds in the conjugated $\pi$ chain. Assuming the electrons are confined to this box, the length of one bond would be $1.54\,\text{nm}/8 = 0.192\,\text{nm}$. Conjugated carbon–carbon bonds are usually about $0.14\,\text{nm}$ long. Our estimate of the bond length is too large by about 37%, which isn't bad considering how simple a model we took for the molecular orbitals.

## Exercise group 2.4

(1) You may recall from your introductory chemistry course that the energy levels of a hydrogen atom satisfy the equation $E_n = -R_H/n^2$, where $n$ is the principal quantum number, and $R_H$ is Rydberg's constant. Rydberg's constant is often given in wavenumbers: $R_H = 109\,677.581\,\text{cm}^{-1}$. Calculate the wavelength of a photon that could cause a transition from the ground state ($n = 1$) to the first excited state ($n = 2$) of a hydrogen atom.

(2) We will see in the next chapter that the vibrational energy levels of a molecule are well approximated by $E_v = h\nu_0\left(v + \frac{1}{2}\right)$, where $\nu_0$ is a constant characteristic of the vibration, and $v$ is a quantum number that can take any of the values $0, 1, 2, \ldots$
   (a) What is the vibrational zero-point energy?
   (b) Develop a general expression for the energies of the photons that could be absorbed in terms of $\Delta v$, the change in the vibrational quantum number of the molecule.
   (c) The lowest energy vibrational transition of a certain molecule is observed at $1200\,\text{cm}^{-1}$. What is the value of $\nu_0$?
   (d) Where should we look for the next-lowest energy transition?

## Key ideas and equations

• For electromagnetic waves: $c = \lambda \nu$, $E = h\nu = hc/\lambda = hc\tilde{\nu}$
• For both photons and particles of matter: $p = h/\lambda$
• Molecular energy is quantized, and there is a minimum molecular energy called the zero-point energy.
• For a particle in a box: $E_n = n^2 h^2 / 8mL^2$
• In absorption spectroscopy, a single photon provides the energy for a molecule to transition from one allowed energy level to another.

## Review exercise group 2.5

(1) What is the condition for a molecule to absorb photons of a particular wavelength?
(2) Most microwave ovens operate at a frequency of $2.45\,\text{GHz}$. Calculate the corresponding photon energy, wavelength and wavenumber.
(3) $^{177}$Au is an unstable isotope that decays by emitting an $\alpha$ particle (a helium nucleus) with a kinetic energy of $6.115\,\text{MeV}$. What is the wavelength of this $\alpha$ particle? The mass of an $\alpha$ particle is $6.644\,656\,2 \times 10^{-27}\,\text{kg}$. Do you expect quantum effects to be important for an $\alpha$ particle with this energy?
(4) The particle-on-a-ring model is a simple variation on the particle-in-a-box model; we imagine that a particle is confined to a ring of radius $r$:
   (a) The main difference between the two models is the quantization condition. A ring doesn't have edges. What would be the correct quantization condition on a ring?

(b) Derive an equation for the energy levels of a particle on a ring.

(c) Unlike the particle-in-a-box problem, $n = 0$ is an allowed value of the quantum number for the particle on a ring. (This is associated with the fact that a ring doesn't have edges, which makes it possible to have a wavefunction which is just constant over the ring.) Also, the positive and negative values of $n$ correspond to different states with the same energy. We say that the states for $n > 0$ are **degenerate**, a concept to which we will return in the next chapter. If we imagine the ring as being perpendicular to the $z$ axis, then positive values of $n$ correspond to counterclockwise circulation around the $z$ axis, while negative values of $n$ correspond to clockwise circulation. Correspondingly, the angular momentum (which you may recall from your first-year physics course) is a vector pointing along the $z$ axis for positive $n$ and along the negative $z$ axis for negative $n$. The $z$ component of the angular momentum vector (the $x$ and $y$ components are zero) is given by $L_z = rp$. Work out a formula for the angular momentum. You will find that your formula is consistent with the rotation directions mentioned above if you allow both positive and negative values of $n$.

(d) We can treat the $\pi$ electrons in benzene approximately as particles on a ring. Sketch the energy level diagram for the lowest few particle-on-a-ring energy levels, and place the $\pi$ electrons in the ground-state configuration. Identify the HOMO and the LUMO.

(e) The HOMO to LUMO transition in benzene is observed near 178 nm. What ring radius does this imply?

(f) If we take the radius as being the distance from the center of the ring to one of the atoms, what is the bond length?

   *Hint*: Draw a regular hexagon, and look at a "slice" bounded by an edge and by the two rays from the center to the ends of the edge. How is the edge length related to the length of the rays (radius)?

# 3

# Spectroscopy

As mentioned in the last chapter, spectroscopy is the study of the interaction of light with matter. We also saw an example of inferring a molecular property (a bond length) from spectroscopic data. In fact, spectroscopy is the principal experimental approach allowing us to study molecules. Our program in this chapter will be to learn about a selected few spectroscopic techniques and what they tell us about molecules. Instructors in your other courses in chemistry and biochemistry will introduce you to additional techniques as you need them. What I hope you will take away from this chapter is that there are a few basic principles that govern spectroscopy which, once mastered, can be applied to all spectroscopy experiments.

## 3.1 Molecular energy

The particles in a molecule are in constant motion:

- The electrons buzz around their orbitals.
- The bonds vibrate, which is to say that the atoms move back and forth repeatedly like balls connected by springs.
- The whole molecule rotates.
- In gases and liquids, the molecules translate, which is to say that they move through space.

Each of these motions has energy associated with it. These motions are not completely independent, but it turns out that we can treat them as if they are for many purposes. We can therefore write the energy of a single molecule as

$$E = E_{\text{elec}} + E_{\text{vib}} + E_{\text{rot}} + E_{\text{trans}},$$

where $E_{\text{elec}}$ is the electronic energy (the energy associated with the electrons), $E_{\text{vib}}$ is the vibrational energy, $E_{\text{rot}}$ is the rotational energy and $E_{\text{trans}}$ is the translational energy. Each of these types of energy has its own energy levels and, as it turns out, these energy levels typically have very distinctive spacings, and so each (except for translational energy, as discussed below) is associated with its own spectroscopy.

23

Let's start by considering the translational energy levels. These might be of interest in a gas. We usually study the spectroscopy of gases in cylindrical sample cells that are at least 10 cm long. Say we have a cell that is this long, and consider the translational energy levels of a nitrogen molecule in this cell, which will be given by the particle-in-a-box equation (2.7). (We could generalize this argument to three dimensions, but we would conclude much the same thing.) Some simple arguments from statistical mechanics show that the average kinetic energy of a gas molecule in one dimension is $\frac{1}{2}k_B T$, where $k_B$ is Boltzmann's constant (to which we will return shortly). At, say, 20 °C, the average kinetic energy works out to $2.02 \times 10^{-21}$ J. Suppose that we have a molecule with this much energy. What would be its quantum number, $n$? The molar mass of a nitrogen molecule is 28.02 g mol$^{-1}$, or $4.65 \times 10^{-23}$ g $= 4.65 \times 10^{-26}$ kg for a single molecule. (Don't forget that the kilogram is the SI unit of mass, not the gram.) From the particle-in-a-box equation, we have

$$n = \sqrt{8m E_n}(L/h)$$
$$= \sqrt{8(4.65 \times 10^{-26}\,\text{kg})(2.02 \times 10^{-21}\,\text{J})}\left[(0.1\,\text{m})/(6.626\,068\,8 \times 10^{-34}\,\text{J Hz}^{-1})\right]$$
$$= 4.138 \times 10^9.$$

Spectroscopy has to do with differences in energy. We might then ask how closely spaced the energy levels are for this value of $n$:

$$\Delta E = E_{n+1} - E_n = \frac{h^2}{8m L^2}\left[(n+1)^2 - n^2\right] = \frac{h^2}{8m L^2}(2n+1).$$

For our values of $n$, $m$ and $L$, we get $\Delta E = 9.76 \times 10^{-31}$ J. This is an extraordinarily small difference in energy, both relative to the average energy of nitrogen molecules, and in terms of the energies of photons (Figure 2.3). Accordingly, no practical spectroscopy experiment could be designed to study the translational energy levels of molecules.

For the other types of molecular energy, the energy levels are in fact sufficiently spaced for spectroscopy. The spacings between energy levels obey the following relationships to each other and to the spectral regions where they are typically observed:

| **Level spacings:** | electronic | > | vibrational | > | rotational |
|---|---|---|---|---|---|
| | ⇓ | | ⇓ | | ⇓ |
| **Spectral region:** | ultraviolet/visible | | infrared | | microwave |

It is important to understand that these relationships do *not* imply that only one type of transition occurs in each of these regions. For example, if the photons in a spectroscopy experiment have enough energy to cause a vibrational transition, they will normally cause rotational transitions at the same time. In this case, we would have $E_{photon} = \Delta E_{vib} + \Delta E_{rot}$. Thus you should think of the correspondences in the boxes above as establishing how much energy is required to cause a certain type of transition, with the understanding that transitions of lower energy can occur at the same time.

## 3.2 The Boltzmann distribution

At any given temperature above absolute zero, not all molecules in a sample will be in their ground states. One of the issues that will be important to our understanding of molecular spectroscopy will be the proportion of molecules that can be expected to occupy a given energy level prior to excitation with a light source. The answer to this question is provided by the **Boltzmann distribution**. Deriving the Boltzmann distribution would take more time and space than we can devote to this topic, so we'll just have to take it as a given. It turns out that there are two factors to consider: the energy of a level, and its **degeneracy**. The degeneracy is the number of different sets of quantum numbers that give the same energy. You would have run into this concept in your introductory chemistry course when the energy levels of atoms were discussed, where you would have learned that there are, for example, three 2p orbitals, and that all three are equal in energy in free atoms. The degeneracy of, for example, a boron atom in its ground state ($1s^2 2s^2 2p^1$) is therefore three.

The Boltzmann distribution is given by the following equation:

$$P(E_i) = g_i \exp\left(-\frac{E_i}{k_B T}\right)/q, \tag{3.1}$$

where $P(E_i)$ is the probability that a molecule has energy $E_i$, $g_i$ is the degeneracy of the energy level, $k_B$ is Boltzmann's constant and $T$ is the absolute temperature. Boltzmann's constant turns out to just be the ideal gas constant expressed on a per-molecule instead of a per-mole basis. In other words, $R = k_B L$, where $L$ is Avogadro's constant.

The constant $q$ appearing in the Boltzmann distribution is the normalization constant for the distribution. In other words, it is chosen so that $\sum_i P(E_i) = 1$, where the sum is taken over the energy levels of the molecule. It is not too hard to see that $q$ must be given by the following formula:

$$q = \sum_i g_i \exp\left(-\frac{E_i}{k_B T}\right). \tag{3.2}$$

$q$ is called the **molecular partition function** and turns out to have a very important role to play in statistical mechanics. To give you just a taste of the importance of the partition function, consider

$$\frac{\partial q}{\partial T} = \sum_i \frac{E_i}{k_B T^2} g_i \exp\left(-\frac{E_i}{k_B T}\right).$$

$$\therefore k_B T^2 \frac{\partial q}{\partial T} = \sum_i E_i g_i \exp\left(-\frac{E_i}{k_B T}\right).$$

$$\therefore \frac{k_B T^2}{q} \frac{\partial q}{\partial T} = \sum_i E_i \left(\frac{g_i e^{-E_i/k_B T}}{q}\right) = \sum_i E_i P(E_i).$$

The symbol $\partial$ indicates a partial derivative, which is a derivative taken holding other variables constant. We use a partial derivative here because $q$ depends on more than one variable. This may not be immediately obvious. However, if you look back at Equation (2.7) for the energy levels of a particle in a box, you will note that these depend on $L$. You can imagine that the three-dimensional generalization of this problem might involve the volume $V$, and you would be right: in general, $q$ depends on $T$ and $V$. The key thing to notice here is the final equation; from basic statistical theory, we know that $\sum_i E_i P(E_i)$ is the average energy, in this case averaged over a large number of molecules. We denote the average by $\langle E \rangle$, so we have

$$\langle E \rangle = \frac{k_B T^2}{q} \frac{\partial q}{\partial T}. \tag{3.3}$$

This average energy can in turn be associated with a thermodynamic quantity called the internal energy. It will turn out that all thermodynamic quantities can be calculated from the partition function.

**Example 3.1 Boltzmann distribution and probabilities** Suppose that we have a very simple system with just two energy levels:

| $i$ | $g_i$ | $E_i/J$ |
|-----|-------|---------|
| 1   | 1     | 0       |
| 2   | 3     | $4 \times 10^{-21}$ |

The molecular partition function is

$$q = g_1 \exp\left(-\frac{E_1}{k_B T}\right) + g_2 \exp\left(-\frac{E_2}{k_B T}\right)$$

$$= 1 + 3 \exp\left(-\frac{4 \times 10^{-21} \, J}{(1.380\,650\,3 \times 10^{-23} \, J\,K^{-1})T}\right)$$

$$= 1 + 3 \exp\left(-\frac{290 \, K}{T}\right). \tag{3.4}$$

We can now write down the probability of finding a randomly chosen molecule in each of the two levels:

$$P(E_1) = g_1 \exp\left(-\frac{E_1}{k_B T}\right)/q = (1 + 3 \exp(-290 \, K/T))^{-1};$$

$$P(E_2) = g_2 \exp\left(-\frac{E_2}{k_B T}\right)/q = \frac{3e^{-290 \, K/T}}{1 + 3 \exp(-290 \, K/T)}. \tag{3.5}$$

These probabilities are plotted against temperature in Figure 3.1. At very low temperatures, $P(E_2) \to 0$ and only level 1 is occupied. This would generalize to systems with more energy levels; as the temperature is lowered toward zero, higher-energy states become less and less probable, until we reach a situation where only the ground state is appreciably occupied. At high temperatures, on the other hand, we see that the probabilities approach the values

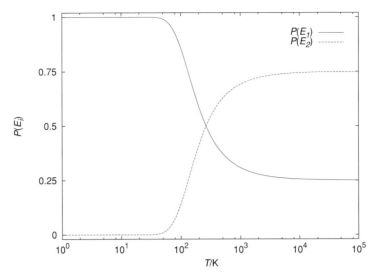

Figure 3.1 Probabilities of the two energy levels from Example 3.1 as a function of temperature. Note the logarithmic scale of the temperature axis.

$P(E_1) = \frac{1}{4}$ and $P(E_2) = \frac{3}{4}$. These probabilities are in direct proportion to the degeneracies. Again, this generalizes to an arbitrary number of states: at very high temperatures, all states are equally probable, so the probability of occupying an energy level is proportional to the number of states of which the energy level consists, i.e. to its degeneracy.

To understand the partition function and what it tells us about a system, let's think a little about the Boltzmann factor $\exp(-E_i/k_B T)$. For convenience, imagine that the ground state has zero energy. We can always arrange for this to be true because the place where we put our zero on an energy scale is arbitrary. Only differences in energy matter in practice. If $E_i \ll k_B T$, then the Boltzmann factor is about 1. The partition function therefore contains a term for this energy level approximately equal to $g_i$, the degeneracy of the level. On the other hand, if $E_i \gg k_B T$, the Boltzmann factor is approximately 0. If we exclude the case where some of the energies are similar to $k_B T$, we get the following:

$$q \approx \sum_{\{i \mid E_i < k_B T\}} g_i$$

In words, this equation says that the partition function counts up the total number of states whose energies are less than $k_B T$. If we consider the case where some of the energies are similar to $k_B T$, we can interpret the partition function similarly, but the number has to be interpreted as a rough indication of the number of accessible states with energies similar to or less than $k_B T$ rather than as an actual count. If you want an example, look back at the molecular partition function (3.4) from Example 3.1: at low temperatures, $q \approx 1$, corresponding to the non-degenerate ground state, while at higher temperatures, $q \to 4$, since all four states (ground state and degenerate excited state) will eventually be accessible.

In between, we get values between 1 and 4, which tells us that there is one low-lying state (the ground state), as well as some states for which $E_i \sim k_B T$.

The Boltzmann distribution is often used to compare the populations of two energy levels. From Equation (3.1), we have

$$\frac{P(E_2)}{P(E_1)} = \frac{g_1 \exp\left(-\frac{E_1}{k_B T}\right)/q}{g_2 \exp\left(-\frac{E_2}{k_B T}\right)/q} = \frac{g_1}{g_2} \exp\left(-\frac{\Delta E}{k_B T}\right),$$

where $\Delta E = E_2 - E_1$. Let's apply this formula to the relative populations of typical ground and excited states of molecules at room temperature.

**Example 3.2 Relative population of an excited electronic state** We have just seen that typical electronic transitions are observed in the UV/visible range. (UV is an abbreviation for ultraviolet.) Let's take a wavelength in that range, say 400 nm, and see what the corresponding energy level spacing implies about the relative populations of the ground and excited state. We first calculate the photon energy, which is the difference in energy between the two electronic states:

$$\Delta E = \frac{hc}{\lambda} = \frac{(6.626\,068\,8 \times 10^{-34}\,\text{J Hz}^{-1})(2.997\,924\,58 \times 10^8\,\text{m s}^{-1})}{400 \times 10^{-9}\,\text{m}}$$
$$= 4.97 \times 10^{-19}\,\text{J}.$$

Assuming that the two electronic states are non-degenerate (i.e. that $g_1 = g_2 = 1$), the relative populations of two states separated by an energy gap of this size at 20 °C is

$$\frac{P(E_2)}{P(E_1)} = \exp\left(\frac{-\Delta E}{k_B T}\right) = \exp\left(\frac{-4.97 \times 10^{-19}\,\text{J}}{(1.380\,650\,3 \times 10^{-23}\,\text{J K}^{-1})(293.15\,\text{K})}\right)$$
$$= 5.16 \times 10^{-54}.$$

This is a really tiny number, which tells us that, really, only the ground state is occupied at room temperature. Note that assuming a higher degeneracy for the excited state than for the ground state would not have altered this conclusion, given the tiny value of the Boltzmann factor.

We can repeat this calculation for typical vibrational (infrared, IR) and rotational (microwave) energy level spacings. We find the following:

| | |
|---|---|
| UV/visible: | only ground state occupied |
| Infrared: | ground state mainly occupied |
| Microwave: | many states occupied |

Knowing this will simplify the interpretation of electronic and vibrational spectra; only transitions starting in the ground state are likely to be observed in typical room-temperature experiments. On the other hand, in experiments where rotational transitions can be observed, we will have to take into account that some molecules are already in excited rotational states at the beginning of the experiment.

One final note about the Boltzmann distribution: throughout this book, you will see equations containing terms that look like $e^{-E/k_BT}$ or $e^{-E_m/RT}$, where $E$ is some quantity (not necessarily symbolized by $E$) with units of energy, and $E_m$ is the corresponding molar quantity. The ubiquitous appearance of such equations in physical chemistry is not a coincidence. It derives from the Boltzmann distribution which governs the statistical distribution of molecular energies.

---

### Exercise group 3.1

(1) Nuclear magnetic resonance is based on the splitting in energy of nuclear spin states in a magnetic field. For a proton in a 7.0 T magnetic field (which you would find in a 300 MHz spectrometer), $\Delta E = 1.97 \times 10^{-25}$ J is the difference in energy between the two spin orientations. What percentage of proton nuclear spins would you expect to be in the higher energy (opposing the field) state at room temperature (20 °C)? What percentage would be in the higher energy state at $-30$ °C?

(2) The Boltzmann distribution is very general, and can be applied to many different kinds of energy. For example, it can be applied to the gravitational potential energy, $E_p = mgh$. The probability of finding a particular molecule at altitude $h$ is of course proportional to the concentration of molecules $n/V$ at that altitude, which in turn is proportional to the pressure.

(a) Put these facts together to derive the **barometric formula**,

$$\frac{p}{p_0} = \exp\left(\frac{-mgh}{k_BT}\right)$$

In this formula, $p$ is the partial pressure of a gas at altitude $h$, and $p_0$ is its partial pressure at sea level.

(b) Show that this formula can be rewritten in the form

$$\frac{p}{p_0} = \exp\left(\frac{-Mgh}{RT}\right)$$

where $M$ is the molar mass of the molecule of interest.

(c) Given that the partial pressure of oxygen at sea level is 0.21 atm, what is the partial pressure at 0 °C at 920 m (the altitude of my house in Lethbridge)? At 1900 m (Val d'Isère, France)? At 3660 m (La Paz, Bolivia)?
*Note*: Don't forget that the SI unit of mass is the kg.

(d) For chemical species that have more than one isotopomer, the dependence of the potential energy on mass leads to **isotopic fractionation**, with heavier isotopes staying, on average, closer to the ground. Given that the abundances given in Appendix G are sea-level abundances, what abundance would you predict for

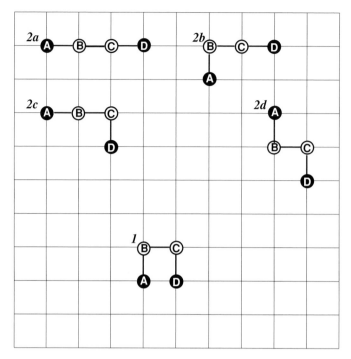

Figure 3.2 The conformations of a small peptide on a square lattice. The dark circles represent hydrophobic residues, while the light circles represent hydrophilic residues.

$^{13}C^{16}O_2$ at 20 000 m (possibly the highest altitude from which viable bacteria have been recovered[1])? Assume a uniform temperature of $0\,^{\circ}C$.

(3) Protein folding is largely driven by the burial of hydrophobic amino acid residues (what is left of the amino acid after the condensation reaction that makes a peptide bond) inside the protein, away from the solvent. A contact between a hydrophobic residue and a water molecule reduces the opportunities for the water to form hydrogen bonds or other energetically favorable contacts. There is therefore an energetic cost to a water–hydrophobic contact. Similarly, there is an energetic cost to a hydrophobic–hydrophilic contact because the hydrophilic group loses opportunities for more favorable contacts with other polar groups (including the solvent). Flipping these statement on their heads, hydrophobic–hydrophobic contacts in a protein, which reduce the number of hydrophobic–water and hydrophobic–hydrophilic contacts, are energetically favorable. One very simple model for protein folding treats residues as beads connected by bonds of unit length placed on a simple lattice.[2] Consider, for example, a small peptide consisting of just four residues (Figure 3.2). We represent hydrophobic residues by dark circles, while light circles represent hydrophilic residues. The bonds between residues are placed along edges of the lattice, and the amino acid residues are placed on lattice

[1] D. W. Griffin, *Aerobiologia* **24**, 19 (2008).   [2] K. A. Dill *et al.*, *Protein Sci.* **4**, 561 (1995).

points. We only count unique conformations, and would leave out any conformations that are just rotations of any we have listed. You should be able to convince yourself that there are just five distinct conformations of the peptide on the lattice, illustrated in Figure 3.2. You might object that conformations 2b and 2c look the same. Because the two ends of a protein are different – there is an amino group at one end of the protein and a carboxy group at the other – we can tell residues A and D apart, even if they are derived from the same amino acid, and so 2b (bent at B, which is nearest to A) and 2c (bent at C, which is nearest to D) are actually different.

Suppose that the energetic payoff for a hydrophobic–hydrophobic contact is $-\epsilon$. Conformation 1, the folded conformation, has one such contact, so its energy is $E_1 = -\epsilon$. Conformations $2a - d$ have no hydrophobic–hydrophobic contacts, so they are degenerate with energy $E_2 = 0$. The latter conformations are unfolded or, as they say in biochemistry, **denatured**.

(a) Write an equation for the probability that the protein is in the denatured state.
(b) A thermal denaturation curve shows the percentage of a protein in the unfolded state as a function of temperature. In this case, we don't know the value of $\epsilon$, so define the reduced temperature $\theta = k_B T / \epsilon$, and draw the thermal denaturation curve as a function of $\theta$. Note that these curves generally look best when plotted against a logarithmic temperature axis.
(c) At what reduced temperature does denaturation start to occur?
(d) What is the minimum percentage of folded protein?

## 3.3 Classes of spectroscopy experiments

We can roughly divide spectroscopy experiments into three classes:

Absorption methods: In an absorption experiment, we measure the amount of light absorbed by a sample as a function of wavelength.

Emission methods: In emission spectroscopy, we measure the amount of light emitted by a sample, typically after the sample has been energized in some way (with photons, heat, electricity etc.).

Scattering methods: Scattering experiments involve the study of photons that have collided with molecules in the sample and been deflected, possibly with some loss or gain of energy.

In any given energy range, we can carry out all three types of experiments. For example, UV/visible spectroscopy usually refers to an absorption experiment. Fluorescence is an emission phenomenon in which molecules first absorb photons, lose some of the energy they absorbed and then re-emit photons of lower energy in the UV/visible range. UV/visible scattering spectroscopy is used as a medical diagnostic test to detect pathologies such as cancers. Each type of spectroscopy gives us different information about the material studied.

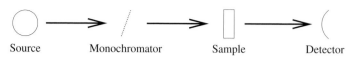

Figure 3.3 Block diagram of a single-beam spectrometer.

## 3.4 Absorption spectroscopy

### 3.4.1 Absorption spectrometers

A simple absorption experiment is diagrammed in Figure 3.3. The light produced by a source (e.g. a lamp) first passes through a **monochromator**, which is a device that selects one wavelength of light. Most commonly, a monochromator would be a diffraction grating, and different wavelengths are selected by rotating the grating. The monochromatic light passes through the sample, and the transmitted light is detected. If the sample absorbs light at a given wavelength, then fewer photons reach the detector than would otherwise be the case. A spectrometer of this design would be called a **single-beam spectrometer**. Single-beam spectrometers are simple, so they are inexpensive and relatively sturdy.

There are two basic types of experiments that can be carried out by a single-beam absorption spectrometer. The simplest type of experiment involves just taking readings at a fixed wavelength. This can be useful in quantifying the amount of substance in a solution, as we will see shortly. The second type of experiment is the measurement of the light intensity as the wavelength is varied. A graph of intensity vs. wavelength is called a **spectrum**.

The problem with single-beam spectrometers is that changes in the signal observed at the detector can have a number of causes. We're interested in absorption of photons by the sample. However, the intensity of the source depends on wavelength, and the detector usually responds more strongly to some wavelengths than to others. Moreover, in solution, the solvent may have an absorption spectrum of its own, which may not be of great interest to us. The solution to these problems is to acquire the spectrum of a **blank**, which is a sample cuvette identical to the one to be used for the sample, but filled with the pure solvent. The spectrum of the blank is then subtracted from the sample spectrum, which in principle gives us the spectrum of the solute(s). The subtraction is typically done automatically by the spectrometer control software.

Another way to solve the problems outlined above is to use a **dual-beam spectrometer** (Figure 3.4). In a dual-beam spectrometer, the monochromated light is split into two beams of equal intensity using a beam splitter. One beam passes through the sample, while the other passes through the blank. The signals from the two detectors are subtracted in hardware by a comparator circuit, and the difference signal is output. Dual-beam spectrometers have one important advantage over single-beam machines: variations in source intensity with time are automatically compensated for. In some designs, rotating mirrors or other mechanical

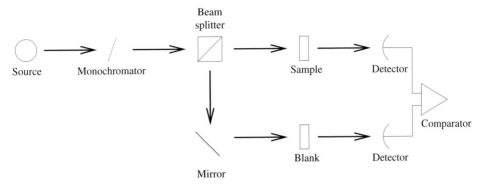

Figure 3.4 Dual-beam spectrometer.

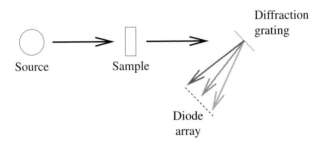

Figure 3.5 The diode array spectrometer.

devices are used to alternately send the transmitted signals from the blank and sample to a single detector. In this case, the subtraction is made between signals that arrive at the detector at different times. It is then not necessary to have two identical detectors, and therefore problems having to do with a relative drift in the responses of the detectors as they age are avoided.

Both single- and dual-beam spectrometers have been around for a long time. A more recent technology is the diode array spectrometer (Figure 3.5), which can be thought of as many single-beam spectrometers working in parallel. In a diode array spectrometer, polychromatic radiation (light containing a range of wavelengths) is sent through the sample. A diffraction grating then separates the transmitted light into its components, which are directed onto a diode array detector. A diode array is the same technology you would find in a digital camera, namely an array of detectors that can be read individually by a computer. Different wavelengths are sent to different parts of the diode array. The entire spectrum is acquired at once, so diode array spectrometers are extremely fast. Moreover, they don't have any moving parts and they tend to be small and sturdy. Typical units fit in the palm of your hand. This makes them ideal for field work, and also for integration into other instruments such as liquid chromatographs.

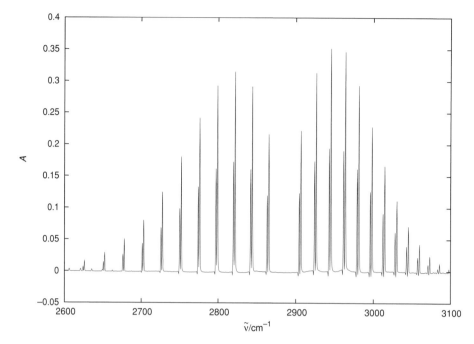

Figure 3.6 Part of the gas-phase infrared spectrum of HCl. This part of the spectrum corresponds to the $v = 0$ to $v = 1$ vibrational transition, with associated rotational transitions. This spectrum was acquired on a Fourier transform instrument, which works on a different principle than the absorption spectrometers discussed here, although the interpretation of the spectrum is not affected. Thanks to Kris Fischer for help in acquiring this spectrum.

### 3.4.2 Absorption spectra

Let's briefly review what happens in an absorption spectroscopy experiment. Photons pass through a sample, and if their energy matches the spacing between two energy levels of the molecule, they may then be absorbed. There are additional conditions for absorption called selection rules so that not every possible absorption is observed, but for now, we just need to focus on the energetics of the absorption process. Since molecules have distinct electronic, vibrational and rotational energy levels, molecular spectra should consist of a set of narrow absorption lines. In the gas phase, this is exactly what we observe. For example, have a look at Figure 3.6, which shows part of the gas-phase IR spectrum of HCl. We will interpret this spectrum in more detail shortly, but for now just notice the narrow lines. These narrow lines are not a feature only of IR spectra. For small molecules, we would see similarly narrow lines in a high-resolution gas-phase UV/visible spectrum, or in any other gas-phase spectroscopy experiment with equipment of appropriate resolution.

In solution though, it's a different story. The energy levels of a molecule depend on its geometry. In solution, frequent collisions between molecules compress and bend bonds. As a result, at any given time, the collection of molecules present in solution all have slightly

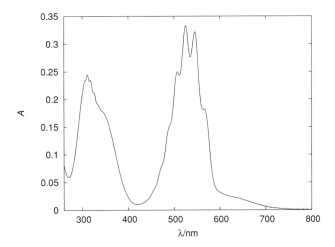

Figure 3.7 UV/visible spectrum of $KMnO_4$ dissolved in water.

different energy levels. The result is that the nice spectral lines typically observed in the gas phase are replaced by broad bands. Figure 3.7 shows the UV/visible spectrum of a solution of potassium permanganate in water. The difference between this spectrum and the gas-phase spectrum of HCl is obvious to the eye, and it's not just a question of the resolution of the instrument.

### 3.4.3 Vibrational spectroscopy

As previously noted, an IR spectrum involves a combination of vibrational and (lower-energy) rotational transitions. You will also recall that, typically, only the vibrational ground state is occupied, but that many rotational states can be occupied at room temperature. Conventionally, the vibrational quantum numbers are denoted $v$, while the rotational quantum numbers are denoted $J$. Both are numbered starting from zero. For the spectrum of HCl shown in Figure 3.6, we are observing a single vibrational transition, from $v = 0$ to $v = 1$, with accompanying rotational transitions. The spectrum has a rough mirror-image symmetry. The lines in the left half of the spectrum, at lower wavenumbers (transition energies) correspond to transitions in which the rotational quantum number decreases, while the lines in the right half of the spectrum correspond to transitions in which the rotational quantum number increases. It turns out to be possible to relate the rotational fine structure to the bond length. However, this is only feasible for small molecules in the gas phase, so this topic is of limited interest to biochemists and we will omit it.

It is however useful to know a little bit of vibrational spectroscopy. For the purpose of analyzing the vibrational spectrum, we can treat molecules as being made of balls (atoms) and springs (bonds), a description known as the **harmonic oscillator approximation**. The vibrational energy levels of a harmonic oscillator are given by

$$E_v = h\nu_0\left(v + \frac{1}{2}\right), \tag{3.6}$$

where $\nu_0$ is a constant known as the **natural frequency**. Stiffer chemical bonds or, in general, vibrational motions (on which more below) correspond to higher natural frequencies. Note once again the phenomenon of zero-point energy: there is a minimum vibrational energy of $\frac{1}{2}h\nu_0$. More importantly, from the point of view of analyzing the spectrum, the energy levels of a harmonic oscillator are equally spaced by $h\nu_0$. In a real molecule, this will only approximately be true, but it does tell us where to look for additional lines in the spectrum; given that the $v = 0$ to $v = 1$ spectrum is centered near $2900\,\mathrm{cm^{-1}}$, then the $v = 0$ to $v = 2$ spectrum should be centered near $2 \times 2900\,\mathrm{cm^{-1}} = 5800\,\mathrm{cm^{-1}}$, the $v = 0$ to $v = 3$ spectrum should be found near $3 \times 2900\,\mathrm{cm^{-1}} = 8700\,\mathrm{cm^{-1}}$ and so on. If we examine the spectrum of HCl over a very wide range of wavenumbers, we indeed find spectra that look qualitatively like the one shown in Figure 3.6 in the expected regions, and nothing in between. This is how we know that the assignment of this vibrational band to the $v = 0$ to $v = 1$ transition is correct.

You may have noticed that each line in the spectrum has a twin: a tall line is accompanied by a shorter copy at slightly lower wavenumber. This is because chlorine has two stable isotopes: $^{35}\mathrm{Cl}$ (abundance 76%) and $^{37}\mathrm{Cl}$ (abundance 24%). The taller lines are therefore associated with the $H^{35}\mathrm{Cl}$ isotopomer, and the shorter lines with the $H^{37}\mathrm{Cl}$ isotopomer. The shift in wavenumber between the two isotopes can be rationalized based on the difference in mass; all other things being equal, heavier species should vibrate more slowly, i.e. have smaller natural frequencies. Given the difference in natural frequency, when analyzing the spectrum, we need to analyze the data for the two isotopomers separately.

There turns out to be a selection rule for asymmetric molecules that forbids a pure vibrational transition, i.e. one with $\Delta J = 0$. The central lines in the spectrum for each isotopomers thus correspond to the rotational transitions $\Delta J = \pm 1$. The average of these two lines would be equivalent to a pure vibrational transition, with $\Delta E = h\nu_0$.

**Example 3.3 Determining the natural frequency**  The center of the spectrum contains the following lines:

| $\tilde{\nu}/\mathrm{cm^{-1}}$ | $A$ |
|---|---|
| 2863.3 | 0.12 |
| 2865.3 | 0.22 |
| 2904.4 | 0.12 |
| 2906.5 | 0.22 |

The two lines at slightly lower wavenumbers with the lower absorbances belong to $H^{37}\mathrm{Cl}$, while the two other lines belong to $H^{35}\mathrm{Cl}$. I will show calculations for $H^{35}\mathrm{Cl}$, and leave the analysis of $H^{37}\mathrm{Cl}$ to an exercise.

The average of the two central lines gives $\tilde{v}_0$, the natural frequency expressed as a wavenumber:

$$\tilde{v}_0 = \frac{1}{2}\left(2865.3 + 2906.5\,\text{cm}^{-1}\right) = 2885.9\,\text{cm}^{-1}.$$

Given the relationship between wavelength and frequency (Equation 2.1) and the fact that wavenumber is just the reciprocal of wavelength, we have

$$v_0 = c\tilde{v}_0 = (2.997\,924\,58 \times 10^8\,\text{m}\,\text{s}^{-1})(2885.9\,\text{cm}^{-1})(100\,\text{cm}\,\text{m}^{-1})$$
$$= 8.6517 \times 10^{13}\,\text{Hz}.$$

In molecules with more than two atoms, there are several ways in which a molecule can vibrate, which we call **vibrational modes**. A water molecule, for example, has three different vibrational modes, which can be described as follows:

(1) The bond angle can vibrate, i.e. the angle will alternately expand and contract around its equilibrium value.
(2) The molecule can undergo a symmetric stretch in which the two bonds stretch and contract together.
(3) The molecule can undergo an antisymmetric stretch in which one bond stretches while the other contracts, followed by a reversal in which the bond that was stretching contracts, and vice versa.

Each of these modes would have its own vibrational frequency.

In general, we can calculate the number of vibrational modes of a molecule as follows. Each atom in a molecule can move in three different ways (corresponding to the three Cartesian axes in space), so a molecule has a total of $3N$ modes of motion. Three of those modes correspond to translational motion of the whole molecule. A molecule also has rotational modes. A non-linear molecule has three rotational modes, again corresponding to the three Cartesian axes. A linear molecule, however, only has two rotational modes; if we pick a coordinate system with the molecule lined up along the $z$ axis, there are rotational modes associated with rotation around the $x$ and $y$ axes, but turning the molecule around the $z$ axis doesn't actually move anything. The number of modes of motion that are not translational or rotational are the vibrational modes. Thus, we have the following formulas:

$$\text{Number of vibrational modes} = \begin{cases} 3N - 6 & \text{(non-linear molecules)} \\ 3N - 5 & \text{(linear molecules)} \end{cases}$$

Not all of these vibrational modes will show up in an infrared absorption experiment, because of an additional selection rule: vibrational absorption is only possible if the dipole moment of the molecule changes during vibration. The dipole moment **p** is a vectorial quantity, calculated by

$$\mathbf{p} = \sum_i q_i \mathbf{r}_i,$$

where $q_i$ is the charge of atom $i$ and $\mathbf{r}_i$ is its position vector. The sum is taken over all atoms. The dipole moment is a time-dependent quantity since atoms move as a result of molecular vibration. In some cases, a particular vibrational mode has no effect on the dipole moment. For example, oxygen and nitrogen have zero dipole moments, and their vibration does not change that. Accordingly, these molecules do not absorb in the infrared. For heteronuclear diatomics like HCl, vibration changes the distance between the atoms and thus the dipole moment. Accordingly, HCl does have a vibrational spectrum. In polyatomic molecules, we have to look at each vibrational mode. Some will satisfy the selection rule and be IR active, while others will not. All three vibrational modes of water, for example, are IR active.

If we turn our attention to carbon dioxide for a moment, we see that it should have $3(3) - 5 = 4$ vibrational modes. These modes are a symmetric stretch, an antisymmetric stretch and two degenerate bending modes. Why two bends? Imagine lining up the molecule along the $z$ axis. It can be bent in the $xz$ plane, and in the $yz$ plane. Of these modes, the symmetric stretch is IR inactive, while the other three modes are IR active. The three active modes make $CO_2$ a greenhouse gas. Of course, so is water, and there is a lot more of that in the atmosphere than there is of carbon dioxide. Why then do we worry so much about carbon dioxide? There are several reasons. First and foremost, by adding a lot of carbon dioxide to the atmosphere, we are disturbing the equilibrium between solar energy reaching the planet, infrared emission from the surface and the greenhouse effect. Even a relatively small disturbance to this equilibrium will have significant effects over time, particularly since warming tends to feed on itself; a small amount of warming due to increased carbon dioxide will lead to more water vapor in the air, which will magnify the initial greenhouse effect of the carbon dioxide itself. Melting of the arctic permafrost will allow additional carbon dioxide to reach the atmosphere due to the decomposition of previously frozen organic matter. These are just a couple of the many complex feedbacks that will result in global warming. There are mitigating effects as well (e.g. increased cloud cover, which reflects some of the Sun's light back into space), but on the whole, the forecast is for warmer weather. The other reason that carbon dioxide is so important is that it absorbs strongly in a wavenumber range where water doesn't absorb, specifically in a band from 2250 to 2400 cm$^{-1}$. It therefore plugs a hole in the planet's emission spectrum where quite a bit of radiation would otherwise be lost to space.

---

### Exercise group 3.2

(1) The central lines in the rotation–vibration spectrum of H$^{37}$Cl are at 2863.3 and 2904.4 cm$^{-1}$. Determine the natural frequency of H$^{37}$Cl.
(2) How many normal modes of vibration would a methane molecule have?
(3) The normal modes of HCN can be roughly described as follows:
   - H–C stretching
   - C–N stretching
   - two degenerate bending modes
   (a) Which of these modes would you expect to be IR active?

(b) The bending modes have a fundamental frequency (in wavenumber equivalents) of $946\,\mathrm{cm}^{-1}$. What is the ratio of HCN molecules in the first excited state relative to the ground state for this mode of motion at $20\,^{\circ}\mathrm{C}$?

### 3.4.4 Electronic spectra

Electronic (UV/visible) spectroscopy will involve a combination of electronic, vibrational and rotational transitions. Because we will typically carry out these experiments in solution, the bands are often widened to the point where even the vibrational structure is lost. Rotational structure is never observed in solution-phase UV/visible spectroscopy. First and foremost, UV/visible spectra give us information about the spacing between electronic energy levels. In the spectrum of $KMnO_4$, for example (Figure 3.7), we see two transitions from the ground electronic state, one with maximum intensity at $311\,\mathrm{nm}$, and the other with maximum intensity at $526\,\mathrm{nm}$. (We typically use the point of maximum intensity to characterize UV/visible bands, even though this involves a mixture of electronic and vibrational excitation.) These wavelengths can be converted to electronic energy level spacings between the ground state and two of the electronic excited states. We can thus start to build the energy level diagram of the permanganate ion. (Potassium ions don't absorb in the UV/visible range.)

**Example 3.4 Electronic energy levels of the permanganate ion** Energy levels of molecules are given in many different kinds of units, depending on the application. Often, electronic energy levels are given in electron-volts (eV), a convenient unit when dealing with phenomena on the molecular scale. Our two transition energies for the permanganate ion are

$$E_1 = \frac{hc}{\lambda} = \frac{(6.626\,068\,8 \times 10^{-34}\,\mathrm{J\,Hz^{-1}})(2.997\,924\,58 \times 10^8\,\mathrm{m\,s^{-1}})}{526 \times 10^{-9}\,\mathrm{m}}$$

$$= 3.78 \times 10^{-19}\,\mathrm{J}$$

$$\equiv \frac{3.78 \times 10^{-19}\,\mathrm{J}}{1.602\,176\,46 \times 10^{-19}\,\mathrm{J\,eV^{-1}}} = 2.36\,\mathrm{eV}.$$

$$E_2 = \frac{hc}{\lambda} = \frac{(6.626\,068\,8 \times 10^{-34}\,\mathrm{J\,Hz^{-1}})(2.997\,924\,58 \times 10^8\,\mathrm{m\,s^{-1}})}{311 \times 10^{-9}\,\mathrm{m}}$$

$$= 6.39 \times 10^{-19}\,\mathrm{J}$$

$$\equiv \frac{6.39 \times 10^{-19}\,\mathrm{J}}{1.602\,176\,46 \times 10^{-19}\,\mathrm{J\,eV^{-1}}} = 3.99\,\mathrm{eV}.$$

There is therefore one excited state about $2.36\,\mathrm{eV}$ above the ground state, and another $3.99\,\mathrm{eV}$ above ground. Note that we can't be sure that there aren't other electronic energy levels in this energy range that we haven't observed. For example, if two electronic states are very close in energy, we wouldn't be able to tell them apart. Moreover, there are selection rules for electronic spectroscopy that might make some states inaccessible in a simple absorption experiment.

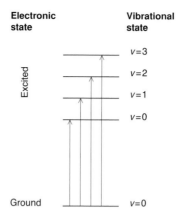

Figure 3.8 Sketch of the electronic/vibrational energy levels of the permanganate ion, with some possible transitions shown.

In many cases, solution UV/visible bands are broadened to the point where vibrational transitions are completely obscured. In others, $KMnO_4$ being an example, we can see vibrational structure in the UV/visible spectrum. These combined vibrational and electronic or **vibronic transitions** are particularly clearly resolved in the lower-energy band of the spectrum of $KMnO_4$, allowing us to recover the vibrational frequency of one of the vibrational modes of the permanganate ion. To understand what happens in this experiment, consider Figure 3.8, which sketches the relevant energy levels. Note that, at room temperature, all of the molecules should be starting out in the ground electronic and vibrational state, as shown in the figure. Since the vibrational energy levels of a given electronic state are roughly equally spaced, the difference between two adjacent transition energies (the arrows in Figure 3.8, corresponding to the vibrational bands in the spectrum) is the vibrational energy level spacing of the excited state.

**Example 3.5 Natural vibrational frequency in permanganate** The following are the wavelengths of the vibrational bands associated with the transition to the 2.36 eV excited electronic state: 507, 526, 546 and 566 nm. Converted to energies, these transitions occur at $3.92 \times 10^{-19}$, $3.78 \times 10^{-19}$, $3.64 \times 10^{-19}$ and $3.51 \times 10^{-19}$ J. The differences between these transition energies should give us the vibrational spacing. Subtracting adjacent terms in this set of transition energies, we get $1.4 \times 10^{-20}$, $1.4 \times 10^{-20}$ and $1.3 \times 10^{-20}$ J. The average vibrational spacing is therefore $1.36 \times 10^{-20}$ J. Note that I kept extra digits in my calculator through the course of this calculation, and used the standard deviation from the mean ($7 \times 10^{-22}$ J) to determine the number of significant figures in the final answer.

From the relationship between the energy level spacing for a harmonic oscillator and the natural frequency, we have

$$\nu_0 = \frac{1.36 \times 10^{-20}\,\text{J}}{6.626\,068\,8 \times 10^{-34}\,\text{J\,Hz}^{-1}} = 2.05 \times 10^{13}\,\text{Hz}.$$

The vibrational selection rules for vibronic transitions are different from those for pure vibrational (or rotation–vibration) transitions. The result is that vibrational transitions that would be forbidden in vibrational spectroscopy are often observed in UV/visible spectra. Again, $KMnO_4$ provides a good example. The permanganate anion has $3(5) - 6 = 9$ vibrational modes. It turns out that the vibrational mode observed in the UV/visible spectrum is the symmetric stretch of the tetrahedral $MnO_4^-$ anion, an IR inactive mode. UV/visible spectroscopy is not the only spectroscopy that allows us to study IR-forbidden vibrational transitions; these may also show up in fluorescence spectra (on which more below) and in Raman spectra (a scattering method we will not study), among other spectroscopies. What this shows us is that various spectroscopies offer complementary information, so if you want to study the structure of a molecule, you often have to pull together information from several spectroscopic techniques.

### 3.4.5 Beer–Lambert law

We now turn to the quantitation of the amount of light absorbed by samples. The absorption of a single photon is a probabilistic event: in order to be absorbed, it has to meet a molecule with the correct energy level spacing. As a photon passes through a thin slice of the sample of thickness $dx$, there is some probability $p\,dx$ that this will occur and that the photon will be absorbed. Here, $p$ is the probability of absorption per unit length. If we have a lot of photons, then we can convert this probability for a single photon into a number of photons absorbed: the number of photons will change by $dn = -np\,dx$ or, since the number of photons is proportional to the intensity of the light, $dI = -Ip\,dx$. We can solve this equation for $I(x)$, where $x$ is the distance traveled through the medium, by **separation of variables**. First, rearrange to put all terms that depend on $I$ on one side of the equation, and all terms that depend on $x$ on the other:

$$\frac{dI}{I} = -p\,dx.$$

Note that the quantity on the left-hand side is the fractional change in intensity over a distance $dx$. Now we integrate both sides of this equation to add up the fractional changes in intensity:

$$\int_{I_0}^{I(x)} \frac{dI'}{I'} = -p \int_0^x dx'.$$

The primed variables are just there to distinguish the variables inside the integrals from those appearing in the limits of integration. The limits of integration correspond to an intensity of $I_0$ as the beam enters the medium at $x = 0$, and to an intensity $I(x)$ at a depth $x$. Evaluating the integrals, we get

$$\ln I' \big|_{I_0}^{I(x)} = -p\,x' \big|_0^x.$$

$$\therefore \ln I(x) - \ln I_0 = \ln\left(\frac{I(x)}{I_0}\right) = -p(x - 0) = -px.$$

$$\therefore px = -\ln\left(\frac{I(x)}{I_0}\right) = \ln\left(\frac{I_0}{I(x)}\right).$$

(If you have trouble following the manipulations of the logarithms here, review the rules in Appendix H.) In a spectroscopy experiment, we measure the intensity of light that is transmitted through the sample. If the path length through the sample is $\ell$, this would be $I(\ell)$:

$$p\ell = \ln\left(\frac{I_0}{I(\ell)}\right).$$

Moreover, we usually compare this intensity to that of the incident radiation, $I_0$. You can see that the equation we have arrived at contains the ratio $I_0/I(\ell)$, which is very convenient. By convention, the **absorbance** is defined as

$$A_\lambda = \log_{10}\left(\frac{I_0}{I(\ell)}\right) = \ln\left(\frac{I_0}{I(\ell)}\right) / \ln 10, \tag{3.7}$$

where the subscript $\lambda$ indicates that the absorbance is measured at a particular wavelength. Thus we have

$$A_\lambda = p\ell / \ln 10.$$

As for the constant $p$, recall that it gives the probability per unit length that a photon is absorbed. This quantity must be proportional to the concentration of the absorbing species: if we double the concentration, we double the probability that a photon will encounter a molecule. Thus, $p = qc$, where $c$ is the concentration of absorbers, and $q$ is another constant. If we combine $q$ with the logarithm, we get

$$A_\lambda = \varepsilon_\lambda c \ell.$$

This equation is known as the **Beer–Lambert law**. The constant $\varepsilon_\lambda$ is called the **molar absorption coefficient**. Path lengths are usually measured in cm and concentrations in mol $L^{-1}$, so $\varepsilon_\lambda$ has units of $L\,mol^{-1}cm^{-1}$ since $A$ is dimensionless. Note that the molar absorption coefficient depends on wavelength, although it is common for people to report just one value, namely the value at the wavelength of maximum absorbance. Another common practice is to report **specific absorption coefficients** rather than the molar coefficients, the difference between the two being that the specific coefficient relates to concentrations in $g\,L^{-1}$ rather than in mol $L^{-1}$. Finally note that spectroscopic studies in which we specifically measure or apply the relationship between absorbance and concentration are sometimes described as **spectrophotometry**, and the instrument used for this purpose would then be referred to as a **spectrophotometer**. In practice, almost any spectrometer can be used as a spectrophotometer, although you may occasionally run into pure spectrophotometers, i.e. machines that record data at a single wavelength selected ahead of time by the experimentalist.

**Example 3.6 Absorbance and the molar absorption coefficient** Ferredoxin is a redox enzyme found in a variety of organisms. Ferredoxin extracted from spinach has an absorption maximum at 485 nm with a molar absorption coefficient of $1.2 \times 10^4\,L\,mol^{-1}cm^{-1}$. What absorbance would you predict at 485 nm for a 60 $\mu$mol $L^{-1}$ solution in a standard cuvette with a 1 cm path length, and what fraction of the photons of this wavelength passing through the sample would be absorbed?

The first part is a straightforward application of the Beer–Lambert law:

$$A_{485} = \varepsilon_{485}c\ell = (1.2 \times 10^4 \, \text{L mol}^{-1}\text{cm}^{-1})(60 \times 10^{-6} \, \text{mol L}^{-1})(1 \, \text{cm}) = 0.72.$$

The absorbance is defined by Equation (3.7):

$$\log_{10}\left(\frac{I_0}{I(\ell)}\right) = 0.72;$$

$$\therefore \frac{I_0}{I(\ell)} = 10^{0.72} = 5.25;$$

$$\therefore \frac{I(\ell)}{I_0} = \frac{1}{5.25} = 0.191.$$

About 19% of the photons would get through, so $100 - 19\% = 81\%$ would be absorbed.

What if there is more than one substance absorbing at a given wavelength? If you look back to our derivation of the Beer–Lambert law, you will see that this would affect the photon absorption probability $p$. There would be some probability of absorption by each species absorbing at the wavelength of interest. Thus, we would get

$$A_\lambda = \ell \sum_i \varepsilon_{\lambda,i} c_i, \tag{3.8}$$

where the sum is taken over all species that absorb at wavelength $\lambda$. We can play all kinds of tricks with this equation. The following example shows how measurements of the absorbance at different wavelengths can be used to calculate concentrations of mixtures if we know the molar absorption coefficients of the components.

**Example 3.7 Concentrations of mixtures from spectroscopic measurements** The molar absorption coefficients at 414 and 425 nm of the oxidized and reduced forms of cytochrome b5 from *Tetrahymena pyriformis* are as follows:[3]

|  | $\varepsilon_\lambda/10^5 \, \text{L mol}^{-1}\text{cm}^{-1}$ | |
|---|---|---|
|  | 414 nm | 425 nm |
| Oxidized | 1.403 | 0.593 |
| Reduced | 0.949 | 2.215 |

During a purification procedure, a spectrum of a solution containing cytochrome b5 is obtained, and the following absorbances are determined in a standard 1 cm cuvette: $A_{414} = 0.25$ and $A_{425} = 0.18$. We would like to know if the enzyme is mostly in the oxidized or mostly in the reduced form, as this might affect further purification steps. To determine the

---

[3] Inferred from the data in H. Fukushima, S. Umeki, T. Watanabe and Y. Nozawa, *Biochem. Biophys. Res. Commun.* **105**, 502 (1982).

concentrations of the two forms, we use Equation (3.8) at each of the two wavelengths:

$$A_{414} = 0.25 = \varepsilon_{414,ox}\ell[ox] + \varepsilon_{414,red}\ell[red]$$
$$= (1.403 \times 10^5 \, L\,mol^{-1})[ox] + (9.49 \times 10^4 \, L\,mol^{-1})[red].$$
$$A_{425} = 0.18 = \varepsilon_{425,ox}\ell[ox] + \varepsilon_{425,red}\ell[red]$$
$$= (5.93 \times 10^4 \, L\,mol^{-1})[ox] + (2.215 \times 10^5 \, L\,mol^{-1})[red].$$

These are two equations in the two unknowns [ox] and [red]. If we solve these two equations, we get

$$[ox] = 1.5 \times 10^{-6} \, mol\,L^{-1},$$
$$[red] = 4.1 \times 10^{-7} \, mol\,L^{-1}.$$

A quick note on solving equations: you can solve these equations by hand using techniques you would have learned in high school. However, most of us now have powerful calculators that can solve these problems for us, if only we learn to use them properly. I strongly suggest that you spend a bit of time with your calculator's manual to learn to use it as a tool to solve equations. In the long run, this will save you a lot of time, particularly since solving equations in multiple unknowns is an error-prone process.

Getting back to our problem, we see that this sample is mostly in the oxidized form, although some is still in the reduced form. If, for example, the next purification step required the protein to be in the oxidized form, we could add an oxidizing agent to shift the remaining reduced protein to the oxidized form, and we could check if we have succeeded by taking another spectrum and repeating this calculation. We might also want to know the total concentration of cytochrome b5, which is just the sum of these two values, so $1.9 \times 10^{-6} \, mol\,L^{-1}$.

---

### Exercise group 3.3

(1) Spectrophotometry is often used to follow chemical reactions in time. Because instrument response can become non-linear at higher absorbances, it's often a good idea to keep the absorbance below 1. Suppose that you are studying an enzyme that converts NADH to its reduced form, $NAD^+$. NADH has a molar absorption coefficient of $6220 \, L\,mol^{-1}cm^{-1}$ at 339 nm, while $NAD^+$ doesn't absorb at this wavelength. What is the maximum concentration of NADH you should use in order to keep the absorbance at 339 nm below 1 in a 1 cm cuvette?

(2) In protein science, instead of giving the molar absorption coefficient, the absorbance of a 1% (by weight) solution is sometimes given. This quantity, denoted $A^{1\%}$, is a particularly useful way to give the absorptivity when the molar mass of the protein is not known. If we do know the molar mass, then we can determine the molar absorption coefficient from $A^{1\%}$.

(a) The molar mass of alkaline phosphatase from calf intestine is $140\,000\,\text{g mol}^{-1}$. (A biochemist would say either $140\,000\,\text{Da}$ or $140\,\text{kDa}$. The dalton (Da) is equal to a gram per mole.) Calculate the concentration (in $\text{mol L}^{-1}$) of a 1% solution of alkaline phosphatase in water at $25\,°\text{C}$.

*Hint*: Assume we have 1 L of water to which enough alkaline phosphatase has been added to make a 1% solution.

(b) $A^{1\%}$ always refers to the absorbance in a cell with a path length of 1 cm. For calf-intestine alkaline phosphatase,[4] $A^{1\%}_{278\,\text{nm}} = 7.6$. Calculate the molar absorption coefficient.

(3) When we want to determine the molar absorption coefficient accurately, we typically measure the absorbance at several concentrations, then plot the results to verify that the Beer–Lambert law is satisfied. If it is, we should get a straight line passing through the origin. This is not always the case. On occasion, we encounter a substance whose $A$ vs. $c$ curve at fixed wavelength is curved.

(a) Suppose that we blindly calculated the molar absorption coefficient by $\varepsilon = A/c\ell$ at each concentration in a case where the $A$ vs. $c$ relationship is non-linear. What would we see in the results? Would the $\varepsilon$ values calculated by this method be useful?

(b) Kageyama has developed a method for determining the concentration of uric acid in blood and urine.[5] The method involves a series of chemical reactions that quantitatively convert uric acid to a yellow product (3,5-diacetyl-1,4-dihydrolutidine). The absorbance at 410 nm is then recorded. He recorded the following data, where [U] represents the concentration of uric acid in the calibration sample:

| $[U]/\text{mg L}^{-1}$ | 50 | 100 | 150 | 200 |
|---|---|---|---|---|
| $A_{410}$ | 0.14 | 0.271 | 0.406 | 0.545 |

If you plot these data, you will see that they fit a straight line through the origin beautifully.

(i) The absorbance is due to the yellow product. Assuming 1:1 conversion of uric acid to the yellow product, calculate the concentration of yellow product, in $\text{mol L}^{-1}$, in each of the samples. The molar mass of uric acid is $168.110\,\text{g mol}^{-1}$.

(ii) If you plot $A_{410}$ vs. the concentration of yellow product, the slope should be $\varepsilon\ell$. As usual, $\ell = 1\,\text{cm}$. Get the slope of the graph by linear regression, and report the molar absorption coefficient.

*Note*: This will not be the last time we need to do a regression in this course. You should learn how to use your calculator to do regressions.

---

[4] M. Fosset, D. Chappelet-Tordo and M. Lazdunski, *Biochemistry* **13**, 1783 (1974).
[5] N. Kageyama, *Clin. Chim. Acta* **31**, 421 (1971).

(4) Chlorophyll *a* and *b* have overlapping spectra. Their specific absorption coefficients at 643 and 660 nm in diethyl ether solution are as follows:

|                | $\varepsilon_\lambda/\mathrm{L\,g^{-1}cm^{-1}}$ | |
|----------------|:--------:|:--------:|
|                | 643 nm   | 660 nm   |
| Chlorophyll *a* | 16.3    | 102      |
| Chlorophyll *b* | 57.5    | 4.50     |

(a) A sample obtained from 0.104 g of a plant leaf extracted using 100 mL of ether and then diluted fivefold gave the following data: $A_{643} = 0.203$ and $A_{660} = 0.625$. Determine the concentrations of chlorophyll *a* and *b* in the samples analyzed, then calculate the masses of the two chlorophylls in the original sample, and finally report the mass of each chlorophyll per unit mass of leaf in $\mathrm{mg\,g^{-1}}$. The latter step normalizes the data to the amount of leaf material present. This facilitates comparisons between samples.

(b) Convert the specific absorption coefficients to molar absorption coefficients. The molar mass of chlorophyll *a* is $893.50\ \mathrm{g\,mol^{-1}}$, while the molar mass of chlorophyll *b* is $907.49\ \mathrm{g\,mol^{-1}}$.

## 3.5 Fluorescence

Consider a molecule that has been excited by absorption of a UV photon. As we have previously seen, this is an amount of energy normally associated with electronic transitions. Moreover, electronic excitation is usually accompanied by vibrational excitation. What happens to the energy gained by the molecule in the absorption process? Conceivably, some or all of it could be immediately re-emitted as a UV photon. Commonly, however, the vibrational excitation is dissipated first. This can happen by a number of mechanisms: redistribution to other vibrational modes, emission of an infrared photon or transfer to other molecules during collisions. Whichever of these mechanisms happens to take away the vibrational energy, the net effect is the same: the molecule ends up in the vibrational ground state of an electronically excited state. It may then go back to the ground electronic state by emitting a photon, which will necessarily be lower in energy than the photon it originally absorbed. This emission process is called **fluorescence**. An energy level diagram showing one possible set of transitions for a fluorescent molecule is shown in Figure 3.9.

Fluorescence spectrometers (Figure 3.10) contain essentially the same parts as absorption spectrometers (Figure 3.3), but these parts are arranged differently. Fluorescent emission is **isotropic**, meaning that the photons are emitted in every direction. That being the case, in order to avoid interference from the light transmitted through the sample, we place the detector in a fluorescence experiment at 90° to the incident beam. Furthermore, we need two monochromators in a fluorescence spectrometer, one for the exciting radiation, and one for the fluorescent emission. In some fluorescence spectrometers, both the excitation and emission wavelengths can be scanned. Scanning the excitation wavelength

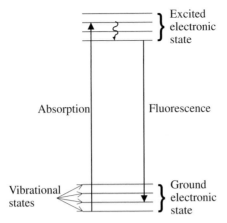

Figure 3.9 Energy level diagram for a fluorescent molecule. Absorption takes us from the ground electronic and vibrational state to an excited state which, in general, will be both electronically and vibrationally excited. The wavy arrow indicates vibrational de-excitation. Fluorescence is the subsequent emission process. It takes the molecule back to the ground electronic state, but may leave the molecule vibrationally excited, as shown.

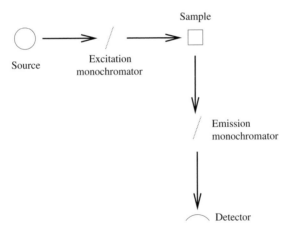

Figure 3.10 Block diagram of a fluorescence spectrometer.

generates a spectrum that is essentially identical to the absorption spectrum, provided an appropriate emission wavelength is selected for observation. A true fluorescence spectrum obtained by scanning the emission wavelength is shifted to longer wavelengths (lower energies) and is often roughly a mirror image of the absorption spectrum. Why? In the absorption process, the molecule goes from the ground-electronic/ground-vibrational state to an excited-electronic/excited-vibrational state corresponding to any one of the possible values of the vibrational quantum number $v$. In the fluorescent emission process, the molecule starts out in the excited-electronic/ground-vibrational state and goes to a ground-electronic/excited-vibrational state, also corresponding to any one of the possible values of $v$. Provided the vibrational levels of the ground and excited electronic states are similar, this produces a mirror-image spectrum. If the fluorescence spectrum turns out not to look like a mirror-image of the absorption spectrum, we then know that the excited state has a different set of vibrational levels than the ground state, which would imply a significantly different excited-state geometry. In other words, the absorption and fluorescence spectra together give us structural information about the electronically excited state.

Some of the most important applications of fluorescence use the fact that emission occurs at a longer wavelength than excitation. In imaging applications, we can, for instance, tag a cellular component with a fluorescent marker, and then filter out both the scattered exciting radiation and other light emanating from the cell (e.g. fluorescence from other sources at different wavelengths), allowing us to see where in the cell a tagged molecule localizes, or perhaps to track its movements through the cell. These techniques have become particularly powerful due to the discovery of green fluorescent protein (GFP) and the development of other-colored fluorescent proteins (yellow fluorescent protein, YFP; enhanced blue fluorescent protein, EBFP; enhanced cyan fluorescent protein, ECFP etc.). Because of the modularity of protein structure, it is often possible to add a sequence for one of these fluorescent proteins to the end of a gene and to get an expressed protein that both fulfills its original function and is fluorescent in a desired spectral range. The original isolation of GFP from a jellyfish by Osamu Shimomura and subsequent contributions to the development of fluorescent proteins into a versatile imaging technology by Martin Chalfie and Roger Tsien were recognized by the award of the 2008 Nobel Prize in Chemistry to these three scientists.

### 3.5.1  *Fluorescence quenching and energy transfer*

The part of a molecule responsible for fluorescence is often relatively simple. For example, the amino acids tryptophan, tyrosine and phenylalanine are fluorescent. In GFP, on the other hand, a triplet of amino acids is responsible for fluorescence. A chemical group that is responsible for fluorescence is called a **fluorophore**. In general, a chemical group that is responsible for the color of a substance, whether because it absorbs visible light or because it fluoresces, is called a **chromophore**.

Now imagine that we put a fluorophore and a chromophore in close proximity to each other, and furthermore that the emission spectrum of the fluorophore overlaps the absorption spectrum of the chromophore. If we excite the fluorophore at a wavelength where it absorbs, it will undergo vibrational de-excitation as usual, but after that, it may transfer its energy to the chromophore rather than fluorescing. This process is called **Förster resonant energy transfer**, or **FRET** for short. As a result of the FRET effect, we would observe reduced fluorescence from the fluorophore, and we would say that its fluorescence has been **quenched**. The chromophore is then known as a **fluorescence quencher**, or simply as a quencher. The fluorophore can also be described as a **donor** and the chromophore as an

**acceptor** in this process. Which terminology we use depends on what we want to emphasize, particularly since there are other mechanisms that lead to fluorescence quenching. We will concentrate here on FRET.

In one common variation on the FRET theme, the acceptor is also fluorescent. After receiving the energy of the donor, the acceptor would go through vibrational de-excitation, and then fluoresce at a longer wavelength than the donor would have. We can therefore detect the energy transfer as fluorescence of the acceptor.

The efficiency of FRET will depend on several factors:

(1) The overlap between the emission (fluorescence) spectrum of the donor and the absorption spectrum of the acceptor. The larger this overlap, i.e. the more similar the emission spectrum of one is to the absorption spectrum of the other, the more efficient the energy transfer process.
(2) If the donor and acceptor chromophores are in separate molecules, then the efficiency of transfer is proportional to the concentration of the acceptor.
(3) The distance between the donor and acceptor. The closer to each other the two are, the more efficient the energy transfer process is. In some cases, we can position a donor and an acceptor on the same molecule. In other cases, we might use the distance dependence of FRET to study the formation of a complex between the molecules containing the donor and acceptor. Either way, we can use FRET to measure the distance between the donor and acceptor, i.e. it can be applied as a molecular ruler.

FRET efficiency is defined as follows:

$$\eta_{\text{FRET}} = 1 - \frac{I_{f,\text{DA}}}{I_{f,\text{D}}}$$

where $I_{f,\text{D}}$ is the fluorescence intensity of the donor in the absence of the acceptor, and $I_{f,\text{DA}}$ is the fluorescence intensity of the donor in the presence of the acceptor. A FRET efficiency of 0 means that none of the excitation is transferred to the acceptor, while a FRET efficiency of 1 means that all of the donor's excitation is transferred to the acceptor. Alternatively, in the variation where the acceptor is fluorescent, the efficiency can be defined as

$$\eta_{\text{FRET}} = \frac{I_{f,\text{AD}}}{I_{f,\text{D}}}$$

where $I_{f,\text{AD}}$ is the fluorescence intensity of the acceptor due to resonant energy transfer from the donor.

The fluorescence intensity is often reported as a **quantum yield**, usually denoted by the symbol $\phi$. The fluorescence quantum yield is the fraction of absorbed photons that are eventually fluoresced. This can be obtained either by comparison with a well-characterized fluorescent standard, or by careful accounting of the photons absorbed and emitted. Both of these are difficult to do correctly. These problems are often circumvented by measuring the FRET efficiency using kinetic methods, which we will study in a later chapter of this book.

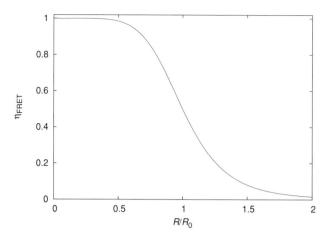

Figure 3.11 FRET efficiency as a function of $R/R_0$.

The FRET efficiency turns out to depend on distance between the donor and acceptor as follows:

$$\eta_{\text{FRET}} = \frac{1}{1 + (R/R_0)^6},\tag{3.9}$$

where $R$ is the distance between the two, and $R_0$ is a constant that depends on the particular donor/acceptor pair we are using. The efficiency is plotted as a function of $R/R_0$ in Figure 3.11. Given $R_0$ and a measurement of the efficiency, we can calculate the distance between the donor and acceptor, subject to some caveats. When $R \ll R_0$, the efficiency approaches 100%, i.e. the donor's excitation is always transferred to the acceptor. For values of $R \gg R_0$, the efficiency of energy transfer tends to zero. FRET will give us the most information about distance when $R$ is of a similar size to $R_0$, i.e. where the efficiency depends most sensitively on the distance. Designing a FRET experiment therefore requires careful selection of a donor/acceptor pair based on an initial structural hypothesis regarding the likely distance between the two chromophores.

**Example 3.8 Distance between A and P sites of the ribosome**  During protein translation, a tRNA (transfer RNA) carrying the next amino acid to be added to the protein enters the ribosome's A (aminoacyl) site. A peptide bond is formed to the amino acid carried by the tRNA in the P (peptidyl) site, at the same time breaking the bond between the P site tRNA and the nascent protein. Then the two tRNAs are pushed along, with the one in the A site moving to the P site, and the one in the P site moving to the E (exit) site. In a classic study, Johnson and coworkers[6] made a short mRNA sequence $\text{AUGU}_n$, with $n = 4$ or 5. AUG is the start codon, and simultaneously incorporates $N$-formylmethionine (fMet) as the first amino acid in the protein in bacteria. UUU encodes phenylalanine (Phe). They

---

[6]  A. E. Johnson *et al.*, *J. Mol. Biol.* **156**, 113 (1982).

labeled the tRNA for fMet with the fluorescent dye AEDANS (donor), and the tRNA for Phe with fluorescein (acceptor). They measured $R_0$ for this donor-acceptor pair, and found it to average about 31 Å. The mRNA, donor, acceptor and ribosomes were then incubated together. This should allow the formation of a complex with tRNA$^{fMet}$ in the P site and tRNA$^{Phe}$ in the A site. In one set of experiments, the quantum yield of the donor, $\phi_{f,D}$, was found to be 0.158, while the quantum yield of the donor in the presence of the acceptor, $\phi_{f,DA}$, was 0.054. We can use these data to determine the distance between the tRNAs in the A and P sites. First, we compute the FRET efficiency:

$$\eta_{FRET} = 1 - \frac{\phi_{f,DA}}{\phi_{f,D}} = 1 - \frac{0.054}{0.158} = 0.66.$$

Note that we used the quantum yield as a measure of fluorescence intensity. Now, from the relationship between efficiency and distance, we have

$$\left(\frac{R}{R_0}\right)^6 = \frac{1}{\eta_{FRET}} - 1 = 0.52;$$

$$\therefore \frac{R}{R_0} = (0.52)^{1/6} = 0.90;$$

$$\therefore R = 0.90 R_0 = 0.90(31 \text{ Å}) = 28 \text{ Å}.$$

---

## Exercise group 3.4

(1) FRET was a key technique in early studies of the structure of actin. The active form, F-actin, is a polymer made using ATP as an energy source. Actin therefore has a binding site for ATP. The FRET studies were greatly aided by the use of fluorescent analogs of ATP, one of which is $\epsilon$-ATP. In one set of experiments, the fluorescence of $\epsilon$-ATP bound to actin was compared to the fluorescence of $\epsilon$-ATP bound to an actin in which one of actin's amino acids, lysine-61, has been labeled with the fluorescent dye FITC (fluorescein-5-isothiocyanate).[7]
  (a) What is the donor and what is the acceptor in this experiment?
  (b) The fluorescence of $\epsilon$-ATP bound to actin with FITC-labeled Lys-61 is 21% of the fluorescence of $\epsilon$-ATP bound to unlabeled actin. For these dyes, $R_0 = 4.74$ nm. What is the distance between the ATP binding site and lysine-61?
(2) Suppose that you want to measure a distance between two domains of a protein. Based on some modeling, you believe this distance to be in the range 40–60 Å. The protein has a tryptophan in one of the two domains and, fortuitously, no other tryptophans. You therefore decide to use FRET, with tryptophan as the donor, to measure the distance between the two domains. To do this, you will need to introduce a fluorescent dye by chemical modification of the other domain. After thinking about the chemistry,

---

[7] M. Miki, C. G. dos Remedios and J. A. Barden, *Eur. J. Biochem.* **168**, 339 (1987).

you decide that you might be able to insert any of the following dyes into the second domain:

| Acceptor | $R_0/\text{Å}$ |
|---|---|
| Nitrobenzoyl | 16 |
| Anthroyl | 25 |
| DPH | 40 |

Which dye(s) could you use, and why?

(3) FRET can be used to detect and quantify conformational changes in proteins. For example, human apolipoprotein E3 (apoE3) undergoes a conformational change when it binds to a lipid. This protein is made of four helices, one of which contains four tryptophans, while another contains the protein's only cysteine. Tryptophan is fluorescent, and cysteine can easily be chemically modified. In one study, a cysteine was tagged with the acceptor AEDANS.[8] The value of $R_0$ was found to be about 23 Å for a tryptophan/AEDANS dye pair. For the labeled protein by itself, the FRET efficiency was 49%. When a lipid was added, the efficiency decreased to 2.7%.

   (a) What are the average distances between the tryptophans and the labeled cysteine in the two cases?

   (b) What is the likely interpretation of these data?

## Key ideas and equations

- Electronic energy levels are much more widely spaced than vibrational energy levels, which are in turn more widely spaced than rotational levels, leading to the appearance of lines or bands associated with transitions between these levels in different spectral regions.

- $P(E_i) = g_i \exp(-E_i/k_\mathrm{B}T)/q$, with $q = \sum_i g_i \exp(-E_i/k_\mathrm{B}T)$; or $\dfrac{P(E_j)}{P(E_i)} = \dfrac{g_j}{g_i} \exp(-\Delta E/k_\mathrm{B}T)$

- The molecular partition function gives us a rough count of the number of states available to a quantum system at temperature $T$.

- Vibrational energies are approximated by $E_v = h\nu_0 \left(v + \frac{1}{2}\right)$. The natural frequency $\nu_0$ measures the stiffness of a vibrational mode.

- A linear molecule has $3N - 5$ vibrational modes, while a non-linear molecule has $3N - 6$ vibrational modes. Only modes whose corresponding motions change the dipole moment of the molecule are IR active.

- $A_\lambda = \varepsilon_\lambda c \ell$ or, in general, $A_\lambda = \ell \sum_i \varepsilon_{\lambda,i} c_i$.

- $\eta_{\mathrm{FRET}} = 1 - \dfrac{I_{f,\mathrm{DA}}}{I_{f,\mathrm{D}}} = \dfrac{I_{f,\mathrm{AD}}}{I_{f,\mathrm{D}}} = \dfrac{1}{1 + (R/R_0)^6}$.

[8] C. A. Fisher and R. O. Ryan, *J. Lipid Res.* **40**, 93 (1999).

## Suggested reading

One of the most thankless and yet essential jobs in science is the collection and organization of data. In a remarkable series of papers spanning more than a decade, Donald Kirschenbaum painstakingly collected absorption coefficients for proteins from the literature:

Donald M. Kirschenbaum, *Int. J. Protein Res.* **3**, 109, 157, 237, 329 (1971).
—— *Int. J. Pept. Protein Res.* **4**, 63, 125 (1972); **5**, 49 (1973); **13**, 479 (1979).
—— *Anal. Biochem.* **55**, 166 (1973); **56**, 237 (1973); **64**, 186 (1975); **68**, 465 (1975); **80**, 193 (1977); **81**, 220 (1977); **82**, 83 (1977); **87**, 223 (1978); **90**, 309 (1978).
—— *Int. J. Biochem.* **11**, 487 (1980); **13**, 621 (1981).
—— *J. Quant. Spectrosc. Radiat. Transfer* **27**, 23, 39 (1982).
—— *Appl. Biochem. Biotechnol.* **7**, 475 (1982); **9**, 187 (1984); **11**, 287 (1985).

To say that this was a huge job is an understatement. To my knowledge, Kirschenbaum's database of molar absorptivities has never been computerized, nor has anyone continued this very important work. The papers cited above therefore remain an invaluable resource to biochemists.

Section 3.2 contains your first glance at statistical mechanics. Some of the ideas from this section will return throughout this book. If you want to get a more systematic treatment, with an emphasis on the connection of statistical mechanics to thermodynamics, there is a nice little book in the Oxford Chemistry Primers series you might want to get:

Andrew Maczek, *Statistical Thermodynamics*; Oxford University Press: Oxford, 1998.

To learn the basic ideas behind FRET, I would recommend two older, but very readable review papers:

Lubert Stryer, *Annu. Rev. Biochem.* **47**, 819 (1978).
Pengguang Wu and Ludwig Brand, *Anal. Biochem.* **218**, 1 (1994).

Among other things, Wu and Brand's paper contains a table of $R_0$ values for common donor–acceptor pairs.

---

## Review exercise group 3.5

(1) The natural vibrational frequency of oxygen, expressed as a wavenumber, is $1580.19\,\mathrm{cm}^{-1}$.

    (a) This frequency was not measured by IR spectroscopy. Why not? What other technique could have been used?

    (b) At what approximate temperature would you expect the population of the first excited vibrational state to become significant?

    *Hint*: Think about the molecular partition function.

(2) The essentials of laser operation can be described using a diagram with three energy levels (Figure 3.12). The material which is to provide the laser emission is placed in a cavity whose length is a multiple of half the wavelength corresponding to the spacing between energy levels **1** and **2**. The ends of the cavity are mirrors, and one mirror will

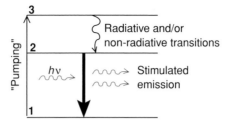

Figure 3.12 Three-level diagram for a laser. Molecules are excited ("pumped") to energy level **3** from which they are rapidly de-excited to energy level **2** by some combination of processes (collisional de-excitation, emission of photons etc.). Some molecules will randomly decay to the ground state, emitting a photon. Stimulated emission occurs when a photon's electromagnetic field induces the emission of an additional photon from an excited molecule.

have a small hole in it. Photons of the appropriate wavelength can therefore form a standing wave in the laser cavity, like the wavefunction for the particle in a box we encountered earlier, but some photons will leak out through the hole. Molecules are excited to energy level **3** from which they are rapidly de-excited to energy level **2**. The initial excitation, called "pumping," can be obtained using a light source of appropriate wavelength, or an electrical discharge, or by any of a number of other means. As photons travel between the two mirrors, their oscillating electromagnetic fields exert forces on the excited molecules which can induce them to emit radiation. This is called stimulated emission. The standing wave therefore tends to build in intensity until the radiation lost through the hole (the business end of the laser) balances the stimulated emission.

(a) Suppose that we have a red laser which emits light with a wavelength of 700 nm. What energy level spacing does this imply for which two levels?

(b) Some authors define temperature using the Boltzmann equation. In other words, they define $T$ such that Equation (3.1) gives the distribution of molecular populations. In lasers, because of the pumping, a population inversion is achieved such that the number of molecules in energy level **2** is larger than the population of energy level **1**. Suppose that for our laser, $N_2/N_1 = 10$. What is the temperature of the lasing medium? What is strange about your answer?

(3) Under given conditions, fluorescence intensities are proportional to concentration. We would therefore define a molar emission coefficient analogously to the molar absorption coefficient: $I_\lambda = \epsilon_\lambda^F c\ell$. Explain why we cannot, however, give a single emission coefficient valid across a range of experiments as we can for the molar absorption coefficient.

# Part Two

## Thermodynamics

# 4

# Thermodynamics preliminaries

Thermodynamics is one of the most useful pieces of science you'll ever learn. Given a little knowledge of thermodynamics, one can understand and explain a broad range of natural phenomena and technological applications. In this book, for instance, we will touch on energy use in living organisms, energy production by fuel-burning engines and spontaneity of chemical processes. This list of topics hardly scratches the surface of the phenomena that can be understood using thermodynamics.

When chemists say "thermodynamics," they usually mean "equilibrium thermodynamics," also known as "classical thermodynamics." This is a theory that emphasizes transformations between equilibrium states. This is important because it is difficult to even define some important quantities like temperature for systems that are out of equilibrium. However, it does mean that we have to be careful how we apply the theory, which requires a certain precision of thought, language and notation. The issues dealt with in the rest of this chapter may seem pedantic or trivial to you, but they are critical prerequisites to understanding anything else in thermodynamics.

## 4.1 The domain of classical thermodynamics

Classical thermodynamics is a theory of equilibrium states and of transformations between these states. What do we mean by equilibrium? A system is in **equilibrium** if its state, under the given conditions, does not change with time. This may seem like an obvious definition, but in thermodynamics, even "state" has a technical meaning: the **state** of a system is the collection of all its observable properties (pressure, volume, chemical composition etc.). Note that the only observable properties we are concerned with in classical thermodynamics are those we can measure with ordinary lab instruments – which we call **macroscopic properties** – and not properties at the atomic or molecular level.

Another limitation of classical equilibrium thermodynamics is that it strictly only applies to **closed systems**, i.e. to systems whose boundaries are impermeable to the flow of matter. This is often an idealization. For example, a beaker containing an aqueous solution is not a closed system since there is a continual exchange of water molecules between the liquid phase and the atmosphere. However, if you are studying a reaction that occurs in solution and your experiment doesn't last too long, it might be a good approximation to pretend that the beaker is a closed system.

## 4.2 Temperature, heat and thermometers

Words that have very precise meanings in science can sometimes mean something subtly different in everyday language. Worse yet, commonly used phrases can sometimes give the wrong impression about a concept. You may be surprised to see a section on temperature, heat and thermometers in this book since you probably think you know what these concepts mean. Perhaps you do, but in my experience, students often have trouble clearly distinguishing between heat and temperature. Anecdotally, I've heard of one graduate student who failed a Ph.D. comprehensive exam because he didn't know the difference between these two concepts. For all I know, this is just a graduate school urban legend, but just to be safe, let's make sure that *you* know what heat and temperature are before we move on.

Heat has a funny, negative definition: it's not work. You will have run across the concept of work in your elementary physics courses. In general, work is force multiplied by distance. This definition of work includes some special cases in which force and distance do not appear explicitly, such as electrical work (charge times electrical potential difference), where we are in fact calculating the work done pushing charges around. Formally, **heat** is an interaction between systems (or between a system and its surroundings) that causes the state of a system to change, but that isn't work. Heat transfer only requires contact (sometimes indirect) between two systems. We're already familiar with this fact from our everyday lives, although you may not have thought of it quite this way before; if you want to heat up some water, you put it in a pot, put the pot on the stove and turn the stove on. The stove is clearly doing something to the pot and water since you can observe various changes in the state of this system, but it's not work since there isn't any pushing involved.

One of the tricky things about heat is that the language we use to describe what happens during heat transfer sometimes suggests that it's some kind of fluid contained in a body ("heat flow," "heat capacity"). The idea of heat as a fluid, the caloric theory, was the dominant theory of heat in the eighteenth century, but turns out to be an incorrect picture, as we will see in Chapter 5. We are nevertheless stuck with this language, which has clearly outlived its usefulness, but just won't go away. The important thing is to remember that heat is an interaction between systems, and not some kind of substance that is stored in a body.

It is very easy to write down definitions of temperature that sound good, but mean little. Temperature is a primitive concept, like space or time. The best we can do is to try to explain what we mean in simple language: **temperature** is a numerical label (e.g. 4 °C) for "hotness." To put it another way, using my sense of touch, I can order objects made from the same material by hotness. (Our perception of hotness is not infallible and is affected by the material of which an object is made, among other factors. I do not claim that we can design thermometers on this principle, only that we would reject temperature scales that grossly violated our innate sense of hotness.) Having done this, we will insist that our temperature scales order the objects the same way, assigning lower temperatures to cooler objects, and higher temperatures to warmer objects. There are of course other definitions of temperature, the best of which probably come from statistical mechanics, specifically from the factor of $T$ appearing in the Boltzmann distribution (Section 3.2). All we really want, however, is to avoid making mental errors when discussing heat and temperature, so this simple definition of temperature as a numerical label for hotness is enough. We can in fact now nicely distinguish between heat and temperature:

> **Heat is an *interaction between* systems.**
> **Temperature is a *property of* a system.**

A **thermometer** is simply a device for assigning temperatures to objects. So how do thermometers work? Suppose we have a system in which one easily measured observable increases monotonically with hotness. Then we could label the temperatures by the values of that observable or, if these values are inconvenient, we could apply some transformation to them to give us a practical temperature scale. A simple example of such a property is volume. For most materials, the volume increases with temperature. This is the basis for the familiar mercury and alcohol thermometers.

When we place objects in contact that have different temperatures, their states change. In other words, heat is transferred between them. This process does not continue indefinitely. Eventually, the two bodies come to the same temperature in a state of thermal equilibrium. Thermometry is based on this principle: if we place a thermometer in thermal contact with another body, it will eventually come to equilibrium with it. At this point, we may read the thermometer to discover the final equilibrium temperature of the composite system that includes both the body under study and the thermometer. Note that the state of the first body, in particular its temperature, has been changed by contact with the thermometer, since heat had to flow between the body and thermometer in order to establish equilibrium. It is impossible to measure the temperature of a body without changing it. However, if our thermometer is sufficiently small and of an intelligent design, this effect can be minimized.

## 4.3 Sign convention

For many thermodynamic quantities such as work it will be necessary to distinguish carefully between positive and negative values. The signs are arbitrary, but a well-chosen and consistently applied sign convention makes the discussion of thermodynamic changes much simpler. The modern convention is as follows:

> **Anything that goes into a system is counted as positive;**
> **anything that comes out is negative.**

All quantities are expressed relative to the system under consideration, not relative to the surroundings, although we will sometimes reverse our perspective and treat the surroundings as a system. Thus, work done on a system is positive, work extracted from a system is negative; heat transferred into a system is positive, heat transferred out is negative, and so on.

Note that the sign convention will require you to clearly identify the system whenever you are working on a thermodynamics problem. In some cases, the system will be obvious. In others, you will have to clearly state what belongs to the system and what belongs to the surroundings.

Some older books (and research papers) use a different convention. Work is singled out for special treatment because the theory was originally devised to treat steam engines and similar devices. In this older convention, work done by the system is positive. Even though we will never use this convention, it's good to know about it in case you ever run across one of these older works. If you do, you will notice that some formulas relating to work differ in sign from those you will see here.

## 4.4 Molar, specific and "total" quantities

Many of the quantities in thermodynamics are **extensive**, which means that, all other things being equal, they increase with the size of the system. For extensive quantities, we sometimes want to talk about the total for a particular system. Often though, we prefer to work in terms of the molar quantity (i.e. the amount per mole), or of the specific quantity (amount per unit mass).

**Notation:** If $X$ is an extensive quantity, then $X_m$ is the corresponding molar quantity, and $x$ is the corresponding specific quantity.

For instance, a system might contain a gas that occupies a certain total volume $V$. On the other hand, we might also talk about the molar volume $V_m$ of the gas, i.e. the volume occupied by one mole of the gas. Because the molar mass relates number and mass, we can easily convert this figure to a specific volume, symbolized $v$. In fact, we have the relationship $V = nV_m = mv$ (here, mass times specific volume, not mass times velocity). Accordingly, we could write every formula relating extensive properties in at least three versions, one for the molar quantities, one for the specific quantities and one for the total. Generally, these formulas look nearly identical, give or take some $m$s and $n$s and the typographical embellishments described above. In this book, I will generally write formulas for the total quantities, letting it be understood that we can divide the formulas by $m$ or $n$ to obtain the corresponding formulas for the specific and molar quantities, except when the distinction is important. When handling data, I try to be consistent and to use the appropriate notation. I find that a little care exerted at this level often reduces confusion as to what various quantities refer to. It also helps with unit analysis: if I think I'm calculating $V_m$, the units had better work out to units of volume per mole.

**Example 4.1 Molar and specific volumes in the ideal gas law** The ideal gas law can be rewritten in terms of the molar volume as follows:

$$pV_m = RT$$

If we divide both sides by the molar mass of the gas, we get

$$pv = RT/M$$

**Example 4.2 Converting between molar and specific quantities**  The molar heat capacity $(C_{p,m})$ of lead is $26.4\,\mathrm{J\,K^{-1}mol^{-1}}$. To calculate the specific heat capacity, we use the molar mass:

$$c_p = \frac{26.4\,\mathrm{J\,K^{-1}mol^{-1}}}{207.2\,\mathrm{g\,mol^{-1}}} = 0.127\,\mathrm{J\,K^{-1}g^{-1}}$$

### Key ideas and equations

• Anything that goes into a system is counted as positive, anything coming out is counted as negative.

### Suggested reading

Thermometry is a surprisingly complicated business. For instance, you may never have thought about the fact that a mercury-in-glass thermometer will give different readings depending on the temperature of the stem, and not just the bulb where we think the measurement occurs, because the glass also expands with temperature. The channel into which the mercury expands therefore has a different diameter depending on the temperature of the stem. Then there's the whole issue of calibration. The best starting point for reading about temperature and thermometers, if this topic catches your fancy, is the web site of the International Temperature Scale of 1990: http://www.its-90.com.

# 5

# The First Law of Thermodynamics

The First Law is about the inter-relationships between a few key quantities: energy, work and heat. These concepts are central to such disparate fields of study as chemistry, engineering and physiology, to name just a few.

In a nutshell, the First Law asserts that energy is conserved. It therefore allows us to do some simple bookkeeping on energy transfers and thus to determine how much energy a system has used or stored. Although this may not sound very exciting, it is remarkably useful.

## 5.1 Differentials

Thermodynamics can't be studied without learning some new mathematics. Central among the mathematical tools of thermodynamics are differentials. **Differentials** are tiny (infinitesimal) increments. They are useful for describing how a change in one quantity is related to a change in another.

There are two kinds of differentials. The first, which are somewhat more straightforward to understand, are exact differentials. **Exact differentials** represent small changes in a property of a system. In thermodynamics, a measurable property is called a **state variable**. A function that only depends on state variables is called a **state function**. Since state functions only depend on the state variables, they also are properties of a system and thus have exact differentials. You have seen exact differentials before, in your calculus course. The differential appearing in an ordinary integral such as

$$\int_{x_1}^{x_2} f(x)\,dx$$

is always an exact differential. In particular, if $dx$ is an exact differential,

$$\int_{x_1}^{x_2} dx = x_2 - x_1 = \Delta x.$$

Only the endpoints matter, and not the way in which $x$ was changed. It doesn't matter if $x$ was changed slowly or quickly, or even if it increased and decreased again before stopping at its final value, only where it started and where it ended up. We say that $x$ is **path independent**. The following three-way correspondence follows:

| | | | | |
|---|---|---|---|---|
| $x$ is a state variable or function | $\Longleftrightarrow$ | $dx$ is an exact differential | $\Longleftrightarrow$ | $\Delta x$ is path independent |

Examples of state variables include pressure, volume and temperature, to name just a few.

The other case, that of **inexact differentials**, is best understood by example. We will often have occasion to talk about $dw$, the amount of work done when some state variable (for instance, the volume) is changed by a small amount. However, $dw$ is not like $dV$; since work is not a state variable (no system is characterized by the amount of work it "has"), it doesn't make sense to talk about the "change in work." Rather, when we integrate $dw$, we will be adding up the small amounts of work done along some path ($\mathcal{P}$) to compute the total work done $w$. Symbolically,

$$\int_{\mathcal{P}} dw = w.$$

In this case, it *may* matter how we do the work. Some forces, like gravitation, are conservative. Work done by or against a conservative force is path independent. However, many forces are non-conservative. For non-conservative forces, it is easy to devise paths that make the same changes in a system (i.e. have the same initial and final states) but for which completely different amounts of work are done. We will see an example of this in the next section. We say that $w$ is a **path function**, and that $dw$ is an inexact differential. For inexact differentials, we therefore have

| | | | | |
|---|---|---|---|---|
| $x$ describes a process | $\Longleftrightarrow$ | $dx$ is an inexact differential | $\Longleftrightarrow$ | $x$ is path dependent |

---

### Exercise group 5.1

(1) Suppose that $F$ is a state function, and consider the following process:

$$(p_1, V_1) \xrightarrow{\Delta F_1 = -1.41\,\text{u}} (p_1, V_2)$$

$$\Delta F_4 \uparrow \qquad\qquad\qquad \downarrow \Delta F_2 = 2.25\,\text{u}$$

$$(p_3, V_1) \xleftarrow{\Delta F_3 = -3.09\,\text{u}} (p_3, V_2)$$

What is $\Delta F_4$? Note that "u" represents the units of $F$, whatever they are.

*Hint*: Imagine starting at $(p_1, V_1)$ and following the arrows around. How has the state changed? What does this imply about the net change in $F$?

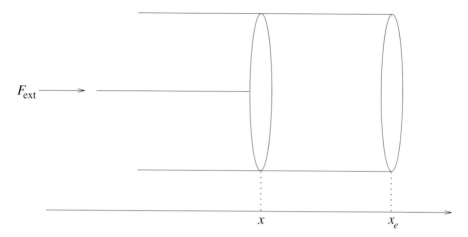

Figure 5.1 Geometry of a piston system used to derive the equation for pressure–volume work. If the cylinder has cross-sectional area $A$, the volume is $V = A(x_e - x)$, where $x_e$ is the position of the end of the cylinder. Accordingly, $dV = -A\,dx$ (area multiplied by change in height) or $dx = -dV/A$. The external force $F_{ext}$ results in an external pressure on the piston of $p_{ext} = F_{ext}/A$ or $F_{ext} = p_{ext}A$. Therefore, the differential of work $dw = F_{ext}\,dx = -p_{ext}\,dV$.

## 5.2 Pressure–volume work

The essence of the First Law of Thermodynamics is the equivalence of heat and work. The type of work that will most concern us in this book is pressure–volume work. This is just ordinary mechanical work, but we will write the equations in a form particularly suited to thermodynamics. Suppose that we have a gas in a piston either being compressed by or expanding against an external force $F_{ext}$ directed perpendicular to the surface of the piston (Figure 5.1). Recall that mechanical work is defined as a force times the distance over which that force is applied, so the work done on or by the system (the gas) during a small displacement $dx$ is

$$dw = F_{ext}\,dx$$

We'll worry about adding up the $dw$s (integrating) later.

Now suppose that the force is distributed over an area $A$ so that we actually measure not force, but pressure $p$. Recall that $F = pA$. Furthermore, a small change $dV$ in the volume $V$ is related to a small change $dx$ in the position of the piston by $dx = -dV/A$ (see Figure 5.1). The negative sign accounts for the fact that if the force (which pushes in) and displacement are in the same direction, the volume decreases. We arrive at the fundamental relationship

$$dw = -p_{ext}\,dV. \tag{5.1}$$

If we want the total work, we just integrate this equation:

$$w = \int_P dw = -\int_P p_{ext}\,dV \tag{5.2}$$

along whatever path $\mathcal{P}$ is followed during the process, i.e. how $p_{\text{ext}}$ changes during the process.

The negative sign in Equation (5.1) is important: since pressures are always positive, the work will always have the opposite sign to the change in volume. Thus, during a compression, the volume decreases and the work is positive. According to our sign convention and in accord with common sense, this means that during a compression work is done *on* the system. Conversely, an expansion represents a volume increase and negative work, i.e. work done *by* the system.

**Example 5.1 Work at constant pressure** Suppose that the external pressure is constant during a process. Then

$$w = -p_{\text{ext}} \int_{V_1}^{V_2} dV = -p_{\text{ext}}(V_2 - V_1) = -p_{\text{ext}}\, \Delta V.$$

One special case of this formula deserves mention. Suppose that a gas expands against a vacuum. This means that the external pressure is zero. Then $w = 0$.

There are some special types of processes (i.e. special paths) that we use all the time in thermodynamics, either because they are theoretically significant, or because they describe (or approximate) some conditions we encounter in real processes. An **isothermal** process is one in which the temperature is constant. A **reversible** process is one in which the system stays in equilibrium with its surroundings at all times. During the expansion of a gas, reversibility would imply that the external and internal pressures are equal during the entire path. This is not something we can really do; if the pressure is the same on both sides of the piston, there is no net force, and the piston doesn't move. In a real (irreversible) expansion, we would have to make the external pressure lower than the internal pressure.

If we can't really do a reversible expansion, why do we care about this kind of process? Consider Figure 5.2, which shows three different paths for an isothermal expansion from $V_1$ to $V_2$. The solid curve is the $pV$ isotherm, i.e. the line along which $T$ is a constant at equilibrium. For an ideal gas, this would be the curve $pV = \text{constant}$. The reversible expansion corresponds to following this curve. According to Equation (5.2), the work done during this expansion is just the area under this curve between $V_1$ and $V_2$ (because $p_{\text{ext}} = p$ in a reversible process), give or take the sign, which will be negative since the system is doing useful work during an expansion. In a real process, we would have to make $p_{\text{ext}}$ smaller than $p$. We could just go directly to the final pressure (the long-dashed line in the figure). The work done would then be (again, ignoring the sign) the area of the rectangle under the long-dashed line. Of course, we could reduce the pressure in steps. The short-dashed line in the figure shows what would happen if we reduced the pressure in three equal steps. The work done in this case would be the area under the "staircase." You can see that taking more steps gives us an amount of work between the reversible work and the result for a single pressure reduction. You should also be able to see that, no matter how many

*The First Law of Thermodynamics*

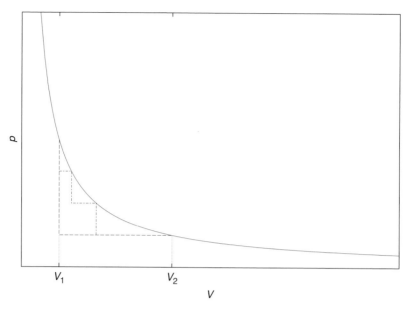

Figure 5.2 Expansion of a gas. The solid curve is the $pV$ isotherm.

pressure-reduction steps we use, we can't produce more work than the reversible process would, we can only approach this limit. This is why reversible processes are important:

> **A reversible process maximizes the work done by a system.**

We can approach reversibility by taking lots of small steps, and we know that the reversible work done by a system is a maximum (i.e. $w$ is maximally negative along a reversible path), so we now have an upper bound on what a machine can accomplish. Note that we haven't for the moment proven this to be a general principle, only observed that it is true for the expansion of a gas. We can do better, and will return to this important principle later in the course.

**Example 5.2 Reversible, isothermal expansion of an ideal gas** Let's calculate the work done during the isothermal, reversible expansion of an ideal gas. Because the process is reversible, $p_{ext} = p$, the pressure of the gas in the piston. Therefore,

$$w = -\int_{V_1}^{V_2} p \, dV.$$

To evaluate the integral, we need to write $p$ as a function of $V$. Because this is an ideal gas, $p = nRT/V$. Furthermore, this is an isothermal expansion so $T$ is constant:

$$w = -nRT \int_{V_1}^{V_2} \frac{dV}{V} = -nRT \ln \frac{V_2}{V_1} = nRT \ln \frac{V_1}{V_2}.$$

In the exercises that follow, you may need to do some slightly harder integrals. If you need a little review, have a look at Appendix I.

---

### Exercise group 5.2

(1) Calculate the work done when a gas expands from a volume of 1 L to 10 L against a constant external pressure of 1 atm. (*Hint* and *caution*: The answer should be in SI units.)

(2) Calculate the work done on or by the gas when 8 mol of nitrogen is compressed isothermally and reversibly at 300 °C from an initial volume of 1 L to a final volume of 30 mL. Assume that nitrogen behaves as an ideal gas. Explain the meaning of the sign of your answer.

(3) 18 mol of nitrogen are held at a pressure of 1.08 atm and a temperature of 20 °C. The external pressure is changed very suddenly to 0.35 atm and is held constant during the expansion. The gas is in good thermal contact with its surroundings, whose temperature is a constant 20 °C so that by the end of the expansion, the temperature of the gas has returned to equilibrium with the surroundings.
   (a) Treat nitrogen as an ideal gas and calculate the work done during the expansion. Explain the meaning of the sign of the work.
   (b) Suppose that the work done on or by the gas results from the raising or lowering of a 2.3 kg mass. The work done when a mass is raised is $w = mgh$ where $h$ is the height to which the mass is raised (negative if lowered). How far does the mass have to move, and in what direction?

(4) 0.10 mol of an ideal gas is expanded isothermally at 50 °C from an initial volume of 1.4 L to a final volume of 3.2 L. Calculate the work done along each of the following paths: (a) reversible, (b) pressure dropped suddenly to the final pressure, (c) a two-step process with two equal pressure drops.

(5) The ideal gas law is only valid for gases at low densities. Real gases are more accurately described by the van der Waals equation:

$$\left( p + \frac{n^2 a}{V^2} \right)(V - nb) = nRT$$

where $a$ and $b$ are constants specific to a particular gas, $R$ is the ideal gas constant, $p$ is the pressure, $V$ the volume and $n$ the number of moles.
   (a) Derive an equation for the work done during a reversible, isothermal expansion of a van der Waals gas.
   (b) For nitrogen, $a = 0.141 \, \text{Pa m}^6 \, \text{mol}^{-2}$ and $b = 3.91 \times 10^{-5} \, \text{m}^3 \, \text{mol}^{-1}$. Calculate the work done when 1 kmol of nitrogen is expanded reversibly and isothermally at a temperature of 300 K from 1 to 3 m³.
   (c) Compare the answer obtained in question 5b to what you would have calculated from the ideal gas equation. Was the extra work worth the effort?

(6) Solids and liquids are often said to be incompressible, which means that changing the pressure does not change their volumes. Because of this, we tend to ignore the effect of pressure on solids and liquids when solving thermodynamic problems. However, solids and liquids are not really incompressible. For copper, for instance, the relationship between pressure and molar volume at constant temperature is

$$p = \frac{1}{\kappa_T} \ln\left(\frac{V_m^\circ}{V_m}\right) + p^\circ$$

where $\kappa_T$, $V_m^\circ$ and $p^\circ$ are constants with the values

$$\kappa_T = 7.25 \times 10^{-12}\,\text{Pa}^{-1}$$
$$V_m^\circ = 7.093 \times 10^{-6}\,\text{m}^3\,\text{mol}^{-1}$$
$$p^\circ = 101\,325\,\text{Pa}$$

Calculate the work done per mole when copper is compressed reversibly and isothermally from an initial pressure of 1 atm to a final pressure of 1000 atm. This is roughly equivalent to the pressure change that a piece of copper would experience if it was dropped from the ocean's surface to the bottom of the Marianas trench, the deepest ocean trench on Earth. Based on your calculation, do you think that ignoring pressure–volume terms (such as the work) is justifiable for solids?

*Hint:* You may need an integral from the table in Appendix I.1.

## 5.3 The First Law of Thermodynamics

The **internal energy** $U$ of a system is the total energy, excluding kinetic energy, associated with the motion of the center of mass or with rotation around the center of mass, as well as the potential energy due to the position of the body in space (e.g. gravitational potential energy). The internal energy is therefore the energy stored in the atoms and molecules of a piece of matter, including the energy associated with the intermolecular forces.

The first law of thermodynamics can be viewed as an extension of classical mechanical results on the conservation of energy. It states that there are two ways to change the internal energy of a system and that these two ways are equivalent: work ($w$) can be done on the system or heat ($q$) can flow into it. Thus the change in internal energy can be written

$$\Delta U = q + w. \tag{5.3}$$

A perpetual motion machine of the first kind is a machine that indefinitely sustains some external energy-using process without external input of energy. The first law forbids such machines; if energy goes out of the system without being replenished from an external source, the internal energy decreases. This process cannot go on indefinitely.

We now come to an important fact:

> **The internal energy is a state function.**

We know this because the internal energy has been defined to exclude external influences so that its value can only depend on the state of the system. On the other hand, as we have seen in Section 5.2, the work is a path function. Since $\Delta U = q + w$, the influence of the path on $w$ must somehow cancel out when we add it to $q$ to give us the change in the state function $U$. It follows that the heat must also be a path function.

> **Work and heat are path functions.**

In fact, these two quantities are the only path functions we will routinely see in this course.

It is often useful to write the First Law in differential form:

$$dU = dq + dw. \tag{5.4}$$

The only hiccup is that we must be clear about what the differentials mean: $dU$, being the differential of a state function, is an exact differential and we can integrate it to get the change in internal energy:

$$\int_{\text{initial state}}^{\text{final state} } dU = U_{\text{final}} - U_{\text{initial}} = \Delta U.$$

On the other hand, $dq$ and $dw$, being differentials of path functions, are inexact differentials. They can only be integrated for a given path to give the total heat and work:

$$q = \int_{\mathcal{P}} dq \quad \text{and} \quad w = \int_{\mathcal{P}} dw.$$

---

### Exercise group 5.3

(1) An induction coil is a device that, when placed in a changing magnetic field, generates electricity. Reasonably large time-varying magnetic fields are produced by high-voltage power lines. Some years ago, a farmer whose property was crossed by high-voltage lines placed a large induction coil under the lines and supplied his farm with electricity from it. He was found guilty of theft. What was he stealing and from whom? Explain (briefly) the scientific basis of the case.

---

### 5.4 Calculus of differentials

Equations written in terms of differentials are just shorthand versions of differential equations (equations relating derivatives). For instance, $dx = \alpha\, dy$ is short for

$$\frac{dx}{dz} = \alpha \frac{dy}{dz}$$

where $z$ is some progress variable (time, amount of reactant used etc.). We use the differential notation when the identity of $z$ doesn't matter. The differential relationship $dx = \alpha\, dy$ can then be thought of as a statement about how a small change in $x$ is related to a small change in $y$.

It follows from the relationship between derivatives and differentials that the usual rules of differentiation apply also to the latter. In particular, if $x$ and $y$ are variables and $a$ is a constant:

$$d(x + y) = dx + dy \qquad \text{(sum rule)}$$
$$d(ax) = a\,dx$$
$$d(xy) = y\,dx + x\,dy \quad \text{(product rule)}$$

## 5.5 Heat and enthalpy

We focus our attention on a system that is only capable of doing pressure–volume work. We do this because that type of system comes up most often in chemistry. Suppose that we heat such a system at constant volume. Then $dw = -p_{ext}\,dV = 0$ because the volume doesn't change. Therefore, the First Law gives us

$$dU = dq_V. \tag{5.5}$$

The subscript in the above equation indicates that the volume is held constant, i.e. it specifies the path. Therefore,

$$\Delta U = q_V.$$

> **At constant volume, the change in internal energy equals the heat transferred to the system.**

It would be nice if there were a state function whose changes corresponded to heat at constant pressure because a number of experiments are performed under those conditions. In thermodynamics, whenever we start a sentence with "it would be nice if there were a state function . . . ," we just make one up that has the desired property. At constant pressure, $p_{ext} = p$ (otherwise the pressure wouldn't stay constant) so, from the First Law and considering only pressure–volume work,

$$dU = dq_p - p\,dV.$$

Let's rearrange this equation slightly:

$$dq_p = dU + p\,dV. \tag{5.6}$$

Note that the last term on the right-hand side is $-dw_p$ for a system that can only perform pressure–volume work. Inspired by this equation, we define the **enthalpy** $H$ by

$$H = U + pV. \tag{5.7}$$

$H$ is a state function because it is the sum of $U$, which is a state function, and of the product of $p$ with $V$, each of which is a state variable. Differentiating this expression, we get

$$dH = dU + p\,dV + V\,dp.$$

At constant pressure, the last term is zero. Considering also Equation (5.6), we conclude that

$$dH = dq_p \implies \Delta H = q_p. \tag{5.8}$$

This is exactly what we were looking for:

> **At constant pressure, the enthalpy change equals the heat transferred to the system.**

When a system is composed only of solids and liquids (called **condensed phases**), an important simplification is possible: since solids and liquids are incompressible and since they are relatively dense, $\Delta(pV)$ is always very small for processes involving only condensed phases provided $\Delta p$ isn't huge. It then follows that $\Delta H \approx \Delta U$.

## 5.6 Heat capacity

You might guess that the larger the temperature change, the larger the amount of heat transferred to (or from) a body. That is essentially correct. Under most conditions, the heat transferred is almost exactly proportional to the temperature change. In other words,

$$q \propto \Delta T.$$

Heat is a path function so we need to specify the path carefully. If we do that, the amount of heat required to cause a given temperature change is very reproducible. At constant volume,

$$q_V = C_V \Delta T,$$

and at constant pressure,

$$q_p = C_p \Delta T.$$

Again note that the notation $q_X$ indicates that $X$ is held constant during the process. $C_V$ and $C_p$ are the **heat capacities** of the system, respectively at constant volume and at constant pressure. In general, these two heat capacities are different.

The above equations are valid provided the relevant heat capacity is approximately constant, which it is for most substances when the change in temperature is not too large. When the heat capacity is not constant, we must rewrite these equations in differential form:

$$dq_X = C_X\,dT \tag{5.9}$$

or

$$C_X = \left. \frac{dq}{dT} \right|_X,$$  (5.10)

where the bar and subscript indicate that $X$ is being held constant. We use the bar rather than just a simple subscript (as we did with $q$, for instance) because holding $X$ constant here applies to the operation of taking the derivative, and not to some particular symbol within the derivative. In other words, we want to know how $q$ changes when we change $T$ while holding $X$ constant. In chemical thermodynamics, $X$ will typically be either $V$ or $p$.

Equation (5.10) tells us how to measure (for instance) $C_V$: we measure the total heat added as a sample is heated at constant volume. This gives us a curve $q(T)$. If we take the derivative of this curve, we get the heat capacity as a function of $T$.

Now recall that $dU$ and $dq_V$ are related by Equation (5.5). If we make this substitution in the appropriate instance of Equation (5.10), we get

$$C_V = \left. \frac{\partial U}{\partial T} \right|_V.$$  (5.11)

The derivative in Equation (5.11) is a partial derivative because $U$ is a function of two variables: $T$ (the variable with respect to which we are differentiating) and $V$ (the variable we are holding constant). The reason that we use a partial derivative here and not in Equation (5.10) again has to do with the distinction between state and path functions. Since the heat added depends on the path, $q$ is not a function of the state variables. It can be thought of as a function of $T$ only in the context of a particular experiment over a well-defined path.

Similarly, using Equation (5.8) we get

$$C_p = \left. \frac{\partial H}{\partial T} \right|_p.$$  (5.12)

A nice way to think of partial derivatives is that they describe experiments. You have no doubt been told that in good experiments, you vary just one thing at a time. $C_V$ would be the slope of a graph of $U$ vs. $T$ in an experiment in which you hold $V$ constant. If the energy depended on other variables (e.g. external magnetic field strength, a key variable in the thermodynamics of magnetic systems), you would hold those constant, too.

We noted earlier that $\Delta H \approx \Delta U$ for processes involving only solids or liquids. It follows that $C_p$ and $C_V$ are also approximately equal to each other for solids or liquids, *but not for gases*. Assuming that a gas obeys the ideal gas law, however, we can easily derive a relationship between $C_{p,m}$ and $C_{V,m}$. Since $H = U + pV$, we have

$$C_{p,m} = \left. \frac{\partial H_m}{\partial T} \right|_p = \left. \frac{\partial U_m}{\partial T} \right|_p + \left. \frac{\partial (pV_m)}{\partial T} \right|_p.$$  (5.13)

For an ideal gas, $pV_m = RT$, so the last derivative is just $R$. The problem is to evaluate $\partial U_m/\partial T|_p$. There is some theory which we won't cover that would let us do this rigorously. We'll settle for a handwave here, although one that brings out an important fact. An ideal gas is, by definition, one whose particles don't interact, i.e. there are no forces acting between particles in an ideal gas. If there are no forces between the particles, then it hardly matters how close together they are. The internal energy of an ideal gas therefore can't depend on the volume. The equation of state for a fixed quantity of an ideal gas can be used to write any one of the variables $(p, V, T)$ in terms of the other two. We can therefore think of the internal energy as being a function of volume and temperature, and then conclude the following useful fact:

---

**The internal energy of an ideal gas only depends on the temperature.**

---

Mathematically, we are saying that for an ideal gas, $U_m = U_m(T)$. Changing variables from $(V, T)$ to $(p, T)$ won't change the form of this equation, so $\partial U_m/\partial T|_p = \partial U_m/\partial T|_V = C_{V,m}$. Putting all the pieces together in Equation (5.13), we get

$$C_{p,m} = C_{V,m} + R.$$

In a lot of problems, we will be able to treat the heat capacity as constant, which simplifies the mathematics a lot. The only thing to watch for is that we are often (but not always) given the molar (or specific) rather than the total heat capacity. This is so common that we often just say "heat capacity" when we mean "molar heat capacity," so you really have to watch the units of any "heat capacities" given to you.

**Example 5.3 Warming an engine** An automobile engine has a constant-pressure heat capacity of $250\,\text{kJ K}^{-1}$. If the engine is initially in equilibrium with its surroundings on a day where the air temperature is $10\,°\text{C}$, how much heat is required to warm the engine to its operating temperature of $110\,°\text{C}$?

The calculation is a straightforward one. First, we need $\Delta T$ in kelvins. However, since the size of a kelvin is exactly the same as the size of a degree Celsius, we can calculate $\Delta T$ directly from the data:

$$\Delta T = 110 - 10\,°\text{C} = 100\,\text{K}.$$

Note that this *only* works because we are taking a difference in temperatures. In any other case, we need to work in kelvins.

To finish the problem, we just multiply the heat capacity of the engine by $\Delta T$:

$$q = C_p \Delta T = 25\,\text{MJ}.$$

**Example 5.4 Warming graphite** The constant-pressure molar heat capacity of graphite is $8.527\,\text{J K}^{-1}\text{mol}^{-1}$. How much heat is required to raise the temperature of a 1 kg block of graphite from 20 to 95 °C?

Here, we are given the *molar* heat capacity rather than the total heat capacity of the sample. Once we have realized this, the calculations are as straightforward as in the last example. 1 kg of graphite corresponds to 83 mol. The heat required is

$$q = nC_{p,m}\Delta T = (83\,\text{mol})(8.527\,\text{J K}^{-1}\text{mol}^{-1})(75\,\text{K}) = 53\,\text{kJ}.$$

Sometimes, a constant heat capacity is not a good approximation. The amount of heat involved in changing the temperature of a system over a constant-$X$ path can then be calculated by

$$q_X = \int_{T_1}^{T_2} C_X(T)\, dT$$

provided we know the equation for the relevant heat capacity. This equation follows from Equation (5.9) simply by adding up (integrating) all the little $dq_X$ contributions along a path.

**Example 5.5 Non-constant heat capacity** The molar heat capacity of copper at constant pressure is given accurately by the following expression over a wide range of temperatures:

$$C_{p,m} = 22.6 + 6.28 \times 10^{-3} T$$

for $C_{p,m}$ in $J\,K^{-1}mol^{-1}$ and $T$ in kelvins.[1] Calculate the heat required to raise the temperature of a 1 kg copper bar from 20 to 500 °C at constant pressure.

If we define $a = 22.6\,J\,K^{-1}mol^{-1}$ and $b = 6.28 \times 10^{-3}\,J\,K^{-2}mol^{-1}$ (to save us carrying these messy constants through the integration process), then $C_{p,m} = a + bT$ and the heat is calculated by

$$q = n\int_{T_1}^{T_2} C_{p,m}\, dT = n\left[aT + \frac{1}{2}bT^2\right]_{T_1}^{T_2} = n\left[a\Delta T + \frac{1}{2}b\left(T_2^2 - T_1^2\right)\right].$$

1 kg of copper is 15.7 mol. Substituting our values into the equation, we get $q = 196\,kJ$. (If you're having trouble getting this value for the heat, you may have made either of two common errors. First of all, a lot of people calculate $(T_2 - T_1)^2$ instead of the correct $T_2^2 - T_1^2$. These are *not* the same. It's also possible that you forgot to convert degrees Celsius to kelvins before evaluating $T_2^2 - T_1^2$. Because we are squaring the temperatures, the difference between kelvins and degrees Celsius can't be ignored here.)

For comparison, suppose we had used the constant "textbook" value of the heat capacity of copper of $24.4\,J\,K^{-1}mol^{-1}$, which is an average valid near 25 °C. We would then have calculated $q = 184\,kJ$. Whether an error of this size is acceptable will very much depend on the purpose of the calculation.

Using our knowledge of heat capacity, we can also solve simple heat balance problems. These can generally only be solved if either the system is in an insulated container so that no heat can escape, or the process is so fast that heat doesn't have time to leak out. In the former case, we would say that the system is **adiabatic**. In the latter case, the process would be **quasi-adiabatic**.

**Example 5.6 Mixing hot and cold water** Suppose that 100 g of water at 20 °C is added to 200 g of water at 80 °C under atmospheric pressure in an insulated enclosure. What is the final temperature of the mixture?

---

[1] *CRC Handbook of Chemistry and Physics*, 66th edn.; CRC Press: Boca Raton, 1985, pp. D-43–D-49.

To solve this problem, note that no heat flows out of this system since it is enclosed in an adiabatic container. Therefore $q = 0$ for the system consisting of the two samples of water. We treat each sample of water as if it were an independent entity. At equilibrium, both samples will reach the same temperature $T_f$. We know the initial temperatures. We write a heat balance equation:

$$q = 0 = m_1 c_p(T_f - T_1) + m_2 c_p(T_f - T_2).$$

In this case, the two substances mixed are both water and have the same value of the specific heat capacity $c_p$. We can therefore simplify this equation:

$$m_1(T_f - T_1) + m_2(T_f - T_2) = 0.$$

Putting in our numbers, we get

$$100(T_f - 20) + 200(T_f - 80) = 0 \Longrightarrow T_f = 60\,°C$$

---

### Exercise group 5.4

(1) A 200 W heating element is submerged in 10 L of water in an insulated bottle whose contents, initially at 25 °C, are held at constant pressure. How long does it take for the water to boil? Assume that the boiling point is 100 °C.

(2) A 1 kg piece of iron is placed in 1 kg of water in a sealed, adiabatic container. The water temperature is initially 20 °C. The final equilibrium temperature of the system is 80 °C. What was the initial temperature of the iron?

(3) The constant pressure heat capacity of the metal indium is given by the equation

$$C_{p,m} = 24.3 + 0.0105T$$

where $T$ is in kelvins and $C_{p,m}$ is in $J\,K^{-1}mol^{-1}$.
   (a) Calculate the amount of heat required to raise the temperature of a 12 g piece of indium from 50 to 100 °C.
   (b) How large an error would result from using a (constant) average value of $28.0\,J\,K^{-1}mol^{-1}$?

(4) X-rays can be generated by firing high-energy electrons into a metal target. Only a small fraction of the kinetic energy of the electrons is converted to X rays however. Most of the energy is transformed into heat.
   (a) Suppose that electrons, each with a kinetic energy of $2.5 \times 10^{-14}$ J, are fired into a copper anode of volume 85 cm³ (volume measured at 25 °C) at a rate of $2.5 \times 10^{17}$ electrons s$^{-1}$. The anode is hollow and contains 30 cm³ of water (measured at 25 °C). The anode and water are in good thermal contact. If all the kinetic energy were transformed to heat and assuming that none of the heat is lost to the surroundings, how long would it take for the anode's temperature to increase to 100 °C from 25? The density of copper at 25 °C is 7.11 g cm$^{-3}$ and its specific heat capacity is 0.38 J K$^{-1}$g$^{-1}$.

(b)  To avoid excessive heating of the anode, the water it contains is constantly replaced
by cool water. Suppose that water with an initial temperature of $4\,^{\circ}$C flows through
the anode at a rate of $6\,\text{L min}^{-1}$. Assuming perfect mixing of the water in the
anode and perfectly efficient heat transfer between the copper and water, what
steady temperature would the anode reach?

*Hint*: How much heat must be removed to hold the temperature steady?

*Note*: The temperature you will calculate here is a lower bound since neither mixing
nor heat transfer will be perfectly efficient in the real system.

---

### 5.6.1  Heat capacity from statistical thermodynamics

In statistical thermodynamics, we combine equations from statistical mechanics, like the
Boltzmann distribution, with equations from thermodynamics in order to predict or rationalize the values of thermodynamic quantities. Here, we will show how the heat capacity
is related to the partition function of a system.

In Section 3.2, we found that the average energy of a system made of identical molecules
was related to the molecular partition function by Equation (3.3). The internal energy, on
the other hand, is the energy stored in a system, and thus in the molecules that make
up that system. This will typically include terms due to intermolecular forces, and not
just to the energy levels of the isolated molecules. However, if we can ignore intermolecular forces, as in an ideal gas, or if we only want to consider the part of the energy
stored by a particular type of molecule, then we can associate the average molecular
energy with the internal energy by $U = N\langle E\rangle$, where $N$ is the number of molecules,
or, on a molar basis, $U_m = L\langle E\rangle$, where $L$ is Avogadro's number. Now, given Equation (5.11), we can calculate the constant-volume heat capacity from the internal energy.
Knowing the partition function of a system therefore gives us its constant-volume heat
capacity, just by taking a couple of derivatives with respect to temperature. This will give
us some insight into the molecular meaning of the heat capacity, as the following example
demonstrates.

**Example 5.7 Heat capacity of a two-level system**  In Example 3.1, we determined the
molecular partition function of a molecule with the following energy levels:

| $i$ | $g_i$ | $E_i/\text{J}$ |
|-----|-------|----------------|
| 1   | 1     | 0              |
| 2   | 3     | $4 \times 10^{-21}$ |

The partition function was

$$q = 1 + 3e^{-290\,\text{K}/T} \tag{3.4}$$

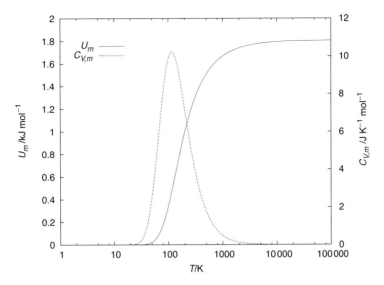

Figure 5.3 Molar internal energy and heat capacity of a system made up of molecules with two energy levels.

Applying Equation (3.3) and the correspondence between average molecular energy and internal energy, we get

$$
U_m = \frac{L k_B T^2}{q} \left.\frac{\partial q}{\partial T}\right|_V = \frac{R T^2}{q} \left.\frac{\partial q}{\partial T}\right|_V
$$
$$
= \frac{3R(290\,\text{K})\exp\left(-290\,\text{K}/T\right)}{1 + 3\exp\left(-290\,\text{K}/T\right)} = \frac{(7.23\,\text{kJ/mol})\exp\left(-290\,\text{K}/T\right)}{1 + 3\exp\left(-290\,\text{K}/T\right)}. \tag{5.14}
$$

The molar heat capacity is therefore

$$
C_{V,m} = \left.\frac{\partial U_m}{\partial T}\right|_V = \frac{(2.09 \times 10^6\,\text{J K mol}^{-1})\exp\left(-290\,\text{K}/T\right)}{T^2 \left(1 + 3\exp\left(-290\,\text{K}/T\right)\right)^2}.
$$

The molar internal energy and heat capacity are plotted in Figure 5.3.

To understand Figure 5.3, compare it to Figure 3.1. Consider first the internal energy. At low temperatures, only the ground state is occupied, so the average energy is the energy of the ground state, in this case zero. As the occupation of the excited state starts to become significant, the internal energy starts to grow because the average energy now includes some molecules in the excited state. Eventually, this curve levels off as the populations of the two levels approach their asymptotic ratio of one ground-state molecule to three excited-state molecules. In fact, the internal energy tends to $\frac{3}{4} L E_2 = 1.8\,\text{kJ mol}^{-1}$. Finally note that because the ground state has zero energy, the molar internal energy (Equation 5.14) is directly proportional to the probability of occupying the excited state (Equation 3.5).

What about the heat capacity? At very low temperatures, the internal energy increases very slowly as we increase $T$ because almost all the molecules are stuck in the ground state. The only way this simple system can store energy is by promoting molecules from the ground to the excited state. At very low temperatures, the probability that a molecule will access the excited state is low, so the heat capacity, which measures this ability to store energy, is correspondingly small. As we approach the temperature where the excited state become significant, there is a rapid increase in the probability that a molecule will reach the excited state as we increase $T$, so the heat capacity starts to rise. The heat capacity is largest where the probability of occupation of the excited state changes most rapidly. Beyond that point, the heat capacity starts to drop because the probability of occupation of the excited state is starting to level off. Eventually, increasing $T$ makes only a negligible difference to the probability of occupying the excited state, and the system is no longer able to store additional energy, at least not while maintaining thermal equilibrium. The heat capacity therefore drops to zero.

---

## Exercise group 5.5

(1) In Exercise group 3.1, problem 3, we considered a simple protein folding problem with the following energy levels:

| $i$ | $g_i$ | $E_i$ |
|-----|-------|-------|
| 1   | 1     | $-\epsilon$ |
| 2   | 4     | 0     |

Plot the molar heat capacity of this model protein, in units of $R$, vs. the reduced temperature $\theta = k_B T / \epsilon$.

---

## 5.7 Phase transitions

Phase transitions represent another easily reproducible set of heating or cooling events. During a phase transition, heat can flow into or out of the system without a change in temperature. The heat evolved in such a process is called the **latent heat**. On a practical level, latent heat is easy to work with since it's just a certain amount of heat per unit of material (mass or moles) that has undergone the phase transition.

**Example 5.8 Melting a solid** Solid cyanamide ($H_2NCN$) has a heat capacity of $78.2\,\mathrm{J\,K^{-1}mol^{-1}}$. It melts at $318.71\,\mathrm{K}$ and its enthalpy of fusion (melting) is $7.272\,\mathrm{kJ\,mol^{-1}}$. Suppose that we want to melt a $14.3\,\mathrm{g}$ sample of cyanamide initially at room temperature ($20\,^\circ\mathrm{C}$). How much heat would that take?

We first have to calculate how much heat it will take to bring the cyanamide from room temperature to the melting point. There are a few data conversions to carry out first:

$$n_{H_2NCN} = \frac{14.3\,g}{42.041\,g\,mol^{-1}} = 0.340\,mol.$$

$$T_i = 20 + 273.15\,K = 293\,K.$$

$$\therefore q_1 = n_{H_2NCN}C_{p,m}\Delta T$$

$$= (0.340\,mol)(78.2\,J\,K^{-1}mol^{-1})(318.71 - 293\,K) = 680\,J.$$

The heat required to melt the sample once it has reached the melting point is just

$$q_2 = n_{H_2NCN}\Delta_{fus}H_m = (0.340\,mol)(7.272\,kJ/mol) = 2.47\,kJ.$$

Note that $q_1$ and $q_2$ are not expressed in the same units. We will have to be careful about that when adding them up:

$$q = q_1 + q_2 = 0.680 + 2.47\,kJ = 3.15\,kJ.$$

The phase transition temperature depends on the pressure, although this dependence is a weak one for phase transitions involving only condensed phases. The two phases that participate in a phase transition can coexist at the phase transition temperature. In other words, the two phases are in equilibrium at that temperature. Taking melting as an example, at the melting (a.k.a. fusion) temperature, the solid and liquid can coexist. If we add a little heat, some of the solid melts, and if we remove a little heat, some of the liquid freezes.

Contrary to popular belief, freezing only exceptionally begins when the system reaches the freezing point. The temperature normally falls somewhat below the freezing point before freezing starts. This is called **supercooling** and it is an example of the creation of a **metastable state**, an unstable state that can persist for a long time if undisturbed. Once freezing has started, the heat released by freezing quickly returns the system to its equilibrium freezing temperature. The ice acts as a template for further freezing and supercooling is no longer possible. Ice is said to **nucleate** freezing. Similarly, steam can be supercooled because condensation is also a process that requires nucleation, this time of water droplets.

Supercooling turns out to be important for understanding frost damage to living tissues. Body fluids can supercool substantially before they freeze. In laboratory experiments, human blood can supercool by 10 degrees or more before it starts to freeze. The degree of supercooling when the blood is still inside the subject is less, but may still be as much as five or six degrees. Because it is in a metastable state, a supercooled liquid freezes quickly, often forming sharp crystals. If this happens in the bloodstream, these sharp crystals will puncture blood vessels, causing hemorrhages. This is one of the things that kills frost-damaged organisms, and also makes it difficult to preserve organs for transplantation by freezing them. There are two possible solutions to this problem. The first is to avoid freezing altogether. This can be done by using antifreeze compounds, some of which work on the principle of freezing-point depression (Section 8.6). Many species of arthropods use this strategy, maintaining high concentrations of glycerol in their body fluids. Arctic fishes also avoid freezing, but using a different bit of physical chemistry. They have antifreeze proteins that bind efficiently to ice microcrystals, thus preventing these from becoming nucleation centers. In other words, these proteins maintain a metastable state by interfering with the ice nucleation process.

The second solution to the problem of frost damage is radically different. Some organisms, notably the wood frog *Rana sylvatica*, have ice-nucleating proteins. As their name suggests, these proteins act as nucleation centers, preventing substantial supercooling. They thus cause water to slowly freeze out of the frog's tissues, including its bloodstream, at temperatures near the equilibrium freezing temperature of frog blood. Ice crystals formed in this way can follow the blood vessels, causing minimal damage, and allowing the frog to, quite literally, hop away after it has been thawed from being frozen solid.

Supercooling is not the only way to create a metastable state. Liquids can be **superheated**. Interestingly, it is not possible to superheat a solid. The difference again has to do with nucleation; bubble formation requires nucleation, but creating a liquid phase from a solid does not. Superheated liquids are dangerous since they can suddenly begin to boil violently, throwing hot liquid and vapor about. This is why we generally use boiling chips (a.k.a. anti-bumping granules) when heating liquids in a laboratory. As a rule, this is much less likely to be a problem at home because the surfaces of household pots are generally sufficiently irregular to nucleate bubble formation so that boiling starts at or very near the normal boiling point. However, superheating is sometimes observed when heating a liquid (water, coffee) in a smooth container (mug, glass measuring cup etc.) in a microwave oven. A surprising demonstration of superheating and of nucleation-induced boiling can be made by adding a drop of cold water to a cup of coffee heated in a microwave oven. If the coffee exceeded its boiling point, the cold water actually causes it to boil. (Be careful if you want to try this. Only fill the mug $\frac{1}{3}$ full. Note that it's possible that the liquid will boil if you bump the mug. Moreover, boiling may be explosive. Again, be careful.) The solubility of gases tends to decrease with increasing temperature, so the dissolved gases in the drop of cold water come out of solution, then becoming nucleation centers for the solvent vapor and thus initiating boiling.

**Example 5.9  Freezing of supercooled water** One mole of water is supercooled to $-10\,^{\circ}\mathrm{C}$, then adiabatically isolated and held at constant pressure. A disturbance (bumping, dust, etc.) causes the water to freeze. Freezing releases heat, which causes the temperature to increase. We want to know the equilibrium composition of the system (quantities of water and ice) and its final temperature.

Since the system is adiabatic, $q = 0$. The heat consists of two offsetting terms: there is a negative contribution from the freezing process and a positive contribution from the temperature rise.

There are two possibilities:

(1) The heat liberated by freezing is sufficient to warm the system to the normal freezing point. We obtain a mixture of ice and water at $0\,^{\circ}\mathrm{C}$.
(2) The heat liberated by freezing *all* of the ice is less than is necessary to raise the temperature to the normal freezing point. We obtain a solid piece of ice at some temperature below zero.

Our first task is therefore to determine which of these two possibilities is the correct one. The enthalpy change on freezing is $-6.007\,\text{kJ mol}^{-1}$ and the heat capacity of liquid water is $75.40\,\text{J K}^{-1}\text{mol}^{-1}$. Therefore, the amount of heat required to raise the temperature of one mole of water from $-10\,°\text{C}$ to $0\,°\text{C}$ is $C_p\Delta T = 754\,\text{J}$. The heat capacity of the water is not sufficient to absorb all of the heat available from freezing without exceeding the melting temperature, so the first of the two possibilities is the correct one: some of the water will freeze, but not all of it. We now have one crucial datum: $T_{\text{final}} = 0\,°\text{C}$.

The direct computation of the heat for simultaneous freezing and warming is difficult. Instead, using the fact that $q_p = \Delta H$ and that $H$ is a state function, we construct a different path for which the calculation is easy: imagine that we first warm the liquid water to $0\,°\text{C}$ and then freeze just the right amount to liberate the heat required to balance the heat we "borrowed" to warm the water:

$$q = 0 = n_{\text{total}}C_{p,m}(\text{H}_2\text{O},\,\text{l})\Delta T + n_{\text{frozen}}\Delta_{\text{freezing}}H_m.$$

Plugging in the data for this problem, we calculate

$$n_{\text{frozen}} = 0.13\,\text{mol}.$$

The rest of the water (0.87 mol) remains liquid.

---

### Exercise group 5.6

(1) One mole of ice at $0\,°\text{C}$ is put in contact with one mole of steam at $100\,°\text{C}$ in an adiabatic enclosure. The pressure is held constant at one atmosphere. What is the final composition and temperature of the system?

(2) Suppose that you pick up a bottle of water at the supermarket on a hot day where the ambient temperature is $32\,°\text{C}$. By the time you get home, you would like to have a drink, but the water has reached ambient temperature. You pour 325 g of water into a large plastic cup with a negligible heat capacity. The ice in your freezer is at a temperature of $-5\,°\text{C}$. How much ice should you add if you want the temperature to equilibrate at $4\,°\text{C}$?

(3) The makers of a new metal alloy wish to determine its thermal properties. In particular, they would like to know the specific heat capacity of the solid and the melting temperature. They proceed as follows:
- A 518.04 g sample is heated to $100.0\,°\text{C}$ and is then transferred to an insulated water-filled calorimeter with a heat capacity of $8403\,\text{J K}^{-1}$ (including the water) at an initial temperature of $22.04\,°\text{C}$. The equilibrium temperature is $24.33\,°\text{C}$.
- A sample is heated beyond its melting point and the temperature is measured as a function of time as it cools. The results are shown in Figure 5.4.

Determine the specific heat capacity of the solid and melting temperature.

(4) A condenser is a device that puts a vapor in thermal contact with a substance (often water) that can absorb heat and therefore cause the vapor to condense. The heat-absorbing substance is usually held at constant temperature by being constantly resupplied from a reservoir held at a constant temperature. In the simplest case, we simply

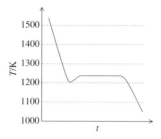

Figure 5.4 Freezing curve for an alloy, showing the temperature as a function of time.

flow tap water through the condenser. Ethanol boils at 78.5 °C and its enthalpy of vaporization is 40.476 kJ mol$^{-1}$. The density of ethanol is 0.79 g mL$^{-1}$.

(a) In a small still, 1 L of liquid ethanol is produced per hour. The condenser is of the usual water-flow type and the ethanol is not cooled appreciably below its boiling point before entering the collection vessel. How much heat is gained by the water flowing through the condenser in an hour?

(b) Tap water enters the condenser at a temperature of 10 °C and flows through the condenser at a rate of 40 L h$^{-1}$. The density of water is approximately constant and is 0.999 g mL$^{-1}$. At what temperature is the water leaving the condenser?

## 5.8 Standard states and enthalpies of formation

We can measure the enthalpy change of a system during a process by measuring heat transfer at constant pressure, and the change in internal energy by measuring heat transfer at constant volume. We can also convert enthalpies into internal energies and vice versa using Equation (5.7). (We will study this conversion in more detail in Section 5.9.) It is an important observation that the total internal energy of a system can't be measured using thermochemical methods. The only things we can measure are *changes* in the internal energy during processes. The same goes for any quantity that, like enthalpy, is derived from the internal energy.

Since, through a combination of measurements and calculations, we can determine changes in enthalpy, we might be tempted to go forth and produce great tables of enthalpies of reaction for every reaction in which we might be interested under a range of different experimental conditions. This turns out to be neither necessary nor desirable for most purposes.

It first makes sense to choose standard reporting conditions such that results from different laboratories can be compared. Most recent tables of thermodynamic data refer to a standard state that is defined as a pressure of 1 bar (100 000 Pa) and a temperature of 25 °C for all components. (Until recently, the standard pressure was one atmosphere (101 325 Pa) and many tables you will encounter were prepared for that pressure. Moreover, the normal boiling (or melting) point of a substance is still generally defined to be the boiling (melting) point at 1 atm. This is particularly important for boiling points, which vary significantly with pressure.) There are other aspects to the definition of a standard state, but for the moment these will suffice. Data obtained under other conditions are generally corrected to the standard conditions before reporting. Quantities measured in or corrected to the standard state are denoted by a superscripted circle, e.g. $\Delta H^\circ$. Tables of thermodynamic data are sometimes produced for other conditions, but in those cases the conditions always need to be specified. Data produced by biochemists should be particularly carefully scrutinized because they often use a slightly different standard state, described in detail later in this book.

One of the reasons that we try to work with state functions, rather than the more directly accessible path functions heat and work, is that state functions are path independent. The importance of this property cannot be overemphasized. It means, among other things, that we can break processes up in our imaginations into steps that may never have happened, but that result in equivalent changes in the system being studied. This leads to an important simplification in the collection and use of thermodynamic data: the **standard enthalpy of formation** $(\Delta_f H^\circ)$ of a chemical compound is defined as the enthalpy change for the process of assembling that compound from its elements in their most stable forms under standard conditions. For instance, the standard enthalpy of formation of methane is the change in enthalpy during the reaction

$$C_{(s)} + 2H_{2(g)} \rightarrow CH_{4(g)}.$$

In this reaction, $C_{(s)}$ would be the most stable form of the element at 25 °C, i.e. graphite. It follows from this definition that the enthalpy of formation of any element in its most stable form is zero. For instance, $\Delta_f H^\circ = 0$ for $O_{2(g)}$, but not for $O_{3(g)}$.

There is one exception to the rule that the element in its most stable form is assigned a standard free energy of formation of zero, namely phosphorus. The form we currently believe to be the most stable form, black phosphorus, is difficult to make in pure form. It was therefore decided to use white phosphorus as the reference form of this element. This exception to our general rule underlines the arbitrary nature of the enthalpy of formation convention.

Solvated ions pose an additional problem: we can't make a solution containing a single type of ion because we make solutions by dissolving neutral ionic compounds in a solvent. We therefore can't measure an enthalpy of formation for a single ion. To put it another way, the partition of the enthalpy of formation of an ion/counterion pair between the two ions is arbitrary. We would still like to have single-ion enthalpies of formation to avoid the need for our tables to contain enthalpies of formation for every possible ionic compound as a solute. The convention adopted is to arbitrarily fix the enthalpy of formation of the hydrogen ion $(H^+)$ to zero. We can then build up tables of enthalpies of formation for ionic compounds by starting with acids, and working our way out from there.

We can write a formation reaction for an aqueous ion by adding enough hydrogen ions to balance for charge, and then adding enough $H_2$ to balance the hydrogen atoms. For

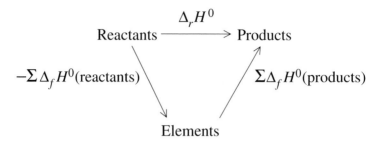

Figure 5.5 Calculation of $\Delta_r H^\circ$ for an arbitrary reaction from $\Delta_f H^\circ$ data. Rearranging the reactants into their elements is the reverse of the corresponding formation reactions, so the change in enthalpy for this step is $-\sum \Delta_f H^\circ$(reactants) (taking into account the stoichiometric coefficients of the reactants). Rearranging the elements into the products involves a change in enthalpy of $\sum \Delta_f H^\circ$ (products). These two terms are added to give the enthalpy of reaction, $\Delta_r H^\circ$.

instance, the formation reaction of a nitrate ion would be

$$\frac{1}{2}N_{2(g)} + \frac{3}{2}O_{2(g)} + \frac{1}{2}H_{2(g)} \rightarrow NO^-_{3(aq)} + H^+_{(aq)}.$$

(Note the use of fractional coefficients in this formation reaction. Formation reactions must *always* be written with a stoichiometric coefficient of 1 in front of the compound or ion being made.) This reaction is the formation reaction for aqueous $HNO_3$. However, since we fixed $\Delta_f H^\circ(H^+$, aq) to be zero, the enthalpy change in this reaction is equal to the enthalpy of formation of the nitrate ion.

Suppose that we have a table of standard enthalpies of formation and we want to know the standard enthalpy change for a balanced reaction A $\rightarrow$ B. We obtain this quantity by the following conceptual decomposition:

(1) Break down A into its elements. The standard enthalpy change for this process is $-\Delta_f H^\circ(A)$ because this step is the reverse of the formation reaction.
(2) Reassemble the elements of A into B. We know this can be done because the reaction is stoichiometrically balanced. The standard enthalpy change here is $\Delta_f H^\circ(B)$.
(3) Put together, the two processes above accomplish the same change as the overall reaction A $\rightarrow$ B, so $\Delta_r H^\circ_m = \Delta_f H^\circ(B) - \Delta_f H^\circ(A)$.

In general, we add the standard enthalpies of formation of the products and subtract the enthalpies of formation of the reactants to get the standard enthalpy change of a reaction, as illustrated in Figure 5.5. Note the use of the subscript $r$ to denote an enthalpy of reaction.

We should pause here to discuss notation. There is a logic to the IUPAC notation for changes in state functions which may not be immediately obvious, but which actually makes good physical sense. Recall that $H$ (for example) is a state function, i.e. a property of a system. During a process, $H$ may change, but it remains a property of a system. When we write $\Delta_r H$, we are denoting the change in the value of $H$ for a system during a chemical reaction. The symbol $\Delta$ indicates a change, and the subscript says what that change is due to. We should not put the subscript $r$ on the $H$ (as was formerly done) because $H$ is a property of the system itself and *not* of a reaction or other process that may affect that system. Subscripts indicating the type of change are therefore always attached to the $\Delta$.

The same logic applies to enthalpies of formation: $\Delta_f H^\circ(A)$ is the (molar) change in enthalpy when a system consisting of an appropriate stoichiometric mixture of the elements in their standard states reacts to form one mole of A in its standard state. The system, which at all times contains the correct number of atoms of each kind to make a mole of A, undergoes a particular type of reaction, and $\Delta_f H^\circ(A)$ is the change in enthalpy of the system during this process. There are two additional points of note here:

(1) Because enthalpies of formation are nearly always given as molar enthalpies, we don't usually bother with the subscript $m$, although it would be more accurate to write $\Delta_f H_m^\circ(A)$.
(2) If we need to specify the state of A, for instance to distinguish two possible phases, we simply add this information to the parenthesis. For example, we can have $\Delta_f H^\circ(H_2O, l)$ for liquid water, or $\Delta_f H^\circ(H_2O, g)$ for the vapor. We usually only include this information if there is any potential for ambiguity.

We say that a reaction is **exothermic** if the system produces heat as a result of the reaction ($\Delta_r H^\circ < 0$). Conversely, a reaction during which the system absorbs heat is said to be **endothermic** ($\Delta_r H^\circ > 0$).

Occasionally, one encounters tables of standard enthalpies of combustion, $\Delta_c H_m^\circ$. These are standard enthalpies for the complete combustion of one mole of a substance by oxygen. Since these tables are given for combustion under standard conditions (1 bar and 25 °C), when water appears as a product, it is assumed to be in the liquid state rather than the vapor. In practical terms, the experiments from which these values are derived are usually carried out in a bomb calorimeter, a device in which the heat generated by the combustion is rapidly dispersed, thus leading to the condensation of the water. Since only the beginning and end states matter in thermodynamics, we don't care that the water was initially generated as vapor, only that it returns to the liquid state before the experiment is over. Given the standard enthalpies of formation of a few key compounds (water, carbon dioxide), we can convert the standard enthalpies of combustion to standard enthalpies of formation, or devise other cycles to compute the standard enthalpy changes during reactions.

**Example 5.10 Enthalpy of formation from enthalpy of combustion** The standard enthalpy of combustion of galactose ($C_6H_{12}O_6$, an isomer of glucose) is $-2804.54$ kJ mol$^{-1}$. What is the standard enthalpy of formation of galactose?

The phrase "enthalpy of combustion of galactose" refers to the enthalpy change in the reaction

$$C_6H_{12}O_{6(s)} + 6O_{2(g)} \rightarrow 6H_2O_{(l)} + 6CO_{2(g)}.$$

For this reaction,

$$\Delta_c H_m^\circ = 6\Delta_f H^\circ(H_2O, l) + 6\Delta_f H^\circ(CO_2) - 6\Delta_f H^\circ(O_2) - \Delta_f H^\circ(\text{galactose}).$$

Using the data from Appendix A.1, we can calculate $\Delta_f H^\circ$(galactose):

$$\Delta_f H^\circ(\text{galactose}) = 6\Delta_f H^\circ(\text{H}_2\text{O}, \text{l}) + 6\Delta_f H^\circ(\text{CO}_2) - \Delta_c H^\circ_m$$
$$= -1271.50 \,\text{kJ mol}^{-1}.$$

**Example 5.11 Reactions and heat balance** What is the temperature change of the solution when 1.5 g of magnesium chloride hexahydrate is dissolved in 100 g of water?

The first assumption in these problems (often unstated) is that the reaction occurs quickly enough that the system can be treated as being quasi-adiabatic. Then we have

$$q = 0 = \Delta_r H^\circ + C_p \Delta T.$$

We will calculate $\Delta_r H^\circ$, which will then allow us to get $\Delta T$ subject to an additional assumption about the heat capacity of the system, discussed later.

The process is

$$\text{MgCl}_2 \cdot 6\text{H}_2\text{O}_{(s)} \rightarrow \text{Mg}^{2+}_{(aq)} + 2\text{Cl}^-_{(aq)} + 6\text{H}_2\text{O}_{(l)}.$$

For this process,

$$\Delta_r H^\circ_m = \Delta_f H^\circ(\text{Mg}^{2+}, \text{aq}) + 2\Delta_f H^\circ(\text{Cl}^-, \text{aq}) + 6\Delta_f H^\circ(\text{H}_2\text{O}, \text{l})$$
$$- \Delta_f H^\circ(\text{MgCl}_2 \cdot 6\text{H}_2\text{O}, \text{s})$$
$$= -17.1 \,\text{kJ mol}^{-1}.$$

The number of moles of magnesium chloride hexahydrate is

$$n = \frac{1.5\,\text{g}}{203.33\,\text{g mol}^{-1}} = 7.4 \times 10^{-3}\,\text{mol}.$$

Therefore $\Delta_r H^\circ = n\Delta_r H^\circ_m = -126\,\text{J}$. The reaction is exothermic, so 126 J of heat is produced, which will go to heating the system (water+solutes). Technically, we need the heat capacity of the solution to complete this problem. However, our system is mostly water, so we assume that the mass and heat capacity are similar to those of the solvent water. Note that the amount of water generated by the reaction (about 0.8 g) is negligible considering the other approximations we are making, particularly with regard to the heat capacity of the solution. Thus,

$$\Delta T = -\frac{\Delta_r H^\circ}{mc_p} = \frac{126\,\text{J}}{(100\,\text{g})(4.184\,\text{J K}^{-1}\text{g}^{-1})} = 0.30\,\text{K}.$$

When we use the heat capacity of water as the heat capacity of a solution, we are living in a state of sin. Why do we do this? There are a few reasons, some of them good ones, and some of them not so good:

• Heat capacity data for solutions are not as readily available as heat capacities for pure substances.

- The heat capacity of a solution depends in a complicated way on the concentration of the solute. For some solutes in some concentration ranges, the heat capacity can increase with the addition of the solute, while the heat capacity may decrease in other concentration ranges for the same solute.
- The purpose of the calculation influences how worried we should be about not knowing the true heat capacity of the solution. Often, we just want to know if mixing A and B will generate enough heat to be dangerous. Having done the previous calculation, I wouldn't worry about adding any amount of magnesium chloride to water. On the other hand, if the calculation had resulted in a predicted $\Delta T$ of several tens of degrees, I might be careful about how I made the addition, particularly since local heating (near the site where the solute is entering solution) can be much higher than the average heating in a well-mixed solution. In cases like this, I don't care if my answer is only accurate to within 20 or 30% (which is pessimistic for the calculation in the example) since I'm only going to rely on the order of magnitude of the answer.

The heat capacity issue isn't the only neglected factor in Example 5.11. Another important effect is that the enthalpy change depends on the concentration of solutes, particularly when these solutes are subject to strong intermolecular forces like the Coulomb force between ions. These forces contribute to the energy of the system, and thus to the enthalpy. To interpret these effects, it is important to understand one additional aspect of the standard state used in thermodynamics tables, namely that the thermodynamic data of solutes are generated by extrapolation to a state of infinite dilution in order to remove the dependence on solute–solute forces. We can quantify the importance of solute–solute forces by measuring the **heat of dilution**. In Example 5.11, we made a $0.074 \, \text{mol kg}^{-1}$ (moles of solute divided by kilograms of solvent) solution of magnesium chloride in water. Now imagine that we diluted this solution with additional water. We could measure the heat released or absorbed as we carried out this dilution. Eventually, the solution would be so dilute that we would, effectively, just be adding water to water, and no further heat would be released. This is the infinite dilution limit. The total heat released in going from $0.074 \, \text{mol kg}^{-1}$ to infinite dilution, called the **integral heat of dilution**, is equal in size and opposite in sign to the amount by which our calculated heat would differ from the value we would actually observe in an experiment. Heats of dilution can vary tremendously in size depending on the ions involved and on the concentration, and they can even change sign for different initial concentrations. In relatively dilute solutions such as the one contemplated in Example 5.11, enthalpies of dilution are usually small (at most $10 \, \text{kJ mol}^{-1}$, and usually much less than $1 \, \text{kJ mol}^{-1}$), but for more concentrated solutions, this effect can add up to several tens of $\text{kJ mol}^{-1}$. Moreover, note the relatively small size of the enthalpy of reaction in the example above. In a case like this, the enthalpy of dilution could easily be similar in size to $\Delta_r H°$, so the temperature change we have calculated could be out by a substantial fraction. Again, whether or not this is a problem will depend on the purpose of the calculation.

## Exercise group 5.7

(1) 1.5 mol of sodium chloride is put into an adiabatic container with 1 L of water. If the water starts out at 298.15 K, what is the final temperature of the solution obtained?

(2) The standard enthalpy of combustion of liquid benzene is $-3268\,kJ\,mol^{-1}$. What is the standard enthalpy of formation of benzene?

(3) Single-use instant cold packs contain water and ammonium nitrate, separated by a membrane. Squeezing the pack breaks the membrane. Dissolving ammonium nitrate in water is an endothermic reaction which absorbs $25.69\,kJ\,mol^{-1}$ of ammonium nitrate dissolved.

    (a) What is the standard enthalpy of formation of the aqueous nitrate ion?

    (b) You have been given a contract to design an instant cold pack which will cool to $2\,°C$ from an initial temperature of $20\,°C$ and which is to contain 250 mL of water. Approximately how much ammonium nitrate should it contain? Assume that the heat capacity of the solution is approximately equal to that of water.

(4) A typical person at rest in a warm room emits heat at a rate of about 100 W. This is a reasonable measurement of a person's resting metabolic rate.

    (a) Suppose that most of the heat originates from oxidizing glucose to carbon dioxide and water. How much glucose (in moles) would this person oxidize per minute? *Note*: Biological oxidations occur in solution phase.

    (b) How many moles of oxygen would be required?

    (c) An average breath brings in about 0.5 L of air containing 21% oxygen. About 25% of the oxygen taken into the lungs is absorbed into the bloodstream. How many breaths per minute would be required by a person at rest to maintain their resting metabolic rate? Assume that the air has a temperature of $20\,°C$ and a pressure of 1 atm.

    (d) How much work is done by the chest muscles each minute expanding the lungs by 0.5 L per breath against a constant external pressure of 1 atm?

### 5.8.1 Adding chemical reactions

Since the change in enthalpy does not depend on the path taken, when a reaction can be broken down into steps for which we know the individual changes in enthalpy, $\Delta_r H°$ for the overall process can be obtained by adding the enthalpy changes of the steps. We now illustrate the application of this principle by a few examples.

**Example 5.12 Calculation of reaction enthalpies from combustion enthalpies** The combustion enthalpies of $\alpha$-D-glucose and of $\beta$-D-glucose are, respectively, $\Delta_c H_m°(\alpha) = -2801.5$ and $\Delta_c H_m°(\beta) = -2807.8\,kJ\,mol^{-1}$. Calculate the enthalpy change during the isomerization of $\alpha$ to $\beta$-D-glucose.

Both compounds have the molecular formula $C_6H_{12}O_6$. The combustion reactions are

$$\alpha\text{-D-glucose} + 6O_2 \rightarrow 6H_2O + 6CO_2$$

and

$$\beta\text{-D-glucose} + 6O_2 \rightarrow 6H_2O + 6CO_2.$$

The isomerization reaction is

$$\alpha\text{-D-glucose} \rightarrow \beta\text{-D-glucose}.$$

If we add the first reaction to the reverse of the second, we get the isomerization reaction. Accordingly,

$$\Delta_{\text{isomer}} H_m^\circ = \Delta_c H_m^\circ(\alpha) - \Delta_c H_m^\circ(\beta) = 6.3 \, \text{kJ mol}^{-1}.$$

**Example 5.13 Changing the state of water in a reaction** The enthalpy of combustion of hydrazine ($N_2H_{4(1)}$) to water vapor and nitrogen gas is $-534 \, \text{kJ mol}^{-1}$. However, standard enthalpies of combustion are normally given for a reaction yielding liquid water as a product. Compute the standard enthalpy of combustion for hydrazine.

The combustion reaction for which we have data is

$$N_2H_{4(1)} + O_{2(g)} \rightarrow 2H_2O_{(g)} + N_{2(g)}.$$

The vaporization reaction is $H_2O_{(1)} \rightarrow H_2O_{(g)}$. We can get the molar enthalpy of vaporization of water at $25\,^\circ$C in one of two ways: Appendix C gives the specific enthalpy of vaporization as $2443 \, \text{J g}^{-1}$, which can be converted to $44.02 \, \text{kJ mol}^{-1}$ using the molar mass; alternatively, we can calculate it as a difference of enthalpies of formation for the vapor and liquid state of water from the data of Appendix A.1. This gives $44.004 \, \text{kJ mol}^{-1}$. The small difference between the two values is due to the limited number of significant figures in Appendix C. Given the precision of the hydrazine combustion data, this difference won't make any difference to the final result.

We want to turn the vaporization reaction around and double it up to cancel the gaseous water in the hydrazine combustion reaction, i.e. $2H_2O_{(g)} \rightarrow 2H_2O_{(1)}$ with $\Delta H_m^\circ = -2(44.004 \, \text{kJ mol}^{-1}) = -88.008 \, \text{kJ mol}^{-1}$. Accordingly, the standard enthalpy of combustion of hydrazine is $-534 + (-88.008) \, \text{kJ mol}^{-1} = -622 \, \text{kJ mol}^{-1}$.

---

### Exercise group 5.8

(1) The following reactions occur in the gas phase:

$$CH_3NO_{2(g)} + HNO_{3(g)} \rightarrow 2NO_{2(g)} + CH_3OH_{(g)} \qquad \Delta_{r(1)} H_m^\circ = 66.48 \, \text{kJ mol}^{-1}$$
$$CH_3NO_{2(g)} + HCl_{(g)} \rightarrow NOCl_{(g)} + CH_3OH_{(g)} \qquad \Delta_{r(2)} H_m^\circ = 8.95 \, \text{kJ mol}^{-1}$$

When $NO_2$ and hydrogen chloride react in the gas phase, the products are $HNO_3$ and NOCl. What is the enthalpy change in this reaction?

(2) Given the following data

$$Cl^-_{(g)} + SO_{2(g)} \rightarrow Cl^-_{(g)} \cdot SO_{2(g)}, \qquad \Delta_{r(1)} H^\circ_m = 91 \text{ kJ mol}^{-1}$$
$$Cl^-_{(g)} \cdot SO_{2(g)} + SO_{2(g)} \rightarrow Cl^-_{(g)} \cdot 2SO_{2(g)}, \qquad \Delta_{r(2)} H^\circ_m = 51.5 \text{ kJ mol}^{-1}$$
$$Cl^-_{(g)} \cdot 2SO_{2(g)} + SO_{2(g)} \rightarrow Cl^-_{(g)} \cdot 3SO_{2(g)}, \qquad \Delta_{r(3)} H^\circ_m = 41.8 \text{ kJ mol}^{-1}$$
$$Cl^-_{(g)} \cdot 3SO_{2(g)} + SO_{2(g)} \rightarrow Cl^-_{(g)} \cdot 4SO_{2(g)}, \qquad \Delta_{r(4)} H^\circ_m = 36 \text{ kJ mol}^{-1}$$

what is the enthalpy change for the complete dissociation of $Cl^-_{(g)} \cdot 4SO_{2(g)}$ into $Cl^-_{(g)}$ and $SO_{2(g)}$?

## 5.9 More on the relationship between internal energy and enthalpy

If we know $\Delta H$, then in principle we can calculate $\Delta U$, or vice versa:

$$\Delta H = \Delta U + \Delta(pV)$$

To do this, we need to know how the product $pV$ has changed during a process. If the process involves only solids or liquids, then $pV$ is very small and consequently its change is insignificant, as we will see in the following example.

**Example 5.14 Difference between $\Delta U$ and $\Delta H$ for the melting of ice**  Suppose that we want to calculate $\Delta U^\circ_m$ for the melting of ice at $0\,^\circ\text{C}$ and $1 \text{ atm} = 101\,325$ Pa. We're going to mind our significant digits carefully during this calculation. Note the table of properties of water in appendix C. The molar enthalpy of fusion of ice at $0\,^\circ\text{C}$ and 1 atm is $6007 \text{ J mol}^{-1}$. At constant pressure, $\Delta U^\circ_m = \Delta H^\circ_m - p\,\Delta V^\circ_m$. Appendix C gives data for water and ice at the standard pressure of 1 bar. For the purpose of this calculation, which only involves condensed phases, the difference between 1 bar and 1 atm is negligible. The molar volumes are computed from the densities and molar masses:

$$V^\circ_m(\text{ice}) = \frac{18.02 \text{ g mol}^{-1}}{0.915 \text{ g cm}^{-3}} = 19.7 \text{ cm}^3 \text{ mol}^{-1}.$$

We want this quantity in SI units:

$$V^\circ_m(\text{ice}) = (19.7 \text{ cm}^3 \text{ mol}^{-1}) \left( \frac{1 \text{ m}}{100 \text{ cm}} \right)^3 = 1.97 \times 10^{-5} \text{ m}^3 \text{ mol}^{-1}.$$

Similarly, for liquid water at $0\,^\circ\text{C}$:

$$V^\circ_m(\text{water}) = \frac{18.02 \text{ g mol}^{-1}}{0.9999 \text{ g cm}^{-3}} \left( \frac{1 \text{ m}}{100 \text{ cm}} \right)^3 = 1.802 \times 10^{-5} \text{ m}^3 \text{ mol}^{-1}.$$

We can now compute the change in the internal energy:

$$\Delta U^\circ_m = 6007 \text{ J mol}^{-1} - (101\,325 \text{ Pa})(1.802 - 1.97) \times 10^{-5} \text{m}^3 \text{ mol}^{-1}$$
$$= 6007 \text{ J mol}^{-1} - (101\,325 \text{ Pa})(-0.17 \times 10^{-5} \text{m}^3 \text{ mol}^{-1})$$
$$= 6007 \text{ J mol}^{-1} + 0.17 \text{ J mol}^{-1} = 6007 \text{ J mol}^{-1}.$$

Note that, within the significant figures of this problem, the $p\Delta V_m^\circ$ term doesn't make any difference. Many chemical processes have enthalpy changes that are known only to a few tens or hundreds of a J mol$^{-1}$ (hundredths or tenths of a kJ mol$^{-1}$), or even just to the nearest kJ mol$^{-1}$, rather than to the nearest J mol$^{-1}$ as in this case. The difference between $\Delta H$ and $\Delta U$ for reactions involving only condensed phases would be even farther removed from the significant figures in these cases.

It is only when processes involve gases that we need to worry about the $pV$ difference between $\Delta H$ and $\Delta U$. In the case of a reaction involving gases, we can approximate the value of $pV$ by $nRT$. If, as will usually be the case, we are interested in changes occurring under standard conditions, $T = 298.15$ K is constant and the number of moles of gas in the system is the only thing that might change during the reaction. Therefore $\Delta_r(pV) \approx RT\Delta_r n_{gas}$, where $\Delta_r n_{gas}$ is the change in the number of moles of gas during the reaction. The relationship between $\Delta_r H$ and $\Delta_r U$ is therefore

$$\Delta_r H \approx \Delta_r U + RT\Delta_r n_{gas}$$

or, if we divide by the number of moles of a reactant,

$$\Delta_r H_m \approx \Delta_r U_m + RT\Delta_r \nu_{gas}. \tag{5.15}$$

Since we divided the change in the number of moles of gas by the number of moles of a reactant, the quantity $\Delta_r \nu_{gas}$ is the difference between the stoichiometric coefficients of *gas-phase* reactants and products.

**Example 5.15 Conversion between enthalpies and internal energies of reaction** What is $\Delta_r U_m^\circ$ for the reaction $H_{2(g)} + \frac{1}{2}O_{2(g)} \rightarrow H_2O_{(g)}$?

This is the formation reaction for gaseous water so $\Delta_r H_m^\circ = \Delta_f H^\circ(H_2O, g) = -241.826$ kJ mol$^{-1}$. For this reaction $\Delta_r \nu_{gas}$ is $1 - \left(1 + \frac{1}{2}\right) = -\frac{1}{2}$. Therefore

$$\Delta_r U_m^\circ \approx \Delta_r H_m^\circ - RT\Delta_r \nu_{gas}$$
$$= -241.826 \text{ kJ mol}^{-1}$$
$$- (8.314\,472 \times 10^{-3} \text{ kJ K}^{-1}\text{mol}^{-1})(298.15 \text{ K})\left(-\frac{1}{2}\right)$$
$$= -240.587 \text{ kJ mol}^{-1}.$$

The transformation between energies and enthalpies most often comes up in calorimetry because a constant-volume ("bomb") calorimeter is the most convenient way to obtain the required data. Figure 5.6 shows the major parts of a typical bomb calorimeter. The calorimeter has three major parts, the calorimeter jacket, which is just an insulated enclosure, the bucket and the bomb. The bucket is filled with a known and carefully measured amount of water. A thermometer (not shown) measures the temperature of the water, which is mechanically stirred during the experiment to ensure rapid equilibration. The bomb is a sealed vessel which is filled with a high pressure of oxygen to ensure complete combustion of the sample. The bomb and bucket are both made of metal and touch each other. The

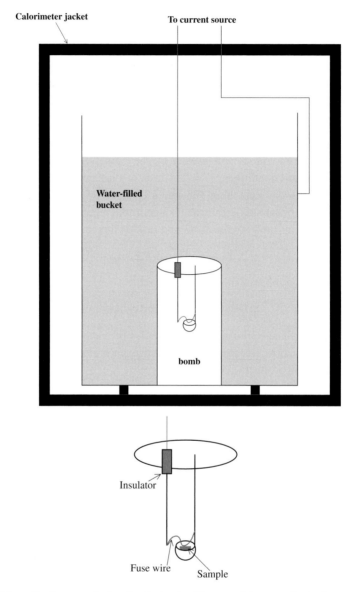

Figure 5.6 Schematic diagram of a bomb calorimeter. The complete apparatus is shown above, while the drawing below shows a blowup of the sample cup and related parts.

bucket sits on top of some insulating posts, both to prevent electrification of the jacket, and to increase the thermal insulation of the bucket by minimizing the contact area with the jacket. The current source is connected to the bucket and to an electrode which is insulated from the bomb. Thus, there is a current path through the fuse wire, a thin wire made of a material (often a magnesium alloy) that burns quickly and completely in air. The fuse wire is installed in such a way that it just touches the sample. When the fuse is ignited by passing a current through it in the oxygen-rich atmosphere of the bomb, it in turn ignites the sample. After a few minutes, the bomb, bucket and water come to thermal equilibrium. If we first burn a sample of known heat of combustion, we can determine the heat capacity of the calorimeter (including the water) from the temperature change. After that, we can burn unknown samples and use the temperature change to calculate the heat of combustion at constant volume, i.e. the change in internal energy.

The bomb calorimetry experiment can be very precise. The key is to minimize the number of variables from experiment to experiment, and to measure everything else carefully: Precisely the same amount of water must be used every time. The amount of fuse wire used is measured, and its heat of combustion is taken into account in the calculations. Finally, we use excellent thermometers which can measure temperatures accurately to three decimal places or better.

Among other applications, bomb calorimeters are used to determine food energies, the Calorie values you see on food packages. The calorie, as you may know, is an older unit of energy based on the amount of heat required to raise the temperature of one gram of water by 1 °C. Unfortunately, the heat capacity of water depends on the temperature, so the value of the calorie depends on the initial temperature. Thermodynamicists use a standard calorie equal to 4.184 J. This is the value I will use throughout the book. Canadian food labels also use this value for the calorie. The International Union of Nutritional Sciences defines a calorie as 4.182 J. In some countries, the calorie is set equal to the International Steam Table calorie, which is 4.1868 J, a value that was convenient for engineers because it's an exact fraction ($\frac{1}{860}$) of a watt-hour (yet another unit of energy). Since nobody can really agree what a calorie is, it's probably better to stick with joules! To add to the confusion, most food energies are given in Calories (with a capital C), which are actually kilocalories.

Getting back to food energies, it's important to understand that the food energy determined by calorimetry is *not* the same as the energy you can extract from your food. The fact that something burns in a calorimeter doesn't mean that you'll be able to digest it, which is a prerequisite to you using a food's energy. A good example is dietary fiber, the indigestible component of plant foods. Dietary fiber is made of cellulose and other indigestible polysaccharides, lignin (the stuff that makes plants woody) and waxes, among other things. All of these compounds burn well in a calorimeter and would register a calorie value. However, since you don't digest them (although the bacteria in your intestinal tract can digest some of them), you certainly don't get any energy out of them. The energy that you can get out of a food is called the catabolizable (or metabolizable) energy, a concept to which we will return in Section 5.11.3.

**Example 5.16  Bomb calorimetry calculations** A bomb calorimeter is set up with a sample of 1.0024 g of octane ($C_8H_{18}$, boiling point 125 °C), 10.2 cm of fuse wire with a constant-volume heat of combustion of $-9.6\,\mathrm{J\,cm^{-1}}$ and 20 atm of pure oxygen. This represents a large excess of oxygen to ensure complete combustion of the octane and fuse wire. The calorimeter has a heat capacity (previously measured) of 12.04 kJ K$^{-1}$. The octane is ignited and the temperature rises from 294.281 to 298.268 K. What is the enthalpy of formation of octane?

We can straightforwardly calculate the heat liberated by the two reactions (combustion of the octane and fuse wire) from the temperature rise. Note that because the reaction is carried out at constant volume and the calorimeter is an (approximately) adiabatic device,

$$\Delta_c U = -C_V \Delta T = -(12.04 \text{ kJ K}^{-1})(298.268 - 294.281 \text{ K})$$
$$= -(12.04 \text{ kJ K}^{-1})(3.987 \text{ K}) = -48.00 \text{ kJ}.$$

Of this amount, the following is due to the fuse wire:

$$\Delta_c U(\text{fuse}) = (10.2 \text{ cm})(-9.6 \text{ J cm}^{-1}) = -98 \text{ J} \equiv -0.098 \text{ kJ}.$$

The heat due to the combustion of octane is therefore

$$\Delta_c U(\text{C}_8\text{H}_{18}) = -48.00 - (-0.098) \text{ kJ} = -47.91 \text{ kJ}.$$

We also know the number of moles of octane because we know the mass used: $n_{\text{octane}} = 1.0024 \text{ g}/114.230 \text{ g mol}^{-1} = 8.7753 \times 10^{-3} \text{mol}$. We can then work out the change in internal energy per mole of octane burned:

$$\Delta_c U_m^\circ = -47.91 \text{ kJ}/8.7753 \times 10^{-3} \text{mol} = -5459 \text{ kJ mol}^{-1}.$$

Note that this is, for all intents and purposes, a *standard* internal energy of combustion since the initial and final temperatures are both close to 298 K.

The standard enthalpy change for the reaction is obtained by $\Delta_c H_m^\circ = \Delta_c U_m^\circ + RT \Delta_c \nu_{\text{gas}}$. We now need a balanced reaction:

$$\text{C}_8\text{H}_{18(l)} + \frac{25}{2}\text{O}_{2(g)} \rightarrow 8\text{CO}_{2(g)} + 9\text{H}_2\text{O}_{(l)}.$$

How do we know that the octane and water should be liquid? The initial temperature is well below the boiling point of octane and the final temperature is well below the boiling point of water. This detail is important because we need to know the states of the compounds to calculate $\Delta_c \nu_{\text{gas}}$. The latter quantity only depends on the stoichiometric coefficients of the gases, in this case oxygen and carbon dioxide, and works out to $\Delta_c \nu_{\text{gas}} = 8 - \frac{25}{2} = -\frac{9}{2}$. Therefore, the standard enthalpy of combustion of octane is

$$\Delta_c H_m^\circ = -5459 \text{ kJ mol}^{-1} + (8.314\,472 \times 10^{-3} \text{ kJ K}^{-1}\text{mol}^{-1})(298.15 \text{ K})\left(-\frac{9}{2}\right)$$
$$= -5470 \text{ kJ mol}^{-1}.$$

We want the standard enthalpy of formation. From the reaction, we have

$$\Delta_c H_m^\circ = 8\Delta_f H_{CO_2}^\circ + 9\Delta_f H_{H_2O_{(l)}}^\circ - \Delta_f H^\circ(C_8H_{18}, l);$$
$$\therefore \Delta_f H^\circ(C_8H_{18}, l) = 8\Delta_f H_{CO_2}^\circ + 9\Delta_f H_{H_2O_{(l)}}^\circ - \Delta_c H_m^\circ$$
$$= 8(-393.51) + 9(-285.83) - (-5470) \text{ kJ mol}^{-1}$$
$$= -250 \text{ kJ mol}^{-1}.$$

Note that the precision of this calculation was limited by the precision in our measurement of $\Delta T$. However, just getting a better thermometer would not, in this case, get us a more accurate number. Bomb calorimeters have some intrinsic limitations, largely associated with the fact that they are not perfectly insulated, and thus that there is a constant heat leak between the calorimeter and lab during the experiment. There are, however, other calorimeter designs for which this is not a problem.

---

### Exercise group 5.9

(1) (a) Compute the standard enthalpy of combustion of acetylene (ethyne) to carbon dioxide and liquid water.

(b) Compute the standard change in internal energy during the combustion of acetylene.

(2) Carbohydrates have the empirical formula $CH_2O$. Show that for the combustion of carbohydrates, $\Delta_c U \approx \Delta_c H$. *Hint*: The temperature in a calorimeter seldom rises very far above room temperature. At these temperatures, carbohydrates are solids and water is liquid.

(3) Catalytic hydrogenation converts unsaturated hydrocarbons into saturated hydrocarbons by reaction with hydrogen gas. How much heat is produced or absorbed per mole of ethene ($C_2H_{4(g)}$) hydrogenated to ethane ($C_2H_{6(g)}$) in a constant-volume reactor at 298.15 K? Indicate clearly whether heat is absorbed or produced.

(4) Formaldehyde (methanal, $CH_2O$) is a gas at room temperature. It has a standard enthalpy of combustion of $-570.78$ kJ mol$^{-1}$.

(a) What is the standard enthalpy of formation of this compound?

(b) Calculate the heat generated by the combustion of 1.0045 g of formaldehyde in a bomb calorimeter assuming that the temperature stays near 25 °C.

(5) It is very important that the amount of sample used in a bomb calorimeter be limited since the heat from the reaction warms the oxygen, which increases the pressure in the bomb. Suppose that we want to use a bomb calorimeter to measure the combustion energy of coal. A typical value for this combustion energy is $-25$ kJ g$^{-1}$. If the manufacturer recommends that the temperature rise in the calorimeter be limited to 6 K and the constant-volume heat capacity of the device is 10 kJ K$^{-1}$, what is the maximum permissible sample size?

(6) (a) During the calibration of a constant-volume calorimeter, a 1.0095 g sample of benzoic acid ($C_6H_5COOH$) is burned completely in an excess of oxygen. The

experiment also used 9.8 cm of fuse wire with a (constant-volume) heat of combustion of $-9.6\,\mathrm{J\,cm^{-1}}$. The temperature rises by $2.579\,^{\circ}\mathrm{C}$. What is the heat capacity of the calorimeter? The change in molar internal energy on combustion of benzoic acid is $-3225.7\,\mathrm{kJ\,mol^{-1}}$.

(b) A 1.1243 g sample of solid sucrose ($C_{12}H_{22}O_{11}$) is then burned in the same calorimeter with 10.1 cm of fuse wire. The temperature rises by $1.805\,^{\circ}\mathrm{C}$. What is the molar energy of combustion of sucrose?

(c) Using the result from question 6 and the standard enthalpies of formation of carbon dioxide and water from Appendix A.1, calculate the standard enthalpy of formation of sucrose. Compare your result with the value given in Appendix A.1.

(7) The bombardier beetle defends itself by spraying a quinone solution at its attackers. Quinone is a noxious substance that repels most of its potential predators. The spray is forcibly expelled because the reaction that forms the quinone

$$C_6H_4(OH)_{2(aq)} + H_2O_{2(aq)} \rightarrow C_6H_4O_{2(aq)} + 2H_2O_{(l)}$$
$$\text{hydroquinone} \qquad\qquad\qquad \text{quinone}$$

is highly exothermic and actually causes the solvent (water) to boil.

(a) Given

$$C_6H_4(OH)_{2(aq)} \rightarrow C_6H_4O_{2(aq)} + H_{2(g)} \qquad \Delta_{r(1)}H_m^{\circ} = 177\,\mathrm{kJ\,mol^{-1}}$$
$$H_2O_{2(aq)} \rightarrow H_2O_{(l)} + \tfrac{1}{2}O_{2(g)} \qquad \Delta_{r(2)}H_m^{\circ} = -94.6\,\mathrm{kJ\,mol^{-1}}$$

calculate the standard enthalpy change for the reaction of hydroquinone with hydrogen peroxide.

(b) Insects are poikilothermic, i.e. they don't generate heat internally and their body temperature is thus roughly the same as the ambient temperature at any given time. Making the dubious assumptions that the enthalpy of reaction is roughly constant with respect to temperature and that the presence of solutes has little effect on the properties of water, what are the minimum concentrations of hydroquinone and of hydrogen peroxide required to bring the solvent to a boil from an ambient temperature of $18\,^{\circ}\mathrm{C}$?

*Hint*: How many moles of the reactants does it take to bring 1 L of water to a boil?

(c) Calculate the standard change in internal energy for the decomposition of hydroquinone to hydrogen and quinone.

## 5.10 The dependence of energy and enthalpy changes on temperature

Not all reactions are carried out at the standard thermodynamic temperature of $25\,^{\circ}\mathrm{C}$, so it is important to be able to adjust data from one temperature to another. To adjust an energy or enthalpy change for temperature, we use the fact that both internal energy and enthalpy are state functions. This means that it doesn't matter which path we take, only what the beginning and end points of the process are. Suppose that we want to use a standard

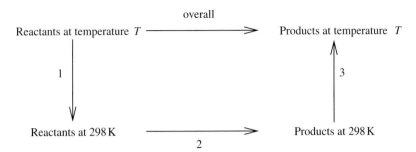

overall

Reactants at temperature $T$ → Products at temperature $T$

1

3

Reactants at 298 K → Products at 298 K

2

Figure 5.7 A thermodynamic cycle used to evaluate the change in a state function at an arbitrary temperature $T$ from data given at 298 K. The overall change is the sum of the changes in steps 1–3.

thermodynamic table to compute the enthalpy change for a given reaction at a different temperature than 298 K. To be definite, imagine that the target temperature, $T$, is higher than 298 K, although this doesn't matter to the overall logic of what we are going to do. We would then break up the process as follows (Figure 5.7):

(1) Imagine that we are starting with reactants at the target temperature $T$. Using the heat capacities of the reactants, we compute the heat removed in cooling them from $T$ to 298 K.
(2) Calculate $\Delta_r H$ at 298 K using the tables.
(3) Calculate the heat added to bring the products from 298 K to $T$.

The overall enthalpy change is the sum of the enthalpy changes in steps 1–3.

**Example 5.17 Calculating an enthalpy of reaction at a non-standard temperature** Calculate the enthalpy change during the reaction

$$SO_{2(g)} + \frac{1}{2}O_{2(g)} \to SO_{3(g)}$$

at 300 °C.

(1) Imagine that we have a stoichiometric mixture of $SO_2$ and $O_2$ at 300 °C. We calculate the heat evolved in cooling this mixture to 298 K:

$$\Delta_{(1)}H_m = \left(C_{p,m}(SO_2) + \frac{1}{2}C_{p,m}(O_2)\right)\Delta T$$

$$= \left(39.8 + \frac{1}{2}(29.35)\,J\,K^{-1}mol^{-1}\right)(25 - 300\,°C)$$

$$= -15.0\,kJ\,mol^{-1}.$$

(2) Use the enthalpies of formation to compute $\Delta_r H_m^\circ$ at 298 K:

$$\Delta_{(2)}H_m = \Delta_f H°(SO_3) - \Delta_f H°(SO_2) - \frac{1}{2}\Delta_f H°(O_2)$$

$$= -395.7 - (-296.81)\,kJ\,mol^{-1} = -98.9\,kJ\,mol^{-1}.$$

(3) Now imagine heating the product back up to 300 °C:

$$\Delta_{(3)} H_m = C_{p,m}(SO_3) \Delta T$$
$$= (50.67 \, \text{J K}^{-1} \text{mol}^{-1})(300 - 25 \, ^\circ\text{C}) = 13.9 \, \text{kJ mol}^{-1}.$$

Therefore,

$$\Delta_r H_m = -15.0 + (-98.9) + 13.9 \, \text{kJ mol}^{-1} = -99.9 \, \text{kJ mol}^{-1}.$$

As you can see from the previous example, the enthalpy of reaction will often vary very little with temperature. The two correction terms (cooling reactants and warming products) will always have opposite signs and will often be similar in size. Exceptions will be found when the reactants and products are very different in nature (e.g. are in different phases of matter). Even then, unless we are looking at very large differences in temperature or are making very precise measurements, we can often treat the enthalpy of reaction as being roughly constant.

---

### Exercise group 5.10

(1) Calculate the enthalpy of combustion of propane ($C_3H_{8(g)}$) at 500 °C.
(2) A thermodynamicist finds that the enthalpy of formation of ozone ($O_{3(g)}$) at 500 °C is 140 kJ mol$^{-1}$. What is the value of the enthalpy of formation at 298 K? The molar heat capacity of ozone is 39.2 J K$^{-1}$mol$^{-1}$.
(3) When water is supercooled, it starts to freeze at a temperature below zero. What is the enthalpy of freezing of water at $-10 \, ^\circ\text{C}$?
(4) Compare the enthalpy of combustion of glucose at 25 °C to the enthalpy of combustion at 37 °C. In your opinion, is the difference sufficiently small to ignore in metabolic calculations?
(5) Industry sometimes uses flow-through reactors, which are basically tubes through which the reactants are pushed and in which they react. The temperature in a flow-through reactor can be made different at the two ends simply by heating one end or cooling the other, or both. Suppose that benzene is hydrogenated to form cyclohexane in a flow-through reactor. The reactants benzene and hydrogen enter the reactor at a temperature of 550 K (well above the boiling point of benzene) and the product cyclohexane exits the reactor at 1290 K (again, well above the b.p. of $C_6H_{12}$). The pressure is constant throughout. How much heat is liberated per mole of benzene hydrogenated under these conditions?

---

## 5.11 Measuring the energy requirements of living organisms

Here's a simple question: how much energy does a given organism use per day? While it's a simple question, figuring out how to answer it takes a bit of thought. There are a number of ways we can measure an organism's energy requirements, some of which we will discuss here. All of these methods (and others we won't consider here) are based on ideas and principles studied in this chapter.

### 5.11.1 Nutritional balance

Suppose that we want to measure the energetic requirements of an animal averaged over a period of several days or weeks. We define the whole animal, including its symbiotic micro-organisms, as our system. A very straightforward application of the First Law then tells us that the difference in energy between inputs (food) and outputs (waste) is the amount of energy used or stored by the animal. If we also know the difference in weight between the beginning and end of the experiment and the energy content of the storage compound (commonly fat), we can determine how much energy was actually used by the organism.

Here is an example, constructed from the data of Max Kleiber:[2] A cow eats 5.27 kg of hay per day. The heat of combustion (at constant volume) of hay is 4.20 Mcal kg$^{-1}$. The heat of combustion of the feces was evaluated by bomb calorimetry and found to average 7.25 Mcal day$^{-1}$. Similarly, her evaporated urine had a heat of combustion of 1.43 Mcal day$^{-1}$. She also produced an average of 191 L of methane (at 25 °C and 1 atm) per day. During the period of the experiment, she didn't gain any weight and didn't lactate.

We have most of the information we need to compute the cow's metabolic requirements. The heats of combustion tell us how much energy is available by burning various substances as fuels. Since metabolism involves a set of reactions equivalent to combustion, and since heat measured at constant volume is a change in internal energy, calorimetric heats of combustion are direct measurements of the energy content of foods and wastes. One of the missing pieces is the change in internal energy during the combustion of methane. A balanced equation for the combustion of methane is

$$CH_{4(g)} + 2O_{2(g)} \rightarrow CO_{2(g)} + 2H_2O_{(l)}.$$

The standard enthalpy of combustion of methane calculated from the enthalpies of formation is $-890.36$ kJ mol$^{-1}$. From Equation (5.15),

$$\begin{aligned}
\Delta_c U_m(CH_4) &= \Delta_c H_m(CH_4) - RT \Delta_c \nu_{gas} \\
&= -890.36 \text{ kJ mol}^{-1} - (8.314\,472 \times 10^{-3} \text{ kJ K}^{-1}\text{mol}^{-1})(298.15 \text{ K})(-2) \\
&= -885.40 \text{ kJ mol}^{-1}.
\end{aligned}$$

The number of moles of methane is determined from the volume by the ideal gas law:

$$n_{CH_4} = \frac{pV}{RT} = \frac{(101\,325 \text{ Pa})(191 \times 10^{-3} \text{ m}^3)}{(8.314\,472 \text{ J K}^{-1}\text{mol}^{-1})(298.15 \text{ K})} = 7.81 \text{ mol}.$$

We therefore have

$$\Delta_c U(CH_4) = (-885.40 \text{ kJ mol}^{-1})(7.81 \text{ mol}) = -6.91 \text{ MJ}.$$

---

[2] Max Kleiber, *The Fire of Life*; Wiley: New York, 1961, pp. 251–265.

We now need to do a bunch of unit conversions and to put the whole thing together. First, the cow took in 5.27 kg of hay with a total energy content of

$$\text{Energy in} = (5.27\,\text{kg day}^{-1})(4.20\,\text{Mcal kg}^{-1})(4.184\,\text{J cal}^{-1}) = 92.6\,\text{MJ day}^{-1}.$$

On the other side of the ledger, the cow excreted energy-containing compounds in her feces and urine, and as methane. Converting the relevant energies to MJ, we get

$$\text{Energy out} = 30.3 + 5.98 + 6.91\,\text{MJ} = 43.2\,\text{MJ}.$$

The metabolic requirements of this particular cow are therefore

$$\text{Metabolic requirements} = \text{Energy in} - \text{Energy out} = 49.4\,\text{MJ day}^{-1}.$$

Installing gas-flow meters on animals; collecting, drying and weighing feces and urine; and subjecting these materials to calorimetric measurements are all procedures that require significant dedication, especially since day-to-day variability will require the data to be collected and averaged over a period of days or weeks. Both because of these difficulties and because more convenient methods are now available, the nutritional balance method is now rarely used for basic physiological studies.

### 5.11.2  Direct calorimetry

According to the First Law, all of the energy used by an organism will either show up as heat or as work. If an organism does no work, everything will show up as heat. As a result, we can measure the resting energetic requirements of animals or the normal metabolic activity of plants or micro-organisms by simply measuring their heat production. There are many possible calorimeter configurations. Here, we'll study just one, a classic design invented by Lavoisier and Laplace.

The Lavoisier–Laplace apparatus is sketched in Figure 5.8. Suppose we place a small animal in an enclosure in good thermal contact with a reserve of ice. Air is allowed to flow in and out. The system (ice+animal) is insulated. There is a provision for collecting and measuring the melt water. Since this is approximately an adiabatic system, all of the heat for melting the ice came from the animal and, therefore, is a direct measure of its metabolic activity, give or take any heat lost as a result of gas exchange.

For example, suppose a guinea pig is placed in such a calorimeter. In 10 hours, 370 g of ice melts.[3] The calculation of the guinea pig's metabolic energy use is then straightforward; since the animal is being held at constant pressure, we are measuring an enthalpy change. However, the system (ice+animal) is largely composed of solids and liquids so that the change in enthalpy is approximately the same as the change in internal energy. The heat of fusion of ice at $0\,^\circ\text{C}$ and 1 atm is $6.007\,\text{kJ mol}^{-1}$. Therefore, the heat required to melt 370 g of ice is

$$q = \frac{(370\,\text{g})(6.007\,\text{kJ mol}^{-1})}{18.02\,\text{g mol}^{-1}} = 123\,\text{kJ}.$$

---

[3]  These are the data of Lavoisier and Laplace, reported by Max Kleiber, *The Fire of Life*; Wiley: New York, 1961, p. 118.

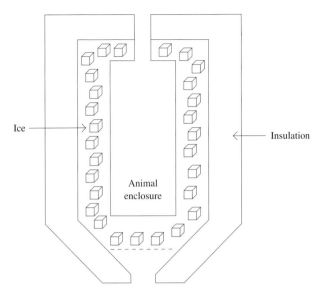

Figure 5.8 Schematic diagram of a Lavoisier–Laplace calorimeter. The animal enclosure is surrounded by an ice jacket which is in turn surrounded, as completely as possible, by insulation. The opening at the top is used to provide the animal with fresh air. The heat generated by the animal melts some of the ice which drips out of the bottom of the apparatus, where it is collected for later measurement.

This energy is provided by the metabolic activity of the animal so that

$$\text{Metabolic requirements} = q/t = 123\,\text{kJ}/10\,\text{h} = 12.3\,\text{kJ}\,\text{h}^{-1}$$
$$\equiv 3.43\,\text{W}$$
$$\equiv 296\,\text{kJ}\,\text{day}^{-1}$$

Direct whole-organism calorimetry has some significant drawbacks:

(1) Because there must be good thermal contact between the organism's enclosure and the ice reservoir, the former will generally be cool. Temperature has profound effects on the metabolism of living organisms.
(2) The heat production by ectotherms ("cold-blooded" animals), plants and micro-organisms, particularly at temperatures near the freezing point of water, is too small to be measured by an ice calorimeter. Calorimeters based on different principles can be designed to overcome both this and the previous problem, but these are generally much more complicated and thus much more expensive.
(3) Calorimetry is generally considered an impractical way to measure the energetic requirements of anything much larger than a guinea pig because of the size of the apparatus required. However, larger whole-body calorimeters have occasionally been built.

Table 5.1 *Physiological and biochemical data required for the indirect calorimetry calculations.*

| Quantity | Value |
| --- | --- |
| Nitrogen content of protein | 17% (by weight) |
| Catabolizable energy from protein | 4.5 kcal g$^{-1}$ |
| $O_2$ needed to catabolize protein | 0.0395 mol g$^{-1}$ |
| $CO_2$ from protein catabolysis | 0.0320 mol g$^{-1}$ |
| Catabolizable energy from fat | 9.5 kcal g$^{-1}$ |
| $O_2$ needed to catabolize fat | 0.0900 mol g$^{-1}$ |
| $CO_2$ from fat catabolysis | 0.0636 mol g$^{-1}$ |
| Catabolizable energy from carbohydrates | 4.0 kcal g$^{-1}$ |
| $O_2$ needed to catabolize carbohydrates | 0.0370 mol g$^{-1}$ |
| $CO_2$ from carbohydrate catabolysis | 0.0370 mol g$^{-1}$ |

### 5.11.3 Indirect calorimetry

Indirect calorimetry isn't calorimetry at all; in fact it's much more closely related in spirit to the nutritional balance method than to calorimetric methods. The idea here is to use a little biochemical knowledge along with measurements of the gas exchange and (sometimes) of urinary composition to calculate the amount of energy-storage compounds (carbohydrates, fat etc.) used, and thus the amount of energy the organism used during the experiment.

In order to do the required calculations, we will require an additional concept, namely that of **catabolizable energy** (sometimes called metabolizable energy). Recall that in the nutritional balance method, we had to know the energy content of both foods and excreta. The reason is that not everything can be used as a fuel in the body. Similarly, when converting certain reserve compounds to work or heat, cells reject certain waste compounds. The catabolizable energy of an energy-storage compound is the average amount of energy that can be extracted from that compound, taking into account the wastes generated. For fat and carbohydrates, the catabolizable energy is virtually the same as the combustion energy. However, the catabolizable energy of protein is much lower than its combustion energy because of the generation of nitrogenous waste.

It is easiest to explain this method by an example: suppose that an experimental subject clears his bladder, then engages in an hour of exercise. During that hour, he uses 3.1 mol of $O_2$ and produces 2.7 mol of $CO_2$. At the end, he again urinates. This urine sample contains 2.0 g of urea (($NH_2)_2CO$), the principal nitrogenous waste produced by mammals. As previously, we want to know how much energy the subject used during this period. Several bits of calorimetric and physiological data are needed to solve this problem, all of which are given in Table 5.1.

Given the amount of urea produced by the subject during the experiment, we can calculate the amount of protein catabolized and thence the amount of oxygen and carbon dioxide used and produced, respectively, in protein metabolism:

$$n_{urea} = 2.0\,g/60.056\,g\,mol^{-1} = 0.033\,mol$$
$$m_N = 2(0.033\,mol)(14.0067\,g\,mol^{-1}) = 0.92\,g$$
$$m_{protein} = 0.92\,g/17\% = 5.4\,g$$
$$O_2 \text{ consumed by protein catabolysis} = (5.4\,g)(0.0395\,mol\,g^{-1}) = 0.21\,mol$$
$$CO_2 \text{ produced by protein catabolysis} = (5.4\,g)(0.0320\,mol\,g^{-1}) = 0.17\,mol$$

This leaves 2.9 mol $O_2$ consumption and 2.5 mol $CO_2$ production to be accounted for. These gases were involved in carbohydrate and fat metabolism, so

$$2.9 = 0.0900m_{fat} + 0.0370m_{carb}$$
$$2.5 = 0.0636m_{fat} + 0.0370m_{carb}$$

Solving these equations simultaneously, we get

$$m_{fat} = 14\,g$$
$$m_{carb} = 45\,g$$

We can now add together the contributions of protein, fat and carbohydrates, using the caloric equivalents of the different energy-storage compounds from Table 5.1, to obtain the energy used by the subject during the experiment:

$$\text{Energy used} = \text{Energy from protein} + \text{Energy from fat} + \text{Energy from carbs}$$
$$= (4.5\,kcal\,g^{-1})(5.4\,g) + (9.5\,kcal\,g^{-1})(15\,g) + (4.0\,kcal\,g^{-1})(42\,g)$$
$$= 333\,kcal \equiv 1.39\,MJ$$

Indirect calorimetry is probably the most accurate way we have of measuring energy use. The principal disadvantage of indirect calorimetry is related to the requirement that gas flow and composition must be measured. This means that the experiment must either be performed in a closed chamber (convenient for small animals, plants and micro-organisms) or (with larger land mammals) that tubes must be connected to the subject's airway. Either way, the range of activities that can be studied in this way is restricted, although portable gas analyzers have been devised.

---

### Exercise group 5.11

(1) How much protein did you metabolize in a day if you excrete 7.0 g of urea during that period? How much energy did you derive from protein metabolism in this time?

(2) A person whose normal metabolic requirements are of about 3000 kcal day$^{-1}$ goes on a hunger strike.

   (a) Assuming that most of the weight loss is initially from fat consumption, how long would it take this person to lose 10 kg?

(b) If this person has 20 kg of muscle (one of the major repositories of protein), how long before it's all gone once protein breakdown becomes the main source of energy?

(3) An important simplification is possible in the indirect calorimetry method described above. The catabolizable energy equivalent of $O_2$ respired is roughly constant across all foodstuffs. It is 110 kcal mol$^{-1}$.

(a) A typical adult takes 17 breaths per minute during the day, each of which brings in 0.0013 mol of $O_2$. At night, the breathing rate drops to about 10 breaths per minute. If a person sleeps for 9 of every 24 hours, how much energy does this person use per day?

(b) Oxygen debt is a condition that arises when the body cannot take in oxygen fast enough to meet the energetic demands of a situation. Anaerobic energy generation mechanisms are then used until the end of the stress. It is necessary afterwards to take in more oxygen than normal ("panting") for a period of time to resupply the body in certain crucial intermediate metabolites. Suppose that a person exercises very hard and, at the end, pants for five minutes. During this period, this person takes in 0.10 mol more $O_2$ than normal. How much energy was required to restore normal metabolic function?

(4) A guinea pig is placed in an ice-jacketed calorimeter with an exercise wheel coupled to an electrical generator. (The generator and wheel are inside the calorimeter.) The current from the generator leaves the calorimeter through some wires to power an electrical load. No food is provided to the guinea pig during the experiment. During a one hour period, 30 g of ice melts and 5 kJ of electrical energy is generated. How much of its stored energy did the guinea pig use in this period?

(5) Recently, a gene has been discovered that causes obesity in rats. It has been named the *ob* gene. Suppose that a rat with the *ob* gene is placed in an ice-jacketed calorimeter while another rat lacking the *ob* gene, but otherwise identical, is placed in a second calorimeter. Which rat do you expect will generate the most heat? Give a thermodynamic reason for your answer.

(6) A shipment consisting of 24 boxes each weighing 10 kg arrives at your office which you must lift to a shelf 1.5 m off the ground. From your elementary physics, you recall that the work done *on* a mass $m$ in raising it to a height $h$ is $mgh$, where $g$ is the acceleration due to gravity. Instead of doing the job right away, you decide to procrastinate by solving a thermodynamics problem first: how much spaghetti (almost a pure carbohydrate) should you eat after carrying out this task to compensate for the work you will have done and avoid losing weight?

(7) (a) An Inuit hunter is out on a day in early spring when the air temperature is $-10\,^\circ$C and the atmospheric pressure is 100 kPa. Due to the intense physical activity of hunting, the hunter's average breathing rate over a six hour period is 15 breaths per minute and each breath has a volume of 0.6 L at body temperature (37 $^\circ$C). The air inspired is warmed to body temperature on the way to the lungs. How much heat is lost in the process of breathing during this six hour period? The molar heat

capacity of air at constant pressure is approximately $29 \, \mathrm{J \, K^{-1} mol^{-1}}$. Note that air may be treated as an ideal gas.

    (b) Seal blubber is almost pure fat. How much seal blubber does the hunter have to eat to offset the energy loss from breathing during the six hours of hunting?

(8) A typical adult male human must consume food with an energy equivalent of about $3000 \, \mathrm{kcal \, day^{-1}}$ to maintain a constant weight. Assume that the food energy is completely converted to heat and, furthermore, that the heat is generated at a constant rate during the day. The human body is mostly water (specific heat capacity $4.184 \, \mathrm{kJ \, K^{-1} kg^{-1}}$). If the body were perfectly insulated, how long would it take before a $70 \, \mathrm{kg}$ adult male reached the normal boiling point of water from an initial temperature of $37 \, °\mathrm{C}$?

(9) A pint of beer delivers approximately $196 \, \mathrm{kcal}$ of metabolizable energy. This figure assumes that the beer is served warm. Cold beer must also be warmed to body temperature ($37 \, ° \mathrm{C}$) and this requires an expenditure of metabolic energy. Calculate the metabolizable energy (in kcal) available from a pint of beer served at $4 \, °\mathrm{C}$. Treat the beer as if its thermal properties were identical to those of water. The density of beer is typically just slightly higher than that of water (about $1 \, \mathrm{g \, mL^{-1}}$) and a pint is $568 \, \mathrm{mL}$.

---

## Key ideas and equations

- A state variable or function has an exact differential and is path independent.
- A path function has an inexact differential and is path dependent.
- $\mathrm{d}w = -p_{\mathrm{ext}} \, \mathrm{d}V$
- During a reversible process, the system and surroundings are constantly in equilibrium.
- First Law of Thermodynamics: $\mathrm{d}U = \mathrm{d}q + \mathrm{d}w$; $U$ is a state function.
- $H = U + pV$
- Heat at constant volume is $\Delta U$; heat at constant pressure is $\Delta H$.
- $C_V = \dfrac{\partial U}{\partial T}\bigg|_V$ and $C_p = \dfrac{\partial H}{\partial T}\bigg|_p$
- Definition of and calculations involving enthalpies of formation.
- Setting up heat balance problems.
- $\Delta_r H_m = \Delta_r U_m + RT \Delta_r \nu_{\mathrm{gas}}$
- How to calculate enthalpy at different temperatures using thermodynamic cycles.

## Suggested reading

Calculating the work done during a process is a surprisingly complex problem in general, so I was careful about the cases I considered in Section 5.2. If you want to delve into these complexities, which may be relevant in physiological studies, for example, I would start with the following paper:

Daniel Kivelson and Irwin Oppenheim, *J. Chem. Ed.* **43**, 233 (1966).

There are many excellent books on animal physiology. I have found the following particularly useful:

Roger Eckert, David Randall and George Augustine, *Animal Physiology: Mechanisms and Adaptations*, 3rd edn.; Freeman: New York, 1988.

For detailed data, experimental protocols and sample calculations on classical physiological measurements, I suggest two old, but still very interesting books:

Max Kleiber, *The Fire of Life: An Introduction to Animal Energetics*; Wiley: New York, 1961.
C. Frank Consolazio, Robert E. Johnson and Louis J. Pecora, *Physiological Measurements of Metabolic Functions in Man*; McGraw-Hill: New York, 1963.

There's more to the story of how *Rana sylvatica* survives freezing than I have explained here. If you want to know more about this, I highly recommend the following articles by Kenneth and Janet Storey:

Kenneth B. Storey and Janet M. Storey, *Annu. Rev. Physiol.* **54**, 619 (1992);
*The Sciences* **39**(3), 33 (1999).

---

### Review exercise group 5.12

(1) Name the thermodynamic quantity corresponding to each of these symbols:
    (a) $\Delta_c U_m$
    (b) $C_V$
    (c) $\Delta_r h$

(2) The bomb calorimetry experiment is generally carried out in two steps: (i) burn a sample of known heat of combustion, and (ii) burn an unknown sample.
    (a) What do you actually measure in a calorimetry experiment?
    (b) What is the purpose of step (i)?
    (c) What quantity can you calculate after step (ii)?

(3) When scientific papers are submitted to scientific journals, they are sent for peer review, i.e. they are sent to experts in the field for an opinion regarding their correctness and novelty. A little while ago, I reviewed a paper in which the following passage appeared:

A part of the biochemical energy becomes heat:

$$\Delta q = \eta \Delta U, \tag{5.16}$$

where $\Delta q$ is the change in heat, and $\eta$ is the efficiency of the energy conversion between $\Delta U$ and $\Delta q$.

What is wrong with this statement?

(4) A 200 g block of steel initially at 100.0 °C is placed in an adiabatic container held at constant pressure with 200 g of liquid water initially at 22.0 °C. The final temperature of the system is 30.3 °C. What is the specific heat capacity of this sample of steel?

(5) (a) A 350 g block of silicon is brought from 20 °C to its melting point, 1693 K. How much heat does this require?

   (b) Do you expect the value calculated in part (a) to be reasonably accurate? Why or why not? If you wanted to do better, what data would you need? How would the calculation differ from that carried out in part (a)?

(6) Suppose that 25 g of sodium hydroxide is dissolved in 200 mL of water at 25 °C. What is the temperature change? According to your calculation, would any special precautions be necessary in preparing this solution?

(7) Suppose that an industrial process generates heat by burning methane in a constant-volume chamber at 1200 K.

   (a) Calculate the enthalpy of combustion of methane at 1200 K.

   (b) Were there any doubtful approximations made in the above calculation?

   (c) Calculate the heat released per mole by the combustion of methane at 1200 K under constant-volume conditions.

(8) Various devices are used to trap, cool and condense steam. One way to condense steam is to bubble it through liquid water. Suppose that 8 g of steam initially at 115 °C is bubbled into 300 g of water initially at 20 °C. All the steam disappears before reaching the surface of the water. Assuming that the vessel in which this occurs is well insulated, what is the final temperature of the water?

(9) During the reversible adiabatic expansion of an ideal gas, the pressure and volume are related by $pV^\gamma = a$ where $\gamma = C_{p,m}/C_{V,m}$, and $a$ is a constant determined by the initial pressure and volume. For gases, $C_{V,m} \approx C_{p,m} - R$, so $\gamma$ can easily be calculated from the constant-pressure heat capacity. Calculate the work done during a reversible adiabatic expansion of 1.85 mol of nitrogen gas from an initial pressure and temperature of 0.94 bar and 225 °C, to a final pressure of 18 mbar. Treat nitrogen as an ideal gas.

(10) 80 mL of 0.50 mol L$^{-1}$ HCl solution is mixed with 45 mL of 0.85 mol L$^{-1}$ NaOH solution. Both solutions have initial temperatures of 22 °C. What is the maximum temperature that the mixture reaches?

   *Note*: Make a reasonable assumption about the solution densities.

(11) At 18 °C, the enthalpy of reaction for

$$C_{(graphite)} + CO_{2(g)} \rightarrow 2CO_{(g)}$$

is 175.52 kJ mol$^{-1}$. The molar heat capacity of carbon monoxide is 29.12 J K$^{-1}$mol$^{-1}$. Accurately calculate the standard enthalpy of formation of carbon monoxide (at 25 °C).

(12) When 10 g of sodium nitrate (NaNO$_{3(s)}$, molar mass 84.99 g mol$^{-1}$) is dissolved in 200 g of water, it completely dissociates and the temperature of the water drops by 2.9 °C. When 10 g of nitric acid (HNO$_{3(l)}$, molar mass 63.01 g mol$^{-1}$) is dissolved in 200 g of water, 93% of the acid dissociates to nitrate and an aqueous proton and the temperature of the water increases by 5.9 °C. The enthalpy of formation of nitric acid is known from other experiments to be −174.1 kJ mol$^{-1}$. Assume that the dissociation

of the acid is responsible for most of the temperature rise in the latter case. What are the enthalpies of formation of sodium nitrate and of the nitrate anion?

(13) The MONJU experimental nuclear reactor in Japan produced heat at a rate of 714 MW. Liquid sodium was used as a coolant. It flowed into the reactor at a temperature of 397 °C and flowed out at a temperature of 529 °C. The molar heat capacity of liquid sodium in this temperature range is $29.26 \, \mathrm{J\,K^{-1}mol^{-1}}$. What flow rate of sodium through the reactor is required to carry away the heat produced by the nuclear reaction given the desired inlet and outlet temperatures? Express your answer in $\mathrm{t\,h^{-1}}$.

(14) Under some conditions, ice crystals sublime rather than melt. If the enthalpies of fusion and of vaporization of water are, respectively, 6.01 and $44.9 \, \mathrm{kJ\,mol^{-1}}$ at 0 °C, what is the enthalpy of sublimation at this temperature?

# 6

# The Second Law of Thermodynamics

The First Law of Thermodynamics tells us that energy is conserved. The Second Law, which was originally formulated to help us understand steam engines, tells us that heat can't be converted into work with 100% efficiency. Our study of the Second Law will lead us to define a new state function known as the entropy. In turn, the entropy will turn out to be pivotal to our understanding of what drives chemical reactions.

## 6.1 The Second Law of Thermodynamics

There are many different statements of the Second Law. My favorite, because it seems to convey most clearly the meaning of this law, is due to Kelvin:

It is impossible for a system to undergo a cyclic process whose sole effects are the flow of an amount of heat from the surroundings to the system and the performance of an equal amount of work on the surroundings.

In more modern language, we might say that it is impossible to continuously convert heat completely into work. The Second Law limits the efficiency of **heat engines**, machines that convert heat into work. Just as the First Law makes a certain kind of perpetual motion machine impossible, the Second Law makes it impossible to build a so-called perpetual motion machine of the second kind, namely one that produces work from heat with perfect efficiency.

There are literally dozens of statements of the Second Law, each of which can be shown to be logically equivalent to Kelvin's. In Section 6.4, we will encounter a modern mathematical statement that is extremely general and is thus well suited to problems that have nothing to do with heat engines. However, all statements of the Second Law have two things in common:

(1) The processes that the Second Law forbids are allowed by the First Law; for instance, the perfect conversion of heat into work does not violate the First Law, which only stipulates the conservation of the total energy.

(2) The reverse of a process forbidden by the Second Law is always allowed. For instance, work can be dissipated to heat with 100% "efficiency." My father-in-law, Walter Oktaba,

a retired electrical power engineer, likes to say that heat is "the most degraded form of energy." What he is saying is that it is easy to turn other forms of energy into heat – it even happens when we don't want it to – but that we face fundamental Second-Law limitations when trying to go in the reverse direction.

The latter point implies that the Second Law is different in nature from the First Law and, indeed, from all of the laws of mechanics: it is asymmetric in time. Consider, for instance, an electric heater, a device that converts electrical work to heat with nearly perfect efficiency. If we had a heat-sensitive camera, we could film this process. If we then ran the film backwards, we would observe a process that, according to the Second Law, can never occur, namely the return of the heat to the heater and its conversion to electricity with perfect efficiency. Classical mechanics, on the other hand, is completely reversible: if we filmed, say, some gas particles flying around a container, and then ran the film backwards, we would observe perfectly reasonable particle trajectories. The First Law is also time-reversible: energy could equally well be conserved by the melting of an ice cube in your hand as by the warming of your hand as water froze in it. However, the latter process never occurs and we will see that the Second Law forbids it. Because it tells us which processes, among all those allowed by mechanics and by the First Law, can and can't be expected to happen, the Second Law has often been said to define an "arrow of time."

The example of running particle trajectories backwards is an interesting one since it brings out an important aspect of the Second Law. Suppose that we filmed a gas being forced out of a nozzle under pressure. The reverse process, the return of the gas molecules to the nozzle against the pressure gradient, does not violate any of the laws of mechanics. If you only knew Newton's laws of motion, you would never suspect that the film was running backwards. However, we know it would never happen because the probability of all the molecules in the gas traveling in just the right direction to make this happen is unimaginably small. As we will see later, the Second Law is essentially statistical in nature. Processes forbidden by the Second Law are not impossible in the strict sense of the word, but are so incredibly unlikely that they might as well be treated as impossibilities.

## 6.2 Intensive and extensive properties

Some properties of a system, like mass, volume, energy and enthalpy, depend on how large a system is. All other things being equal, if we double the size of a system, the values of these properties double too. We say that such quantities are **extensive**.

**Intensive** properties, on the other hand, are those whose values are independent of the size of the system. These include temperature and pressure. Doubling the size of a system while holding the density constant would have no effect on the pressure, for instance.

If you're having trouble figuring out what the difference is, imagine the following experiment: take two identical copies of your system separated by a removable wall. Now remove the wall and ask yourself what the value of a given property for the two copies put together is. If a property is twice as large as in the original system, then it's extensive. If it hasn't changed, it's intensive.

## 6.3 Reversible processes and entropy

A modern statement of the Second Law requires the definition of a state function called the **entropy**. Defining the entropy rigorously requires a detailed consideration of heat engines, the subject of Section 6.8. Since heat engines are not particularly important in biochemistry, to do so would be a distraction at this point. Rather, I will only show that the existence of such a state function is plausible.

We need to reconsider reversible processes, which we first encountered in Section 5.2. Recall that a reversible process is one during which the system is constantly at equilibrium. During reversible pressure–volume work, for instance, $p_{ext} = p$ so that

$$dw_{rev} = -p\,dV.$$

Since there is no pressure difference driving the change, it could go either way with only an infinitesimal change in pressure in one direction or the other, hence the name "reversible."

Note that $dw_{rev}$ is of the form of an intensive state variable ($p$) multiplied by an exact differential of an extensive state function ($-dV$, the differential of the extensive state function $-V$).

What about reversible heating? The variable that determines whether or not two systems are in thermal equilibrium is the temperature. Thus, during a reversible heating process, $T_{ext} = T$. Since $T$ plays the same role for heat as $p$ does for work, being an intensive state variable whose gradient determines the direction in which heat will flow, we would like to write

$$dq_{rev} = T\,dS,$$

where $S$ is a extensive state function. It turns out that this is exactly what we need to do: the entropy $S$ is a state function such that

$$\Delta S = \int_{rev.\ path} \frac{dq}{T} \tag{6.1}$$

or, in differential form,

$$dS = \frac{dq_{rev}}{T}. \tag{6.2}$$

We will see very shortly that $S$ is deeply connected to the Second Law. In the meantime, let's learn how to calculate changes in entropy. The main problem we will face will be to find reversible paths over which to evaluate the integral in Equation (6.1).

**Example 6.1 Entropy of heating at constant pressure** Reversible heat transfer means that the heat is provided from one body at temperature $T$ to another body at the same temperature. We will see just a bit later how we can at least imagine doing this. For now, just note that, whether it is provided reversibly or not, we can calculate the heat in the absence of phase transitions from the heat capacity. Specifically, $dq_{rev} = C_p\,dT$. If $C_p$ is constant over the temperature range of interest,

$$\Delta S = \int_{T_1}^{T_2} \frac{C_p\,dT}{T} = C_p \ln\left(\frac{T_2}{T_1}\right).$$

Note that $\Delta S > 0$ if $T_2 > T_1$.

**Example 6.2 Entropy of expansion** What is the change in entropy for the isothermal expansion of an ideal gas? In this example, the issue of calculating the reversible heat is a bit trickier.

Note that the description of the process (isothermal expansion) isn't enough to determine the heat or work, since the path is incompletely specified. However, it is enough to compute the entropy change. Initially, we have a certain amount of gas at a certain temperature, pressure and volume. The final state is the same amount of gas at the same temperature, but at a higher volume and thus, by the ideal gas law, at a lower pressure. Because the process is isothermal,

$$\Delta S = \int \frac{dq_{rev}}{T} = \frac{1}{T} \int dq_{rev} = \frac{q_{rev}}{T}.$$

To compute the entropy, we must use a reversible path. Specifically, the correct reversible path in this case is a reversible, isothermal expansion of an ideal gas. As we saw on page 73, the internal energy of an ideal gas only depends on $T$. Thus, for an isothermal process on an ideal gas, $\Delta U = 0$. Therefore,

$$q_{rev} = -w_{rev}.$$

$$w_{rev} = -\int_{V_1}^{V_2} p\,dV = -nRT \int_{V_1}^{V_2} \frac{dV}{V} = -nRT \ln\frac{V_2}{V_1}.$$

We conclude that

$$\Delta S = nR \ln\frac{V_2}{V_1}.$$

---

### Exercise group 6.1

(1) What is the entropy change during a reversible, adiabatic process?
(2) What is the entropy change during a reversible, isothermal process?
(3) The enthalpy of vaporization of helium at its normal boiling point of 4.22 K is 84 J mol$^{-1}$. What is the entropy of vaporization of helium at constant pressure?
(4) Calculate the change in entropy of a 50 g sample of potassium bromide that is heated from 12 to 53 °C.
(5) The standard enthalpies of formation of liquid and gaseous TiCl$_4$ are, respectively, $-804.16$ and $-763.2$ kJ mol$^{-1}$. TiCl$_4$ boils at 408 K. The heat capacities of the two

phases can both be expressed in the form

$$C_{p,m} = A + BT + CT^2 + DT^3 + E/T^2$$

where $C_{p,m}$ is in $\mathrm{J\,K^{-1}mol^{-1}}$ and $T$ is in kelvins. For the liquid, $A = 143.048$, $B = 7.600\,362 \times 10^{-3}$, $C = 1.530\,575 \times 10^{-6}$, $D = -5.383\,76 \times 10^{-10}$ and $E = -20\,638$. For the vapor, $A = 106.8573$, $B = 1.049\,482 \times 10^{-3}$, $C = -2.843 \times 10^{-7}$, $D = 2.4257 \times 10^{-11}$ and $E = -1\,043\,516$.

(a) Calculate the enthalpy of vaporization at the boiling point.

(b) Calculate the entropy of vaporization at the boiling point.

## 6.4 The Second Law of Thermodynamics and entropy

In terms of entropy, the Second Law can be very simply stated:

> **The entropy of an adiabatic system never decreases.**

Since we certainly believe that the Universe is adiabatic – with what could the Universe be in thermal contact? – it is often said that the entropy of the Universe increases. This is a consequence of the more general statement above.

Processes with a positive $\Delta S_{\text{universe}}$ will be said to be **thermodynamically allowed**. Many books use the word "spontaneous" for "thermodynamically allowed." The trouble is that, in everyday language, "spontaneous" implies that a process *will* happen, when all that thermodynamics tells us is whether it *could* happen. Some things that could happen don't, for reasons that have to do with kinetics rather than thermodynamics. I therefore try to only use "spontaneous" for things that are known to occur spontaneously according to our usual understanding of this word.

In this section, examples of the increase in entropy during thermodynamically allowed (and actually spontaneous) processes will be studied.

**Example 6.3 Spontaneous heat flow** We already know that heat flows spontaneously from hotter to colder bodies. The Second Law should of course agree with this observation.

Suppose that we have two bodies, one at temperature $T_1$ and the other at temperature $T_2$. An amount of heat $|q|$ flows from the first to the second. This amount of heat is supposed to be so small that the temperatures of the two bodies are not sensibly affected by the flow. If we want to consider larger heat flows, we would only have to make minor adjustments to the mathematics below, but the principle would be the same.

In order to compute the change in entropy, we have to invent a reversible path, because that's how entropy is defined. We need some way to move heat from body 1 to body 2 such that heat flows out of body 1 into a body at temperature $T_1$, and into body 2 from a body at temperature $T_2$. This requires a third body whose temperature can be adjusted and which will ferry the heat from body 1 to body 2. Specifically, we imagine a heat exchanger whose temperature can be adjusted adiabatically. (An ideal gas will do nicely, since an adiabatic

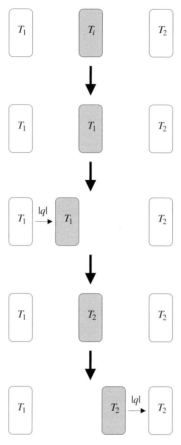

Figure 6.1 A reversible path for transferring heat from one body to another. A heat exchanger (shaded), starting from an arbitrary temperature $T_i$, is brought to temperature $T_1$ by a reversible adiabatic process. Heat $|q|$ is reversibly transferred from body 1 to the heat exchanger, reversibility being ensured by having body 1 and the exchanger at the same temperature. The heat exchanger is then brought to temperature $T_2$ by a second reversible adiabatic process. Finally, heat $|q|$ is transferred reversibly to body 2. If the amount of heat transferred is such as to result in a significant change in the temperatures of either or both of bodies 1 and 2, we could simply operate this process in a cycle, each time transferring a small amount of heat $dq$, resulting in infinitesimally small changes in the temperatures of the two bodies at each step.

compression of an ideal gas increases its temperature and adiabatic expansion decreases it.) The strategy is illustrated in Figure 6.1. We start by performing a reversible adiabatic process to bring the heat exchanger to $T_1$. We then allow $|q|$ to flow out of the first body reversibly at $T_1$ into the heat exchanger. We perform a second reversible adiabatic process on the heat exchanger to bring it to $T_2$. The heat extracted from the first body is allowed to flow reversibly into the second. The only non-zero contributions to the change in entropy are from the two heat flow steps:

$$\Delta S_{universe} = \int \frac{dq}{T_1} + \int \frac{dq}{T_2} = \frac{-|q|}{T_1} + \frac{|q|}{T_2}.$$

The two terms in this equation correspond to the change in entropy of the first body and to the change in entropy of the second body. We don't count the change in entropy of the heat exchanger because it's just an imaginary device needed to make the path reversible; it's not part of the physical Universe.

From the last equation, we can read off our result: the change in entropy of the Universe is positive if and only if $T_1 > T_2$. In other words, $\Delta S_{universe}$ is positive when heat flows from hotter to colder bodies.

**Example 6.4 The melting of ice at room temperature** We again use our heat exchanger trick. This time, we want to remove heat from the room (assumed to be large enough to release it isothermally) at $25\,°C$ and dump it into the ice at $0\,°C$ (the melting temperature). The ice–water system will remain at this temperature as long as we don't melt all of the ice. The heat of fusion of ice at $0\,°C$ (273 K) and 1 atm is $6.007\,kJ\,mol^{-1}$. We then have

$$\Delta_{fus} S_m = \frac{6.007\,kJ\,mol^{-1}}{273.15\,K} = 21.99\,J\,K^{-1}mol^{-1};$$

$$\Delta S_{room,m} = \frac{-6.007\,kJ\,mol^{-1}}{298.15\,K} = -20.15\,J\,K^{-1}mol^{-1};$$

$$\therefore \Delta S_{universe,m} = \Delta_{fus} S_m + \Delta S_{room,m} = 1.84\,J\,K^{-1}mol^{-1} > 0,$$

in accord with our everyday experience that this process is spontaneous.

---

### Exercise group 6.2

(1) Using your knowledge of thermodynamics, prove that steam at $100\,°C$ will condense in a room at $20\,°C$.

(2) Prove, by a calculation based on thermodynamic principles, that a glass of water initially at $20\,°C$ with a heat capacity of $1200\,J\,K^{-1}$ in a room at $20\,°C$ cannot spontaneously extract heat from the room and warm itself to $50\,°C$.

(3) An ideal gas is compressed isothermally from $V_1$ to $V_2$. Is the change in entropy positive, negative or zero? What about the change in internal energy?

(4) 20 g of iron metal at $85\,°C$ is placed in an insulated container with 80 g of water at $4\,°C$.
   (a) Calculate the final temperature of the system.
   (b) Calculate the entropy change for the iron–water system. Is the result in accord with the Second Law of Thermodynamics?

(5) At constant temperature and pressure, the reaction

$$Na_{(s)} + H_2O_{(l)} \rightarrow Na^+_{(aq)} + OH^-_{(aq)} + \frac{1}{2}H_{2(g)}$$

is spontaneous and exothermic. For simplicity, suppose that the reaction is carried out in an adiabatic container. What can you say (rigorously) about size and/or sign of the entropy change for the reaction?

*Hint*: Break up the entropy change into a part that is due to the reaction itself, and a part that is associated with the heating.

---

### 6.4.1 Entropy of mixing

Any spontaneous process, whether or not it involves heat flow, should lead to a positive change in entropy. A particularly important example, and one that requires a sneaky method of attack, is the entropy change on mixing two gases.

Suppose that two different ideal gases are separated from one another by a removable barrier. Each gas is held at identical temperatures and pressures $T$ and $p$. The first is in a compartment of volume $V_1$ and the other in a compartment of volume $V_2$. Then, the wall is suddenly removed. We have a pretty good idea what will happen: the two gases will diffuse and mix. Does the Second Law predict this behavior?

Ideal gases are made up of non-interacting particles. Therefore, the two gases in our system are completely insensible to each other. Thus we should ask ourselves what effect removing the barrier has on *each* gas, as if the other weren't there. The answer to that question is that, after the system comes to equilibrium, the partial pressure of each gas will have decreased, the temperature remaining constant and the volume increasing. Thus, each gas undergoes an isothermal expansion. To compute the entropy change, we need to calculate the entropy change during the reversible isothermal expansion of an ideal gas. We have already made this calculation, in Example 6.2. The result was

$$\Delta S = nR \ln \frac{V_f}{V_i}.$$

In the case of our mixing problem, gas 1 expands from $V_1$ to $V_1 + V_2$ and gas 2 expands from $V_2$ to $V_1 + V_2$:

$$\Delta_{\text{mix}} S = R \left( n_1 \ln \frac{V_1 + V_2}{V_1} + n_2 \ln \frac{V_1 + V_2}{V_2} \right). \tag{6.3}$$

We have proven what we set out to prove: since the arguments of the logarithms are both greater than unity, the logarithms are positive, so $\Delta_{\text{mix}} S$ is positive. Once again, the Second Law agrees with our intuition about what should happen. The reverse process, spontaneous unmixing, would have a negative entropy change, so it is not thermodynamically allowed.

Note that we haven't used the fact that the pressures of the two gases are initially the same. Thus, Equation (6.3) is correct even when the two gases start out at different pressures.

The entropy of mixing is usually written in a slightly different form for the special case where the two gases *are* initially at the same pressure. For this case, it shouldn't take much effort to convince yourself using the ideal gas law that the mole fraction of gas 1 is

$$X_1 = \frac{n_1}{n_1 + n_2} = \frac{V_1}{V_1 + V_2}$$

and that a similar expression holds for the second gas. Therefore,

$$\Delta_{\text{mix}} S = -nR (X_1 \ln X_1 + X_2 \ln X_2) \tag{6.4}$$

where $n = n_1 + n_2$ is the total number of moles of gas. If we divide both sides by $n$, we get the molar entropy of mixing:

$$\Delta_{\text{mix}} S_m = -R (X_1 \ln X_1 + X_2 \ln X_2).$$

This calculation leads us to Gibbs's paradox: the entropy of mixing does not depend on the identities of the gas molecules. However, in one special case, the entropy of mixing is zero; If the two gases are the same, then the "before" and "after" states of the system are identical and, because entropy is a state function, $\Delta S = 0$. Imagine that we could "morph" one gas into another, slowly changing the properties of one until they were identical to the properties of the other. The thermodynamic arguments we have just seen predict that the entropy of mixing would be constant, regardless of how similar the two gases were, right up to the point where we actually made them identical, at which point the entropy of mixing would discontinuously drop down to zero. This is Gibbs's paradox. The resolution of this paradox requires the recognition that one always has either one case or the other; because of the particulate nature of matter, it is impossible to slowly morph one atom or isotope into another. Two gases are either identical (in which case $\Delta S = 0$) or they are not (in which case Equation (6.4) applies).

**Example 6.5 Entropy of mixing** Suppose that two 2 L flasks are connected by a stopcock. One flask contains neon at a pressure of 50 kPa. The other contains nitrogen at a pressure of 90 kPa. Both are held at 19 °C. The stopcock is opened. Since the two gases are initially at different pressures, we must use Equation (6.3) to calculate the entropy of mixing.

$$n_{\text{Ne}} = \frac{pV}{RT} = \frac{(50 \times 10^3 \text{ Pa})(2 \times 10^{-3} \text{ m}^3)}{(8.314\,472 \text{ J K}^{-1}\text{mol}^{-1})(292 \text{ K})} = 0.041 \text{ mol}.$$

The other flask contains

$$n_{\text{N}_2} = \frac{pV}{RT} = \frac{(90 \times 10^3 \text{ Pa})(2 \times 10^{-3} \text{ m}^3)}{(8.314\,472 \text{ J K}^{-1}\text{mol}^{-1})(292 \text{ K})} = 0.074 \text{ mol}.$$

The entropy of mixing is therefore

$$\Delta_{\text{mix}} S = (8.314\,472 \text{ J K}^{-1}\text{mol}^{-1}) \left[ (0.041 \text{ mol}) \ln \left( \frac{2+2\text{L}}{2\text{L}} \right) \right.$$

$$\left. + (0.074 \text{ mol}) \ln \left( \frac{2+2\text{L}}{2\text{L}} \right) \right]$$

$$= 0.66 \text{ J K}^{-1}.$$

---

## Exercise group 6.3

(1) Some substances don't mix. While the above derivation is not completely general, the entropy of mixing is always positive. So why don't (for instance) oil and water mix?

(2) Note that the sum of the mole fractions of all the components of a system must equal 1. Therefore, for a two-component system, $X_2 = 1 - X_1$. Find the mole fractions $X_1$ and $X_2$ that maximize the entropy of mixing.

---

## 6.5 A microscopic picture of entropy

The problem with entropy is that it's difficult to build a mental picture of what it measures from the thermodynamic definition alone. The other key concepts in thermodynamics (energy, heat, work) are more familiar to us and can be related to everyday experience. Entropy isn't like that. In this section, we will develop the molecular theory of entropy. The emphasis will be on insight rather than rigor.

In thermodynamics, we specify the macroscopic state of a system by the values of the system's state variables or state functions. This macroscopic state, however, is compatible with many microscopic states (**microstates** for short), i.e. with many different arrangements of the molecules and of their energy. As a simple example, consider the vibrational energy of a gas of harmonic oscillators (a simple model for molecular vibrations; Section 3.4.3). Because of the equal spacing of harmonic oscillator energy levels, we can think of the vibrational energy as being stored in units of $h\nu_0$, the spacing between the energy levels. A total of $k$ units of excess vibrational energy (beyond the zero-point energy) could be spread over the molecules in any of a number of ways. For the sake of argument, suppose that $k < N$, the number of molecules. There could be $k$ molecules with one unit of excess vibrational energy each. Alternatively, there could be $k - 1$ molecules with one unit each, and one with two. At the other extreme, all $k$ units could be piled into one molecule. Each of the possible ways to distribute the energy is a microstate.

We will consider a system with fixed values of $N$, $U$ and $V$. The Boltzmann distribution (Section 3.2) is very general. In particular, it can be applied to the total energy of a system. For a system of fixed internal energy $U$, because each of the microstates has the same energy, then each microstate has the same probability. Let $\Omega$ be the number of distinguishable microstates corresponding to the macroscopic state $(N, U, V)$. To put it another way, the single energy level $U$ of the system has degeneracy $\Omega$. Now recall that the partition function counts the number of states accessible to a system. Thus, for a system at fixed $(N, U, V)$, the partition function for the system as a whole has the value $\Omega$.

Many of the equations we have derived thus far for the molecular partition function apply equally well to the partition function of a system, which we denote by $Q$. We will now derive an equation for the connection between the entropy and the system partition function.

From the thermodynamic definition of entropy, we have $dS = dq_{rev}/T = C_V \, dT/T$ since we consider a system at constant volume. We can calculate the entropy by integrating this equation. If we also substitute for $C_V$ using Equation (5.11), we get

$$S = S(0) + \int_0^T \frac{1}{T} \frac{\partial U}{\partial T}\bigg|_{V,N} dT. \tag{6.5}$$

The integral on the right-hand side can be evaluated by integration by parts. We get

$$S = S(0) + \frac{U(T) - U(0)}{T} + \int_0^T \frac{U}{T^2} dT.$$

The internal energy at $0\,K$, $U(0)$, is just the zero-point energy of the system. The value of this energy is arbitrary as we can always redefine the zero of our energy scale such that $U(0) = 0$. We will therefore drop this term from our equation.

Noting that the average energy of the system is just the internal energy, and using Equation (3.3), which relates the average energy to the partition function $Q$, we get

$$U = \frac{k_B T^2}{Q} \frac{\partial Q}{\partial T}\bigg|_{V,N} = k_B T^2 \frac{\partial \ln Q}{\partial T}\bigg|_{V,N}.$$

If you can't quite see how I got the last equality in this equation, apply the chain rule to the result. Substituting this expression into the integral in Equation (6.5), we have

$$S = S(0) + \frac{U}{T} + k_B \int_0^T \frac{\partial \ln Q}{\partial T}\bigg|_{V,N} dT$$

$$= S(0) + \frac{U}{T} + k_B \ln Q - k_B \ln Q(0).$$

We can combine $S(0) - k_B \ln Q(0)$ into a single constant. We will see later that this constant can arbitrarily be set to zero, and that $S(0) = k_B \ln Q(0)$ is a sensible choice consistent with the equation we are about to derive. This leaves us with

$$S = \frac{U}{T} + k_B \ln Q. \tag{6.6}$$

Now we will take a shortcut in order not to get too bogged down in mathematical details. Our derivation so far treats $Q$ as a function of $(N, T, V)$. (Note the derivatives with respect to $T$, in particular in the statistical definition of the internal energy.) If that is the case we want to treat, then Equation (6.6) gives us the entropy, provided we have the appropriate partition function (Equation (3.2) with the energies interpreted as allowed energies of the system of $N$ molecules). However, we started out talking about a system with constant $(N, U, V)$. We can approach such a system from one with fixed $(N, T, V)$ by imagining that the spacing between the energy levels is much larger than $k_B T$ and that the latter quantity is such that it falls somewhere in the gap between two energy levels. In such a range of temperatures, $Q$ is roughly constant, so $U$ is roughly constant (which is the condition we want to impose) and much smaller than $k_B T$. The energy becomes exactly constant as the energy level spacing and $k_B T$ go to infinity. If we divide our equation for the entropy by $k_B T$ and take the limit as $U/k_B T \to 0$, then recall that $Q = \Omega$ for a system at constant $U$, we get **Boltzmann's equation**:

$$S = k_B \ln \Omega. \tag{6.7}$$

The entropy calculated by this equation is either called the Boltzmann or the statistical entropy.

We can now provide a statistical interpretation of the entropy. When we specify the macroscopic state, we remain ignorant of the precise microscopic state the system occupies. Indeed, this microstate can change from moment to moment as molecules move around and exchange energy. The entropy is larger for systems with more microstates, so it measures our degree of ignorance of the microstate.

The Second Law says that the entropy of an adiabatic system increases in a thermodynamically allowed process. According to the Boltzmann definition of entropy, this would mean that thermodynamically allowed processes favor an increase in the number of microscopic states available to an adiabatic system. Consequently, thermodynamically allowed processes result in a loss of information about adiabatic systems or, in general, about the Universe. To put it another way, as the Universe evolves, there is a tendency for its energy to be spread over more and more microstates.

### 6.5.1 Vibrational entropy

Let's treat an example to see the statistical entropy in action. Suppose that our system consists of just two identical molecules and that we focus on the part of the energy stored in a particular molecular vibration. Recall again that vibrational energy levels are roughly equally spaced. Additionally, suppose that the total energy of the system is two units (in multiples of $h\nu_0$). How many different ways can the energy be shared between the two molecules? Both units of energy can be stored in molecule A, both in molecule B one in each of molecules A and B. Apparently then, $\Omega = 3$, and the molar entropy of our system is $S = k_B \ln 3$. This is the correct value for the statistical entropy if the two molecules are bound to a surface so that we can tell them apart. However, if the molecules can move around, then we can't really tell which is which. Accordingly, we should only count *one* of the two states in which both units of energy are stored in the same molecule. The entropy in that case would be $S = k_B \ln 2$.

Let's generalize this example. Suppose that I have $n$ units of energy shared among $m$ indistinguishable molecules. There's a trick for figuring out how many different ways I can share out $m$ units of energy among $n$ molecules. Imagine that we have $n$ balls (the units of energy) arranged on a line with two fixed walls at the ends defining a "room" that contains all the balls, and $m - 1$ movable walls. The movable walls can divide the room into $m$ "cubicles," each cubicle representing a molecule. Imagine that we start with an empty room and insert the walls and balls randomly, one by one, into the room. How many different arrangements of balls and walls are there? This will be the number of microstates. In total, we have $n + m - 1$ objects to arrange. For now, suppose that the balls are distinguishable

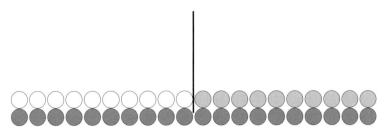

Figure 6.2 An experiment to study diffusive mixing of molecules on a surface. A wall separates two parts of the surface. On the left, a single layer of molecules of species A (represented by open circles) has been adsorbed onto the surface, whose atoms are represented by darker gray circles. The right side of the surface is covered with molecules of species B (light gray circles).

from each other, as are the walls (colored or numbered, say). We will place them from left to right in the room by randomly picking one ball or wall, placing it, then going back to pick another one, over and over again until they are all gone. There are $n + m - 1$ ways to pick the first object. That leaves us $n + m - 2$ objects to pick from the next time, then $n + m - 3$ objects, and so on, until there is only one object left to place. There are therefore $(n + m - 1)(n + m - 2)(n + m - 3)\ldots(2)(1) = (n + m - 1)!$ different ways to order the objects, assuming that each ball is distinguishable from the next, and similarly for the walls. However, they are not distinguishable. By a similar argument, the number of different ways of ordering the $n$ balls is $n!$, so that many arrangements are indistinguishable because they differ only in the ordering of the balls, which we can't actually tell apart. To make this clear: picking ball 1, then ball 2 is the same as picking ball 2, then ball 1. Similarly, $(m - 1)!$ arrangements differ only by the ordering of the walls. The total number of ways of sharing $n$ units of energy among $m$ molecules is therefore

$$\Omega_{n,m} = \frac{(n + m - 1)!}{n!(m - 1)!}.$$

This number can be colossal, even for what we would consider to be tiny numbers of molecules. Suppose, for example, that $n = 100$ and $m = 100$. Then $\Omega_{n,m} = 4.53 \times 10^{58}$.

### 6.5.2 Configurational entropy of adsorbed molecules

Let's consider another example, namely the diffusion of adsorbed molecules over a surface. Adsorbed molecules are simply molecules that have become stuck to a surface. These molecules will generally only stick to certain sites, sometimes the atoms of the surface, sometimes the spaces between the atoms. They can hop from site to site or switch places with their neighbors, which results in diffusion, which we can picture as a random wandering of the molecules over the surface.

Now consider the experimental preparation shown in Figure 6.2. A wall separates two halves of the surface. On each side, we have covered the surface with a single layer of a different adsorbate (the molecules stuck to the surface). Call the adsorbate on the left A and that on the right B. The entropy of this original arrangement is $S = k_B \ln 1 = 0$ since there's only one way to put one atom at each available site on each side. For simplicity, suppose that there are exactly $N$ sites (and thus $N$ atoms of A and B) on each side of the wall. At $t = 0$, we remove the wall. If the adsorbed atoms can trade places with their neighbors (generally the case), then they will start to mix. If the forces between two As or two Bs or between an A and a B are the same (not guaranteed), then we would expect that the As and Bs would eventually be well mixed. What do we mean by that? We could mean one of at least two things: (a) The probability of finding an atom of A or of B at any given site eventually tends to $\frac{1}{2}$ (since there are equal numbers of atoms of A and B); (b) If we take a portion of the surface that is large so that it contains many adsorption sites, then there should be about the same number of A and B molecules in this portion. The first of these ideas tells us how to calculate the entropy: there are $2N$ lattice sites. The number of different ways that I can put $N$ atoms of A on $2N$ lattice sites is

$$\Omega = \frac{(2N)!}{N!N!}.$$

The rationale for this formula is as follows: I have $2N$ ways of choosing the site for the first atom, $2N - 1$ remaining sites for the second atom and so on. Multiplying these out I get $(2N)(2N - 1)\ldots(N + 1) = (2N)!/N!$. However, the As are indistinguishable, so I have to divide by $N!$, the number of different ways I could order the $N$ molecules of A, to account for the different orders in which I could populate the same sites. Once I have decided where the As go, the Bs are placed in the remaining sites so there are no choices left to make. Using the statistical theory you may have learned in high school, this quantity is $C(2N, N)$, the number of different combinations of $N$ objects chosen from a set of $2N$ objects.

If we plug $\Omega$ into the Boltzmann formula, we get an equation for the entropy. Unfortunately, this equation is very hard to use because of the factorials. In any reasonable experiment, $N$ is large (of the order of a mole), and we simply can't calculate factorials of very large numbers directly. The following formula, known as Stirling's approximation, is useful in these cases:

$$\ln N! \approx N \ln N - N.$$

The entropy becomes

$$S = k_B \ln \left( \frac{(2N)!}{N!N!} \right) = k_B \left[ \ln(2N)! - 2 \ln N! \right]$$
$$\approx k_B \left[ 2N \ln(2N) - 2N - 2 (N \ln N - N) \right]$$
$$= 2Nk_B \ln \frac{2N}{N} = 2Nk_B \ln 2.$$

Note that $2N$ is the total number of atoms adsorbed on the surface. Expressing this number in moles, we have $n = 2N/L$ or $2N = nL$. Since $k_B = R/L$, the entropy can be rewritten $S = nR \ln 2$. As the entropy of the original (ordered) system was zero, the change in entropy on mixing the two adsorbates is $\Delta_{\mathrm{mix}} S = nR \ln 2 - 0 = nR \ln 2$. Note that we get exactly the same result from Equation (6.4) with $X_1 = X_2 = \frac{1}{2}$, despite the entirely different means by which these results were obtained!

We calculate the entropy of mixing by assuming that any given site is equally likely to be occupied by an A or by a B. At first glance, this is vaguely unsettling; our second description of what we meant by a well-mixed system suggests that the As and Bs should be spread out evenly, and yet microscopic states in which all the As and Bs are neatly organized (e.g. the original configuration with all the As on the left) count in the total. These two ideas do not contradict each other because the configurations in which the As and Bs are well mixed by our second criterion are overwhelmingly more likely than those in which the As and Bs are segregated, again assuming that there are no significant differences in the interatomic forces acting on the two types of molecules. To see this, let's ask ourselves how many of the $\Omega$ microscopic states place exactly $N_l$ molecules of A on the left and the rest on the right. Again, we won't have to worry about the Bs since they will simply be placed wherever we haven't put an A. There are

$$\omega_l = C(N, N_l) = \frac{N!}{N_l!(N - N_l)!}$$

ways of arranging $N_l$ molecules of A on the $N$ adsorption sites on the left. Similarly, the remaining $N - N_l$ molecules of A can be placed in

$$\omega_r = C(N, N - N_l) = \frac{N!}{N_l!(N - N_l)!}$$

ways on the right. The total number of arrangements of $N$ molecules of A such that $N_l$ are the on left and $N - N_l$ are on the right is just the product of these two quantities:

$$\omega(N, N_l) = \frac{N!^2}{N_l!^2 (N - N_l)!^2}.$$

How does this function behave? Figure 6.3 shows the number of configurations as a function of $N_l$ for $N = 50$. Note the logarithmic scale. Configurations with $N_l \approx \frac{1}{2}N$ are *overwhelmingly* more numerous, and so are overwhelmingly more likely to be observed. Table 6.1 shows the probability that the number of molecules of A on the left differs from the most probable value $(N/2)$ by more than 1% as a function of $N$. Note how rapidly this probability decreases. One can imagine that, for a truly macroscopic system containing a number of molecules of the order of Avogadro's number, the probability of significant fluctuations would be unmeasurably tiny. This is a good thing; a very similar statistical argument could be applied to the mixing of gases. Accordingly, the probability that we would ever encounter an unmixed pocket of air in the open atmosphere in which the oxygen content is either significantly higher or lower than average (for the altitude) is vanishingly small. The laws of mechanics allow for the possibility of the atmosphere unmixing itself into regions of high and low oxygen concentrations, but the probability of observation of such a state is just too small to even calculate.

Table 6.1 *Probability of a 1% fluctuation from the mean in the number of particles on the left or right of the center line as a function of the total number of particles N.*

| $N$ | $\Pr\left(\left|\frac{N}{2} - N_l\right| > \frac{1}{100}\frac{N}{2}\right)$ |
|---|---|
| 100 | 0.7774 |
| 1000 | 0.5871 |
| 10 000 | 0.1424 |
| 20 000 | 0.0365 |
| 30 000 | 0.0075 |
| 35 000 | 0.0019 |

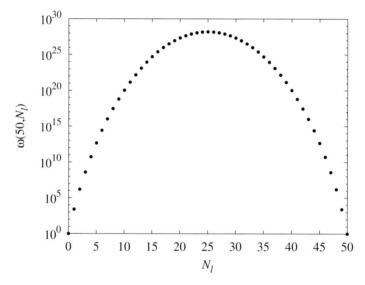

Figure 6.3 Number of microscopic states with exactly $N_l$ of the 50 atoms of A on the left half of a 100-site surface. Note the logarithmic scale for the number of microstates.

### 6.5.3 *Qualitative arguments about entropy*

Given the qualitative definition that entropy measures the number of microstates available to a system, we can often determine the sign of the entropy change in a reaction without doing any calculations.

**Example 6.6** The entropy change for the reaction

$$C_{(s)} + 2S_{(s)} \rightarrow CS_{2(l)}$$

is positive. The reason is straightforward: we are taking two orderly solids (very few microstates) and converting them to a molecular liquid. Molecules in the liquid state have much more freedom of motion than atoms locked into solids, and so have more microstates available to them. Accordingly, the entropy of the liquid product is higher than the entropy of the solid reactants.

**Example 6.7** It is tempting to think that the entropy change for the dissociation of an ionic crystal into its ions such as

$$MgF_{2(s)} \rightarrow Mg^{2+}_{(aq)} + 2F^-_{(aq)}$$

is positive. After all, we break up a nice orderly crystal (not a lot of microstates) and turn it into free-moving ions (apparently, lots of microstates). However, there is another factor to consider, namely the fact that the solvent molecules organize themselves around the ions, producing a negative contribution to the entropy. Moreover, the ions in solution retain some of their organization due to their mutual attractions and repulsions so that the entropy contribution from breaking up the solid lattice is not as large as might be expected. It turns out that the loss of microstates of the solvent is the more important effect in the case of magnesium fluoride (and in many others, although not for all ionic compounds) so that the entropy of the system actually *decreases* in this process. The punch line of this example is this: if you're trying to predict the sign of the entropy change, you have to make sure that you considered all contributions to the entropy of the system, and sometimes you have to be prepared to admit that you don't know a priori what the sign will be.

In your previous courses in chemistry, you may have run across discussions of entropy that suggested that entropy was a measure of disorder or randomness. You may have noticed that I have so far carefully avoided any language of the sort. Why? Because it's misleading. In *some* cases we can relate the number of microstates available to a system to intuitive notions of disorder. In many others, it's not so clear, and trying to do so leads to errors, often because we haven't taken all sources of "disorder" into account, as in the above example, but sometimes for more subtle reasons. For example, suppose that I asked you to say whether a mole of helium or a mole of krypton has the higher entropy at $25\,^\circ$C. You might reason that helium is much lighter, so at the same temperature, its atoms have a much higher average speed. You might then conclude that these fast-moving helium atoms are more "disordered" than the slower-moving krypton atoms, and thus that helium should have the higher entropy. This argument is in fact wrong. Because of the dependence of the translational energy levels on the mass of a particle (which we can get from the particle-in-a-box model, Section 2.4), there are more translational microstates available to a heavy atom at a given temperature, so krypton actually has a higher entropy than helium. It's much better in general to think in terms of microstates than disorder. The microstate approach at least leads you to ask the right questions, even if you might not always know how to answer them.

## Exercise group 6.4

(1) The entropy of money: suppose that someone tells you that he has a total of 10¢ in his pocket. There are a few different combinations of coins that could add up to 10¢. If we think of these different possibilities as microstates, we can calculate an entropy.

    (a) What is the entropy of 10¢? Give your answer in terms of $k_B$.

    (b) Explain what the entropy calculated in question 1 measures.

    (c) If you were told that one of the coins was a 5¢ piece, would the entropy increase or decrease? Why?

(2) Does the entropy of the system increase or decrease during the reaction $N_2O_{4(g)} \rightarrow 2NO_{2(g)}$?

(3) Do you expect $\Delta_r S$ to be positive or negative for the reaction

$$Na_{(s)} + \frac{1}{2}Cl_{2(g)} \rightarrow NaCl_{(s)}?$$

(4) When we adsorb gas onto a surface, we often express the amount adsorbed as a percentage of the surface sites occupied, which we call the percent coverage. Suppose that we have 10% coverage in one experiment, and 90% coverage in another. How would the entropy of the gas compare in the two experiments? In other words, in which case, if any, is the entropy higher?

(5) A protein is made up of a string of amino acids or, more precisely, of the residues obtained as a result of the condensation reaction that makes a peptide bond. Two of the bonds along the "backbone" of a protein can rotate freely, as can some of the bonds in the side chains. As a result, proteins can adopt many conformations. At physiological temperatures, many proteins have a unique overall fold, called the native fold, which includes a small number of closely related conformations. For the sake of argument in this question, assume that the native fold consists of just one conformation. Proteins can be unfolded, resulting in more-or-less random conformations, by heating to high temperatures.

    (a) In a series of investigations, Makhatadze and Privalov measured the entropy of unfolding of proteins at $125\,°C$.[1] All of these values tend to fall into a narrow range, with an average of $52\,J\,K^{-1}(\text{mol residues})^{-1}$. How many conformations per residue does this value imply for the unfolded proteins?

    (b) If the native fold corresponded to many conformations, what effect would this have on the calculation of part 5?

---

## 6.6 Entropy and evolution

The idea that entropy is a measure of disorder has given it a certain popular appeal. Since entropy is a hard quantity to reason about, even for experienced scientists, this has

---

[1] These investigations were summarized in G. I. Makhatadze and P. L. Privalov, *Protein Sci.* **5**, 507 (1996).

sometimes led to the circulation of fallacies. One of the more annoying ones has been the recurring argument that evolution somehow contradicts the Second Law.

Now it's difficult to give a good version of a bad argument, but I will try to put the argument against evolution based on the Second Law in its best possible light. I apologize in advance for failing to do something that can't be done, namely to make this argument truly plausible. With this caveat, here is my attempt to present the argument:

The Second Law of Thermodynamics says that disorder increases with time. Evolutionary theory, if we are to believe it, says that the complex, highly organized life forms we see today evolved from simpler life forms. This is a clear contradiction to the Second Law, therefore evolution can't have happened.

There are lots of problems with this argument. We could talk about the fact that the Second Law applies to adiabatic systems, so that we have to consider living organisms *and* their surroundings, a point to which we will return below. We could question the use of the equilibrium theory, which is where these particular ideas about entropy come from, to a biosphere that is very far from equilibrium. It turns out though that the higher-order refinements of non-equilibrium thermodynamics are not necessary to understand the intellectual error made in this argument against evolution. We could also point out that disorder is a meaningless concept, and that the Second Law says nothing about it. If we want to rephrase the Second Law argument against evolution in terms of microstates, we run into a problem: what are the microstates we should consider? Evolution is a process of, at the very least, an ecosystem. How do we define, never mind count, the microstates of an ecosystem? Even that wouldn't be quite right because the entropy of the solar radiation received by the Earth is an important part of the equation. The "disorder" approach to entropy, or its more correct microstate formulation, doesn't get us anywhere unless we can say what microstates we should be counting and account for external inputs correctly.

Ultimately though, the real problem with this argument is the following: evolution is not a process in the sense we would normally use the term in physical science. It is something that happens as a result of several other processes, most saliently the reproduction and death of living organisms. All of the processes carried out by living organisms result in a net increase in entropy. Let's take the key process of reproduction, for example. Reproduction requires a lot of energy. That energy comes from metabolism, which generates a large amount of high-entropy waste products. Simply put, life is a highly refined entropy generator. The objection raised against evolution on the basis of the Second Law therefore seems to rest on a faulty premise; living organisms, singly or as groups, increase the entropy of the Universe. There is therefore no decrease in entropy caused by evolution to explain away.

Another way to look at the error made by the people who advance the Second Law argument against evolution is that they are trying to isolate one aspect of the temporal evolution of the biosphere and to apply the Second Law to this one isolated aspect. Specifically, they want to look at how species change with time as an isolated phenomenon. This is wrong. Even if we knew how to define the entropy change associated with biological evolution, there is no guarantee in physical theory that any one, arbitrary partition of the entropy of a complex system will necessarily increase with time. It is, again, the entropy of the biosphere, along with key inputs such as solar radiation, that we need to consider. An analogy would be to look at the transformations undergone by an ethylene molecule during the manufacture of polyethylene and to claim that the Second Law has been violated because the eventual product is more ordered than the starting material. We would reject such an argument because it leaves out the substantial energy input to the process, and concomitant increase in entropy due to the dissipation of a portion of that energy. We should similarly reject the one-dimensional view of the temporal evolution of the biosphere required by the Second-Law argument against evolution. Biological evolution is something that happens as a byproduct of the normal day-to-day entropy-generating activities of living organisms. It can't be separated out and asked to satisfy the Second Law all by itself.

## 6.7 The Third Law of Thermodynamics

The Second Law is closely associated with the Third Law of Thermodynamics:

The entropy change during an isothermal process between phases whose initial and final states are true thermodynamic equilibrium states approaches zero as the temperature approaches absolute zero.

This implies that the entropy of a regular crystalline solid approaches a universal constant (which we may as well call zero) as the temperature decreases to absolute zero. Note that this is in accord with our microscopic picture of entropy; there's just one way to arrange the atoms in a perfect crystal ($\Omega = 1$), so there is no entropy associated with the arrangement of the atoms in space. As we approach absolute zero, the energy of the system approaches the zero-point energy, the absolute minimum possible. There is no excess energy to share among the atoms, and so no entropy associated with this sharing. Accordingly, a perfect crystal at absolute zero would have just one microstate, and its entropy would be zero. This is not true for every material. Glasses, for instance, are not crystalline, and thus not directly subject to the Third Law. Also, some crystalline materials (notably $CO$) seem to have a certain amount of irreducible disorder as the temperature is decreased. However, conceptually, the Third Law remains very useful since it defines an experimental zero point for entropy, namely the low-temperature entropy of a perfect crystal.

This property of entropy allows us to define **absolute entropies** (unlike enthalpy where we have to tabulate differences from a reference state). Since the entropy of a perfect crystal at absolute zero is zero, if a substance forms a regular crystal at low temperature, we can compute its entropy at any other temperature by adding the changes in entropy during the heating of the substance from zero to, for example, the standard temperature. Note that the absolute entropies of normal substances are *always positive* since the entropy change on heating is positive and the lowest possible value, defined by the entropy of a perfect crystal, is zero. In the case that a substance does not form a perfect crystal at low temperatures, we can infer the absolute entropy from entropies of reaction, given the entropies of the other substances involved in a particular reaction. Entropies of reaction are obtained from enthalpies and free energies of reaction, the latter being the topic of the next few chapters of this book.

Note again that thermodynamic tables will contain *absolute entropies*, not entropies of formation.

> **The standard entropy of an element always has a positive value.**

If we want to calculate the entropy change in a reaction from a table of standard entropies, we therefore have to consider all reactants and products, including any elements participating in the reaction.

As usual, ions in solution pose special problems. The difficulty is that we can't measure the entropy of a single ion since we always create them in ion/counterion pairs when we dissolve an ionic compound in a solvent. In this case, the absolute entropy of the aqueous proton is arbitrarily assigned to zero, and all the other entropies for aqueous ions are given relative to this reference substance. Consequently, the entropies of aqueous ions are actually relative (not absolute) entropies, and some of them are negative.

You may be surprised that a well-studied science like thermodynamics still gives rise to some controversies. In fact, there are several controversial questions in thermodynamics, one of which follows. The Third Law is often stated in an entirely different way than the statement we discussed above:

It is not possible to lower the temperature of a system to absolute zero.

There have been many attempts over the years to prove one statement from the other, as well as attempts to show that this version of the Third Law is a consequence of some combination of the three Laws of Thermodynamics, but to my knowledge, no entirely satisfactory proof of the equivalence of the two statements of the Third Law has been obtained. We believe, based on a number of lines of evidence, that this statement of the Third Law is also correct. It is possible that it is in fact an independent law, i.e. a Fourth Law of Thermodynamics.

**Example 6.8 Entropy of formation** Although we have tables of absolute entropies, we can calculate standard entropies of formation when required. Consider the standard entropy of formation of liquid water, for example. The formation reaction is

$$H_{2(g)} + \frac{1}{2}O_{2(g)} \rightarrow H_2O_{(l)}.$$

We can look up the standard entropies of the reactants and products in the table in Section A.2. For this reaction,

$$\Delta_f S^\circ = S_m^\circ(H_2O) - S_m^\circ(H_2) - \frac{1}{2}S_m^\circ(O_2)$$

$$= 69.95 - 130.680 - \frac{1}{2}(205.152)\, J\,K^{-1}mol^{-1} = -162.31\, J\,K^{-1}mol^{-1}.$$

**Example 6.9 Residual entropy** CO molecules adopt more or less random orientations (oxygen to the left, oxygen to the right) in the crystal. Estimate the molar entropy of a CO crystal at low temperature.

At low temperatures, only the disorder of the CO molecules contributes to the entropy. The associated entropy is sometimes called the **residual entropy** since it's the entropy we can't account for by pure Third-Law behavior. Each CO molecule has two possible orientations. The number of different ways of arranging the CO molecules is therefore

$$\Omega = 2^N$$

where $N$ is the number of molecules in the crystal. The Boltzmann entropy is therefore

$$S = k_B \ln 2^N = N k_B \ln 2.$$

To convert the result to a molar basis, first note that the entropy per molecule is $S/N = k_B \ln 2$. Multiply both sides by Avogadro's number. Since $R = L k_B$, we get

$$S_m = R \ln 2 = 5.76 \, \text{J K}^{-1} \text{mol}^{-1}.$$

The measured residual entropy is slightly lower (about $5 \, \text{J K}^{-1} \text{mol}^{-1}$), indicating that CO isn't fully disordered at low temperature, but nearly so.

If we want to calculate the absolute entropy of a substance at some temperature $T$ given its entropy at a reference temperature (usually $25\,°\text{C}$), all we have to do is to add the changes in entropy in going from the reference temperature to $T$. The following two examples show how this is done.

**Example 6.10 Absolute entropy as a function of temperature** The entropy of gold at $298\,\text{K}$ is $47.4 \, \text{J K}^{-1} \text{mol}^{-1}$ and its specific heat capacity at constant pressure is $25.42 \, \text{J K}^{-1} \text{mol}^{-1}$. To compute the entropy of gold at (for instance) $273\,\text{K}$, we first calculate the change in entropy during the process of cooling gold to this temperature from $298\,\text{K}$:

$$\Delta S = (25.42 \, \text{J K}^{-1} \text{mol}^{-1}) \ln\left(\frac{273}{298}\right) = -2.23 \, \text{J K}^{-1} \text{mol}^{-1}.$$

The entropy at $273\,\text{K}$ is therefore

$$S(273\,\text{K}) = S(298\,\text{K}) + \Delta S = 45.2 \, \text{J K}^{-1} \text{mol}^{-1}.$$

**Example 6.11 Absolute entropy and phase changes** Suppose that we want to calculate the entropy of liquid sodium at $800\,\text{K}$. The entropy of sodium at $298\,\text{K}$ is $51.30 \, \text{J K}^{-1} \text{mol}^{-1}$. The melting temperature is $371\,\text{K}$ and the latent heat of fusion is $2.6 \, \text{kJ mol}^{-1}$. We can use a constant heat capacity for the solid, which is only to be heated by $73\,\text{K}$, but not for the liquid whose temperature will rise several hundred degrees from the melting point to the target temperature. For liquid sodium,[2]

$$C_{p,m}(\text{l}) = 37.46 - 0.019\,15 \, T + 1.063 \times 10^{-6} \, T^2$$

where the heat capacity is in $\text{J K}^{-1} \text{mol}^{-1}$ and $T$ is in kelvins.

---

[2] *CRC Handbook of Chemistry and Physics*, 66th edn; CRC Press: Boca Raton, 1985, pp. D-43–D-49.

We first deal with the increase in temperature from 298 K to the melting point. Since the solid is assumed to have a constant heat capacity, this is straightforward:

$$\Delta_{(1)} S_m = (28.2 \, \text{J K}^{-1} \text{mol}^{-1}) \ln \left( \frac{371 \, \text{K}}{298 \, \text{K}} \right) = 6.18 \, \text{J K}^{-1} \text{mol}^{-1}.$$

A phase transition is a reversible isothermal process so

$$\Delta_{(2)} S_m = \frac{2.6 \, \text{kJ mol}^{-1}}{371 \, \text{K}} = 7.0 \, \text{J K}^{-1} \text{mol}^{-1}.$$

To calculate the change in entropy of the liquid, we start from the differential form of the definition of entropy: $dS = dq_{\text{rev}}/T$. For a heating process, $dq_{\text{rev}} = C_p \, dT$ so we have

$$dS_m = \frac{C_{p,m}}{T} dT = dT \left( \frac{37.46}{T} - 0.019 \, 15 + 1.063 \times 10^{-6} T \right);$$

$$\therefore \Delta_{(3)} S_m = \int_{371}^{800} dT \left( \frac{37.46}{T} - 0.019 \, 15 + 1.063 \times 10^{-6} T \right)$$

$$= \left[ 37.46 \ln T - 0.019 \, 15 T + 5.315 \times 10^{-7} T^2 \right]_{371}^{800}$$

$$= 37.46 \ln \left( \frac{800}{371} \right) - 0.019 \, 15(800 - 371) + 5.315 \times 10^{-7}(800^2 - 371^2)$$

$$= 20.84 \, \text{J K}^{-1} \text{mol}^{-1}.$$

The entropy at 800 K is therefore

$$S_m(800 \, \text{K}) = S_m^\circ + \Delta_{(1)} S_m + \Delta_{(2)} S_m + \Delta_{(3)} S_m = 85.3 \, \text{J K}^{-1} \text{mol}^{-1}.$$

---

### Exercise group 6.5

(1) Calculate the entropy of nitrogen at 400 K.

(2) The standard entropy of liquid methanol at 298.15 K is 126.8 J K$^{-1}$mol$^{-1}$ and its heat capacity is 81.6 J K$^{-1}$mol$^{-1}$. Methanol boils at 337.8 K with an enthalpy of vaporization at that temperature of 35.21 kJ mol$^{-1}$. The heat capacity of the vapor is 43.89 J K$^{-1}$mol$^{-1}$. Calculate the entropy of methanol vapor at 800 K.

(3) The heat capacity of solid silver in J K$^{-1}$mol$^{-1}$ is given by the equation

$$C_{p,m} = 21.2 + 4.27 \times 10^{-3} T + 1.51 \times 10^5 / T^2$$

where $T$ is the temperature in kelvins. What is the change in entropy per mole of silver when this substance is heated from 298 K (room temperature) to its melting point, 1234 K, at constant pressure? Compare this to the entropy change during the melting process. The heat of fusion for silver is 11.95 kJ mol$^{-1}$.

(4) When a carbonate solution is acidified, carbon dioxide is produced.

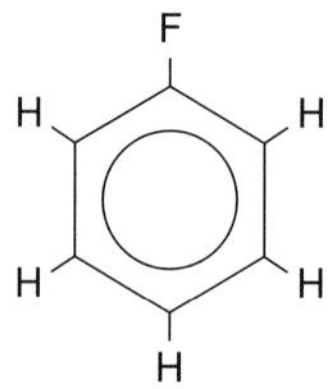

Figure 6.4 Fluorobenzene.

(a) Can we say anything about the sign of the entropy change without doing any calculations?

(b) Calculate the entropy change for this reaction.

(5) In crystals, flat aromatic molecules (e.g. benzene) typically stack up with the carbon atoms of one ring directly above the carbon atoms of the ring below due to the interactions between their delocalized $\pi$ orbitals. For small substituents that lie in the plane of the ring, like fluorine, there is often only a very small energy difference between lining up one fluorine above another, or having a fluorine atom lying above a hydrogen atom. The relative positions of the fluorine atoms in the different layers are therefore essentially random. Calculate the residual molar entropy of fluorobenzene (illustrated in Figure 6.4).

---

## 6.8 Heat engines and the Carnot cycle

Entropy was discovered by Sadi Carnot, an engineer who was trying to understand steam engines and their fundamental limitations. Historically then, putting heat engines at the end of a chapter on entropy is backwards. However, as pointed out earlier, we don't really need to know about heat engines to study biochemistry. This section, and the next on refrigerators, are included to expand your scientific culture given the pivotal importance of heat engines in the history of thermodynamics.

We know, from Kelvin's statement of the Second Law, that a heat engine must allow some heat to flow through it in order to function. Suppose that the engine is in contact with two constant-temperature reservoirs, respectively at temperatures $T_h$ (high) and $T_l$ (low). The engine contains a compressible fluid and operates in a reversible cycle. Working in a cycle is important because an engine that doesn't eventually return to its original state can't run for very long: the state (e.g. the temperature and positions of the pistons) can't keep changing in the same direction without putting the engine into a state where it can no longer perform work (too hot, all pistons out, etc.). Reversibility makes this a maximally efficient engine. (We will prove this in Section 7.1.)

The Carnot engine cycle consists of four steps (Figure 6.5):

(1) The fluid starts the cycle at the higher temperature $T_h$ and reversibly draws a quantity of heat from the corresponding heat bath.

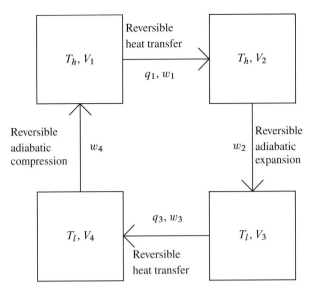

Figure 6.5 The Carnot cycle.

(2) The fluid is adiabatically (and reversibly) expanded until it reaches the lower tempera-
    ture $T_l$.
(3) A quantity of heat is ejected reversibly into the lower-temperature heat bath.
(4) The fluid is adiabatically (reversibly) compressed until it has returned to its original
    state.

The total work done on the fluid in all four steps is $w$. (For a useful engine, this will
be negative.) There are only two steps in which heat enters or leaves the engine, namely
steps 1 and 3. Since the engine returns to its original state after the fourth step, all state
functions return to their original values. Therefore

$$\Delta U_{\text{engine}} = 0$$

and

$$\Delta S_{\text{engine}} = 0.$$

$\Delta U_{\text{engine}}$ and $\Delta S_{\text{engine}}$ can easily be written down:

$$\Delta U_{\text{engine}} = w + q_1 + q_3 = 0. \tag{6.8}$$

$$\Delta S_{\text{engine}} = \frac{q_1}{T_h} + \frac{q_3}{T_l} = 0. \tag{6.9}$$

Using these two equations, we find

$$q_3 = -\frac{T_l q_1}{T_h}$$

and

$$w = -q_1 \left( \frac{T_h - T_l}{T_h} \right).$$

The **efficiency** $\eta$ of a heat engine is the amount of work obtained per unit heat extracted. The work obtained, remembering that all the symbols used so far are relative to the engine, is $-w$. The heat extracted is $q_1$. We obtain $\eta$ directly from the last equation:

$$\eta = -\frac{w}{q_1} = \frac{T_h - T_l}{T_h} = 1 - \frac{T_l}{T_h}. \tag{6.10}$$

Note that the efficiency is less than 100% unless $T_l = 0$ (or $T_h \to \infty$). Since the Third Law says that we can't reach absolute zero, the efficiency of a Carnot engine is always less than 100%, in accord with Kelvin's statement.

It can be proven that, if the Second Law is correct, no heat engine can be more efficient than a reversible Carnot engine. The proof is by contradiction: you assume the existence of an engine that is more efficient than a Carnot engine working between the same two heat reservoirs, and use it to operate a Carnot engine in reverse, moving heat from a cool body to a warm one. In other words, the Carnot engine becomes a Carnot refrigerator. You then show that the two engines together violate the Second Law by producing work in a cyclic process with no net heat transfer from the warmer to the colder body. If the Second Law is correct, it then follows that no engine operating between $T_h$ and $T_l$ can be more efficient than a reversible Carnot engine. The proof is not very difficult, but, like most of these things, it takes time and space to develop, so this outline will have to suffice.

Real engines won't reach the maximum efficiency given by Equation (6.10) since they won't be working reversibly. Moreover, most engines don't operate on the Carnot cycle. Analyses like the one we have just seen for the Carnot cycle can be carried out for other reversible engine cycles, with analogous results. For real, irreversible engines, we typically find that higher efficiencies are reached at higher ratios of $T_l/T_h$, just as for a reversible Carnot engine, but the exact relationship may be different from Equation (6.10).

**Example 6.12 Robot tug-of-war** Two robots which derive their power from heat engines are pitted against each other in a tug-of-war. The first robot's engine operates between the temperatures of 200 and 400 K. Robot number 2 extracts work from an engine whose working temperatures are 700 and 900 K. Both robots consume the same fuel at the same rate. On thermodynamic grounds, which robot do you expect to win the contest?

For robot 1,

$$\eta_1 = 1 - \frac{200}{400} = 0.50.$$

For robot 2,

$$\eta_1 = 1 - \frac{700}{900} = 0.22.$$

Thus, robot 1 has a clear advantage.

**Example 6.13 The internal combustion engine** Modern internal combustion engines are also heat engines. They operate on a different cycle than the Carnot cycle, but for the sake of this example, we will treat these engines as Carnot engines. Since many engine parts are now made of aluminum, the melting point of this metal (932 K) sets an upper limit on the combustion temperature. The exhaust temperature is generally around 400 K. The efficiency of this engine can't be any greater than

$$\eta = 1 - \frac{T_l}{T_h} = 1 - \frac{400}{932} = 0.571.$$

A typical modern car consumes $6\,L\,(100\,km)^{-1}$ at highway speeds. The enthalpy of combustion of gasoline is $-34.7\,MJ/L$ so the heat energy required to go 1 km is $2.1\,MJ\,km^{-1}$. Since $\eta = -w/q_1$, the maximum amount of work done per kilometer is

$$w = -(0.571)(3.5\,MJ\,km^{-1}) = -1.2\,MJ\,km^{-1}.$$

This is of course just the theoretical (Carnot) efficiency. The real efficiency is quite a bit lower, so a real engine actually produces substantially less than 1.2 MJ of work for every kilometer traveled, a greater amount of the heat obtained from combustion being wasted.

---

## Exercise group 6.6

(1) A heat engine takes in heat at a rate of 10 MW and ejects heat at a rate of 8 MW.
   (a) How much work does it perform?
   (b) If the second quantity of heat is ejected at 400 K, what is the minimum upper operating temperature of this engine?
(2) An electrical generator is a type of heat engine. Suppose that a generator burns ethanol at a rate of $10\,L\,h^{-1}$. The generator's combustion chamber runs at 450 °C and the temperature of the exhaust manifold is 130 °C. What is the maximum electrical power (measured in W, i.e. in $J\,s^{-1}$) that such a generator might produce? The density of ethanol is $790\,g\,L^{-1}$ and its molar mass is $46.069\,g\,mol^{-1}$.
(3) A coal-fired power plant is designed to produce 12 MW of electricity. (This is a small unit that might be used to power a pulp-and-paper mill or other relatively energy-intensive factory.) The coal is used to heat steam to a temperature of 560 °C. After running through the turbines, the steam has been condensed and the liquid water cooled to 38 °C.
   (a) Calculate the minimum amount of heat that must be generated to produce the desired amount of electricity.
   (b) Calculate the amount of heat required to bring 1 kg of liquid water from 38 °C up to steam at 560 °C.
   (c) How fast must water be circulated through the heating unit to generate the required amount of electricity?
   (d) The enthalpy of combustion of coal is $29\,MJ\,kg^{-1}$. What is the minimum rate at which coal must be burned to sustain the required level of power production?

(4) Cogeneration is a process in which a heat engine is used to drive an electrical generator, and the waste heat from the engine is used either in industrial processes, for the heating of buildings or for hot water.

(a) TransAlta operates a natural gas cogeneration plant in Sarnia, Ontario that generates 440 MW of electrical power. This plant provides power to the electrical grid, and steam to several local industries. The generating plant uses a combination of gas turbines (analogous to jet engines, but configured to turn a generator rather than to produce thrust) and steam turbines, with the steam produced using the hot exhaust gases from the gas turbines. The cooled steam that emerges from the steam turbines is then distributed to the plant's industrial customers. For the purpose of thermodynamic analysis, we can treat the combination of gas and steam turbines as one engine. Suppose that the gas turbines operate at 800 °C and that the steam exhausted from the steam turbines is at a temperature of 150 °C (typical values for cogeneration plants). Calculate the amount of natural gas (in m$^3$) that must be burned per day to sustain the electrical output of this cogeneration plant, assuming that the generator is an ideal Carnot engine. The specific heat of combustion of natural gas is 37.0 kJ m$^{-3}$.

(b) At what minimum rate is heat ejected by the generator?

(c) The outlet steam is used as a source of heat at various industrial sites, as mentioned above. If this steam was not available, it would be necessary to generate heat for the industrial plants some other way, perhaps by burning more natural gas. Assuming an ideal Carnot generator and no heat losses, how much natural gas is saved per day?

(d) A real steam turbine will function at a somewhat lower efficiency than that calculated in question 4a. Is this as serious a problem for a cogeneration plant as it is for a regular power plant which makes no use of the waste heat?

---

## 6.9 Refrigerators

A refrigerator is simply a heat engine running backward; instead of producing work from the transfer of heat it transfers heat under the action of work. Just as a reversible Carnot engine is the most efficient heat engine possible, a reversible Carnot refrigerator is the most efficient refrigerator possible, i.e. the one that uses the least work to transfer a given amount of heat.

Ordinary household electric refrigerators operate on a Carnot cycle. The only difference between a real refrigerator and the one we are going to analyze is that a real refrigerator can't operate reversibly. Imagine running the Carnot cycle (Figure 6.5) *backward* starting in the state $(T_h, V_1)$.

(4) The fluid is adiabatically and reversibly expanded until it reaches the lower temperature $T_l$ (the temperature of the refrigerated compartment).

(3) A quantity of heat is drawn reversibly from the refrigerated compartment into the working fluid.
(2) The fluid is adiabatically and reversibly compressed until it reaches the higher temperature $T_h$ (the temperature of the room).
(1) A quantity of heat is ejected reversibly from the fluid into the room.

Note that reversing the cycle turns compressions into expansions, and vice versa. Precisely the same equations apply as before. In particular, Equations (6.8) and (6.9) are still valid. However, $\eta$ is the wrong measure of "efficiency" for a refrigerator. Instead, we want to know the amount of heat removed ($q_3$) per unit work performed (just $w$ this time since work is done on the engine). This is called the **coefficient of performance** $\eta^*$. Using the energy- and entropy-balance equations, we get

$$w = q_3 \left( \frac{T_h}{T_l} - 1 \right) \tag{6.11}$$

so that

$$\eta^* = \frac{q_3}{w} = \frac{T_l}{T_h - T_l}.$$

**Example 6.14 Work of refrigeration** Refrigerators and freezers are insulated. However, their insulation isn't perfect and, in any event, the door must occasionally be opened to retrieve or add items so that they are not truly adiabatic systems. Suppose that a freezer is designed to maintain a temperature of $-5\,°C$ in a room at $20\,°C$ against an average heat leak of 100 W. A watt is a joule per second so this figure is just the amount of heat we need to remove per second. The coefficient of performance of the refrigerator is

$$\eta^* = \frac{268}{293 - 268} = 10.7.$$

Since $\eta^* = q_3/w$, the minimum work required to combat the heat leak is

$$w = \frac{100\,\text{W}}{10.7} = 9.35\,\text{W}.$$

Note that this is the average power consumption for an ideal refrigerator operating under the given conditions. A real refrigerator cycles on and off so that its peak power consumption is much higher than this figure, but of course it consumes very little power while its compressor is off. Furthermore, a real refrigerator is less efficient than a reversible Carnot refrigerator so even its average power consumption would be somewhat higher than calculated here.

**Example 6.15 Heat generated by a refrigerator** A refrigerator ejecting heat into a room at $20\,°C$ removes 200 kJ of heat from a $2\,°C$ compartment in an hour. By how much does it heat the room?

The coefficient of performance of this refrigerator, assuming that it is an ideal Carnot refrigerator, is

$$\eta^* = \frac{275\,\text{K}}{293 - 275\,\text{K}} = 15.3.$$

The work performed to remove 200 kJ from the cold compartment is

$$w = \frac{q_3}{\eta^*} = \frac{200\,\text{kJ}}{15.3} = 13.1\,\text{kJ}.$$

The energy-balance equation for the cycle says that

$$q_1 + q_3 + w = 0;$$
$$\therefore q_1 = -(w + q_3) = -213\,\text{kJ}.$$

This equation implies that the work done is also converted to heat and ejected into the room during the refrigeration cycle. In a real refrigerator, a larger amount of work would be done to extract the same amount of heat, and therefore a larger amount of waste heat would be ejected into the room.

---

## Exercise group 6.7

(1) Most of the foods we eat are mostly water. Thus, refrigerating or freezing them is approximately equivalent to refrigerating or freezing an equal mass of water. Suppose that an 8 kg turkey, initially at 20 °C, is placed in a large industrial freezer held at −5 °C. The freezer's coils are on the roof of the building and it's a nice warm day, with an exterior temperature of 15 °C. What is the minimum electrical work required to bring the turkey into equilibrium with the rest of the freezer's contents? Assume that the freezer is sufficiently large that its temperature is not significantly affected by the presence of the turkey.

(2) During a hot summer's day, an air-conditioned apartment has a steady-state heat leak of 2 MJ h⁻¹ when the interior temperature is 25 °C and the external temperature is 35 °C. What is the minimum electrical work (in Watts, i.e. in J s⁻¹) that the air conditioner is expending?

---

## 6.10 Thermodynamics: the cynic's view

We have now seen the three Laws of Thermodynamics and some of their consequences. A cynic might paraphrase these laws as follows:

(1) At best, you might break even.
(2) You can't break even, except at absolute zero.
(3) You can't get to absolute zero.

Some slight hope is offered by cosmological models in which the expansion of the Universe is eventually halted and reversed by gravity. In some of these models, the entropy of the Universe actually decreases during the collapse, which would make things behave quite differently than they do now! Unfortunately, the best data currently available suggest that gravity isn't strong enough to halt the expansion, so it looks as though the cynics are right again.

## Key ideas and equations

- $dS = dq_{rev}/T$
- Second Law of Thermodynamics:
  - It is impossible to continuously convert heat completely into work.
  - $\Delta S \geq 0$ for any thermodynamically allowed process of an adiabatic system.
- $S = k_B \ln \Omega$. Entropy measures the number of microstates available to a system or, equivalently, our degree of ignorance of the microscopic state.
- The Third Law allows us to define an absolute entropy scale. We calculate the entropy of reaction as a difference of standard entropies.
- For a heat engine, $\eta = -w/q_{in} = 1 - T_l/T_h$.
- For a refrigerator, $\eta^* = q_{removed}/w = T_l/(T_h - T_l)$.

## Suggested reading

For a colorful qualitative discussion of the Second Law and its implications, see

P.W. Atkins, *The Second Law*; Scientific American Books: New York, 1984.

There is a nice overview of the importance of entropy in biological systems in the following paper:

Jayant B. Udgaonkar, *Resonance* **6**(9), 61–66 (2001).

The choices of the symbols $U$ and $S$ for the internal energy and entropy might seem somewhat arbitrary. In fact, they are. These choices were made by Clausius (Figure 6.6) in the course of his investigations of thermodynamics in the mid-1800s. He started to study energy and entropy before he had names for these concepts, and he picked letters for these quantities that weren't being used for something else. Because of his influence, these symbols were retained. The story is told in the following paper:

Irmgard K. Howard, *J. Chem. Ed.* **78**, 505–508 (2001).

---

## Review exercise group 6.8

(1) Which of the quantities $p$, $q$, $S$ and $U$ are state functions?
(2) Calculate the entropy of water vapor at 300 °C and 1 atm. The entropy of liquid water at 0 °C and 1 atm is 63.2 J K$^{-1}$mol$^{-1}$.
(3) Critique the following argument: The entropy change during a phase transition is calculated by $\Delta S = \Delta H/T$. Since $\Delta H < 0$ for the freezing of a liquid, $\Delta S < 0$. According to the Second Law, processes for which $\Delta S < 0$ can't happen so freezing is never spontaneous.

Figure 6.6 Clausius-Strasse, Zurich. Rudolf Clausius who, among other contributions, was the orig-inator of the concept of entropy, was a Professor of Physics at the famous ETH (Eigenössische Technische Hochschule, the Swiss Federal Institute of Technology) in Zurich from 1855 to 1867. There is now a street ("Strasse") named in his honor on the campus of the ETH. Unfairly, this street is a dead end.

(4) The entropy of vaporization of most substances at their normal boiling points is usually between 85 and 95 $J\,K^{-1}mol^{-1}$. This observation is known as Trouton's rule. There are exceptions to Trouton's rule. Water, for instance, has a very high entropy of vaporization at its boiling point: 108.951 $J\,K^{-1}mol^{-1}$. Why?

(5) (a) Use a qualitative argument to predict the sign of the entropy change associated with the combustion of ethanol at 25 °C.

(b) Calculate the standard entropy of combustion of ethanol. Hopefully your reasoning in part (a) agrees with your answer to this question!

(c) Suppose that the reaction were carried out at 500 °C instead of 25 °C. Outline the calculations you would have to carry out to obtain the entropy of combustion at this temperature. Clearly list all thermodynamic data you would need.

*Note*: Ethanol boils at 79 °C.

# 7

# Free energy

Determining whether or not a process is thermodynamically allowed based on entropy is a laborious process because we need to calculate the entropy change both of the system and of its surroundings. Fortunately, for commonly encountered experimental conditions, we can simplify the calculations at the cost of defining two more state functions.

## 7.1 The Clausius inequality

Suppose that we are studying a real (irreversible) isothermal process in which both system and surroundings are held at temperature $T$. In order to maintain the system at this temperature, the surroundings must give up any heat absorbed by the system during any process that occurs. If $dq$ is the heat absorbed by the system during an infinitesimal part of the process, the surroundings must lose the same amount of heat, i.e. the heat relative to the surroundings is $-dq$. We can imagine removing the heat from the surroundings using a heat exchanger and reversibly putting it back into the system as described in Section 6.3. Because the surroundings are assumed to be large enough to be able to exchange heat with the system without any significant temperature change, this transfer of heat is reversible with respect to the surroundings. The entropy change of the surroundings is then just $-dq/T$. The Second Law guarantees that the entropy change of the Universe is non-negative, i.e. that

$$dS = dS_{\text{system}} + dS_{\text{surroundings}} = dS_{\text{system}} - \frac{dq}{T} \geq 0$$

or

$$dS \geq \frac{dq}{T}. \tag{7.1}$$

This is the Clausius inequality. We dropped the subscript "system" because all quantities in the inequality are relative to the system. Every isothermal thermodynamically allowed process will satisfy this inequality.

We derived the Clausius inequality from the Second Law. We can also derive the Second Law from the Clausius inequality; if the system is adiabatic, then $dq = 0$ so $\Delta S_{\text{system}} \geq 0$. Since either can be used to prove the other, the Clausius inequality is just another form of the Second Law.

The Clausius inequality has an extremely interesting implication:

$$dS = \frac{dq_{rev}}{T} \geq \frac{dq}{T}.$$

$$\therefore dq_{rev} \geq dq$$

$$\text{or} \quad q_{rev} \geq q.$$

However, since $\Delta U = q + w$ and $U$ is a state function, then for a reversible and an irreversible path that accomplish the same overall change, we must have

$$w_{rev} \leq w.$$

Since work done *by* a system is negative, the maximum useful work is done by a reversible process. In other words, reversible processes are maximally efficient. A reversible path is necessarily infinitely slow so thermodynamics confirms the old adage that "haste makes waste."

---

### Exercise group 7.1

(1) Show that the entropy change for any system that undergoes an endothermic process is positive.

(2) Is the entropy change for a process undergoing an exothermic process positive or negative?

(3) (a) Calculate the work done per mole when an ideal gas is expanded reversibly by a factor of 2 at $115\,°C$.

   (b) A friend thinking of investing in a company that makes engines shows you the company's prospectus. Analyzing the machine, you realize that it works by harnessing the expansion of a gas. The machine's operating temperature is approximately constant at $115\,°C$ and the gas doubles in volume during the power extraction phase of operation. A quick calculation shows that the machine is claimed to produce about $3.5\,kJ\,mol^{-1}$ of work during the expansion. What advice would you give your friend? Give a brief thermodynamic rationale.

---

### 7.2 Free energy functions

Suppose that a process occurs at constant temperature and volume. Then, $q = \Delta U$ (Equation 5.5). The Clausius inequality can then be rearranged to

$$\Delta U - T\Delta S \leq 0.$$

Again, this seems like an ideal opportunity to define a new state function. Let

$$A = U - TS.$$

This is called the Helmholtz free energy, or sometimes the Helmholtz function. By differentiation, we get

$$dA = dU - T\,dS - S\,dT.$$

At constant temperature ($dT = 0$) this gives, after integration,

$$\Delta A = \Delta U - T\Delta S.$$

> **For a thermodynamically allowed process at constant temperature and volume, $\Delta A < 0$.**

It should be emphasized that this derivation of the Helmholtz free energy makes it an accounting trick; we have turned questions about the entropy of the Universe into questions about a state function of the system under special conditions (constant volume and temperature).

Living organisms typically function under conditions that closely approximate constant temperature and pressure, at least over time scales of a few hours. Under these conditions, $q = \Delta H$ (Equation 5.8) and Inequality (7.1) becomes

$$\Delta G = \Delta H - T\Delta S \leq 0. \tag{7.2}$$

The state function $G$ that satisfies this inequality under the stated conditions is the Gibbs free energy

$$G = H - T S.$$

Like the Helmholtz free energy, the Gibbs free energy provides us with an easy way to determine the consequences of the Second Law for a special class of systems, namely those held at constant temperature and pressure:

> **For a thermodynamically allowed process at constant temperature and pressure, $\Delta G < 0$.**

Historically, the understanding that chemical reactions were driven by a balance of enthalpic (heat) and entropic effects came late, with the work of Gibbs in fact. For a long time, chemists believed that heat was the major driver of chemical reactions, i.e. that only exothermic reactions are thermodynamically allowed. This belief was sustained by the fact that most thermodynamically allowed reactions are in fact exothermic, the entropic term often being relatively small. However, there are many examples of spontaneous endothermic chemical reactions, clear evidence that both heat and entropy are important in determining what is and isn't thermodynamically allowed.

---

### Exercise group 7.2

(1) Calculate the change in Gibbs free energy of an ideal gas during an isothermal expansion from $V_1$ to $V_2$.

*Hint*: The internal energy of an ideal gas only depends on its temperature.

---

## 7.3 Free energy as maximum work

Suppose that a system undergoes a thermodynamically allowed, isothermal process. Then,

$$dA = dU - T\,dS.$$

We can evaluate $dU$ along any path (since it's a state function). Therefore

$$dU = dq_{\mathrm{rev}} + dw_{\mathrm{rev}}.$$

From the definition of entropy, $dq_{\mathrm{rev}} = T\,dS$. Putting everything together, we get

$$dA = T\,dS + dw_{\mathrm{rev}} - T\,dS = dw_{\mathrm{rev}}$$

or, after integration,

$$\Delta A = w_{\mathrm{rev.}}$$

Since $w_{\mathrm{rev}} \leq w$, a negative $\Delta A$ has a simple interpretation:

$-\Delta A$ **is the maximum work that can be done by an isothermal process.**

We can apply a similar treatment to the Gibbs free energy:

$$dG = dH - T\,dS - S\,dT,$$

but since

$$dH = dU + p\,dV + V\,dp$$

and using the expression for $dU$ derived above, we get

$$dG = T\,dS + dw_{\mathrm{rev}} + p\,dV + V\,dp - T\,dS - S\,dT = dw_{\mathrm{rev}} + p\,dV + V\,dp - S\,dT.$$

At constant temperature and pressure, we can drop the terms in $dp$ and $dT$:

$$dG = dw_{\mathrm{rev}} + p\,dV.$$

Now we split the work into two terms, namely the pressure–volume work which, under reversible conditions, is $-p\,dV$, and the rest, which we will call $dw'_{\mathrm{rev}}$. This gives

$$dG = dw'_{\mathrm{rev}}$$

which integrates to

$$\Delta G = w'_{\mathrm{rev}}.$$

**For isothermal, constant pressure processes,**
$-\Delta G$ **is the maximum non-pressure–volume work.**

**Exercise group 7.3**

(1) Derive a relationship between the change in Helmholtz and Gibbs free energies at constant temperature. Use your relationship to show how the expansion work and the two types of free energy are related for a process occurring at constant pressure.

## 7.4 Standard states and tabulated values of the state functions

We noted in Section 5.8 that the standard state for a gas was a pressure of 1 bar (100 000 Pa) and that the standard temperature was 25 °C. Since the precise definition of the standard state will now become much more important, we return to this subject.

In defining a standard state for solutes, we need to decide how we will measure concentrations. Molarities are convenient and many thermodynamic tables are based on this concentration scale. However, for high-accuracy work, molarities present a problem: the volume of the solution is temperature dependent, so the same solution can have different molarities at different temperatures. Mole fractions and molalities (moles of solute per kilogram of solvent) are better behaved. Nevertheless, we will express concentrations of solutes with molarities. For dilute solutions, any concentration measure will do. Furthermore, extensive tables of data based on a molarity scale are available so that this is perhaps the most convenient concentration scale for most purposes.

The standard states for each possible state of matter are:

**Solids:** pure solid held at the standard temperature and pressure.
**Liquids:** pure liquid held at the standard temperature and pressure.
**Gases:** held at the standard temperature and pressure, and ideally behaving.
**Solutes**: in a solution held at the standard temperature and pressure, at a concentration of 1 mol L$^{-1}$ (or 1 mol kg$^{-1}$ for the molal standard), and ideally behaving.

The condition that a gas or solute be "ideally behaving" is imposed in order to assist in reproducibility of tabulated thermodynamic data. In practice, it means that dilute gases and solutions are to be used for establishing thermodynamic properties. What precisely is meant by "dilute" may vary from one gas to another or from one solute to another. In general, this requirement conflicts with the concentration and pressure specifications, which implies that tabulated thermodynamic data will normally be obtained by extrapolation from experimental data. For now, our working assumption will be that all substances behave ideally in the concentration and pressure ranges considered. We will study non-ideal behavior in Chapter 9.

The standard pressure was changed just a few decades ago from 1 atm to 1 bar. You will therefore still find many tables of thermodynamic data based on the atmosphere. The difference between a bar and an atmosphere is small, so this makes only a small, but not negligible, difference to standard free energy data. Just be aware of this minor historical hiccup, and see the suggested reading on page 154 for further information.

One of the side effects of our definition of solute standard states is that it produces solutions with very odd acid–base concentrations; the concentrations of both hydrogen and hydroxide ions is 1 M, but we know that neutralization would quickly take place in such a solution. This isn't a problem since the standard state is an abstraction intended to give us common conditions for data presentation. More seriously, it requires us to know the protonation states of all acids and bases, since each ion is treated individually as an entity whose standard concentration is $1 \, mol \, L^{-1}$. Biochemists, who don't always have all the necessary acid and base dissociation constants to sort out exactly what ions are in their solutions and at what concentrations, therefore generally modify their standard state in the following manner: the standard state for all normal solutes is a $1 \, mol \, L^{-1}$ solution; in the standard state, the pH is 7 (which fixes the hydrogen and hydroxide ion concentrations as well as the ratios of acids to their conjugate bases). Only the total concentration of any given ion is considered, irrespective of its protonation state. For example, the biochemists' standard state for the phosphate ion is a pH 7 solution with a total phosphate concentration ($[H_3PO_4] + [H_2PO_4^-] + [HPO_4^{2-}] + [PO_4^{3-}]$) of $1 \, mol \, L^{-1}$ (in the molar standard state convention, which is almost universally used in biochemistry). To distinguish the two quantities, standard free energies evaluated in the normal convention are denoted $G^\circ$ while biochemists' standard free energies are denoted $G^{\circ'}$.

The definition of the biochemists' standard state has some interesting side-effects for balancing reactions; since we don't know the protonation states of the acids (only that the appropriate mixture is present for the pH), biochemists are often deliberately vague about which protonation state is involved in a reaction, and the reactions may sometimes appear not to be balanced with respect to hydrogen ions. For example, they will write $P_i$ for what they call "inorganic phosphate" rather than write one of $PO_4^{3-}$, $HPO_4^{2-}$ etc. Again, the idea is that the reaction occurs in a pH-appropriate mixture of phosphate species, so their thermodynamic data averages over this mixture rather than giving us values for the reaction involving a specific ion.

Some tables of biochemical standard data define the reaction conditions further. Technically, the chemists' standard state and the simple biochemists' standard state described above both refer to an infinitely dilute solution containing no other solutes. However, biomolecules often function only in solutions with a high ionic strength (high concentrations of charged solutes). Just as is the case for pH, some biochemical standard data are given at an ionic strength corresponding to a physiologically realistic situation. Magnesium ions are of particular importance to the function of a number of biomolecules. Some biochemical standard data accordingly also refer to a particular, near-physiological magnesium concentration, commonly $pMg = 3$, i.e. $[Mg^{2+}] = 10^{-3} \, mol \, L^{-1}$. The only way to tell for sure what conditions the data refer to is to look at the caption of the table from which they were taken.

There is one additional thing to look out for in tables. The free energy, like the enthalpy, needs to be referred to some common zero, and precisely the same solution is adopted here as for enthalpy: $\Delta_f G^\circ$, the standard free energy of formation from the elements (and the hydrogen ion, for ionic species) is tabulated.

**Example 7.1 Standard free energy change from tables** Suppose that we want to compute the change in the Gibbs free energy for the reaction

$$C_6H_{12}O_{6(aq)} + 6O_{2(aq)} \rightarrow 6CO_{2(aq)} + 6H_2O_{(l)}$$

under standard conditions. This is just

$$
\begin{aligned}
\Delta_r G^\circ &= 6\Delta_f G^\circ(CO_2, aq) + 6\Delta_f G^\circ(H_2O, l) \\
&\quad - \left[\Delta_f G^\circ(C_6H_{12}O_6, aq) + 6\Delta_f G^\circ(O_2, aq)\right] \\
&= 6(-386.05) + 6(-237.140) - [(-914.25) + 6(16.35)] \text{ kJ mol}^{-1} \\
&= -2922.99 \text{ kJ mol}^{-1}.
\end{aligned}
$$

The negative sign tells us that this reaction would be thermodynamically allowed under standard conditions (1 mol L$^{-1}$ glucose, 1 mol L$^{-1}$ dissolved oxygen and 1 mol L$^{-1}$ dissolved carbon dioxide in pure water, an unrealizable state). Also, under standard conditions, we can obtain up to 2923 kJ mol$^{-1}$ of non-$pV$ work from this reaction. This could be muscle work, for instance: the change in volume of a working muscle is minimal so essentially all the work can be used to lift weights.

**Example 7.2 Entropies from enthalpies and free energies** Given a table of standard enthalpies and free energies of formation, we can compute the standard entropy change during a reaction. For the oxidation of glucose in solution (from the above example), $\Delta_r H^\circ = -2858.94$ kJ mol$^{-1}$. Since $\Delta_r G^\circ = \Delta_r H^\circ - T\Delta_r S^\circ$,

$$
\begin{aligned}
\Delta_r S^\circ &= (\Delta_r H^\circ - \Delta_r G^\circ)/T \\
&= \frac{-2858.94 - (-2922.99) \text{ kJ mol}^{-1}}{298.15 \text{ K}} = 214.8 \text{ J K}^{-1}\text{mol}^{-1}.
\end{aligned}
$$

We can then infer absolute entropies if we know the entropies of all but one of the reactants or products. Using the data from Appendix A.2, for instance, we see that we have all the entropies, except for the entropy of $CO_{2(aq)}$. We can calculate the latter quantity as follows:

$$
\Delta_r S^\circ = 6S^\circ(CO_2, aq) + 6S^\circ(H_2O, l) - \left[S^\circ(C_6H_{12}O_6, aq) + 6S^\circ(O_2, aq)\right].
$$

$$
\begin{aligned}
\therefore S^\circ(CO_2, aq) &= \frac{1}{6}\left[\Delta_r S^\circ - 6S^\circ(H_2O, l) + S^\circ(C_6H_{12}O_6, aq) + 6S^\circ(O_2, aq)\right] \\
&= \frac{1}{6}\left[214.8 - 6(69.95) + 264.01 + 6(110.88) \text{ J K}^{-1}\text{mol}^{-1}\right] \\
&= 120.7 \text{ J K}^{-1}\text{mol}^{-1}.
\end{aligned}
$$

## Exercise group 7.4

(1) Is the combustion of ethanol thermodynamically allowed under standard conditions?
(2) (a) What is the standard entropy of vaporization of water at 298 K?
    (b) Calculate the absolute entropy of water vapor under standard conditions.
(3) What is the standard entropy change during the reaction

$$CO_{(g)} + \frac{1}{2}O_{2(g)} \rightarrow CO_{2(g)}$$

at 700 K?
(4) Differences in free energies for reactions are fixed by experimental data, but our decision to assign the free energies of formation of certain substances to be zero is purely arbitrary. Suppose that we had decided instead that the free energies of formation of the elements and of the aqueous hydrogen ion should each be 13 kJ mol$^{-1}$. What would the free energy of formation of the hydroxide ion then have to be?

## 7.5 Activity: expressing the dependence of Gibbs free energy on concentration

Suppose that we have a table of standard Gibbs free energies of formation but that we want the free energy of formation of a substance at a different pressure or concentration than standard. We can easily deal with one case, namely that of ideal gases, and will treat the others by analogy.

Consider this two-step process:

$$\left\{\begin{array}{c} \text{elements} \\ \text{in their} \\ \text{standard states} \end{array}\right\} \xrightarrow{1} \left\{\begin{array}{c} \text{Gaseous compound} \\ \text{in its} \\ \text{standard state} \end{array}\right\} \xrightarrow{2} \left\{\begin{array}{c} \text{Gaseous compound} \\ \text{at standard temperature} \\ \text{and pressure } p \end{array}\right\}$$

$\Delta G_m = \Delta_{(1)}G_m + \Delta_{(2)}G_m$ for this overall process would be the molar free energy of formation of the compound at pressure $p$ from elements in their standard states. Let us call this quantity $\Delta_f G_m$. (Note the absence of the superscripted naught.) $\Delta_{(1)}G_m$ is just the standard free energy of formation of the compound, $\Delta_f G_m^\circ$. $\Delta_{(2)}G_m$ is the change in free energy when we adjust the pressure of the compound isothermally from $p^\circ$ to $p$.

To calculate $\Delta_{(2)}G_m$, we need to figure out how the free energy depends on $p$. Recall

$$G = H - TS.$$

Therefore

$$dG = dH - T\,dS - S\,dT.$$

Also,

$$H = U + pV$$

so that

$$dH = dU + p\,dV + V\,dp.$$

By definition,

$$dU = dq + dw.$$

If we take a reversible path and only pressure–volume work is possible,

$$dU = T\,dS - p\,dV.$$

We now substitute this expression for $dU$ into $dH$, and $dH$ into $dG$ to obtain

$$dG = V\,dp - S\,dT. \tag{7.3}$$

Equation (7.3) is a very important equation: it tells us how the Gibbs free energy depends on $p$ and $T$. It is very general, and can be applied to any system.

For an isothermal process, $dG = V\,dp$. We can calculate $\Delta_{(2)}G_m$ directly from this equation:

$$\Delta_{(2)}G_m = \int_{p^\circ}^{p} dG_m = \int_{p^\circ}^{p} V_m\,dp = RT \int_{p^\circ}^{p} \frac{dp}{p} = RT \ln \frac{p}{p^\circ}.$$

In this derivation, we used the ideal gas law in the form $pV_m = RT$ to eliminate $V_m$.

We can now put the pieces together. The free energy of formation of an ideal gaseous compound at pressure $p$ is

$$\Delta_f G_m = \Delta_f G_m^\circ + RT \ln \frac{p}{p^\circ}. \tag{7.4}$$

The fraction $p/p^\circ$ measures how far from the standard state the gas is. We call this quantity the activity $a$. When a gas is in the standard state, $a = 1$ and the logarithm in Equation (7.4) is zero. If $a \neq 1$, the gas is out of the standard state, and we have the logarithmic correction to the free energy of formation shown above. By analogy, for any substance, we write

$$\Delta_f G_m = \Delta_f G_m^\circ + RT \ln a. \tag{7.5}$$

The activity is a dimensionless quantity that gives the extent of the deviation from the standard state.

For ideally behaving substances, the activity is defined as follows:

**Solids:** $a = 1$.
**Liquids:** $a = X$, the mole fraction. The mole fraction is the number of moles of a particular substance (in this case, a liquid, often the solvent) divided by the total number of moles of all species in a system.
**Gases:** $a = p/p^\circ$ where $p^\circ$ is the standard pressure.
**Solutes:** $a = c/c^\circ$ where $c^\circ$ is the standard concentration for that species. In the chemists' standard state, this would be $1\ mol\,L^{-1}$ if our standard state is defined on the molarity scale, and $1\ mol\,kg^{-1}$ for the molality-scale standard state. In the biochemists'

standard state, the standard concentration would be $1\,\mathrm{mol\,L^{-1}}$ for most substances (with species differing only in their protonation state treated as one substance), but $10^{-7}\,\mathrm{mol\,L^{-1}}$ for hydrogen and hydroxide ions.

We have already seen why $p/p^\circ$ is a sensible definition for the activity of an ideal gas. Solids don't have a property corresponding to concentration; they are either there or they are not, which is why the activity of a pure solid is just 1. We will see later why the other choices are correct for dilute solutions.

For non-ideal substances, Equation (7.5) will remain valid, but we will have to modify our definitions of the activities. This important topic will be discussed in Chapter 9.

To get the dependence of the free energy change during a reaction on the activities of the participating species, we simply add the free energies of formation of the products and subtract the free energies of formation of reactants in the usual way. Using the logarithm manipulation rules, we get, for any reaction,

$$\Delta_r G_m = \Delta_r G_m^\circ + RT \ln \left( \frac{\displaystyle\prod_{i\,\in\,\text{products}} a_i}{\displaystyle\prod_{j\,\in\,\text{reactants}} a_j} \right). \tag{7.6}$$

(The symbol $\prod$ represents a product, in the same way that $\sum$ represents a sum of terms. Thus, $\prod_{i=1}^{N} x_i = x_1 x_2 x_3 \ldots x_N$.) The quantity following the logarithm is the **reaction quotient**, usually denoted $Q$:

$$\Delta_r G_m = \Delta_r G_m^\circ + RT \ln Q. \tag{7.7}$$

**Example 7.3 Calculating the free energy of reaction under non-standard conditions** In Example 7.1 we computed the standard change in Gibbs free energy for the reaction

$$C_6H_{12}O_{6(aq)} + 6O_{2(aq)} \rightarrow 6CO_{2(aq)} + 6H_2O_{(l)}.$$

Now suppose that we want to calculate the actual change in free energy when the concentrations of glucose, oxygen and carbon dioxide are, respectively, 0.005, 0.001 and $0.03\,\mathrm{mol\,L^{-1}}$ at $25\,^\circ\mathrm{C}$. For this reaction,

$$Q = \frac{(a_{CO_2})^6 (X_{H_2O})^6}{(a_{C_6H_{12}O_6})(a_{O_2})^6}.$$

The activities of the solutes are easy to calculate. For example, $a_{CO_2} = [CO_2]/c^\circ = (0.03\,\mathrm{mol\,L^{-1}})/(1\,\mathrm{mol\,L^{-1}}) = 0.03$. We need the mole fraction of water to compute $Q$. The mole density of water at $25\,^\circ\mathrm{C}$ is $55.33\,\mathrm{mol\,L^{-1}}$. If there are no other solutes than those listed above, the total number of moles of all species in one liter of water is $[H_2O] + [C_6H_{12}O_6] + [O_2] + [CO_2] = 55.33 + 0.005 + 0.001 + 0.03 \, mol\,L^{-1} = 55.37 \, mol\,L^{-1}$. Thus,

$$X_{H_2O} = \frac{55.33 \, mol\,L^{-1}}{55.37 \, mol\,L^{-1}} = 0.999.$$

$$\Delta_r G_m = \Delta_r G_m^\circ + RT \ln Q$$
$$= -2922.99 \, kJ \, mol^{-1}$$
$$+ (8.314\,472 \times 10^{-3} \, kJ\,K^{-1}mol^{-1})(298.15 \, K) \ln \frac{(0.03)^6 (0.999)^6}{(0.005)(0.001)^6}$$
$$= -2859.28 \, kJ \, mol^{-1}.$$

Since the mole fraction of the solvent is usually very close to 1, we generally leave it out. If we had done this here, we would have obtained $-2859.27 \, kJ \, mol^{-1}$, a negligible difference from the value computed above.

---

### Exercise group 7.5

(1) Consider the following reaction occurring in aqueous solution:

$$H_2O_{(l)} + A_{(aq)} \rightarrow P_{(aq)}.$$

Suppose that this reaction is not allowed under some particular conditions, but only just. In other words, suppose that the free energy change is just slightly positive. Holding all other conditions constant, we add an inert salt to the solution. Can this make the reaction thermodynamically allowed? Why or why not?

(2) Is the reaction

$$2HgS_{(s)} + Cl_{2(g)} \rightarrow Hg_2Cl_{2(s)} + 2S_{(s)}$$

thermodynamically allowed if the pressure of chlorine gas is 2 bar and the temperature is 25 °C?

(3) 20 mL of a 0.04 mol/L solution of lead (II) nitrate is mixed with 15 mL of a 0.003 mol/L solution of ammonium sulfate at 25 °C. Is a lead (II) sulfate precipitate formed?

(4) The overall reaction in an alkaline battery is

$$Zn_{(s)} + 2MnO_{2(s)} \rightarrow ZnO_{(s)} + Mn_2O_{3(s)}.$$

(a) Calculate the maximum electrical work that an alkaline battery can perform per kilogram of zinc at 25 °C.

(b) If an alkaline battery initially contains 10 g of zinc and 15 g of manganese (IV) oxide, what is the maximum **total** electrical work that can be produced at 25 °C?

(5) A fuel cell is a device that generates electricity from an oxidation reaction at constant temperature and pressure (i.e. the net reaction is combustion but no flame or explosion is involved; this is similar to biological oxidations). Calculate the maximum electrical

work that can be obtained from the oxidation of 1 kg of hydrogen by gaseous oxygen in a hydrogen fuel cell under the following conditions: $p_{H_2} = 1$ bar, $p_{O_2} = 0.2$ bar and pure liquid water is produced at 298 K.

## 7.6 Adjusting $\Delta G$ to different temperatures

If we want to calculate $\Delta_r G$ at a non-standard temperature, we typically just apply Equation (7.2), i.e. $\Delta_r G = \Delta_r H - T \Delta_r S$. We can use the methods described in previous chapters to first adjust $\Delta_r H$ and $\Delta_r S$ to the desired temperature. More commonly, however, we simply assume that $\Delta_r H$ and $\Delta_r S$ are temperature independent. We can treat these two quantities as being independent of temperature over a significant range of temperatures, because they both depend on $\Delta C_{p,m}$, the difference in heat capacities of reactants and products, which is typically small, and certainly much smaller than $\Delta_r S^\circ$ for most reactions. We are also helped by the fact that $\Delta_r H^\circ$ and $\Delta_r S^\circ$ change in the same direction (either both increase or both decrease, depending on the sign of $\Delta C_{p,m}$) as we change the temperature, so the negative sign in Equation (7.2) tends to result in some cancellation of the error we introduce by neglecting the temperature variation of these quantities.

**Example 7.4 ATP hydrolysis at 37 °C** Muscle contraction is powered by the hydrolysis of ATP:

$$ATP + H_2O \rightarrow ADP + P_i$$

where $P_i$ represents any phosphate ($PO_4^{3-}$, $HPO_4^{2-}$ etc.), as discussed on page 146. A muscle is an isothermal free energy machine. It does not produce work from heat, but by harnessing the free energy change of a chemical reaction.

The standard Gibbs free energy change for this hydrolysis in a physiological biochemical standard state[1] is $\Delta_r G^{\circ\prime} = -32.49$ kJ mol$^{-1}$, and the standard enthalpy change is $\Delta_r H^{\circ\prime} = -30.88$ kJ mol$^{-1}$. These data are at 298.15 K, but a muscle works at 37 °C, i.e. 310.15 K. From the basic definition of $G$, we have

$$\Delta_r S^{\circ\prime} = \frac{\Delta_r H^{\circ\prime} - \Delta_r G^{\circ\prime}}{298.15 \text{ K}} = 5.40 \text{ J K}^{-1}\text{mol}^{-1}$$

Neither $\Delta_r H^{\circ\prime}$ nor $\Delta_r S^{\circ\prime}$ vary dramatically with temperature so their values at 310.15 K are approximately the same as their values at 298.15 K. Thus

$$\Delta_r G_{310}^{\circ\prime} = -30.88 \text{ kJ mol}^{-1} - (310.15 \text{ K})(5.40 \times 10^{-3} \text{ kJ K}^{-1}\text{mol}^{-1}) = -32.55 \text{ kJ mol}^{-1}.$$

Note that this value isn't terribly different from the value at 25 °C. Of course, 37 °C isn't very far removed from 25 °C, either.

The approximate physiological concentrations of phosphate, ATP and ADP are, respectively, 2 mmol L$^{-1}$, 4 mmol L$^{-1}$ and 0.3 mmol L$^{-1}$.[2] Thus, under physiological conditions,

---

[1] $T = 298.15$ K, pH = 7, pMg = 3, $I = 0.25$ mol L$^{-1}$. Data from Alberty and Goldberg, *Biochemistry* **31**, 10610 (1992).
[2] Bernard L. Oser (Ed.), *Hawk's Physiological Chemistry*, 14th edn.; McGraw-Hill: New York, 1965, pp. 217–218.

the change in free energy during hydrolysis is

$$\Delta_r G_m = \Delta_r G_{310}^{\circ}{}' + RT \ln Q$$
$$= -32.55 \text{ kJ mol}^{-1}$$
$$+ (8.314\,472 \times 10^{-3} \text{ kJ K}^{-1} \text{mol}^{-1})(310.15 \text{ K}) \ln \left( \frac{(0.3 \times 10^{-3})(2 \times 10^{-3})}{4 \times 10^{-3}} \right)$$
$$= -55.26 \text{ kJ mol}^{-1}.$$

This is the maximum amount of non-pressure–volume work available. Of course, muscles mainly do non-$pV$ work, since their volume is relatively constant, being made primarily of incompressible components. Therefore, this is the maximum work that can be done by muscle, per mole of ATP hydrolyzed.

Note that in the last calculation of the example, we *must* use a standard free energy of reaction adjusted to the temperature of the calculation. We can't use $\Delta_r G^{\circ}$ (or $\Delta_r G^{\circ}{}'$) from 25 °C and a different $T$ in this equation.

---

### Exercise group 7.6

(1) Calculate the maximum work available from the oxidation of glucose at 37 °C and solute concentrations of 0.005 mol L$^{-1}$ of glucose, 0.001 mol L$^{-1}$ of dissolved oxygen and 0.03 mol L$^{-1}$ of dissolved carbon dioxide. Compare your answer to the calculation of Example 7.3.

(2) (a) Based on thermodynamic considerations alone, can liquid boron trichloride be made by reacting diborane ($B_2H_{6(g)}$) with chlorine gas if the pressures of the two reactants are both 0.5 bar and the pressure of hydrogen gas is 0.03 bar at 25 °C?

    (b) What if the reaction was run at $-10$ °C instead of 25 °C? Assume that $\Delta_r H_m^{\circ}$ and $\Delta_r S_m^{\circ}$ are independent of temperature.

(3) Creatine phosphate is an alternative energy storage compound found in many cells. Just as in the case of ATP, considerable free energy can be released by the hydrolysis of creatine phosphate:

$$\text{creatine phosphate} + H_2O \rightarrow \text{creatine} + P_i$$

with a $\Delta_r G^{\circ}{}'$ of $-37.7$ kJ mol$^{-1}$ at 310 K. Suppose that, in a certain muscle with a volume of 1 L, the concentrations of creatine and creatine phosphate are both approximately 1 mmol L$^{-1}$ and that the concentration of phosphate is 2 mmol L$^{-1}$. What is the maximum work that can be done by such a muscle if all the concentrations are held constant by homeostatic mechanisms when the entire store of creatine phosphate is turned over once? How far would this amount of work lift a 1 kg mass?

## Key ideas and equations

- Reversible processes are maximally efficient.
- $A = U - TS$ is the Helmholtz free energy.
  - $\Delta A < 0$ for a thermodynamically allowed process at constant $T$ and $V$.
  - $-\Delta A$ is the maximum work that can be done by a machine operating isothermally.
- $G = H - TS$ is the Gibbs free energy.
  - $\Delta G < 0$ for a thermodynamically allowed process at constant $T$ and $p$.
  - $-\Delta G$ is the maximum non-$pV$ work that can be done by a machine operating isothermally and at constant pressure.
- Definition of standard state (1 bar, 25 °C, solutes at $1\,\mathrm{mol\,L^{-1}}$, everything behaving ideally).
  - Biochemists' standard state: lump together species that differ by protons, pH $= 7$.
- $\Delta_r G_m = \Delta_r G_m^\circ + RT \ln Q$
- Activities of ideal substances:

| Phase | $a$ |
|---|---|
| Solid | 1 |
| Liquid | $X$ |
| Gas | $p/p^\circ$ |
| Solute | $c/c^\circ$ |

- To adjust $\Delta G$ to different temperatures, it is often adequate to treat $\Delta H$ and $\Delta S$ as being independent of temperature.

## Suggested reading

For detailed explanations of the biochemists' standard state, see

Robert A. Alberty, *Biochim. Biophys. Acta* **1207**, 1 (1994).
Robert A. Alberty, *Biochem. Ed.* **28**, 12 (2000).

The latter paper includes a large table of thermodynamic data in the biochemists' standard state.

If you ever need to convert data from the old standard pressure of 1 atm to the new standard pressure of 1 atm, see the following paper:

Richard S. Treptow, *J. Chem. Ed.* **76**, 212 (1999).

Because the formation reactions for most compounds involve some gaseous elements, this conversion can affect free energies of formation for compounds in any state of matter.

---

## Review exercise group 7.7

(1) Many reactions involve water as a reactant or product. When treating such reactions, we generally ignore the activity of the water, i.e. set its value to unity. Why is it often reasonable to do this?

(2) What is wrong with the following statement?

$\Delta G$ gives the maximum work that a system can perform at constant temperature and pressure.

(3) For processes involving only solids or liquids, what is the relationship between $\Delta G$ and $\Delta A$?

(4) A common dry cell ("battery"), also known as a Leclanché cell, has the following overall reaction:

$$Zn_{(s)} + 2NH^+_{4(aq)} + 2MnO_{2(s)} \rightarrow Zn^{2+}_{(aq)} + 2NH_{3(aq)} + H_2O_{(l)} + Mn_2O_{3(s)}.$$

Suppose that a dry cell operating at 25 °C starts off with the following concentrations: $[NH^+_4] = 1.5 \, \text{mol L}^{-1}$, $[Zn^{2+}] = 0.03 \, \text{mol L}^{-1}$ and $[NH_{3(aq)}] = 0.12 \, \text{mol L}^{-1}$. The solid reactants zinc and manganese(IV) oxide are present in excess. The water exists in a concentrated paste rather than as a free solution (hence the name "dry cell"). For the sake of argument, assume that $a_{H_2O}$ is constant and has the value 0.5. Calculate the maximum work that can be performed by this cell per mole of zinc consumed:

(a) initially, and
(b) when half the ammonium ion available has reacted.

(5) While ATP is the compound used to directly power most metabolic processes, it is not a very good long-term energy storage compound because of its low stability. ATP is generally made from ADP by harnessing the energy obtained when sugars are oxidized. A muscle can therefore be considered to be powered, indirectly at least, by the oxidation of sugars. A teaspoon of sucrose (ordinary table sugar) weighs about 4 g. To what maximum height would the oxidation of a teaspoon of sugar allow an organism working at 298 K to raise a 1 kg mass under standard conditions? The molar mass of sucrose is $342.299 \, \text{g mol}^{-1}$. Note that when this process occurs in a cell, aqueous oxygen and carbon dioxide are involved instead of the gaseous molecules.

(6) CODATA, the Committee on Data for Science and Technology, produces a table that gives generally accepted values of thermodynamic properties for a broad range of substances. Their table is a little different from the one in most textbooks. Here is an excerpt from that table, with notations adjusted to match the ones used in this book:

| Substance | State | $\frac{\Delta_f H^\circ \,(298.15 \, \text{K})}{\text{kJ mol}^{-1}}$ | $\frac{S^\circ_m \,(298.15 \, \text{K})}{\text{J K}^{-1}\text{mol}^{-1}}$ | $\frac{H_m(298.15 \, \text{K}) - H_m^\circ(0 \, \text{K})}{\text{kJ mol}^{-1}}$ |
|---|---|---|---|---|
| Al | s | 0 | 28.30 | 4.540 |
| $Al^{3+}$ | aq | $-538.4$ | $-325$ | |
| $H^+$ | aq | 0 | 0 | |
| $H_2$ | g | 0 | 130.680 | 8.468 |

(a) Calculate the standard free energy of formation of $Al^{3+}_{(aq)}$.
(b) To which thermodynamic property of a substance is the last column of the table related? Give the precise mathematical relationship.

(7) Industrially, methanol ($CH_3OH$) is made by reaction of carbon monoxide with hydrogen. The reaction is carried out at high temperatures in the presence of a catalyst. Suppose, however, that you wanted to carry out the reaction at $25\,°C$, the pressures of CO and of $H_2$ both being held constant at 0.5 bar. Is there any *thermodynamic* reason why this could not be done?

*Note*: Methanol boils at $65\,°C$.

(8) The equation

$$\Delta_r G_m = \Delta_r G_m^° + RT \ln Q$$

is used to compute the Gibbs free energy of reaction under particular conditions. The enthalpy change depends extremely weakly on the reaction conditions. Use this fact to derive an equation relating the entropy change to the standard entropy change and to $\ln Q$.

(9) While ATP is the principal short-term energy storage compound in living organisms, many other compounds are used for this purpose. Glucose-6-phosphate (G6P) is an intermediate in glucose metabolism, but it can be and sometimes is used as a source of work. Its hydrolysis reaction is

$$G6P + H_2O \rightarrow glucose + phosphate.$$

For this reaction, $\Delta_r G_m^{°\prime} = -16.7\,kJ\,mol^{-1}$ and $\Delta_r H_m^{°\prime} = -35.1\,kJ\,mol^{-1}$ at 298 K. Suppose that the concentration of G6P in an organelle at $37\,°C$ is $50\,\mu mol\,L^{-1}$, the concentration of glucose is $20\,\mu mol\,L^{-1}$ and the total concentration of phosphates is $2\,mmol\,L^{-1}$. What is the maximum work available from this reaction per mole of G6P?

(10) The following reactions have been used in voltaic cells (batteries):

$$2Na_{(s)} + 5S_{(s)} \rightarrow Na_2S_{5(s)} \qquad \Delta_r G_m^° = -401\,kJ\,mol^{-1}$$
$$VH_{(s)} + NiOOH_{(s)} \rightarrow V_{(s)} + Ni(OH)_{2(s)} \qquad \Delta_r G_m^° = -130\,kJ\,mol^{-1}$$

(a) What is the standard free energy of formation of $Na_2S_5$?

(b) Battery designers don't generally care about the amount of electrical work per mole. What they really care about is the work per unit mass, heavy batteries having a number of obvious disadvantages. Assuming a stoichiometric mixture of the reactants in each case, batteries based on which of the above reactions will store the most energy for a given mass?

(c) The reactions shown above involve solid reactants and products only. This makes the calculations required for the last question much more straightforward than they would otherwise be. Why? What additional complication(s) would arise if some of the reactants or products were solutes?

# 8

# Chemical equilibrium and coupled reactions

Life is a far-from-equilibrium phenomenon, so one might reasonably question the utility of equilibrium theory to biochemists. There are a few reasons why you should care about equilibrium. One is that we still do a lot of experiments in test tubes, and under those conditions, systems eventually go to equilibrium. Another reason is that some reactions, like acid–base reactions, equilibrate much faster than others, so they can be treated as being in equilibrium. There is another, more subtle reason: we saw earlier that reversible processes, those in which the system is constantly in equilibrium, are the most efficient ones possible. Equilibrium therefore sets a limit to the efficiency of chemical processes, an idea we will use when we discuss how one reaction can drive another forward.

## 8.1 What does $\Delta_r G_m$ mean?

The molar free energy of reaction, $\Delta_r G_m$ (as well as other $\Delta$ quantities in thermodynamics), is a funny quantity. We typically think of it as the difference in free energy between reactants and products, since that's how we calculate it. However, its meaning is a bit more subtle than that. To make things specific, imagine that we have a reaction

$$\text{reactants} \rightarrow \text{P} + \text{some other products}$$

in a system held at constant temperature and pressure. $\Delta_r G_m$ represents the change in free energy when we convert one molar equivalent of reactants to one mole of P *under specified conditions* (concentrations, pressures etc.). In other words, the composition of the system should not change as our mole of P is made. However, if we carry out this conversion, we *will* change the quantities of reactants and products, and thus the composition. There is a way out of this apparent contradiction: instead of making one mole of P, imagine that we convert some very small fraction of the reactants to products. Let $n$ be the number of moles of P at some particular point in time. If we know the initial composition and the stoichiometry of the reaction, then we can calculate the number of moles of any component of the reaction if we know $n$. We can therefore think of $G$ as a function of the one composition variable $n$. (This assumes there is only one reaction going on in our system. In general, we need

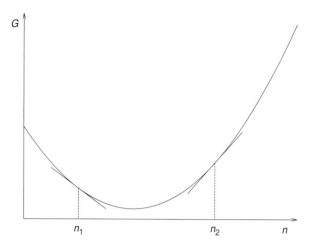

Figure 8.1 Schematic showing how the free energy of a system, $G$, varies with the number of moles of a product, $n$. $\Delta_r G_m$ is the slope of the graph of $G$ vs. $n$. This sketch shows two tangent lines at two different values of $n$. At $n = n_1$, the slope of the tangent is negative, so $\Delta_r G_m < 0$, indicating that the reaction is thermodynamically allowed, i.e. that it will proceed to make more product (increase $n$), provided kinetic factors don't prevent it from proceeding. At $n = n_2$, $\Delta_r G_m$ is positive, so the reverse reaction, consuming products and decreasing $n$, is thermodynamically allowed.

as many variables as there are reactions.) Now suppose that, over some period of time, the reaction causes a change in $n$ of $\Delta n$, where $\Delta n$ is sufficiently small that it makes only a negligible difference to the overall composition of the system. We divide the change in the free energy during this period of time, $\Delta_r G$, by the number of moles of P made, $\Delta n$, to get $\Delta_r G_m$:

$$\Delta_r G_m = \frac{\Delta_r G}{\Delta n}. \tag{8.1}$$

Since we converted only a very small fraction of the reactants to products, the reaction conditions didn't change significantly during this operation. The quantities on the right of the equality are ordinary differences: $\Delta_r G$ is the change in the total free energy of the system when $\Delta n$ moles of P were produced or, if you prefer, $G(n + \Delta n) - G(n)$. Now if you think back to your calculus course, you may recognize that if $\Delta n$ is small, then the right-hand side of Equation (8.1) is in fact a *derivative* of $G$ with respect to $n$. In other words, the quantity we refer to as $\Delta_r G_m$ is actually

$$\Delta_r G_m = \frac{dG(n)}{dn}.$$

Figure 8.1 shows the relationship of $G(n)$ and $\Delta_r G_m$, which is actually the slope of the tangent line to $G(n)$ at a particular $n$.

## 8.2 Free energy and equilibrium

For thermodynamically allowed processes at constant pressure and temperature, $\Delta_r G_m < 0$. This implies that the Gibbs free energy of a system held under these conditions will decrease until it hits a minimum. Referring to Figure 8.1, we could start to the left of the equilibrium, where $\Delta_r G_m < 0$, and move to increasing values of $n$ (more product), until we hit the minimum of $G$, where $\Delta_r G_m = 0$. If, on the other hand, the initial composition was such that we were to the right of the equilibrium, where $\Delta_r G_m > 0$, then the reverse reaction, consuming product, would be thermodynamically allowed, and the system would again move toward the minimum in $G$. At the minimum, $\Delta_r G_m = 0$, meaning that neither the forward nor the reverse process is thermodynamically favored. The system is then at **equilibrium**.

Equation (7.6) gives the dependence of $\Delta_r G_m$ on the reaction quotient $Q$. At equilibrium, $\Delta_r G_m = 0$ so that

$$\Delta_r G_m^{\circ} = -RT \ln K \qquad (8.2)$$

or

$$K = \exp(-\Delta_r G_m^{\circ}/RT) \qquad (8.3)$$

where $K$ is the value of $Q$ at equilibrium. Given that $Q$ is a ratio of activities of products and reactants, you will probably recognize $K$ as the **equilibrium constant** which you have encountered in previous chemistry courses.

There are a couple of things to note:

(1) The equilibrium constant is a ratio of activities. Activities are dimensionless, hence the equilibrium constant is dimensionless.
(2) $\Delta_r G_m^{\circ}$ depends on the standard state, so the value of the equilibrium constant also depends on the standard state. This may seem a little strange because the standard state is arbitrary. However, the equilibrium constant is constructed from activities which themselves depend on the standard state. The dependence of $\Delta_r G_m^{\circ}$ and of the activities on the standard state offset each other in such a way that any calculation of observable quantities from equilibrium relationships will yield identical results, regardless of the standard state chosen, provided it is consistently used.

**Example 8.1 Gas solubility**  For the process

$$O_{2(g)} \rightleftharpoons O_{2(aq)}$$

$$\Delta_r G_m^{\circ} = \Delta_f G^{\circ}(aq) - \Delta_f G^{\circ}(g)$$
$$= 16.35 - 0 \,\text{kJ mol}^{-1} = 16.35 \,\text{kJ mol}^{-1}$$

We can use this to compute the equilibrium constant for this process at $25\,^\circ$C:

$$K = \exp(-\Delta_r G_m^\circ / RT) = \exp\left(\frac{-16.35 \times 10^3 \, \text{J mol}^{-1}}{(8.314\,472\,\text{J K}^{-1}\,\text{mol}^{-1})(298.15\,\text{K})}\right)$$
$$= 1.37 \times 10^{-3}.$$

The equilibrium constant for this process corresponds to the ratio

$$K = a_{\text{aq}}/a_{\text{g}} = [O_2]/p_{O_2}$$

(give or take the standard concentration and standard pressure, both of which are 1 in appropriate units). This can be rearranged to the form

$$[O_2] = K p_{O_2},$$

which is **Henry's law** of gas solubility. Henry's law is valid for dilute solutions. Note that the choice for the activity of a solute made on page 149 is required to make ideal solution theory consistent with Henry's law. This is in fact why this choice was made.

The experimental value[1] of the Henry's law constant for oxygen is $1.26 \times 10^{-3}$. Because of the exponential function that appears in Equation (8.3), small errors in the thermodynamic data are amplified. The difference between the calculated and experimental equilibrium constants is therefore not terribly disturbing. In fact, if we work backwards and calculate $\Delta_f G^\circ(O_2, \text{aq})$ from the experimental Henry's law constant, we get $16.55\,\text{kJ mol}^{-1}$ which, considering the typical spread of values for measurements of this kind, is probably within the statistical uncertainty of this quantity.

The partial pressure of oxygen in air at sea level is about 21 kPa. The activity of oxygen would be $p_{O_2}/p^\circ = 21\,\text{kPa}/100\,\text{kPa} = 0.21$. Using our calculated Henry's law constant, we would calculate a dissolved oxygen concentration of

$$[O_2] = (1.37 \times 10^{-3})(0.21) = 2.9 \times 10^{-4}\,\text{mol L}^{-1}$$

at sea level.

**Example 8.2 Solvent vapor pressure** Suppose that we want to know the vapor pressure of a solution made by dissolving 0.2 mol of sodium chloride in 100 mL of water at $25\,^\circ$C. Since sodium chloride is involatile, the vapor pressure of the solution is entirely due to the liquid–vapor equilibrium of water:

$$H_2O_{(l)} \rightleftharpoons H_2O_{(g)}.$$

To determine the equilibrium constant for this process, we calculate $\Delta_r G_m^\circ$:

$$\Delta_r G_m^\circ = \Delta_f G_m^\circ(\text{g}) - \Delta_f G_m^\circ(\text{l})$$
$$= -228.582 - (-237.140)\,\text{kJ mol}^{-1} = 8.558\,\text{kJ mol}^{-1}$$
$$\therefore \quad K = \exp\left(\frac{-8.558 \times 10^3 \, \text{J mol}^{-1}}{(8.314\,472\,\text{J K}^{-1}\,\text{mol}^{-1})(298.15\,\text{K})}\right) = 0.03167.$$

[1] J. C. Kotz and P. Treichel, *Chemistry & Chemical Reactivity*, 3rd edn; Harcourt Brace: Fort Worth, 1996; p. 664.

The equilibrium constant for this process is $K = a_g/a_l = p_{H_2O}/p^\circ X_{H_2O}$, so we have $p_{H_2O} = Kp^\circ X_{H_2O}$, which is a form of **Raoult's law**: for the pure solvent, $X_{H_2O} = 1$, so if we call the vapor pressure of the pure solvent $p^\bullet_{H_2O}$, we get $p^\bullet_{H_2O} = Kp^\circ$. We can now use this equation to eliminate $K$ from our equation for the vapor pressure of the solution, and we get

$$p_{H_2O} = p^\bullet_{H_2O} X_{H_2O}$$

which now looks like Raoult's law as you may have seen it in your introductory chemistry course. Raoult's law is known to be valid for dilute solutions. Again, it was necessary to choose $a = X$ for solvents on page 149 in order to make the dilute solution theory agree with Raoult's law.

We now need to calculate the mole fraction of water in this problem: the mole density of water at 25 °C is 55.33 mol L$^{-1}$ so the number of moles of water is

$$n_{H_2O} = (55.33 \text{ mol L}^{-1})(0.100 \text{ L}) = 5.53 \text{ mol}.$$

Since NaCl dissociates into its ions in solution,

$$X_{H_2O} = \frac{n_{H_2O}}{n_{H_2O} + n_{Na^+} + n_{Cl^-}} = \frac{5.53}{5.53 + 2(0.2)} = 0.93.$$

Thus,

$$p_{H_2O} = (0.031\,67)(0.93) = 0.030 \text{ bar}.$$

**Example 8.3 The solubility product** Suppose that we want to calculate the **solubility product** of silver sulfide at 25 °C. This is the equilibrium constant for the reaction

$$Ag_2S_{(s)} \rightleftharpoons 2Ag^+_{(aq)} + S^{2-}_{(aq)}.$$

All we have to do is calculate $\Delta_r G^\circ_m$ and then $K_{sp}$:

$$\Delta_r G^\circ_m = 2\Delta_f G^\circ(Ag^+, aq) + \Delta_f G^\circ(S^{2-}, aq) - \Delta_f G^\circ(Ag_2S)$$
$$= 2(77.11) + 79 - (-40.7) \text{ kJ mol}^{-1} = 274 \text{ kJ mol}^{-1}$$

$$\therefore \quad K_{sp} = \exp\left(\frac{-274 \times 10^3 \text{ J mol}^{-1}}{(8.314\,472 \text{ J K}^{-1}\text{mol}^{-1})(298.15 \text{ K})}\right) = 1.0 \times 10^{-48}.$$

**Example 8.4 Standard free energies of formation from solubility measurements** The solubility of iron (III) hydroxide in water is $1.1 \times 10^{-15}$ mol L$^{-1}$ at 25 °C. If the standard free energy of formation of $Fe^{3+}_{(aq)}$ is $-4.6$ kJ mol$^{-1}$ and the free energy of formation of the hydroxide ion is $-157.220$ kJ mol$^{-1}$, what is the standard free energy of formation of solid $Fe(OH)_3$?

First note that the solubility product is the equilibrium constant for the reaction

$$Fe(OH)_{3(s)} \rightleftharpoons Fe^{3+}_{(aq)} + 3OH^-_{(aq)}.$$

We can use the solubility to compute the solubility product: at equilibrium, $a_{Fe^{3+}} = 1.1 \times 10^{-15}$. While it is tempting to say that $a_{OH^-} = 3a_{Fe^{3+}}$, we must consider the autoionization of water which produces $a_{OH^-} = 10^{-7}$ by itself. Since the amount of hydroxide generated by the latter process is much, much greater than that generated by the dissociation of iron (III) hydroxide, we need not consider the simultaneous equilibria, the autoionization of water being by far the more important process. Therefore $a_{OH^-} = 10^{-7}$, and

$$K_{sp} = (a_{Fe^{3+}})(a_{OH^-})^3 = 1.1 \times 10^{-36}.$$

The free energy change for the reaction is

$$\Delta_r G_m^\circ = -RT \ln K_{sp} = -(8.314\,472\,J\,K^{-1}mol^{-1})(298.15\,K) \ln(1.1 \times 10^{-36})$$
$$= 205\,kJ\,mol^{-1}.$$

However,

$$\Delta_r G_m^\circ = \Delta_f G^\circ(Fe^{3+}, aq) + 3\Delta_f G^\circ(OH^-, aq) - \Delta_f G^\circ(Fe(OH)_3)$$
$$\therefore \Delta_f G^\circ(Fe(OH)_3) = \Delta_f G^\circ(Fe^{3+}, aq) + 3\Delta_f G^\circ(OH^-, aq) - \Delta_r G_m^\circ$$
$$= -4.6 + 3(-157.220) - 205\,kJ\,mol^{-1} = -682\,kJ\,mol^{-1}.$$

Equilibrium problems often involve acids and bases. You will already be familiar with the pH scale, but you were likely told that $pH = -\log_{10}[H^+]$. This is wrong. First of all, taking the logarithm of a dimensioned quantity ($[H^+]$ has units of $mol\,L^{-1}$) is not mathematically sensible. [What are the units of $\log(mol\,L^{-1})$?] Second, pH is usually measured electrochemically (Chapter 10). Electrochemical measurements, as we will see, are equivalent to measuring the change in free energy, and the latter quantity is related to activities, and not directly to concentrations. The correct definition of pH is

$$pH = -\log_{10} a_{H^+}.$$

**Example 8.5 Acid dissociation constants** are also equilibrium constants. For instance, for acetic acid, $K_a$ is the equilibrium constant for the reaction

$$CH_3COOH_{(aq)} \rightleftharpoons CH_3COO^-_{(aq)} + H^+_{(aq)}.$$

$$\Delta_r G_m^\circ = \Delta_f G^\circ(CH_3COO^-, aq) - \Delta_f G^\circ(CH_3COOH, aq)$$
$$= -369.31 - (-396.39)\,kJ\,mol^{-1} = 27.08\,kJ\,mol^{-1}.$$

$$\therefore \quad K_a = \exp\left(\frac{-27.08 \times 10^3\,J/mol}{(8.314\,472\,J\,K^{-1}mol^{-1})(298.15\,K)}\right) = 1.80 \times 10^{-5}.$$

Given the dissociation constant, we can compute the pH of a solution of known formal concentration. If for instance we have a $0.20\,mol\,L^{-1}$ solution of acetic acid in water, we are saying that

$$a_{CH_3COOH} + a_{CH_3COO^-} = 0.20.$$

It is often the case in these problems that the autoionization of water is a negligible source of protons. If we make this assumption, which we can check later, the number of moles of protons in solution will be equal to the number of moles of $CH_3COO^-$ so

$$K_a = 1.80 \times 10^{-5} = \frac{a_{H^+}^2}{0.20 - a_{H^+}}.$$

After some rearrangement, this gives the quadratic equation

$$a_{H^+}^2 + 1.80 \times 10^{-5} a_{H^+} - 3.60 \times 10^{-6} = 0.$$

The physically reasonable solution of this quadratic equation is

$$a_{H^+} = 1.89 \times 10^{-3}.$$

Note that $a_{H^+} \gg 10^{-7}$, which confirms that we were justified in ignoring the autoionization of water. The pH is therefore

$$pH = -\log_{10} a_{H^+} = 2.72.$$

---

### Exercise group 8.1

(1) Calculate the equilibrium constant ($K_w$) at 298.15 K for the autoionization of water $H_2O \rightarrow H^+ + OH^-$.

(2) Predict the solubility of nickel (II) chloride in water at 25 °C. Comment on the magnitude of the value computed.

(3) Calculate the pH of a solution with a formal concentration of $2 \times 10^{-7} \, mol \, L^{-1}$ of the strong acid HI at 25 °C.

(4) A sample of 25 g of sodium acetate ($CH_3COONa$) is dissolved in 1 L of water at 25 °C. What is the pH of the solution? The base ionization constant ($K_b$) of acetate at this temperature is $5.6 \times 10^{-10}$.

(5) The ozone layer is approximately 20 km above sea level. At this altitude, the mean temperature is 210 K and the partial pressure of oxygen is 1.2 kPa.

(a) Accurately calculate $\Delta_r G_m^\circ$ for the reaction

$$2O_{3(g)} \rightarrow 3O_{2(g)}$$

at 210 K, taking into account the temperature dependence of $\Delta_r H_m^\circ$ and $\Delta_r S_m^\circ$.

(b) Calculate $\Delta_r G_m^\circ$ for the above reaction at 210 K under the assumption that neither $\Delta_r H^\circ$ nor $\Delta_r S^\circ$ vary significantly with temperature. Is the result significantly different?

(c) Calculate the equilibrium pressure of ozone under the conditions prevailing at an altitude of 20 km.

(d) The above calculation gives a value that is several orders of magnitude smaller than the actual partial pressure of ozone at 20 km. Why?

## 8.3 Catalysts and equilibrium

A catalyst is a substance that speeds up a chemical reaction without being consumed by it. Enzymes are biological molecules, usually proteins, with catalytic activity. In this section, we will focus on enzymes, but what we will say applies equally to other types of catalysts. Indeed, the mechanism we will study is essentially the same one used to understand catalysis at surfaces, give or take some differences of interpretation of the symbols.

The simplest enzyme reaction mechanism is the Michaelis–Menten mechanism:

$$\text{E} + \text{S} \overset{1}{\rightleftharpoons} \text{C} \overset{2}{\rightleftharpoons} \text{E} + \text{P}$$

where E is the enzyme, S is the substrate (reactant), C is the enzyme–substrate complex and P is the product.

Let's analyze this mechanism stepwise. Under arbitrary conditions,

$$\Delta_{r(1)}G_m = \Delta_{r(1)}G_m^\circ + RT \ln \frac{a_C}{a_E a_S}$$

with

$$\Delta_{r(1)}G_m^\circ = \Delta_f G^\circ(\text{C}) - \Delta_f G^\circ(\text{E}) - \Delta_f G^\circ(\text{S})$$

for the first step, and

$$\Delta_{r(2)}G_m = \Delta_{r(2)}G_m^\circ + RT \ln \frac{a_E a_P}{a_C}$$

with

$$\Delta_{r(2)}G_m^\circ = \Delta_f G^\circ(\text{E}) + \Delta_f G^\circ(\text{P}) - \Delta_f G^\circ(\text{C})$$

for the second. The total free energy change when one equivalent of S is converted to one equivalent of P by the enzyme is the sum of the free energy changes of the two steps:

$$\Delta_r G_m = \Delta_{r(1)}G_m + \Delta_{r(2)}G_m = \Delta_f G^\circ(\text{P}) - \Delta_f G^\circ(\text{S}) + RT \ln \frac{a_P}{a_S}.$$

Note that the thermodynamic properties of E and C drop out. Also, since $\Delta_f G^\circ(\text{P}) - \Delta_f G^\circ(\text{S})$ is $\Delta_r G_m^\circ$ for the reaction S → P, the equation for $\Delta_r G_m$ given above is just the free energy change for the overall reaction. Perhaps we shouldn't be surprised; since $G$ is a state function, the path taken doesn't matter. The net change in free energy must be the same whether the enzyme was involved or not.

We conclude the following:

---
**Catalysts do not affect the equilibrium ratio of products to reactants.**
---

In speeding up a reaction, a catalyst must speed up both directions by the same factor.

**Exercise group 8.2**

(1) What is the equilibrium ratio of $\alpha$- to $\beta$-lactose in the enzyme-catalyzed reaction

$$\text{enzyme} + \beta\text{-lactose} \rightleftharpoons \text{complex} \rightleftharpoons \text{enzyme} + \alpha\text{-lactose}$$

at 25 °C?

## 8.4 Coupled reactions

Living organisms sometimes find it necessary to alter the position of equilibrium. For instance, in order to maintain reasonable internal supplies of nutrients, they must pump them in from the outside against the normal tendency of diffusion to equalize concentrations. If enzymes can't accomplish these tasks, how are they achieved?

Suppose that a cell (or a chemical engineer) needs to make a particular compound, C, using the reaction

$$A + B \overset{1}{\rightleftharpoons} C + M$$

but that under the conditions normally prevailing in the cell (or in an industrial reactor), $\Delta_{r(1)}G > 0$. If, however, the cell also has a supply of another substance, say X, that reacts with M:

$$X + M \overset{2}{\rightleftharpoons} Y$$

then Le Chatelier's principle tells us this will shift equilibrium 1 to the right because reaction 2 removes M from the system. Approaching the matter from a different angle, we have produced a new overall reaction, namely

$$A + B + X \rightleftharpoons C + Y$$

with an overall free energy change $\Delta_r G = \Delta_{r(1)}G + \Delta_{r(2)}G$. If $\Delta_{r(2)}G < -\Delta_{r(1)}G$, then $\Delta_r G_m$ for the overall reaction is negative and the reaction can proceed to make the desired product C. Reactions 1 and 2 are said to be *coupled* by M.

Note that M doesn't appear in the eventual overall reaction. However, it is crucial that a species M exist that is a product of the unfavorable reaction and a reactant in the favorable reaction.

**Example 8.6 Displacement of oxygen from titanium** For the reaction

$$\text{TiO}_{2(s)} + 2\text{Cl}_{2(g)} \rightleftharpoons \text{TiCl}_{4(l)} + \text{O}_{2(g)}$$

$\Delta_r G_m^\circ = 152.3\,\text{kJ mol}^{-1}$. Titanium (IV) chloride is an important precursor of titanium metal so, in an industrial setting, we would want to maximize its yield. Since $\Delta_r G_m^\circ$ is large and positive, this reaction will have a tiny equilibrium constant:

$$K = \exp(-\Delta_r G_m^\circ/RT) = \exp\left(\frac{-152.3 \times 10^3 \text{ J mol}^{-1}}{(8.314\,472 \text{ J K}^{-1} \text{ mol}^{-1})(298.15 \text{ K})}\right)$$

$$= 2.08 \times 10^{-27}.$$

In order to make this reaction more favorable, we need to couple it to another reaction that will remove a product. The trick used in titanium refining is to couple this reaction to the oxidation of graphite:

$$C_{(s)} + O_{2(g)} \rightleftharpoons CO_{2(g)}$$

for which $\Delta_r G_m^\circ = -394.36$ kJ mol$^{-1}$. Thus, for the coupled reaction

$$TiO_{2(s)} + 2Cl_{2(g)} + C_{(s)} \rightleftharpoons TiCl_{4(s)} + CO_{2(g)}$$

$\Delta_r G_m^\circ = -242.1$ kJ mol$^{-1}$. The equilibrium constant for the coupled reaction is then

$$K = \exp(-\Delta_r G_m^\circ/RT) = \exp\left(\frac{242.1 \times 10^3 \text{ J mol}^{-1}}{(8.314\,472 \text{ J K}^{-1} \text{ mol}^{-1})(298.15 \text{ K})}\right)$$

$$= 2.55 \times 10^{42}.$$

Now that's a big number!

It is tempting to think that this would also work if we had a thermodynamically allowed reaction that made one of the reactants with a negative overall free energy change for the coupled process. For instance, we could try to add a reaction $Z \rightleftharpoons B$ to our original $A + B \rightleftharpoons C + M$. In ordinary chemical reactions, this doesn't work because the amount of B we can make is limited by stoichiometry, i.e. by the initial amount of Z. Typically, the equilibrium constant for an unfavorable reaction like reaction 1 will be extremely small, so making a limited amount of one of the reactants just won't do us much good. To put it another way, having a large equilibrium constant for the reaction $Z \rightleftharpoons B$ means that Z will be stoichiometrically converted to B. If the equilibrium constant for reaction 1 is very small, only a very small amount of the A and B present, whatever their origin, will be converted into products.

In biological systems, we do sometimes observe coupling through a reactant. Phosphorylation reactions are an important class of reactions where this is commonly seen:

$$X + P_i \rightarrow \text{phospho-X} + H_2O$$

These are often very unfavorable ($\Delta_r G^\circ \gg 0$) and are only made feasible by coupling them to ATP hydrolysis, which has a significantly negative free energy change under the conditions that prevail in a cell:

$$ATP + H_2O \rightarrow ADP + P_i, \qquad \Delta_r G^{\circ\prime} = -37.64 \text{ kJ mol}^{-1}. \qquad (8.4)$$

Note that water *cannot* be the coupling agent because the activity of water will be constant during these reactions due to its abundance. Phosphate might be the coupling agent, except that it's a reactant of the phosphorylation reaction, and we just argued that in general coupling reactions in this manner doesn't work. How then, are these reactions coupled? These reactions are always catalyzed by enzymes. A typical mechanism might be

$$E + ATP \rightarrow [E \cdot ATP]$$
$$[E \cdot ATP] + X \rightarrow [E \cdot ATP \cdot X]$$
$$[E \cdot ATP \cdot X] \rightarrow [E \cdot ADP \cdot phospho\text{-}X]$$
$$[E \cdot ADP \cdot phospho\text{-}X] \rightarrow [E \cdot ADP] + phospho\text{-}X$$
$$[E \cdot ADP] \rightarrow E + ADP$$

Many details will depend on the reaction and enzyme (e.g. order of binding, dissociation), but generally the mechanism is at least roughly as shown here. In some cases, an intermediate phosphorylated enzyme is involved. Whatever the exact mechanistic details, note that the enzyme actually catalyzes the combined reaction, i.e. transfer of a phosphate group to X from ATP, and not the two separate reactions involving $P_i$. In a case like this, thinking in terms of coupling two simpler reactions is a convenience since the "overall reaction" is the only reaction that actually happens.

**Example 8.7 Glutamine synthetase** In living cells, the amino acid glutamine [$CONH_2(CH_2)_2CHNH_2COOH$] is made from ammonium and glutamate [$^{-}CO_2(CH_2)_2$ $CHNH_2COOH$] by the following reaction

$$NH_4^+ + glutamate \rightarrow glutamine + H_2O, \qquad K_1' = 0.003. \qquad (8.5)$$

Note that $K_1'$ is the equilibrium constant for this reaction in the biochemists' standard state, i.e. at pH 7 and 25 °C.

Glutamine synthetase, the enzyme responsible for this reaction, has binding sites both for the reactants and for ATP. The reaction that occurs *in vivo* is as follows:[2]

$$E + ATP \rightarrow C_1$$
$$C_1 + glutamate \rightarrow C_2$$
$$C_2 + NH_4^+ \rightarrow C_3 \rightarrow C_4 + glutamine$$
$$C_4 \rightarrow E + ADP + P_i$$

The overall reaction is simply

$$ATP + NH_4^+ + glutamate \rightarrow ADP + P_i + glutamine. \qquad (8.6)$$

We can think of this reaction as being a result of coupling reaction (8.5) with the hydrolysis of ATP (Equation 8.4). As in the case of phosphorylation reactions, the enzyme itself couples the two reactions. Note in this case, however, that the reaction does not involve the transfer of a phosphate group to glutamate. Instead, the binding of ATP to the enzyme and the subsequent hydrolysis reaction change the conformation of the enzyme in ways that favor the synthesis of glutamine. A simple way of describing this is that when the

---

[2] This mechanism was adapted from H. S. Gill *et al.*, *Biochemistry* **41**, 9863 (2002), who studied a glutamine synthetase from *Mycobacterium tuberculosis*.

ATP "fires" it gives the other reaction a "kick" through the enzyme that pushes it toward the product state. The final state (ADP, phosphate and glutamine, in this case) has a lower overall free energy than the initial state (ATP, ammonium and glutamate), so once it's done, although the reverse reaction is possible, it isn't very likely.

A living cell exists out of equilibrium, so we can't say exactly what concentration of the product glutamine will be formed under *in vivo* conditions. We can however fix a theoretical maximum: to synthesize glutamine by Reaction (8.6), the free energy change $\Delta_r G$ cannot be positive or the reaction would not occur. If $\Delta_r G$ for the overall reaction is negative, there is in some sense some "left over" free energy which could be used to drive the reaction further toward products. This means that the maximum ratio of products to reactants will be reached when $\Delta_r G = 0$, i.e. when the reaction is at equilibrium.

The equilibrium constant for Reaction (8.5) is related to the activities of its reactants and products by

$$K_1' = 0.003 = \frac{(a_{\text{glutamine}})(a_{H_2O})}{(a_{\text{glutamate}})(a_{NH_4^+})}. \tag{8.7}$$

Similarly, for Equation (8.4),

$$K_2' = \exp(-\Delta_r G_m^{\circ\prime}/RT) = 3.93 \times 10^6 = \frac{(a_{\text{ADP}})(a_{P_i})}{(a_{\text{ATP}})(a_{H_2O})}. \tag{8.8}$$

Multiplying Equations (8.7) and (8.8) together, we get the equilibrium constant for the overall Reaction (8.6):

$$K = 1.18 \times 10^4 = \left. \frac{(a_{\text{glutamine}})(a_{\text{ADP}})(a_{P_i})}{(a_{\text{glutamate}})(a_{NH_4^+})(a_{\text{ATP}})} \right|_{\text{max}}.$$

Typically in living cells, $[\text{ATP}]/[\text{ADP}] \approx 10$, $[P_i] \approx 2 \, \text{mmol} \, L^{-1}$, and the ammonium concentration is around $30 \, \mu\text{mol} \, L^{-1}$. It follows that the maximum ratio of glutamine to glutamate is approximately

$$\left. \frac{(a_{\text{glutamine}})}{(a_{\text{glutamate}})} \right|_{\text{max}} \approx 1.18 \times 10^4 \frac{10(30 \times 10^{-6})}{2 \times 10^{-3}} = 1.8 \times 10^3.$$

---

### Exercise group 8.3

(1) Suppose that you want to make a compound C by the reaction

$$A \rightarrow B + C, \qquad\qquad \Delta_r G^\circ = 204 \, \text{kJ} \, \text{mol}^{-1} \tag{0}$$

Your lab also has the supplies necessary for the following reactions:

$$X + C \rightarrow P, \qquad\qquad \Delta_r G^\circ = -509 \, \text{kJ} \, \text{mol}^{-1} \tag{A}$$

$$Y + B \rightarrow Q, \qquad\qquad \Delta_r G^\circ = -50 \, \text{kJ} \, \text{mol}^{-1} \tag{B}$$

$$W + Z \rightarrow 2A, \qquad\qquad \Delta_r G^\circ = -550 \, \text{kJ} \, \text{mol}^{-1} \tag{C}$$

(a) Provide a brief argument (possibly using a simple calculation) showing that the original reaction (0) will not produce much C under reasonable experimental conditions.

(b) Which of the supplementary reactions (A–C) could be used to increase the yield of C relative to what you would get from reaction (0)? Outline your reasoning for each reaction in a few words.

(2) The first step in glycolysis is the formation of glucose-6-phosphate from glucose:

$$glucose + P_i \rightarrow glucose\text{-}6\text{-}phosphate + H_2O$$

(a) Calculate the equilibrium ratio at $25\,^\circ C$ of glucose-6-phosphate to glucose if the phosphate concentration is $2.5\,\mathrm{mmol\,L^{-1}}$.

| Species | $\Delta_f G^{\circ\prime}/\mathrm{kJ\,mol^{-1}}$ |
| --- | --- |
| glucose | $-436.42$ |
| glucose-6-phosphate | $-1325.00$ |
| $H_2O$ | $-157.28$ |
| $P_i$ | $-1058.56$ |

(b) If the phosphate ion is transferred from ATP, what is the equilibrium ratio of glucose-6-phosphate to glucose? The *in vivo* concentrations of ATP and ADP are, respectively, $5\,\mathrm{mmol\,L^{-1}}$ and $0.5\,\mathrm{mmol\,L^{-1}}$ and the standard free energy change of hydrolysis of ATP in the biochemical convention is $-37.64\,\mathrm{kJ\,mol^{-1}}$.

(3) A key reaction in the synthesis of fats in living organisms is

$$glycerol + ATP \rightarrow glycerol\text{-}1\text{-}phosphate + ADP.$$

The ratio of ATP to ADP is held fixed at a value of 10 by homeostatic mechanisms. The maximum ratio of glycerol-1-phosphate to glycerol observed is 770 at $25\,^\circ C$. $\Delta_r G^{\circ\prime}$ for ATP hydrolysis

$$ATP + H_2O \rightarrow ADP + P_i$$

is $-37.64\,\mathrm{kJ\,mol^{-1}}$ at $25\,^\circ C$. What is $\Delta_r G^{\circ\prime}$ for the phosphorylation of glycerol?

## 8.5 Active transport

Among the more biologically important examples of coupled reactions are active transport processes used to "pump" substrates and products in and out of cells or organelles. One of the most famous examples of active transport is the cotransport of sodium and potassium ions in red blood cells. Cells pump sodium out and potassium in, powering this process by hydrolyzing ATP. They do this for a number of reasons: they need potassium for certain internal processes; they also couple the spontaneous inward sodium flow to other transport systems.

Table 8.1 *Some data relevant to erythrocytes and their transport properties.*

| | |
|---|---|
| Temperature in the human body | 310 K |
| Physiological $\Delta_r G^\circ_{310}{}'$ for ATP hydrolysis[3] | $-32.55$ kJ mol$^{-1}$ |
| [ATP] in erythrocytes | 5 mmol L$^{-1}$ |
| [ADP] in erythrocytes | 0.5 mmol L$^{-1}$ |
| [$P_i$] in erythrocytes | 2.5 mmol L$^{-1}$ |
| [K$^+$] in plasma | 4 mmol L$^{-1}$ |
| [Na$^+$] in plasma | 145 mmol L$^{-1}$ |
| [K$^+$] inside erythrocytes | 140 mmol L$^{-1}$ |
| [Na$^+$] inside erythrocytes | 12 mmol L$^{-1}$ |

The ATP-powered cotransport of sodium and potassium is mediated by a transmembrane protein. This protein can best be thought of as an ATPase (an enzyme that hydrolyzes ATP) that undergoes a set of conformational changes that affect its ability to bind sodium and potassium, as well as the side of the membrane on which the bound ions are exposed. It is not clear exactly how this transporter functions, but it suffices (for our purposes, at least) to know that it exists. It provides the mechanism whereby ATP hydrolysis is coupled to the transport process.

Three sodium ions are transported out for every two potassium ions transported in, and one ATP molecule is hydrolyzed in the process. Additional data are provided in Table 8.1, for the special case of erythrocytes (red blood cells).

Let's work out the algebra for a general case. This will let us ask later whether the 3:2:1 ratio of sodium ions to potassium ions to ATP reflects some fundamental physical limitation or whether some other principle is at work. The overall process is

$$\nu_{Na}\,\mathrm{Na^+_{(in)}} + \nu_K\,\mathrm{K^+_{(out)}} + \mathrm{ATP} + \mathrm{H_2O} \rightarrow \nu_{Na}\,\mathrm{Na^+_{(out)}} + \nu_K\,\mathrm{K^+_{(in)}} + \mathrm{ADP} + \mathrm{P_i}$$

where $\nu_{Na}$ and $\nu_K$ are the numbers of each of the cations transported per molecule of ATP hydrolyzed. The standard free energy change for this process is just the standard free energy change for ATP hydrolysis because the standard free energies of formation of the ions cancel out. (Potassium is potassium, whether it's on one side of the membrane or the other.)

We can calculate $\Delta_r G_m$ under physiological conditions from the data in Table 8.1:

$$\Delta_r G_m = \Delta_r G_m^\circ + RT \ln \left( \frac{[\mathrm{Na^+_{(out)}}]^{\nu_{Na}} [\mathrm{K^+_{(in)}}]^{\nu_K} [\mathrm{ADP}][P_i]}{[\mathrm{Na^+_{(in)}}]^{\nu_{Na}} [\mathrm{K^+_{(out)}}]^{\nu_K} [\mathrm{ATP}]} \right) \tag{8.9}$$

$$= -32.55\,\mathrm{kJ\,mol^{-1}} + (8.314\,472 \times 10^{-3}\,\mathrm{kJ\,K^{-1}mol^{-1}})(310\,\mathrm{K}) \tag{8.10}$$

$$\times \ln \left( \frac{(145 \times 10^{-3})^3 (140 \times 10^{-3})^2 (0.5 \times 10^{-3})(2.5 \times 10^{-3})}{(12 \times 10^{-3})^3 (4 \times 10^{-3})^2 (5 \times 10^{-3})} \right)$$

$$= -16.33\,\mathrm{kJ\,mol^{-1}}$$

---

[3] Computed from data in Alberty and Goldberg, *Biochemistry* **31**, 10610 (1992). This value corresponds to physiological conditions, including appropriate values for the ionic strength and magnesium concentration, both of which have a significant effect on the thermodynamics of this reaction.

Table 8.2 *The ratio* $[K_{in}^+]/[K_{out}^+]$ *as a function of the stoichiometric coefficients of sodium and potassium in sodium/potassium active transport.*

|  |  | $\nu_{Na}$ | | | | |
|---|---|---|---|---|---|---|
|  |  | 2 | 3 | 4 | 5 | 6 |
| $\nu_K$ | 1 | 14 814 | 1226 | 101 | 8 | 0.7 |
|  | 2 |  | 35 | 10 | 3 | 0.8 |
|  | 3 |  |  | 5 | 2 | 0.9 |
|  | 4 |  |  |  | 2 | 0.9 |
|  | 5 |  |  |  |  | 0.9 |

The transporter uses only about half of the available free energy. (Compare $\Delta_r G_m$ to $\Delta_r G_m^\circ$.) There are a couple of reasons why the transporter can't use all of the available free energy:

- The transporter creates an electrical gradient, with more positive charge outside than inside the cell. Some of the free energy is doubtless used to combat this electrical gradient which, once established, opposes further export of sodium ions.
- Using more of the free energy would take us closer to equilibrium. However, at equilibrium, the ion transporter no longer has a preferred direction, i.e. it can run forwards and use ATP to pump the ions around, or it can run backwards and synthesize ATP as sodium and potassium ions move down their respective gradients. Staying away from equilibrium keeps the transporter biased in the forward direction.

Since there are good reasons why the transporter can't use all of the free energy made available by the hydrolysis of ATP, let's suppose that the difference $\Delta_r G_m - \Delta_r G_m^\circ = 16.22 \, \text{kJ mol}^{-1}$ is fixed. Unlike sodium ions, potassium ions can diffuse through the cell membrane, which is important to their role in polarizing the membrane (Section 10.5.2). We could then ask what ratio of $[K_{in}^+]/[K_{out}^+]$ could be reached for various values of the stoichiometric coefficients $\nu_{Na}$ and $\nu_K$ at fixed $[Na_{in}^+]/[Na_{out}^+]$. While these two concentration ratios are not independent, this calculation has an interesting lesson to teach us. Solving Equation (8.9) for the potassium ion ratio, we get

$$\frac{[K_{in}^+]}{[K_{out}^+]} = \left\{ \frac{[Na_{in}^+]^{\nu_{Na}}[ATP]}{[Na_{out}^+]^{\nu_{Na}}[ADP][P_i]} \exp\left( \frac{\Delta_r G_m - \Delta_r G_m^\circ}{RT} \right) \right\}^{1/\nu_K}.$$

Table 8.2 shows the results of these calculations. I only considered cases where $\nu_{Na} > \nu_K$ since this system exists in part to polarize the cell membrane, which requires that a different number of ions be pumped out than in. You can see from the results that the smaller the stoichiometric coefficients of the ions, the larger the ratio of potassium inside the cell to outside can be maintained. However, with smaller stoichiometric coefficients, more ATP is expended to move a given number of ions across the membrane. For $\nu_{Na} = 6$, it is no longer possible to maintain a higher concentration of potassium inside the cell than out. The 3:2

ratio may be the best compromise that can be reached between maintaining a high ratio of potassium inside to outside while economizing ATP.

**Example 8.8 Transport coupled to substrate phosphorylation** There are many variations on the theme of active transport. In some cases, a molecule is covalently modified during the transport process. For instance, in some organisms, glucose is actively transported into mitochondria while being phosphorylated by ATP. Glucose phosphorylation is the first step in glycolysis so two purposes are being served by a single process. The two processes coupled here are ATP hydrolysis and simultaneous glucose transport and phosphorylation:

$$\text{ATP} + \text{H}_2\text{O} \to \text{ADP} + \text{P}_i, \qquad\qquad \Delta_r G^{\circ\prime} = -37.64 \,\text{kJ mol}^{-1}$$
$$\text{glucose}_{(\text{out})} + \text{P}_i \to \text{glucose-6-phosphate}_{(\text{in})} + \text{H}_2\text{O}, \ \Delta_r G^{\circ\prime} = 21.16 \,\text{kJ mol}^{-1}$$

(All data are given for the usual biochemists' standard state at $25\,^\circ$C.) Suppose that we want to know the maximum ratio of glucose-6-phosphate inside the mitochondria to glucose outside at $25\,^\circ$C if the ATP to ADP ratio is 100.

The overall reaction is

$$\text{ATP} + \text{H}_2\text{O} + \text{glucose}_{(\text{out})} \to \text{ADP} + \text{glucose-6-phosphate}_{(\text{in})}$$

with $\Delta_r G^{\circ\prime} = -16.48 \,\text{kJ mol}^{-1}$. The maximum ratio is obtained when all of the free energy is used in transport. Then we have

$$Q = \frac{(a_{\text{ADP}})(a_{\text{G6P}})}{(a_{\text{ATP}})(a_{\text{gluc}})} = \exp\left(\frac{16.48 \times 10^3 \,\text{J mol}^{-1}}{(8.314\,472 \,\text{J K}^{-1}\text{mol}^{-1})(298.15 \,\text{K})}\right) = 771;$$

$$\therefore \ \frac{a_{\text{G6P}}}{a_{\text{gluc}}} = 771\frac{a_{\text{ATP}}}{a_{\text{ADP}}} = 771(100) = 7.71 \times 10^4.$$

---

### Exercise group 8.4

(1) Suppose that a glucose transporter in a cell is powered by ATP hydrolysis at $310\,\text{K}$. Two glucose molecules are transported into the cell for every ATP hydrolyzed. The ratio of ATP to ADP in the cell is 15 and the concentration of phosphate is $5 \,\text{mmol L}^{-1}$. If the extracellular glucose concentration is $3 \,\text{mmol L}^{-1}$, what is the maximum glucose concentration that can be obtained in the cell? Any additional data required can be taken from Table 8.1.

(2) The chemiosmotic theory of ATP synthesis suggests that mitochondria use the oxidation of carbohydrates and fats to operate a proton pump resulting in an internal pH of 7.0 when the pH of the surrounding cytoplasm is 5.5. When these protons spontaneously flow back in across the mitochondrial membrane, their free energy is harnessed to synthesize ATP. The concentrations of ATP, ADP and phosphates in the mitochondria are 1, 1 and 2.5 mM, respectively. For the ATP synthesis reaction $\text{ADP} + \text{P}_i \to \text{ATP} + \text{H}_2\text{O}$, $\Delta_r G^{\circ\prime} = 37.64 \,\text{kJ mol}^{-1}$ at $25\,^\circ$C.

 (a) How much free energy is required to synthesize 1 mol of ATP in a mitochondrion at $25\,^\circ$C?

(b) How much free energy is produced by moving 1 mol of protons into the mitochon-drion from the outside?

(c) What is the minimum number of protons that must cross the membrane to synthe-size one molecule of ATP?

*Note*: Report the smallest sufficiently large whole number.

## 8.6 Temperature and equilibrium

We now seek to work out the temperature dependence of the equilibrium constant of a reaction. First note that

$$\ln K = -\frac{\Delta_r G_m^\circ}{RT}.$$

Therefore, for two different temperatures $T_1$ and $T_2$, we have

$$\ln \frac{K_1}{K_2} = \ln K_1 - \ln K_2 = -\frac{\Delta_r G_{1,m}^\circ}{RT_1} + \frac{\Delta_r G_{2,m}^\circ}{RT_2}.$$

$\Delta_r H_m^\circ$ and $\Delta_r S_m^\circ$ vary much more slowly with temperature than $\Delta_r G_m^\circ$, so provided the temperatures $T_1$ and $T_2$ aren't too different, we have

$$\ln \frac{K_1}{K_2} = \frac{\Delta_r H_m^\circ - T_2 \Delta_r S_m^\circ}{RT_2} - \frac{\Delta_r H_m^\circ - T_1 \Delta_r S_m^\circ}{RT_1};$$

$$\therefore \ln \frac{K_1}{K_2} = \frac{\Delta_r H_m^\circ}{R} \left( \frac{1}{T_2} - \frac{1}{T_1} \right). \tag{8.11}$$

Note that this equation agrees with Le Chatelier's principle: if the reaction is exothermic ($\Delta_r H_m^\circ < 0$), then the equilibrium constant decreases as the temperature increases. Con-versely, if it is endothermic, the equilibrium constant increases with increasing temperature.

**Example 8.9 pH of neutral water at 37 °C**  The autoionization constant for water is the equilibrium constant for the process

$$H_2O_{(l)} \rightleftharpoons H^+_{(aq)} + OH^-_{(aq)}.$$

At 25 °C, $\Delta_r G_m^\circ$ for this reaction is 79.920 kJ mol$^{-1}$. The equilibrium constant is therefore

$$K_{298} = \exp \left( \frac{-79.920 \times 10^3 \text{ J mol}^{-1}}{(8.314\,472 \text{ J K}^{-1}\text{mol}^{-1})(298.15 \text{ K})} \right) = 9.97 \times 10^{-15}.$$

(Note that this equilibrium constant is normally called $K_w$. We omit the $w$ here to avoid excessive subscripting.) To adjust this equilibrium constant to (for instance) 37 °C, we also need the enthalpy of reaction, $\Delta_r H_m^\circ = 55.815$ kJ mol$^{-1}$. Therefore

$$\ln \left( \frac{K_{310}}{K_{298}} \right) = \frac{55.815 \times 10^3 \text{ J mol}^{-1}}{8.314\,472 \text{ J K}^{-1}\text{mol}^{-1}} \left( \frac{1}{298.15 \text{ K}} - \frac{1}{310.15 \text{ K}} \right) = 0.871;$$

$$\therefore K_{310} = (9.97 \times 10^{-15}) \exp(0.871) = 2.38 \times 10^{-14}.$$

This means that the pH of neutral water at body temperature (37 °C) is not 7 but 6.81.

**Example 8.10 Boiling point and altitude** The boiling point is the temperature at which the vapor pressure equals the atmospheric pressure; if the partial pressure were lower than the atmospheric pressure, bubbles couldn't form because the atmosphere would crush them out of existence. If the partial pressure were significantly higher, an explosion would result. (This can happen when a liquid is superheated.) Since the atmospheric pressure varies with altitude, it follows that the boiling temperature does too. Suppose that I wanted to calculate the boiling point of water where I live in Lethbridge, Canada. The average atmospheric pressure here, approximately 920 m above sea level, is 90 kPa or 0.90 bar.

Boiling involves the phase transition $l \to g$. For this process, $K = a_g/a_l$ but $a_l = X = 1$ for a pure solvent and $a_g = p/p^\circ$. If we express pressures in bars, $a_g = p$ (since $p^\circ = 1$ bar) so that $K = p$.

Now we run into a slight bit of awkwardness. While the thermodynamic standard pressure is 1 bar, the so-called normal boiling point is still quoted at 1 atm $= 1.013\,25$ bar. At precisely this pressure, water boils at $100\,^\circ$C. We therefore take $K_1 = 1.013\,25$ at $T_1 = 373.15\,$K in Equation (8.11). We want to calculate the temperature $T_2$ at which $K_2 = 0.90$. (Note that it doesn't matter whether we associate the normal boiling point with $T_1$ and $K_1$ or with $T_2$ and $K_2$, provided we are consistent.) The enthalpy of vaporization at the normal boiling point is $40.66\,$kJ mol$^{-1}$. Substituting all these quantities in Equation (8.11), we get

$$\ln\left(\frac{1.013\,25}{0.90}\right) = \frac{40\,660\,\text{J mol}^{-1}}{8.314\,472\,\text{J K}^{-1}\text{mol}^{-1}}\left(\frac{1}{T_2} - \frac{1}{373.15\,\text{K}}\right)$$

$$\therefore \frac{1}{T_2} = 2.704 \times 10^{-3}\,\text{K}^{-1}$$

$$\therefore T_2 = 369.81\,\text{K} \equiv 96.7\,^\circ\text{C}$$

The fact that water boils at a lower temperature at higher altitudes than at sea level has observable consequences. For instance, the cooking times given on packages of pasta are appropriate to sea-level cooking since most of the population of North America lives much closer to sea level than I do. Since chemical reactions, including those responsible for cooking, speed up with increasing temperature, the estimated cooking times given on packages of pasta are always too short for those of us who live more than a few hundred meters above sea level. This issue affects any recipes (or laboratory procedures) that involve boiling, but not other methods of cooking, although other pressure-related issues arise in baking, particularly of leavened breads.

**Example 8.11 Boiling-point elevation** You have probably studied freezing-point depression and boiling-point elevation in your previous courses in chemistry. In short, solutes tend to reduce the freezing point and to increase the boiling point. These phenomena can also be understood via Equation (8.11).

We don't cook pasta in pure water; we salt the water, which increases its boiling point. Does that make a significant difference to cooking times? A rough rule of thumb is that we want to make the cooking water roughly as salty as sea water, which contains about

0.5 mol L$^{-1}$ of salt. We will do the calculation for sea level, but we could easily do it for another altitude by adjusting $p$. The basic reasoning is exactly as in the last example, except that $a_l = X_l$. The mole density of water near 100 °C is 53.19 mol L$^{-1}$. The mole fraction of the water is therefore

$$X_{H_2O} = \frac{n_{H_2O}/V}{n_{H_2O}/V + [Na^+] + [Cl^-]} = \frac{53.19 \text{ mol L}^{-1}}{53.19 + 2(0.5 \text{ mol L}^{-1})} = 0.982$$

Our process is l $\rightarrow$ g with equilibrium constant $K = a_g/a_l = p/X_{H_2O}$. We know that pure water ($X_{H_2O} = 1$) boils at 100 °C at 1 atm, so we take $T_1 = 373.15$ K and $K_1 = 1.013\,25$. At sea level, $p = 1$ atm, so $K_2 = 1.013\,25/0.982 = 1.03$ at the unknown $T_2$. Proceeding as in the previous example, we have

$$\ln\left(\frac{1.013\,25}{1.03}\right) = \frac{40\,660 \text{ J mol}^{-1}}{8.314\,472 \text{ J K}^{-1}\text{mol}^{-1}}\left(\frac{1}{T_2} - \frac{1}{373.15 \text{ K}}\right)$$

$$\therefore \frac{1}{T_2} = 2.676 \times 10^{-3} \text{ K}^{-1}$$

$$\therefore T_2 = 373.68 \text{ K} \equiv 100.5 \text{ °C}$$

This certainly isn't enough of an increase in the boiling point to have much of an effect on cooking times. The real reason we add salt to water when we're cooking starches (pasta, potatoes, rice) is of course for flavor.

There is another way to use Equation (8.11). Instead of $(K_1, T_1)$, let's write $(K, T)$. Rearranging the equation, we get

$$\ln K = -\frac{\Delta_r H_m^\circ}{R}\frac{1}{T} + c \tag{8.12}$$

where $c$ is a constant that depends on $K_2$, $T_2$ and $\Delta_r H_m^\circ$. If we have values of the equilibrium constant at several different temperatures, we can therefore get $\Delta_r H_m^\circ$ from the slope of the graph of $\ln K$ vs. $T^{-1}$. Specifically, the slope is $-\Delta_r H_m^\circ/R$. This procedure is better than using pairs of equilibrium constants with Equation (8.11) and averaging at the end because it uses *all* the data at once.

The slope of a graph is best obtained by **linear regression**, sometimes also known as *least-squares fitting*. Fortunately, all modern scientific calculators, as well as spreadsheet programs, provide linear regression routines, so this is something we can easily do.

**Example 8.12 Partition coefficient and temperature** The partition coefficient is the equilibrium constant for the transfer of a solute between two immiscible solvents. For instance, the hexane–water partition coefficient of nicotine refers to the equilibrium

$$\text{nicotine}_{(aq)} \rightleftharpoons \text{nicotine}_{(hexane)}$$

Partition coefficients between a hydrophobic solvent like hexane and water are often used to assess the ease with which a solute is likely to penetrate a cell membrane, since the inside of a cell membrane is hydrophobic. The hexane–water partition coefficient of nicotine has

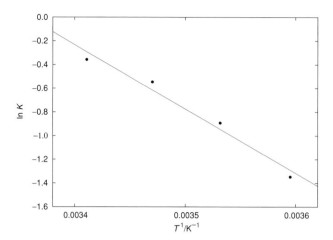

Figure 8.2 Temperature variation of the hexane–water partition coefficient of nicotine (Example 8.12).

been measured as a function of temperature:

| $T/^\circ C$ | 5 | 10 | 15 | 20 |
|---|---|---|---|---|
| $K$ | 0.26 | 0.41 | 0.58 | 0.70 |

We want to calculate both $\Delta_r H^\circ_m$ and $\Delta_r G^\circ_m$ for this process.

To get $\Delta_r H^\circ_m$, we plot $\ln K$ vs. $T^{-1}$, remembering to convert $T$ to kelvins before we generate the plot. The graph is shown in Figure 8.2. It's not exactly a great fit, although we don't have enough points to say for sure if this is just scatter or some kind of systematic experimental error. We will assume the former. The slope of the graph is $-5425$ K and its intercept is 18.22.

$$\text{slope} = -\Delta_r H^\circ_m / R.$$
$$\therefore \Delta_r H^\circ_m = -R(\text{slope})$$
$$= -(8.314\,472\,\text{J K}^{-1}\text{mol}^{-1})(-5425\,\text{K})$$
$$= 45.1\,\text{kJ mol}^{-1}.$$

What about $\Delta_r G^\circ_m$? We could calculate this from the partition coefficient at $25\,^\circ$C. However, $25\,^\circ$C is outside of the experimental range, so we can't just pick out the value of $K$ at this temperature from the data set. Rather, we use the equation of the line we just fit to estimate $K$ at this temperature. Note that even if we had data at $25\,^\circ$C, it might still be a good idea to use the equation of the line to estimate $K$ at this temperature. Why? Our data set displays significant scatter (Figure 8.2). That being the case, we could not be sure that the measurement at $25\,^\circ$C was all that accurate. The line represents our best fit of the entire data set, so our best estimate of $K$ at any given temperature comes from the equation of the line.

The equation of the line of best fit is

$$\ln K = (-5425 \text{ K})\frac{1}{T} + 18.22.$$

$$\therefore \ln K(25\,^\circ\text{C}) = \frac{-5425 \text{ K}}{298.15 \text{ K}} + 18.22 = 0.0185.$$

$$\therefore \Delta_r G^\circ_m = -RT \ln K$$

$$= -(8.314\,472 \text{ J K}^{-1}\text{mol}^{-1})(298.15 \text{ K})(0.0185)$$

$$= -46 \text{ J mol}^{-1}.$$

(Yes, that is J mol$^{-1}$, not kJ mol$^{-1}$ this time. It's a very small $\Delta_r G^\circ_m$.)

---

### Exercise group 8.5

(1) The solubility of salts generally goes up with temperature. What does this tell us about the thermodynamics of these processes?

(2) Calculate the vapor pressure of sodium metal at 50 °C.

(3) In chemical laboratories, solvents are often boiled away by reducing the pressure in the reaction vessel rather than by heating. This makes it possible to dry heat-sensitive compounds. Calculate the pressures at which the following solvents boil at 298 K:
   (a) water ($T_b = 373$ K, $\Delta_{vap} H^\circ = 45.02$ kJ mol$^{-1}$ at 298 K)
   (b) ethanol ($T_b = 352$ K, $\Delta_{vap} H^\circ = 40.48$ kJ mol$^{-1}$ at 298 K)
   (c) benzene ($T_b = 353$ K, $\Delta_{vap} H^\circ = 33.9$ kJ mol$^{-1}$ at 298 K)
   All other things being equal, which of these solvents would be easiest to remove by pumping?

(4) (a) Using the data in Appendix A.1, predict the normal boiling point of water.
   (b) What are the possible reasons that your answer might differ from the accepted value of 100 °C?

(5) Relate the formula

$$\ln\left(\frac{K_2}{K_1}\right) = \frac{\Delta_r H^\circ_m}{R}\left(\frac{1}{T_1} - \frac{1}{T_2}\right)$$

to Le Chatelier's principle. Consider both endothermic and exothermic reactions.

(6) (a) The equilibrium constant for dissolving hydrogen sulfide in water

$$H_2S_{(g)} \rightleftharpoons H_2S_{(aq)}$$

   is 0.087 at 298.15 K. What is the standard free energy of formation of an aqueous hydrogen sulfide molecule?
   (b) The slope of a graph of $\ln K$ vs. $T^{-1}$ for the above solubility process is 2100 K. What is the standard enthalpy of formation of aqueous hydrogen sulfide?
   (c) The standard entropy of gaseous $H_2S$ at 298.15 K is 205.77 J K$^{-1}$mol$^{-1}$. What is the standard entropy of aqueous $H_2S$?

(7) Calcium carbonate decomposes under certain conditions to calcium oxide and carbon dioxide.

(a) Is calcium carbonate stable with respect to calcium oxide and carbon dioxide at $25\,^\circ$C under normal atmospheric conditions? The pressure of carbon dioxide in normal air is $3.3 \times 10^{-4}$ bar.

(b) Calcium oxide is an enormously important industrial chemical used in steel production, in water treatment and in the pulp and paper industry, among other applications. It has traditionally been produced by heating calcium carbonate to a high temperature. Assuming that the partial pressure of carbon dioxide is $3.3 \times 10^{-4}$ bar, what is the minimum temperature (in degrees Celsius) at which the decomposition reaction becomes thermodynamically allowed?

*Hint:* Although the temperature you will calculate will seem high, an approximation that we have used for temperatures near standard still gives very good results.

(8) Carbonated drinks are made by forcing carbon dioxide into an aqueous solution under high pressure. Typical soft drinks are made at $4\,^\circ$C under a pressure of $120$ kPa of carbon dioxide. Once the carbon dioxide is dissolved in the aqueous solution, some of it reacts with water to form $H^+$ and $HCO_3^-$. (Carbonic acid itself is present in such small concentrations that it is undetectable in aqueous solutions.)

(a) Assuming there are no other acids present, calculate the pH of the solution formed under the conditions given above.

(b) Based on your calculations, why do you think that carbonated drinks are usually made at $4\,^\circ$C rather than at room temperature?

(9) Consider the following equilibrium vapor pressure data for benzene:

| $T/^\circ$C | 3.0 | 35.3 | 72.6 | 80.1 |
|---|---|---|---|---|
| $p/$mmHg | 30 | 150 | 600 | 760 |

Determine the enthalpy of vaporization of benzene from these data.

(10) Xanthine is an intermediate in purine metabolism.

It is only slightly soluble, and so tends to precipitate out of solution at higher concentrations, leading to urinary stones. (This is just one of the insoluble compounds that can be involved in the formation of stones.) The hydroxyl protons are ionizable so xanthine is a diprotic acid. Two processes are important in the solubilization of xanthine:

$$H_2Xan_{(s)} \overset{K_1}{\rightleftharpoons} H_2Xan_{(aq)}$$

$$H_2Xan_{(aq)} \overset{K_2}{\rightleftharpoons} H^+_{(aq)} + HXan^-_{(aq)}$$

(The second proton is too weakly acidic to be physiologically relevant.)

(a) The equilibrium constants have recently been measured.[4] Their values at 298.15 K are $K_1 = 1.29 \times 10^{-4}$ and $K_2 = 2.95 \times 10^{-8}$. What is the solubility of xanthine in aqueous solution at a fixed (buffered) pH of 6.5 (a typical pH for urine)? *Hint*: The solubility includes all forms of xanthine in solution.

(b) At 310.15 K, $K_1 = 2.04 \times 10^{-4}$. The standard enthalpy of formation of solid xanthine is $-379.6 \, kJ \, mol^{-1}$. Estimate the standard enthalpy of formation of aqueous xanthine.

(c) Estimate the standard free energy of formation of solid xanthine.

(d) Estimate the standard free energy of formation of aqueous xanthine ($H_2Xan_{(aq)}$).

### 8.6.1 Relative humidity and the dew point

Relative humidity is the fraction of the equilibrium vapor pressure of water in a sample of air. Thus, when we say that the relative humidity is 50%, we mean that the partial pressure of water is 50% of its equilibrium value. Since the equilibrium vapor pressure varies with temperature, the relative humidity does too. Accordingly, the relative humidity can change during the day even if the actual partial pressure of water in the atmosphere does not change.

**Example 8.13** Suppose that the relative humidity is 30% when the temperature is 25 °C on a still day. The equilibrium vapor pressure of water at this temperature is 0.031 67 bar (Example 8.2). Thus, the partial pressure of water is 0.30(0.031 67 bar) = 0.0095 bar. If the temperature drops overnight to 10 °C, the relative humidity will have increased because the equilibrium vapor pressure of water has decreased. We can calculate the equilibrium vapor pressure; the enthalpy of vaporization at 25 °C is 44.004 kJ mol$^{-1}$. Thus

$$\ln\left(\frac{K_{10}}{K_{25}}\right) = \frac{44\,004 \, J \, mol^{-1}}{8.314\,472 \, J \, K^{-1}mol^{-1}} \left(\frac{1}{298.15 \, K} - \frac{1}{283.15 \, K}\right) = -0.94;$$

$$\therefore K_{10} = 0.031\,67 \, \exp(-0.94) = 0.0124.$$

The equilibrium partial pressure is therefore 0.0124 bar at 10 °C. If the vapor pressure of water has not changed, the relative humidity is then

$$\text{rel. humidity} = \frac{0.0095}{0.0124} \times 100\% = 77\%.$$

---

[4] E. Königsberger *et al., J. Chem. Thermodyn.* **33**, 1 (2001).

It follows from these considerations that if the temperature goes on dropping, the relative humidity will eventually reach 100%. This is the **dew point**. If the temperature drops slightly below the dew point, water will tend to condense on surfaces that efficiently nucleate condensation. To calculate the dew point, we work out the temperature at which the current partial pressure of water in the air will equal the equilibrium vapor pressure.

**Example 8.14** Carrying on with our previous example, the dew point would be reached when $K = 0.0095$. Therefore

$$\ln\left(\frac{0.031\,67}{0.0095}\right) = \frac{44\,004\,\text{J mol}^{-1}}{8.314\,472\,\text{J K}^{-1}\text{mol}^{-1}}\left(\frac{1}{T_{\text{dew}}} - \frac{1}{298.15\,\text{K}}\right);$$

$$\therefore T_{\text{dew}} = 279.2\,\text{K} \equiv 6.1\,^\circ\text{C}.$$

---

### Exercise group 8.6

(1) Suppose that the partial pressure of water vapor in the air is $1.0 \times 10^{-2}$ bar. What is the dew point, in degrees Celsius?
(2) Ice fog occurs when the condensation of water vapor into ice is thermodynamically allowed. Ice crystals then grow on surfaces that are good nucleators. If the partial pressure of water is 0.004 bar, below what temperature would you predict ice fog to form? Report your final answer in degrees Celsius. The vapor pressure of ice at $0\,^\circ\text{C}$ is $6.025 \times 10^{-3}$ bar.

---

### Key ideas and equations

- $\Delta_r G_m^\circ = -RT \ln K$
- Important types of equilibria:
  - $g \rightleftharpoons$ aq: Henry's law
  - $l \rightleftharpoons g$: Raoult's law and vapor pressure/boiling
  - Ionic compound solubility, equilibrium constant $K_{\text{sp}}$
  - Acid dissociation, equilibrium constant $K_a$
  - Base ionization, equilibrium constant $K_b$
- $\text{pH} = -\log_{10} a_{\text{H}^+}$
- Catalysts do not affect the thermodynamics of a reaction, including the equilibrium ratio of products to reactants.
- Coupled reactions
- $\ln\left(\dfrac{K_1}{K_2}\right) = \dfrac{\Delta_r H_m^\circ}{R}\left(\dfrac{1}{T_2} - \dfrac{1}{T_1}\right)$

  This can be rewritten in the form $\ln K = -\dfrac{\Delta_r H_m^\circ}{R}\dfrac{1}{T} + C$, so $\Delta_r H_m^\circ$ can be recovered from the slope of a graph of $\ln K$ vs. $T^{-1}$.

## Review exercise group 8.7

(1) The solubility of gases in water generally decreases with increasing temperature. Explain this observation using thermodynamics.

(2) Suppose that you want to make compound C by the gas-phase reaction

$$A_{(g)} + B_{(g)} \rightarrow C_{(g)}$$

but the above reaction produces a very poor yield under normal conditions. How might you try to improve the yield of C? Describe two different methods.

(3) The maximum allowed occupational exposure to lead vapor is $0.05 \, \text{mg m}^{-3}$. Is it safe to keep a lead bar in a poorly ventilated laboratory at $25\,^{\circ}\text{C}$? The molar mass of lead is $207.2 \, \text{g mol}^{-1}$.

(4) What is the solubility product of iron(II) sulfate at 350 K?

(5) Suppose that the Earth had a much lighter atmosphere, generating a surface pressure of 0.2 bar. Could liquid water exist on Earth at this pressure?

(6) The solubility of cobalt (II) fluoride is $0.15 \, \text{mol L}^{-1}$ at $25\,^{\circ}\text{C}$. Estimate the free energy of formation of solid cobalt (II) fluoride.

(7) A $1.2 \times 10^{-4} \, \text{mol L}^{-1}$ solution of NaOH is prepared. What is the pH of this solution at $25\,^{\circ}\text{C}$ and at $60\,^{\circ}\text{C}$?

(8) The enthalpy of vaporization of toluene at its normal boiling point of $111\,^{\circ}\text{C}$ (at 1 atm) is $33 \, \text{kJ mol}^{-1}$. Calculate the boiling point of toluene at an atmospheric pressure of 0.90 bar.

(9) The acid dissociation constant of phosphoric acid is $7.5 \times 10^{-3}$ at $25\,^{\circ}\text{C}$. What is the standard free energy of formation of the aqueous $H_3PO_4$ molecule?

(10) Ribonuclease has two conformations, only one of which is active. At $50\,^{\circ}\text{C}$, the equilibrium concentrations of the active and inactive forms are, respectively, $9.97 \times 10^{-4}$ and $2.57 \times 10^{-6} \, \text{mol L}^{-1}$ in a solution with a total molarity of $1 \, \text{mmol L}^{-1}$. At $100\,^{\circ}\text{C}$, the equilibrium concentrations of the active and inactive forms are $8.6 \times 10^{-4}$ and $1.4 \times 10^{-4} \, \text{mol L}^{-1}$.

(a) What is $\Delta_r H^{\circ}$ for the process that converts the active form to its inactive conformation?

(b) What is the ratio of the active to the inactive form at $37\,^{\circ}\text{C}$?

(11) Hemoglobin Howick (HH) is a naturally occurring mutant form of hemoglobin. It normally exists in solution as a dimer (two copies of the protein associated together), but can also form a tetramer:

$$2 \, \text{dimer} \rightleftharpoons \text{tetramer}$$

The equilibrium constant for this reaction is $3.3 \times 10^5$ (standard state $= 1 \, \text{mol L}^{-1}$) for deoxygenated HH. If a solution of $100 \, \mu\text{mol L}^{-1}$ of HH monomers is made, what are the equilibrium concentrations of dimer and tetramer?

(12) The boiling point decreases at decreasing atmospheric pressure, but increases due to the presence of solutes. Calculate the approximate concentration of sodium chloride required to raise the boiling point of water held at a pressure of 0.90 bar back to 100 °C. What mass of salt would have to be dissolved in 1 L of water to reach this concentration?

*Note*: The answer does not depend on the enthalpy or free energy of vaporization of water. Try to answer this question without these pieces of data, if you can.

(13) In glycolysis, glucose is metabolized in steps to produce ATP and pyruvate. The reactions of glycolysis are as follows:

|  | | $\dfrac{\Delta_r G^{\circ\prime}}{\text{kJ mol}^{-1}}$ |
|---|---|---|
| (1) | D-glucose + ATP → D-glucose-6-phosphate + ADP | −16.7 |
| (2) | D-glucose-6-phosphate → D-fructose-6-phosphate | 1.7 |
| (3) | D-fructose-6-phosphate + ATP → | |
| | D-fructose-1,6-diphosphate + ADP | −14.2 |
| (4) | D-fructose-1,6-diphosphate → | |
| | dihydroxyacetone phosphate + glyceraldehyde-3-phosphate | 23.8 |
| (5) | dihydroxyacetone phosphate → glyceraldehyde-3-phosphate | 7.5 |
| (6) | glyceraldehyde-3-phosphate + $P_i$ + $NAD_{ox}$ → | |
| | 1,3-diphosphoglycerate + $NAD_{red}$ + $H^+$ | 6.3 |
| (7) | 1,3-diphosphoglycerate + ADP → 3-phosphoglycerate + ATP | −18.8 |
| (8) | 3-phosphoglycerate → 2-phosphoglycerate | 4.6 |
| (9) | 2-phosphoglycerate → 2-phosphoenolpyruvate + $H_2O$ | 1.7 |
| (10) | 2-phosphoenolpyruvate + ADP → pyruvate + ATP | −31.4 |

The pyruvate is further oxidized in the citric acid cycle.

(a) What is the overall reaction for glycolysis? Also calculate the biochemists' standard free energy change for the overall reaction.

(b) If this reaction were carried out in isolation (i.e. in the absence of other reactions involving the reactants and products, which is *not* the case in a living cell) in a solution in which [glucose] = 0.01, [pyruvate] = $6 \times 10^{-5}$ [$P_i$] = 0.002, all in mol L$^{-1}$, [$NAD_{ox}$]/[$NAD_{red}$] = 0.25, and the pH is 7, what is the maximum ratio of ATP to ADP that could be obtained at 25 °C?

(14) The octanol–water partition coefficient of a compound is the equilibrium constant for the process $A_{(aq)} \rightleftharpoons A_{(octanol)}$. This partition coefficient correlates reasonably well with the tendency of a substance to enter cell membranes and thus to bioaccumulate. Bleaching pulp for the production of paper produces a number of environmentally undesirable compounds, including chlorinated cymenes. Tetrachloro-*p*-cymene (TCPC, illustrated below) has a molar mass of 274.1 g mol$^{-1}$, a solubility

in water of $0.15\,\mathrm{g\,m^{-3}}$ and an octanol–water partition coefficient of $6.8 \times 10^6$ at $25\,^{\circ}\mathrm{C}$.[5]

The high solubility of TCPC in hydrophobic phases such as cell membranes poses a serious toxicological problem, but it also presents opportunities for removal of this substance from aqueous waste. If $10.0\,\mathrm{m^3}$ of an aqueous solution containing $0.98\,\mathrm{g}$ of TCPC is put in contact with $2.0\,\mathrm{L}$ of octanol, what mass of TCPC remains in the aqueous phase once the system has come to equilibrium?

*Note*: To get a reasonable answer for this problem, you'll have to keep a lot of digits (as many as possible) in intermediate steps of the calculation.

(15) It has recently been proposed that there are two different forms of the nitrate ion in aqueous solution, which have tentatively been named $\alpha$-nitrate and $\beta$-nitrate.[6] The equilibrium constant for the interconversion

$$\alpha\text{-NO}_{3(\mathrm{aq})}^{-} \rightleftharpoons \beta\text{-NO}_{3(\mathrm{aq})}^{-}$$

has been measured at several temperatures over the range $10–70\,^{\circ}\mathrm{C}$. In particular, $K = 0.122$ in dilute sodium nitrate solution at $25\,^{\circ}\mathrm{C}$, and $K = 0.132$ at $30\,^{\circ}\mathrm{C}$. In the discussion section of their paper, Simeon and coworkers argue that the reaction shown above involves the breaking of a hydrogen bond between nitrate and water, i.e. that the reaction is something along the lines of

$$[\mathrm{NO}_3(\mathrm{H}_2\mathrm{O})_{n+1}]^{-} \rightleftharpoons [\mathrm{NO}_3(\mathrm{H}_2\mathrm{O})_n]^{-} + \mathrm{H}_2\mathrm{O}.$$

Calculate the standard entropy change for the reaction. Is the result consistent with the breaking of a hydrogen bond as suggested by Simeon and coworkers?

(16) As the temperature of a kernel of popcorn increases, two things happen:

- The starch and protein that make up most of the kernel cook.
- The water heats up, increasing its vapor pressure.

The increase in the vapor pressure of water eventually causes the kernel to explode. The cooked starch and protein dehydrate rapidly, making the popped corn crispy and fluffy.[7] The explosive release of steam is what makes the popcorn jump, based on exactly the same physics as rocket propulsion. In essence, a popcorn kernel is a tiny little rocket!

[5] R. Lun *et al.*, *J. Chem. Eng. Data* **42**, 951 (1997).
[6] V. Simeon *et al.*, *Phys. Chem. Chem. Phys.* **5**, 2015 (2003).
[7] H. McGee, *On Food and Cooking*; Collier: New York, 1984, p 242. Incidentally, this book makes for excellent pleasure reading.

(a) It is known that popcorn has to reach a temperature of at least $200\,°C$ before it will pop. Estimate the yielding pressure of a kernel of popcorn, assuming that the water in the kernel behaves like pure water.

(b) Estimate the heat of vaporization of water at $200\,°C$.

(c) Suppose that a $15\,g$ serving of popcorn starts out at $20\,°C$. Popcorn is approximately $14\%$ water (by mass). How much heat is required to bring all the moisture in the $15\,g$ serving to $200\,°C$ and then to evaporate it completely?

(d) Unlike most reactions involving solids, popcorn expands spectacularly when it explodes. Suppose that you pop $20\,mL$ (initial volume) of a variety that expands by a factor of 40 when popped.[8] How much expansion work does the popcorn do if the atmospheric pressure is 1 atm?

(e) Suppose that you wanted to calculate $\Delta U$ for the process of popping corn. Are heat or work effects more important? What additional data would you need to obtain a reasonably accurate estimate of $\Delta U$ for this process?

(17) (a) The Henry's law constant for nitrogen in aqueous solution is $6.2 \times 10^{-4}\,mol\,L^{-1}bar^{-1}$ at $298.15\,K$. Determine the free energy of formation of the aqueous nitrogen molecule.

(b) The Henry's law constant ($k_H$) for aqueous nitrogen depends on temperature according to

$$\frac{d(\ln k_H)}{d(1/T)} = 1300\,K$$

Determine the enthalpy of formation of the aqueous nitrogen molecule.
*Hint*: Look at Equation (8.12).

(c) Calculate the Henry's law constant for nitrogen at $37\,°C$.

(d) Scuba divers can dive down to a depth of $40\,m$ using ordinary air. At this depth, the diver experiences a pressure of about $5\,bar$. Predict the solubility of nitrogen in an aqueous solution (e.g. blood) at this pressure and at $37\,°C$.

(e) The volume of blood in a typical male adult human is about $5.7\,L$. Suppose that a diver makes a rapid ascent from $40\,m$ to the surface. How much nitrogen comes out of solution? Assume that nitrogen behaves like an ideal gas and that the pressure at the surface is $1\,bar$, and express your answer in cubic centimeters.

(f) The sudden release of dissolved gases after a rapid ascent causes a syndrome known as decompression sickness. In extreme cases, decompression sickness can be fatal. Based on your calculation in the last part of this question, do you think that a rapid ascent from $40\,m$ could be dangerous? Explain briefly.

---

[8] This expansion factor is at the low end of typical values. Source: www.popcorn.org. (I'm not making this up.)

# 9

# Non-ideal behavior

We will start this chapter with a straightforward calculation, namely that of the solubility of lead (II) iodide in water at $20\,°C$ from the thermodynamic data in Appendix A.1. From the free energies of formation, we first calculate the standard free energy change for the reaction

$$PbI_{2(s)} \rightarrow Pb^{2+}_{(aq)} + 2I^-_{(aq)}$$

at $25\,°C$: $\Delta_r G^{\circ}_{298} = 45.89\,kJ\,mol^{-1}$. The equilibrium constant for this reaction at $25\,°C$ is therefore $K_{298} = 9.13 \times 10^{-9}$. To find the equilibrium constant at $20\,°C$, we use Equation (8.11). This requires $\Delta_r H^{\circ}_m$, which we calculate from the standard enthalpies of formation, yielding $\Delta_r H^{\circ}_m = 62.75\,kJ\,mol^{-1}$. Thus, $K_{293} = 5.93 \times 10^{-9}$. According to the ideal solution theory we have been using, this equilibrium constant is related to the concentrations by

$$K_{293} = \frac{[Pb^{2+}]}{c^{\circ}} \left(\frac{[I^-]}{c^{\circ}}\right)^2 .$$

Furthermore, if $s = [Pb^{2+}]$, then $[I^-] = 2s$, where $s$ is the solubility in $mol\,L^{-1}$. Therefore

$$K_{293} = 4 \left(\frac{s}{c^{\circ}}\right)^3 .$$

Solving this equation for $s$, we get $s_{293} = 1.14 \times 10^{-3}\,mol\,L^{-1}$ or, using the molar mass of $PbI_2$ of $461.0\,g\,mol^{-1}$, $s_{293} = 0.526\,g\,L^{-1}$. The experimental solubility at $20\,°C$ is $0.63\,g\,L^{-1}$. Our calculation therefore gives too low a solubility, missing the correct value by more than 15%. Why is this, if thermodynamics is supposed to be such an exact science? So far, we have assumed that all solutions behave ideally. They do not. We are going to need a way to treat systems with non-ideal components. In this chapter, we will deal with one particularly important case in detail, namely that of electrolyte solutions, i.e. solutions containing ions. Because they contain charged particles which attract and repel, electrolyte solutions tend to deviate from ideal behavior at much lower concentrations than solutions of uncharged molecules.

Before we do that, we need to develop some theory we can use to incorporate non-ideal substances into thermodynamics.

## 9.1  Activity coefficients

We know how the activity depends on pressure, concentration or mole fraction for ideally behaving substances and mixtures. All substances behave ideally when they are sufficiently dilute or, in the case of a solvent, when the solutes are sufficiently dilute. We therefore do not want to throw away what we have learned, but rather modify it for non-ideal systems. We do this by introducing an **activity coefficient** $\gamma_i$ for each component of the system, which will multiply the ideal-substance activity expression in our corrected equations for the activity. It is easiest to see what this means by example: suppose that component $i$ is a solute in a solution. We will now write the activity as

$$a_i = \gamma_i \frac{c_i}{c^\circ}.$$

We recognize the activity of an ideal solute, $c_i/c^\circ$, as a factor in this expression. Note that since $a_i$ is dimensionless and the ratio $c_i/c^\circ$ is also dimensionless, $\gamma_i$ must itself have no units. When $c_i$ is very small, the solution should behave ideally and thus $\gamma_i \to 1$. As the concentration increases and non-ideal effects become important, $\gamma_i$ will start to deviate from unity. Depending on the substance and on the concentration range, $\gamma_i$ may be greater than or less than unity, indicating, respectively, that the solute is either more or less active than it would be at the same concentration if it behaved ideally. The same solute may in fact have an activity coefficient that is less than unity in some solutions and greater than unity in others. In general, the activity coefficient depends not only on the concentration of the solute itself, but on the concentrations and identities of all other solutes in the solution and, for that matter, on the identity of the solvent. Thus, the activity coefficient of a potassium ion is different in a solution that only contains $1\,\mathrm{mol\,L^{-1}}$ of potassium chloride than it is in a solution that also contains $0.5\,\mathrm{mol\,L^{-1}}$ of sodium nitrate; the activity coefficient of a potassium ion in a $1\,\mathrm{mol\,L^{-1}}$ solution of potassium hydroxide in water would be different than the activity coefficient of the same ion in a $1\,\mathrm{mol\,L^{-1}}$ solution of KOH in ethanol, and so on.

We could make similar comments about the activity coefficients of other phases of matter. Table 9.1 shows how the activity is constructed for a variety of substances and the limits in which $\gamma_i$ approaches unit magnitude. Note that in order to get ideal behavior, we should also require other components (e.g. other solutes in solution) to be dilute; even at very low concentrations of a solute $i$, $\gamma_i$ may be very different from unity if there are other solutes present at significant concentrations.

Ideal substances are ones in which intermolecular forces between molecules, excluding interactions with the solvent, are negligible. With respect to the solvent, ideal behavior is obtained when the solutes are either present at extremely low concentrations such that they do not disrupt the overall solvent structure too much, or when the intermolecular forces between two solvent molecules are similar to the forces between solvent and solute molecules. The activity coefficient therefore measures the extent to which intermolecular forces deviate from those that would act in an extremely dilute solution or mixture.

Table 9.1 *Activities of some common substances.
The final column of this table gives the limits in
which we recover ideal behavior.*

| Substance type | $a_i$ | $\gamma_i \rightarrow 1$ when |
|---|---|---|
| Gas | $\gamma_i \dfrac{p_i}{p^\circ}$ | $\dfrac{p_i}{p^\circ} \rightarrow 0$ |
| Solvent | $\gamma_i X_i$ | $X_i \rightarrow 1$ |
| Solute | $\gamma_i \dfrac{c_i}{c^\circ}$ | $\dfrac{c_i}{c^\circ} \rightarrow 0$ |
| Pure solid | 1 | |

If we want to measure an activity coefficient, what we need to do is to find a measurable quantity that depends on this coefficient. For instance, suppose we want to measure the activity coefficient of a solvent as a function of the concentration of a particular solute. We can use the vapor pressure of the solvent to measure its activity coefficient. Vapor pressure is related to the equilibrium $1 \rightleftharpoons g$. For this process, at equilibrium, $K = a_g/a_1$. Non-ideal behavior of gases is typically only observable at very high pressures. In most situations of interest in chemistry, it is therefore an excellent approximation to treat all gases as ideal. Certainly, vapor pressures are almost always small enough for the vapor to behave ideally, so $\gamma_g = 1$. The equilibrium constant is known from experiments with the pure solvent: $K = p^\bullet/p^\circ$, where $p^\bullet$ is the vapor pressure of the pure solvent. Therefore

$$\frac{p^\bullet}{p^\circ} = \frac{p}{p^\circ \gamma_l X}$$

or

$$\gamma_1 = \frac{p}{p^\bullet X}, \tag{9.1}$$

where $p$ is the vapor pressure and $X$ is the mole fraction of the solvent. Measuring the vapor pressure as a function of mole fraction therefore tells us how the activity coefficient varies with $X$ for a certain type of solute. Note that Equation (9.1) involves a comparison between the measured vapor pressure $p$ and the vapor pressure we would predict from Raoult's law (page 161), i.e. from ideal solution theory.

**Example 9.1 Activity coefficient of water in a salt solution** An aqueous solution containing 1.641 mol kg$^{-1}$ LiI, 7.384 mol kg$^{-1}$ LiBr and 3.282 mol kg$^{-1}$ LiCl has a vapor pressure of 2.27 kPa at 318.36 K. The vapor pressure of pure water at this temperature is 9.566 kPa. We want to calculate the activity coefficient of water in this solution.

We haven't worked with molalities yet, although hopefully you have seen concentrations expressed in these units in your other courses. The molality is the number of moles of solute per kilogram of solvent. We're going to use Equation (9.1) to calculate the activity coefficient. We need the mole fraction of water. Imagine that we have a solution that contains exactly 1 kg of water. This solution would also contain 1.641 mol of LiI, 7.384 mol of LiBr and 3.282 mol of LiCl. The mole fraction of water would therefore be calculated as follows:

$$n_{H_2O} = \frac{1000\,g}{18.02\,g\,mol^{-1}} = 55.49\,mol$$

$$X_{H_2O} = \frac{n_{H_2O}}{n_{H_2O} + 2n_{LiI} + 2n_{LiBr} + 2n_{LiCl}}$$

$$= \frac{55.49\,mol}{55.49 + 2(1.641) + 2(7.384) + 2(3.282)\,mol} = 0.6927$$

$$\therefore \gamma_{H_2O} = \frac{p}{p^\bullet X} = \frac{2.27\,kPa}{(9.566\,kPa)(0.6927)}$$

$$= 0.343$$

This is a *very* non-ideal solution ($\gamma_{H_2O}$ is very different from unity).

---

### Exercise group 9.1

(1) The freezing point of a $130.0\,g\,kg^{-1}$ solution of ammonium chloride in water is $-9.47\,^\circ C$. What is the activity coefficient of water in this solution?

---

### 9.1.1 Physical interpretation of the activity coefficient

So far, an activity coefficient just measures the deviation from ideal behavior. It turns out however that the activity coefficient has a physical interpretation that is conceptually useful. To be specific, consider the case of a solute. The free energy of formation of a solute at concentration $c_i$ can be written

$$\Delta_f G_m = \Delta_f G_m^\circ + RT \ln a_i = \Delta_f G_m^\circ + RT \ln \left( \gamma_i \frac{c_i}{c^\circ} \right)$$

$$= \Delta_f G_m^\circ + RT \ln \left( \frac{c_i}{c^\circ} \right) + RT \ln \gamma_i.$$

Now recall that the free energy is the reversible non-$pV$ work. The free energy of formation would be the reversible work required to create the solute in a solution at concentration $c_i$ from the elements in their standard states. We can break down this work into three terms, corresponding to the three terms in the equation for $\Delta_f G_m$:

- $\Delta_f G_m^\circ$ is the work required to create the solute at standard concentration in an ideal solution from the elements.
- $RT \ln(c_i/c^\circ)$ is the work required to transfer the solute from an ideal solution at standard concentration to an ideal solution at concentration $c_i$. Recall that ideal solutions are defined by a lack of solute–solute forces.

- $RT \ln \gamma_i$ is the work done against solute–solute forces in the transfer to a real solution. This work will clearly depend on the concentrations of all solutes in the solution.

We could repeat this analysis for any of the states of matter. The important point to note is that $RT \ln \gamma_i$ has a clear physical interpretation as the extra work required to introduce a substance into a particular environment, resulting from the intermolecular forces acting, excluding interactions with a solvent which are already accounted for in the standard free energy of formation.

## 9.2 Electrolyte solutions

An electrolyte is a compound that releases ions in solution. This category includes both compounds that dissociate completely when they dissolve (strong electrolytes), such as ionic compounds, and those that dissociate only partially (weak electrolytes), such as weak acids. The reason that an electrolyte solution can't generally be treated as ideal is that the energy contains significant electrostatic repulsion and attraction terms. These terms contribute to the internal energy and therefore, following the chain of definitions, to the enthalpy and free energy functions. Ideal solution theory does not take solute–solute forces into account. These forces will become more and more significant as the solution becomes more concentrated. Also, since electrostatic potential energy depends on the product of the charges involved, the corrections will be larger for solutions involving highly charged ions.

If we place charged particles in a solution, we do not obtain a random distribution of solutes. Rather, negative charges tend to stay closer to positive charges than they do to other negatively charged solutes and vice versa. There are many molecules of solvent between two typical ions of opposite charge. However, if we ignore the solvent molecules, the following structure becomes apparent: we can think of each negative ion as being more or less surrounded by positive ions, each of which is also more or less surrounded by negative ions, and so on. Relative to any given ion, the solution has approximately the structure of an onion, with alternating shells of positive and negative charge. This arrangement of ions relative to any given ion is called an **ionic atmosphere**. Of course, due to diffusion, the precise positions of all the ions in solution are constantly changing, but as ions move, because of the electrostatic force, their ionic atmospheres tend to readjust. In a sense, an ion drags along its ionic atmosphere as it moves through the solvent. Diffusion does mix things up to some extent, and from time to time, two cations or two anions will meet, but if you could take snapshots of the environment of an ion in solution, you would find that its nearest neighbors were oppositely charged most of the time.

The existence of ionic atmospheres means that an electrolyte solution is more organized than a non-electrolyte solution of similar total solute concentration. This in turn implies that, at moderate concentrations, removing an ion from a solution (through, e.g., a chemical reaction) requires a rearrangement of the other ions to reform the ionic atmospheres of the remaining ions. The activity of an ion in a moderately concentrated solution, which is a measure of its availability to participate in reactions, must therefore be lower than the activity of an ideal solute. In other words, $\gamma_i \leq 1$ for ions in moderately concentrated solutions. In very concentrated solutions, like-charged ions are often close to each other, and their mutual repulsion may cause $\gamma_i$ to be greater than unity.

### 9.2.1 Mean ionic activity coefficients

There's one little problem: $\gamma_i$ is difficult to measure for an individual ion since we can't make a solution that contains an ion without a counterion. (There are electrochemical methods for measuring $\gamma_i$, but to this day, it remains controversial whether these methods really measure a single-ion activity coefficient.) Instead, we use a **mean ionic activity coefficient** $\gamma_\pm$ which is the geometric mean of the activity coefficients of the individual ions. Specifically, for an ionic compound $M_\alpha X_\beta$,

$$\gamma_\pm^{\alpha+\beta} = (\gamma_{M^{z+}})^\alpha (\gamma_{X^{z-}})^\beta \tag{9.2}$$

where $z_+$ and $z_-$ are the charges of the two ions. (If you have trouble deciphering the the formulas in this section, try writing down the equations for a specific compound like $MgCl_2$.) To see why we should use a geometric mean, consider the solubility equilibrium

$$M_\alpha X_{\beta(s)} \rightleftharpoons \alpha M_{(aq)}^{z+} + \beta X_{(aq)}^{z-}.$$

For this equilibrium,

$$K_{sp} = (a_{M^{z+}})^\alpha (a_{X^{z-}})^\beta$$

$$= \left(\gamma_{M^{z+}} \frac{[M^{z+}]}{c^\circ}\right)^\alpha \left(\gamma_{X^{z-}} \frac{[X^{z-}]}{c^\circ}\right)^\beta$$

$$= (\gamma_{M^{z+}})^\alpha (\gamma_{X^{z-}})^\beta \left(\frac{[M^{z+}]}{c^\circ}\right)^\alpha \left(\frac{[X^{z-}]}{c^\circ}\right)^\beta. \tag{9.3}$$

We would like the mean activity coefficient to be defined such that we can use it instead of $\gamma_{M^{z+}}$ or $\gamma_{X^{z-}}$. If we make this substitution, we get

$$K_{sp} = \gamma_\pm^\alpha \gamma_\pm^\beta \left(\frac{[M^{z+}]}{c^\circ}\right)^\alpha \left(\frac{[X^{z-}]}{c^\circ}\right)^\beta$$

$$= \gamma_\pm^{\alpha+\beta} \left(\frac{[M^{z+}]}{c^\circ}\right)^\alpha \left(\frac{[X^{z-}]}{c^\circ}\right)^\beta. \tag{9.4}$$

If we now compare Equations (9.3) and (9.4), we see that they imply Equation (9.2).

We rarely use the definition (9.2) directly. The important thing to remember is that we can use $\gamma_\pm$ instead of the activity coefficient of a particular ion in solution when we have a reaction involving an ion–counterion pair.

It is also important to note that $\gamma_\pm$ is a property of a particular *pair* of ions at a particular concentration. If we change either ion, the value of $\gamma_\pm$ may change. For example, $\gamma_\pm$ is different in a $1 \, mol \, L^{-1}$ solution of sodium fluoride ($\gamma_\pm \approx 0.57$) than it is in a $1 \, mol \, L^{-1}$ solution of sodium iodide (0.74).

### 9.2.2 Ionic strength

We need another definition before we carry on. As you can imagine from the discussion so far, because highly charged ions are subject to stronger intermolecular forces than ions with smaller charges, non-ideal effects will start to show up at lower concentrations for the former than for the latter. Suppose, for example, that we have equimolar solutions of NaCl and $MgSO_4$. In the first solution, we have only singly charged ions so the potential energy due to ion–ion attractions and repulsions is proportional to the square of the elementary charge. On the other hand, in the second solution, the ions are doubly charged so the electrostatic energy terms are proportional to four times the square of the elementary charge. We therefore expect the second solution to deviate much more significantly from ideality than the first. The **ionic strength** is a concentration measure that corrects for the fact that ions carrying larger charges experience larger intermolecular forces. It is defined as

$$I_c = \frac{1}{2} \sum_i z_i^2 c_i \tag{9.5}$$

where $z_i$ is the charge on ion $i$ and $c_i$ is its molarity. The sum is taken over *all* ions in a solution, including any spectator ions. Spectator ions are often added to solutions to control the ionic strength since this turns out to be an important variable to hold constant in many experiments.

**Example 9.2** The ionic strength of a $1 \, mol \, L^{-1}$ solution of sodium sulfate is

$$
\begin{aligned}
I_c &= \frac{1}{2} \left( z_{Na^+}^2 [Na^+] + z_{SO_4^{2-}}^2 [SO_4^{2-}] \right) \\
&= \frac{1}{2} \left[ (1)^2 (2 \, mol \, L^{-1}) + (-2)^2 (1 \, mol \, L^{-1}) \right] = 3 \, mol \, L^{-1}
\end{aligned}
$$

### 9.2.3 Debye–Hückel theory

We will study here a theory for the activity coefficients of ions in dilute solutions known as the **Debye–Hückel limiting law**. This theory appears as the low-ionic-strength limit of a more general theory which is, however, not that much more accurate than the limiting law outside of the range where the limiting law applies. We will therefore focus on the limiting law, although I will comment on more accurate theories later.

We won't derive the Debye–Hückel limiting law. The derivation is long and, frankly, there are only one or two points in it that are really enlightening. The key ideas behind Debye–Hückel theory and its intellectual descendents can, however, be summarized. Recall that $RT \ln \gamma_i$ is the work required to introduce a solute into a particular solution resulting from solute–solute forces. In the case of an electrolyte solution, the forces involved are electrostatic forces. The technique used to calculate the work due to electrostatic forces is to imagine that we initially create an uncharged solute, and then gradually charge it up. In principle, we should consider the force between the ion we are adding and all the other ions in solution. To do that, we need to know the distribution of the ions in solution. Debye and Hückel assumed that the electrostatic potential energies of the ions followed a Boltzmann distribution (Section 3.2). This introduces an exponential fall-off $\exp(-r/r_D)$ in the effective potential of an ion, which can be understood as being due to the ionic atmosphere. To picture this, imagine two ions, A and B, that are separated by several layers of ionic atmosphere. The ions in the ionic atmospheres of ions A and B also exert forces on these two ions. In the overall balance of forces, this diminishes the importance of the term due to the specific force between ions A and B. The parameter $r_D$ is called the Debye screening length. Among other things, the screening length is inversely proportional to the square root of the ionic strength. In solutions with a large ionic strength, $r_D$ is small, and the effective potential of an ion falls off rapidly with distance. When we calculate the work required to charge up an ion in such an environment, we find that this work is inversely proportional to $r_D$, and thus directly proportional to $\sqrt{I_c}$. This leads to the Debye–Hückel limiting law:

$$\ln \gamma_i = -\frac{z_i^2 e^3}{8\pi} (\varepsilon k_B T)^{-3/2} \sqrt{2000 L I_c}$$

where $z_i$ is the charge of the ion (in elementary charge units), $e$ is the elementary charge (in coulombs), $\varepsilon$ is the permittivity of the solvent, $k_B$ is Boltzmann's constant, $L$ is Avogadro's number and $I_c$ is the ionic strength in $mol\,L^{-1}$.

The permittivity is a constant source of confusion because of conflicting terminology and notation in the literature. The symbol $\varepsilon$, which I use here for the permittivity itself, is sometimes used for the relative permittivity, which is in turn is also known as the dielectric constant. IUPAC recommends $\varepsilon_r$ for the relative permittivity, although $\kappa$ has often been used in the past. The relationship between the permittivity $\varepsilon$ and the relative permittivity $\varepsilon_r$ is $\varepsilon = \varepsilon_r \varepsilon_0$, where $\varepsilon_0$ is the permittivity of the vacuum. The easiest way to tell the permittivity and relative permittivity apart is that the former has units of $C^2 N^{-1} m^{-2}$ or $C^2 J^{-1} m^{-1}$, while the latter is dimensionless.

Many of the terms appearing in the Debye–Hückel equation are constants. We can combine them to obtain the simpler expression

$$\ln \gamma_i = -A z_i^2 (\varepsilon T)^{-3/2} \sqrt{I_c} \tag{9.6}$$

where $A = 1.107 \times 10^{-10}$. The constant $A$ has strange units which I won't write down here. You just have to be careful to use the correct units for everything else in this equation, and $\ln \gamma_i$ will come out as a dimensionless quantity. Also note that the permittivity is strongly temperature dependent, which is why it is better to leave it as an explicit constant in this equation.

At zero ionic strength, $\ln \gamma_i = 0$, which implies that $\gamma_i = 1$. This is, as expected, the ideal solution limit. At non-zero ionic strength, $\ln \gamma_i$ is negative, implying that $\gamma_i < 1$. This is in accord with our earlier argument that ions should generally be less available than ideal solutes in the concentration range where the Debye–Hückel limiting law applies.

Debye–Hückel theory has been experimentally confirmed to work extremely well for very dilute solutions. However, significant deviations are observed when the ionic strength is greater than about $0.01 \, \text{mol} \, \text{L}^{-1}$. Thus, it gives us a good first approximation for dilute solutions, but it cannot be applied in high-ionic-strength solutions.

Equation (9.6) will be useful when we are studying reactions in which there are combinations of ions that cannot be thought of as a binary compound. In the opposite case, when ions appear in combinations that do correspond to a binary compound ($H^+$ and $F^-$, $Pb^{2+}$ and $2I^-$ etc.), we prefer to use mean ionic activity coefficients for simplicity. We previously worked out an equation for the mean ionic activity coefficient of the general ionic compound $M_\alpha X_\beta$. Taking a natural logarithm of Equation (9.2), we get

$$(\alpha + \beta) \ln \gamma_\pm = \alpha \ln (\gamma_{M^{z+}}) + \beta \ln (\gamma_{X^{z-}}).$$

If we now use Equation (9.6) for the activity coefficients of each ion and collect like terms, we get

$$(\alpha + \beta) \ln \gamma_\pm = -A(\varepsilon T)^{-3/2} \sqrt{I_c} \left( \alpha z_+^2 + \beta z_-^2 \right). \tag{9.7}$$

Since we must have charge neutrality,

$$\alpha z_+ + \beta z_- = 0.$$

If we use this equation to eliminate (e.g.) $\beta$ from Equation (9.7), we get the following, after a lot of rearrangements and simplifications:

$$\ln \gamma_\pm = A z_+ z_- (\varepsilon T)^{-3/2} \sqrt{I_c}. \tag{9.8}$$

Because the ions in an ion–counterion pair are oppositely charged, we see that $\ln \gamma_\pm$ is negative, as we expect.

In the following example, we calculate the mean ionic activity coefficient of an ionic solute.

**Example 9.3** For a solution of $5 \, \text{mmol} \, \text{L}^{-1} \, Fe(NO_3)_2$ in water at $25 \, ^\circ\text{C}$,

$$I_c = \frac{1}{2} \left( (2)^2(5) + (-1)^2(10 \, \text{mmol} \, \text{L}^{-1}) \right) = 15 \, \text{mmol} \, \text{L}^{-1}.$$

This is just beyond the range of applicability of Debye–Hückel theory, so the mean activity coefficient obtained will only be approximately correct. At $25 \, ^\circ\text{C}$, the relative permittivity of water is 78.37, so the permittivity is $\varepsilon = \varepsilon_r \varepsilon_0 = 78.37(8.854 \, 187 \, 817 \times 10^{-12} \, \text{C}^2 \text{J}^{-1} \text{m}^{-1}) = 6.939 \times 10^{-10} \, \text{C}^2 \text{J}^{-1} \text{m}^{-1}$.

$$\ln \gamma_\pm = (1.107 \times 10^{-10})(2)(-1) \left[ (6.939 \times 10^{-10} \, \text{C}^2 \text{N}^{-1} \text{m}^{-2})(298.15 \, \text{K}) \right]^{-3/2}$$
$$\times \sqrt{15 \times 10^{-3} \, \text{mol} \, \text{L}^{-1}}$$
$$= -0.288;$$
$$\therefore \gamma_\pm = 0.750.$$

Once we know how to calculate ionic activity coefficients, we can calculate quantities that depend on the activity of an ion, as the following examples demonstrate.

**Example 9.4** The solubility of barium sulfate in water at $25\,^{\circ}\mathrm{C}$ is $1.05 \times 10^{-5}\,\mathrm{mol\,L^{-1}}$. Solubility refers to the reaction

$$\mathrm{BaSO_{4(s)}} \rightleftharpoons \mathrm{Ba^{2+}_{(aq)}} + \mathrm{SO^{2-}_{4(aq)}}.$$

This means that the ionic strength is

$$I_c = \frac{1}{2}[2^2(1.05 \times 10^{-5}) + (-2)^2(1.05 \times 10^{-5})] = 4.20 \times 10^{-5}\,\mathrm{mol\,L^{-1}}.$$

We can now compute $\gamma_{\pm}$ from the ionic strength:

$$\ln \gamma_{\pm} = (1.107 \times 10^{-10})(2)(-2)\left[(6.939 \times 10^{-10}\,\mathrm{C^2 N^{-1} m^{-2}})(298.15\,\mathrm{K})\right]^{-3/2}$$
$$\times \sqrt{4.20 \times 10^{-5}\,\mathrm{mol\,L^{-1}}}$$
$$= -0.0305;$$
$$\therefore \gamma_{\pm} = 0.970.$$

Knowing this, we can compute $K_{\mathrm{sp}}$:

$$K_{\mathrm{sp}} = \gamma_{\pm}^2 s^2 = (0.970)^2(1.05 \times 10^{-5})^2 = 1.04 \times 10^{-10}.$$

This in turn allows us to compute the standard free energy change for this reaction:

$$\Delta_r G_m^{\circ} = -RT \ln K_{\mathrm{sp}} = 57.0\,\mathrm{kJ\,mol^{-1}}.$$

If, in addition, we know the standard free energies of formation of the aqueous barium and sulfate ions, we can calculate the standard free energy of formation of solid barium sulfate:

$$\Delta_r G_m^{\circ} = \Delta_f G^{\circ}(\mathrm{Ba^{2+}}, \mathrm{aq}) + \Delta_f G^{\circ}(\mathrm{SO_4}^{2-}, \mathrm{aq}) - \Delta_f G^{\circ}(\mathrm{BaSO_4})$$
$$\therefore \Delta_f G^{\circ}(\mathrm{BaSO_4}) = \Delta_f G^{\circ}(\mathrm{Ba^{2+}}, \mathrm{aq}) + \Delta_f G^{\circ}(\mathrm{SO_4}^{2-}, \mathrm{aq}) - \Delta_r G_m^{\circ}$$
$$= -560.7 + (-743.6) - 57.0\,\mathrm{kJ/mol} = -1361.3\,\mathrm{kJ\,mol^{-1}}$$

This number is within $0.7\,\mathrm{kJ\,mol^{-1}}$ of the accepted value.

The fact that $\gamma_i$ (or $\gamma_{\pm}$) decreases with increasing ionic strength, at least for small to moderate ionic strengths, is responsible for the **salting in** effect: the solubility of ionic compounds tends to increase with increasing ionic strength, provided the ionic strength is not too high.

**Example 9.5** Suppose that we want to do an accurate calculation of the pH of a $0.001\,\mathrm{mol\,L^{-1}}$ sodium hydroxide solution at $298\,\mathrm{K}$. The ionic strength of this solution is

$$I_c = \frac{1}{2}\left[(-1)^2(0.001) + (1)^2(0.001)\right] = 0.001\,\mathrm{mol\,L^{-1}}.$$

The mean activity coefficient of the hydroxide ions is calculated by

$$\ln \gamma_\pm = (1.107 \times 10^{-10})(1)(-1)\left[(6.939 \times 10^{-10}\, C^2 N^{-1} m^{-2})(298\, K)\right]^{-3/2}$$
$$\times \sqrt{0.001\, mol\, L^{-1}}$$
$$= -0.0372;$$
$$\therefore \gamma_\pm = 0.963.$$

The activity of the hydroxide ions is therefore

$$a_{OH^-} = \gamma_\pm c_{OH^-}/c^\circ = 9.63 \times 10^{-4}.$$

Using $K_w = 10^{-14} = a_{H^+} a_{OH^-}$ we get $a_{H^+} = 1.038 \times 10^{-11}$ or a pH of 10.98. Because the pH scale is a logarithmic scale, the difference in pH from what we would calculate using ideal solution theory is not very large in this case.

---

## Exercise group 9.2

(1) Taking non-ideal effects into account, what is the pH of a $0.0058\, mol\, L^{-1}$ aqueous solution of HCl at $25\,^\circ C$?

(2) The solubility of barium fluoride in water at $25\,^\circ C$ is $6.8 \times 10^{-3}\, mol\, L^{-1}$. Using Debye–Hückel theory, calculate the solubility product of this compound.

(3) Uric acid is a diprotic acid. The first acid dissociation constant of uric acid has been measured as a function of temperature in a solution to which NaCl had been added to maintain a constant ionic strength of $0.15\, mol\, L^{-1}$:[1]

| $T/^\circ C$ | 25 | 32 | 37 | 42 |
|---|---|---|---|---|
| $K_a/10^{-6}$ | 5.50 | 6.17 | 6.46 | 7.41 |

The calculations used by the authors of this paper to obtain the equilibrium constants given above assumed ideal behavior. Note that the dissociation of the second proton is not relevant to us here.

(a) Activity coefficients of ions vary weakly with temperature. Using this piece of information, explain why we should be able to get $\Delta_r H_m^\circ$ accurately from these data without considering corrections for non-ideal behavior.
    *Hint*: Think about using just two of the above points. What formula would you use?

(b) Calculate $\Delta_r H_m^\circ$ using *all* the data.

(c) Once we get outside of the range of applicability of the Debye–Hückel limiting law, the mean ionic activity coefficients depend mostly on the charges of the ions and on their sizes. The size dependence is a little less strong than the charge dependence, although we always have to be careful with the hydrogen ion. Tables of mean ionic

---

[1] Z. Wang and E. Königsberger, *Thermochim. Acta* **310**, 237 (1998).

activity coefficients are available for many common acids, for instance in the *CRC Handbook of Chemistry and Physics*. We can approximate the mean ionic activity coefficient of dissociated uric acid (i.e. of an $H^+$/hydrogen urate ion pair) using the mean ionic activity coefficient of $HNO_3$. The nitrate ion is a little smaller than the hydrogen urate ion, but the two are planar and have similar hydrogen-bonding patterns in water. The mean activity coefficient of nitric acid has been measured and has the value $\gamma_\pm = 0.773$ at an ionic strength of $0.15\,mol\,L^{-1}$ at $25\,°C$. Assuming that uric acid itself behaves ideally, correct the measured value of $K_a$ at $25\,°C$ for the non-ideal behavior of $H^+$ and of the hydrogen urate ion.

(d)  Calculate $\Delta_r G_m^\circ$ at $25\,°C$ for the acid dissociation equilibrium.

(e)  What value of $\Delta_r G_m^\circ$ would you have calculated if we had not taken the non-ideal behavior of uric acid into account?

---

## 9.3 The solubility of ionic compounds in aqueous solution

The method for calculating the solubility of ionic compounds is best described by example. Suppose that we want to calculate the solubility of $PbI_2$ in water at 293 K, just as we set out to do in the opening of this chapter. The reaction of interest is

$$PbI_{2(s)} \rightarrow Pb^{2+}_{(aq)} + 2I^-_{(aq)}.$$

The equilibrium constant for this reaction is

$$K_{sp} = a_{Pb^{2+}}(a_{I^-})^2 = \left(\gamma_\pm \frac{[Pb^{2+}]}{c^\circ}\right)\left(\gamma_\pm \frac{[I^-]}{c^\circ}\right)^2 = \gamma_\pm^3 \left(\frac{[Pb^{2+}]}{c^\circ}\right)\left(\frac{[I^-]}{c^\circ}\right)^2$$

$$= 4\gamma_\pm^3 \left(\frac{s}{c^\circ}\right)^3. \tag{9.9}$$

To get this equation, we used $[Pb^{2+}] = s$ and $[I^-] = 2s$, where $s$ is the solubility of the salt. We calculated $K_{sp}$ at $20\,°C$ on page 185 and found $K_{sp} = 5.93 \times 10^{-9}$.

If we want to calculate the solubility, we need to know $\gamma_\pm$. However, this depends on the ionic strength, which depends on the concentrations of lead and iodide ions, which in turn depends on the solubility. Your first impression might be that we're stuck. These equations can be solved by iteration, which is to say that you take a guess at the value of $\gamma_\pm$ (e.g. $\gamma_\pm = 1$), then use that to calculate $s$, then calculate the ionic strength and $\gamma_\pm$ from $s$, then repeat the process using the improved estimate of $\gamma_\pm$ until the answer converges to the desired number of decimal places. This works well, and is still useful in some cases, but most of us have better technology at our disposal now, since most modern scientific calculators have a numerical equation solver, as do most numerically oriented computer programs, including spreadsheets like Excel. The trick is to reduce our equations to a single equation in one unknown, preferably the solubility $s$, although in a spreadsheet you wouldn't even have to do that. We start by writing the ionic strength in terms of $s$:

$$I_c = \frac{1}{2}\left[(2)^2[Pb^{2+}] + (-1)^2[I^-]\right] = \frac{1}{2}(4s + 2s) = 3s.$$

We can now express the mean ionic activity coefficient in terms of $s$: at $20\,^\circ$C, the relative permittivity of water is 80.18, so the permittivity is $\varepsilon = \varepsilon_r \varepsilon_0 = 80.18(8.854\,187\,817 \times 10^{-12}\,\text{C}^2\text{J}^{-1}\text{m}^{-1}) = 7.099 \times 10^{-10}\,\text{C}^2\text{J}^{-1}\text{m}^{-1}$. Thus,

$$\gamma_\pm = \exp\left\{ \frac{(1.107 \times 10^{-10})(2)(-1)}{[(7.099 \times 10^{-10}\,\text{C}^2\text{J}^{-1}\text{m}^{-1})(293.15\,\text{K})]^{3/2}} \sqrt{3s} \right\}$$

$$= \exp(-4.039\sqrt{s}). \tag{9.10}$$

We can isolate $s$ from Equation (9.9):

$$s = \frac{1.14 \times 10^{-3}\,\text{mol}\,\text{L}^{-1}}{\gamma_\pm}. \tag{9.11}$$

Now substitute Equation (9.10) into Equation (9.11) and rearrange to the form

$$s - \frac{1.14 \times 10^{-3}\,\text{mol/L}}{\exp(-4.025\sqrt{s})} = 0.$$

We now have an equation involving only the unknown $s$. It's time to go to our calculator. The exact instructions will depend on the calculator model, but roughly you will have to do the following: enter the above equation in your equation solver. Your calculator will also need an initial guess for $s$. In this case, we have a very good place to start: the ideal solution value, which we get by setting $\gamma_\pm = 1$, i.e. $s = 1.14 \times 10^{-3}$. If you now run the equation solver, you will get $s = 1.32 \times 10^{-3}\,\text{mol}\,\text{L}^{-1}$, or $0.61\,\text{g}\,\text{L}^{-1}$. This value is in excellent agreement with the experimental value of $0.63\,\text{g}\,\text{L}^{-1}$.

**Example 9.6** We can try to use the same reasoning to calculate the solubility of barium carbonate at $20\,^\circ$C. The reaction is

$$\text{BaCO}_{3(s)} \rightarrow \text{Ba}^{2+}_{(aq)} + \text{CO}^{2-}_{3(aq)}$$

You should verify the following data and equations for this reaction:

$$\Delta_r G^\circ_{298} = 49\,\text{kJ}\,\text{mol}^{-1}$$
$$\Delta_r H^\circ_m = 3.2\,\text{kJ}\,\text{mol}^{-1}$$
$$K_{298} = 2.6 \times 10^{-9}$$
$$K_{293} = 2.5 \times 10^{-9}$$
$$I_c = 4s$$
$$s = 5.0 \times 10^{-5}/\gamma_\pm$$
$$\gamma_\pm = \exp(-9.328\sqrt{s})$$

If we solve these equations for $s$, we get $s = 5.35 \times 10^{-5}\,\text{mol}\,\text{L}^{-1}$. The experimental value of the solubility is $1.0 \times 10^{-4}\,\text{mol}\,\text{L}^{-1}$, so our final answer is off by a factor of almost 2. What went wrong? It's not the Debye–Hückel theory, which works perfectly well in this range of ionic strengths. The problem is that we have neglected some important chemistry, namely the reaction of the carbonate ion with water:

$$CO_{3(aq)}^{2-} + H_2O_{(l)} \rightleftharpoons HCO_{3(aq)}^{-} + OH_{(aq)}^{-}.$$

Since this reaction removes carbonate ions from solution, by Le Chatelier's principle, this will increase the solubility of barium carbonate. If we do a (much more complex) calculation taking this last reaction into account, we again get essentially quantitative agreement with the experimental value.

Solubility problems are notorious for giving the wrong answer when solved naively, as we did in the previous example. Aqueous chemistry is complicated, and ions often enter into multiple equilibria, including acid–base equilibria like the one in our last example, and complex ion formation. The lesson is this: if you don't get the chemistry right, it doesn't matter how sophisticated your theory of non-ideal solutions is, you'll still get the wrong answer.

---

## Exercise group 9.3

(1) Acetic acid ($CH_3COOH$) has an acid dissociation constant of $1.75 \times 10^{-5}$ at 298 K.
  (a) Calculate the pH of a $0.1 \, \text{mol L}^{-1}$ solution of acetic acid in water.
  (b) Calculate the concentration of acetate ions in this solution. Do a full Debye–Hückel calculation.
  
  *Hints*: Treat the acetic acid as ideally behaving. The method is analogous to that used to compute solubilities.
(2) The solubility product of silver iodide (AgI) is only $8 \times 10^{-17}$ at 298 K. This is tiny, so it is not generally necessary to use Debye–Hückel theory to compute the solubility of this compound. However, if other ions are present, they affect the ionic strength and so change the solubility. Compare the solubility of silver iodide in pure water to its solubility in a $0.008 \, \text{mol L}^{-1}$ solution of sodium nitrate ($NaNO_3$) in water.
  *Note*: All sodium and nitrate salts are extremely soluble.
(3) The acid dissociation constant of hydrofluoric acid is $6.6 \times 10^{-4}$ at 25 °C.
  (a) Calculate the standard free energy of formation of aqueous HF.
  (b) Using Debye–Hückel theory, calculate the concentrations of the $H^+$ and $F^-$ ions in a $0.10 \, \text{mol L}^{-1}$ aqueous solution of HF. Assume a unit activity coefficient for molecular HF.
  (c) Calculate the pH of the above solution.

---

## 9.4 Beyond the limiting law

What do we do if Debye–Hückel theory isn't good enough? After all, $0.01 \, \text{mol L}^{-1}$ is a very small ionic strength. The first thing that causes Debye–Hückel theory to break down is that, as the solution becomes more concentrated, we need to take into account the fact that ions occupy space. This introduces an extra factor of $(1 + R/r_D)^{-1}$ in the activity coefficient, where $R$ is the radius of the ion, and $r_D$ is the Debye screening length. As previously mentioned, the Debye screening length is inversely proportional to the ionic strength, so we get

$$\ln \gamma_i = -\frac{Az_i^2(\varepsilon T)^{-3/2}\sqrt{I_c}}{1 + BR\sqrt{\frac{I_c}{\varepsilon T}}}$$

where $B$ is another combination of constants with the value $B = 1.058 \times 10^6$. As mentioned previously, this extended Debye–Hückel theory isn't much better than the limiting law.

The basic Debye–Hückel expressions assume that all the non-ideal effects are due to direct solute–solute interactions. However, solutes affect the solvent, and this in turn affects other solutes. Hückel showed that if we take into account interactions between ions mediated by the solvent, a linear term is added to the Debye–Hückel expressions:

$$\ln \gamma_i = -\frac{Az_i^2(\varepsilon T)^{-3/2}\sqrt{I_c}}{1 + BR\sqrt{\frac{I_c}{\varepsilon T}}} + CI_c$$

where $C$ is a positive constant determined by fitting this expression to experimental data. This version of Debye–Hückel theory is useful over a much broader range of concentrations. Among other things, it correctly predicts that activity coefficients will eventually increase with ionic strength, and can even eventually exceed unit magnitude. This leads to the **salting out** effect, where the solubilities of ionic compounds decrease with increasing ionic strength at high ionic strengths. Salting in and salting out are general phenomena not restricted to simple ionic compounds. In particular, they are also observed with proteins.

### Key ideas and equations

- To treat non-ideal effects, we introduce activity coefficients as detailed in Table 9.1.
- $I_c = \frac{1}{2}\sum_i z_i^2 c_i$
- Definition and use of the mean ionic activity coefficient
- For ionic solutes in solutions with $I_c \lesssim 0.01 \text{ mol L}^{-1}$,

$$\ln \gamma_i = -Az_i^2 (\varepsilon T)^{-3/2} \sqrt{I_c}$$

or

$$\ln \gamma_\pm = Az_+z_-(\varepsilon T)^{-3/2}\sqrt{I_c}$$

The mean ionic activity coefficient is used when the reactants or products of a reaction represent a neutral binary compound.
- How to set up and solve solubility problems using Debye–Hückel theory.

## Suggested reading

Size matters. In crowded environments, such as the inside of a cell, non-ideal behavior correlates to size of the molecule. The activity coefficients in these cases can be huge, with $\gamma_i$ for large molecules in the cell often reaching values in excess of 10, and possibly as high as $10^2$. These effects have been discussed by Minton in the following short review:

Alex P. Minton, *J. Biol. Chem.* **276**, 10577 (2001).

Many molecules of biological interest are charged, including many biological macromolecules. Modern refinements of Debye–Hückel theory have been reviewed by Pitzer:

Kenneth S. Pitzer, *Acc. Chem. Res.* **10**, 371–377 (1977).

For a review of electrostatic effects on the free energy of biological macromolecules, although not one phrased in terms of activity coefficients, see:

Jianzhong Wu and Dimitrios Morikis, *Fluid Phase Equilib.* **241**, 317–333 (2006).

---

## Review exercise group 9.4

(1) Tris [short for tris(hydroxymethyl)aminomethane] is a base that is used to make buffer solutions, especially in biochemistry. A solution is made that contains $5.2 \, \mathrm{mmol \, L^{-1}}$ of tris and $8.0 \, \mathrm{mmol \, L^{-1}}$ of sodium chloride. The $K_b$ of tris at $25\,^\circ$C is $1.1 \times 10^{-6}$.
  (a) What is the concentration of the acid form of tris in this solution at $25\,^\circ$C? Use Debye–Hückel theory to answer this question, and assume that the base has an activity coefficient of unity.
  (b) What is the pH of this solution?

(2) The inside of a cell is sometimes described as a gel. The cytoplasm is crowded with proteins and other macromolecules, which increases the viscosity of the solution and reduces the volume available for solutes. Crowding tends to result in large activity coefficients for macromolecules. These effects can be simulated *in vitro* by dissolving polymers in a solvent along with the molecule(s) of interest. Minton has calculated the dependence of the activity coefficient on the molar mass of a spherical protein in a solution of the polymer dextran in which the dextran occupies 3% of the total volume.[2] For a $100\,000 \, \mathrm{g \, mol^{-1}}$ (biochemists would say dalton, abbreviated Da) protein, the activity coefficient in this particular solution is 2.7, while for a $200\,000 \, \mathrm{g \, mol^{-1}}$ protein, it is 4.2. Suppose that we have a protein of molar mass $100\,000 \, \mathrm{g \, mol^{-1}}$ that associates into a dimer with an equilibrium constant, measured in dilute solution, of $5 \times 10^5$.

---

[2] A. P. Minton, *J. Biol. Chem.* **276**, 10577 (2001).

(a) Calculate the concentrations of the protein and of its dimer in an aqueous solution with a protein concentration of $0.203 \text{ g L}^{-1}$, neglecting non-ideal effects.

(b) Calculate the concentrations of the protein and of its dimer in the dextran solution described above with $0.203 \text{ g L}^{-1}$ of protein. Assume that the protein and its dimer are both roughly spherical so that you can use Minton's estimates of the activity coefficients.

# 10

# Electrochemistry

I have to admit that I wasn't fond of electrochemistry when I was a student. It has since grown on me. Looking back, I think that I was told too much in my first exposure to the subject. Electrochemistry is a fussy science, with all kinds of complications that can make it difficult to obtain reproducible data from experiments. If your professors tell you about all these complications up front, then it becomes difficult to appreciate the utility of the subject. You just lose track of the big picture in a haze of contact potentials and transference numbers. That being said, it's possible to go too far the other way and to leave out important details you ought to know about when studying electrochemistry. I'm going to try to steer a middle course, one that emphasizes that electrochemistry is one of the key ways to get thermodynamic data, but also one that points out some of the things you need to think about when you do electrochemistry. Hopefully, you'll come through this experience with more enthusiasm for electrochemistry than I had when I was in your place.

## 10.1 Free energy and electromotive force

In several of the problems in Chapter 7, we used the fact that the maximum electrical work can be calculated from the Gibbs free energy, the latter representing the maximum (reversible) non-$pV$ work. Electrochemical cells convert chemical into electrical energy, or vice versa. In this section, we will develop a relationship between free energy and the **electromotive force (emf)**, also known as the reversible voltage or electric potential, which is the voltage generated by a cell under reversible conditions. This will lead us to the fundamental equation of electrochemistry.

We first note that measuring the electromotive force generated by a cell is easy; reversibility implies equilibrium. If we use a calibrated voltage source and set it to oppose our cell, the electromotive force is just the voltage produced by the calibrated source when the two are exactly balanced such that no current flows. In practice, especially for work requiring less precision, we measure the emf using a voltmeter. A voltmeter measures the current passing through a large resistance, so it's not really making an equilibrium measurement. However, a good quality voltmeter uses a very large resistance, so the current is very small, which gives us a near-reversible measurement.

The electrical work performed when a charge $q$ moves through an electric potential difference $E$ is

$$w = qE.$$

Therefore, under reversible conditions (where $E$ is the emf), at constant pressure and temperature,

$$\Delta G = w_{\text{rev}} = qE.$$

This is not the most convenient form of the relationship between $\Delta G$ and $E$. Chemists normally count units of matter in moles, and charge in electrical circuits is normally carried by electrons. Therefore write

$$\Delta G = -nFE \tag{10.1}$$

where $n$ is the number of moles of electrons and $F$ is the charge of a mole of electrons, a quantity referred to as Faraday's constant. The negative sign comes from the fact that $q$ is negative for electrons. Note that $E = -\Delta G/nF$ is an intensive property, unlike $G$ which is an extensive property. When the electrons come from a chemical reaction, we can divide Equation (10.1) by the number of moles of product formed, and we get

$$\Delta_r G_m = -v_e FE$$

where $v_e$ is the number of electrons exchanged per reaction, i.e. the stoichiometric coefficient of the electrons in a redox reaction. We will see how to identify $v_e$ a little later. Right now, recall that thermodynamically allowed reactions are those for which $\Delta_r G_m < 0$. Therefore, the emf $E$ will be *positive* for a thermodynamically allowed process.

Since the electromotive force is directly proportional to $\Delta G$, everything we know about $\Delta G$ can be transferred to this quantity. For instance,

$$\Delta_r G_m = \Delta_r G_m^\circ + RT \ln Q = -v_e FE.$$

If we divide this equation by $-v_e F$, we get

$$E = -\frac{\Delta_r G_m^\circ}{v_e F} - \frac{RT}{v_e F} \ln Q.$$

Define the standard emf $E^\circ$ by

$$E^\circ = -\frac{\Delta_r G_m^\circ}{v_e F}$$

and conclude that

$$E = E^\circ - \frac{RT}{v_e F} \ln Q. \tag{10.2}$$

This is the **Nernst equation**, the central equation of electrochemistry. It tells us how the electromotive force varies with the activities of the reactants and products of a reaction.

## 10.2  Reduction and oxidation

The reason that it is fruitful to talk of an electric potential or emf in relation to chemical thermodynamics is that many reactions can be separated into two subreactions, called **half-reactions**, one in which electrons are used and one in which electrons are produced. When a substance gains electrons we say that it is reduced (GER: gain of electrons = reduction). When a chemical species loses electrons, it is oxidized (LEO: loss of electrons = oxidation). A substance that can oxidize another is an oxidizing agent and one that can reduce another is a reducing agent. Because an oxidizing agent causes another substance to lose electrons, it must be accepting electrons itself and therefore be reduced. A mirror-image statement can be written for reducing agents.

For example, consider the reaction of oxygen and iron to form rust, iron(III) oxide:

$$2Fe_{(s)} + \frac{3}{2}O_{2(g)} \rightarrow Fe_2O_{3(s)}.$$

We can separate the overall reaction into the half-reactions

$$2Fe_{(s)} + 3H_2O_{(l)} \rightarrow Fe_2O_{3(s)} + 6H^+_{(aq)} + 6e^-$$

and

$$\frac{3}{2}O_{2(g)} + 6H^+_{(aq)} + 6e^- \rightarrow 3H_2O_{(l)}$$

where $e^-$ represents an electron. Iron has been oxidized and oxygen has been reduced in the process. The stoichiometric coefficient of the electrons which appears in the Nernst equation, $\nu_e$, can be determined from the half-reactions. In this case, $\nu_e = 6$.

You may wonder how the half-reactions given above for the oxidation of iron were obtained. Especially in solution, balancing redox (reduction–oxidation) reactions can be tricky. There are many methods available, of which one is presented here. The method consists of a set of steps you must carry out in the correct order:

(1)  Identify the atoms other than oxygen and hydrogen participating in the reaction. Typically, there are only a few. Separate the reaction into half-reactions based on the non-oxygen and non-hydrogen atoms. The main rule of thumb here is that you have to set it up so that eventually you will be able to balance the non-oxygen and non-hydrogen atoms. There are a few small exceptions to this rule, in reactions involving molecular oxygen and/or molecular hydrogen. In these cases, one of the half-reactions may involve $O_2$ or $H_2$.
(2)  Balance each half-reaction for all atoms except O and H.
(3)  Balance each half-reaction for O by adding $H_2O_{(l)}$ as appropriate.
(4)  Balance each half-reaction for H by adding $H^+_{(aq)}$ as appropriate. At this point, your reactions should be balanced for all atoms, but not for charge. Check that they are properly balanced for the atoms.

(5) Balance each half-reaction for charge by adding electrons as appropriate. At this point, one of the half-reactions should have electrons on the reactant side and the other on the product side. If this is not the case, something went wrong earlier. Go back and check your work.

(6) You now want to prepare to add the half-reactions. When you are done, you shouldn't have any electrons left. If necessary, multiply one or both half-reactions by appropriate factors such that the number of electrons in each half-reaction is the same. This common number of electrons is $\nu_e$. Add the results. This reaction is balanced for acidic conditions.

(7) If the reaction occurs in neutral medium, there are (at least initially) no excess protons in solution, so it is OK to have $H^+$ as a product, but not as a reactant. If the reaction occurs in basic medium, there shouldn't be *any* $H^+$ in the overall reaction. If you have protons on either side of a reaction in basic medium, or if you have protons on the reactant side of a reaction in neutral medium, add an equal number of $OH^-_{(aq)}$ to each side of the reaction to "neutralize" excess $H^+$, converting the latter to $H_2O_{(l)}$.

**Example 10.1 Balancing redox reactions in acid**  Acidified solutions of $Na_2Cr_2O_7$ and of KBr are mixed. The two react, forming $Br_2$ and $Cr^{3+}$. We first focus on the bromine and chromium atoms, separating and balancing the half-reactions with respect to these atoms:

$$Cr_2O_7^{2-} \rightarrow 2Cr^{3+}$$
$$2Br^- \rightarrow Br_2$$

We balance for oxygen by adding water molecules as appropriate:

$$Cr_2O_7^{2-} \rightarrow 2Cr^{3+} + 7H_2O$$
$$2Br^- \rightarrow Br_2$$

We next balance for hydrogen by adding $H^+$ ions:

$$Cr_2O_7^{2-} + 14H^+ \rightarrow 2Cr^{3+} + 7H_2O$$
$$2Br^- \rightarrow Br_2$$

Balance for charge by adding electrons:

$$Cr_2O_7^{2-} + 14H^+ + 6e^- \rightarrow 2Cr^{3+} + 7H_2O$$
$$2Br^- \rightarrow Br_2 + 2e^-$$

Finally, arrange for the electrons to cancel by multiplying the second half-reaction by 3 (such that $\nu_e = 6$), and add:

$$Cr_2O_7^{2-} + 14H^+ + 6Br^- \rightarrow 2Cr^{3+} + 7H_2O + 3Br_2.$$

Since the reaction occurs in acidic medium, we're done.

**Example 10.2 Neutralizing hydrogen ions in redox balancing**  If the reaction of the last example had occurred in a neutral (or basic) environment, there would not have been many protons available to participate in the reaction. In these cases, add a sufficient number of hydroxide ions to each side to convert all the protons appearing to water and obtain, after cancellation of excess water molecules,

$$Cr_2O_7^{2-} + 7H_2O + 6Br^- \rightarrow 2Cr^{3+} + 14OH^- + 3Br_2.$$

## Exercise group 10.1

(1) Carbon dioxide reacts with calcium metal in a basic aqueous environment to produce ethanedioate ions (a.k.a. oxalate, $C_2O_4^{2-}$) and calcium ions. Balance the reaction.

(2) Permanganate ($MnO_4^-$) reacts with sulfur dioxide ($SO_2$) to form manganese(II) ions and $HSO_4^-$ in acidic solution. Balance the reaction.

(3) Solid bismuth(III) hydroxide ($Bi(OH)_3$) reacts with the $SnO_2^{2-}$ anion to form $SnO_3^{2-}$ and metallic bismuth in basic solution. Balance the reaction.

(4) Pyruvate ions ($C_3H_3O_3^-$) react with glucose ($C_6H_{12}O_6$) to form lactate ($C_3H_5O_3^-$) and gluconate ($C_6H_{11}O_7^-$) ions. The unbalanced half-reactions are $C_3H_3O_3^- \rightarrow C_3H_5O_3^-$ and $C_6H_{12}O_6 \rightarrow C_6H_{11}O_7^-$. The reaction medium is slightly acidic. Balance the reaction.

## 10.3 Voltaic cells

If we allow the different species in a redox reaction to freely intermix, it will prove impossible to harness their free energy to derive electrical work. In order to produce electricity, it is necessary to separate the reduction and oxidation half-reactions. This is done by building a voltaic cell, a cell that converts free energy into electricity. Although there are many different designs of voltaic cells, every one has three essential components:

(1) spatial separation between the two half-reactions;
(2) a conductive connection that allows electrons to flow between the electrodes and
(3) a means for electrical neutrality to be maintained in all parts of the system.

In one half-cell, oxidation occurs. The electrons generated by the oxidation reaction are carried by the conductive connection to the other half-cell where reduction (gain of electrons) occurs. This would normally lead to accumulation of negative charge in the reduction half-cell counteracting the normal tendency for current to flow, hence the need for a mechanism to maintain electrical neutrality. For example, consider the reaction

$$Zn_{(s)} + Cu_{(aq)}^{2+} \rightarrow Cu_{(s)} + Zn_{(aq)}^{2+}.$$

If we just put a piece of solid zinc into a copper(II) solution, this spontaneous reaction occurs without any possibility of extracting work from it. Instead, we put together an apparatus like the one shown in Figure 10.1. We have two beakers. In one, we place a solid zinc electrode bathing in a zinc sulfate solution. In the other, we place a solid copper electrode in a copper(II) sulfate solution. We connect the two electrodes by an electrical

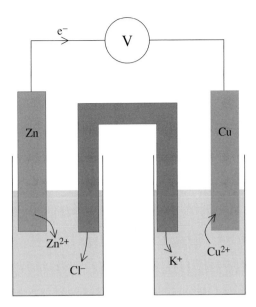

Figure 10.1 A simple copper–zinc voltaic cell. The gray U-tube is the salt bridge.

circuit running through a piece of electrical machinery (shown as a voltmeter in the figure). We connect the fluids in the two beakers by a **salt bridge**, a device we describe in detail below. For now, we just need to know that a salt bridge provides ions to balance the overall charge without allowing the two solutions to mix. The zinc is oxidized, producing $Zn^{2+}$ and two electrons. These electrons travel through the external circuit where they can be used to operate electrical equipment, and eventually reach the copper electrode. There, they combine with copper(II) ions at the surface of the electrode to produce solid copper.

As you may have learned in your earlier chemistry courses, the two electrodes in an electrochemical cell have specific names. The electrode where oxidation occurs (zinc in Figure 10.1) is the **anode**. The electrode where reduction occurs (copper) is the **cathode**.

There are many ways to provide for the maintenance of electrical neutrality between two half-cells, some as simple as a porous glass disc that allows for the movement of ions from one solution to the other. If the ions on either side of a simple junction such as a porous glass disc have different mobilities (average velocity per unit electric field strength), then diffusion can create a small separation of charge across the junction, which leads to a potential difference between the two solutions known as the liquid junction potential. The liquid junction potential may not matter much if we're making a battery, but it varies from one cell to another, so it hampers reproducibility of results for high-precision electrochemical studies. The solution to this problem is to have two liquid junctions which generate potentials of roughly equal size, but opposite signs, such that they cancel out. This is what a salt bridge does. A salt bridge can be as simple as a tube filled with a saturated potassium chloride solution. The ends of the tube are tightly plugged with cotton wool to minimize transfer of material from the salt bridge to the half-cells. Alternatively, the salt bridge may be made by packing a tube with a gel saturated with potassium chloride. Potassium chloride is a deliberate choice, since the two ions in this compound have very similar mobilities. Accordingly, roughly equal amounts of potassium and chloride leak out of opposite ends of the tube, generating very small liquid junction potentials. Better still, since this happens at opposite ends of the tube, the liquid junction potentials are opposite in sign with respect to the circuit. Overall, the salt bridge contributes to the cell emf a term which is a difference of two small numbers of similar magnitude, thus minimizing the effect of liquid junctions. In the case that potassium chloride cannot be used, other ionic compounds whose ions would have similar mobilities can be substituted. Ammonium nitrate is a popular choice.

There is a standard notation for electrochemical cells, which is best demonstrated by example. In the case of the cell described in the last paragraph, we would write

$$Zn_{(s)}|Zn^{2+}_{(aq)}||Cu^{2+}_{(aq)}|Cu_{(s)}.$$

A solid bar represents an interface between two phases. A double bar represents a junction (e.g. a salt bridge) between two miscible phases (i.e. two phases that would mix if we put them in direct contact). Depending on what we want to emphasize, we can either show only essential species, as above, or we can write down a more complete description of the contents of the cell. We can also indicate concentrations of species in our diagram when they matter, e.g.

$$Zn_{(s)}|ZnSO_{4(aq)}(0.52\,mol\,L^{-1})||CuSO_{4(aq)}(0.28\,mol\,L^{-1})|Cu_{(s)}.$$

In one common convention which you may have seen in your introductory chemistry course, we put the anode on the left. However, we don't always know a priori which of the two half-reactions will provide the anode, and it makes no real difference to the way we do calculations or interpret the results, so we won't bother with that convention here.

## 10.4 Standard reduction potentials

Just as with free energies, we do not want to tabulate the cell emf for every possible electrochemical cell. Instead, we tabulate half-cell potentials relative to a standard half-cell, namely the formation of hydrogen gas from hydrogen ions:

$$H^+_{(aq)} + e^- \rightarrow \frac{1}{2}H_{2(g)}$$

The reduction potential of this half-cell is arbitrarily set to zero. Every other half-cell is measured relative to this one, either directly or indirectly. By convention, we report half-cell *reduction* potentials. Again, as with the other thermodynamic quantities we have encountered, *standard* reduction potentials, i.e. reduction potentials for the half-reactions in which all reactants and products are in their standard states, are tabulated.

Because the number of electrons is divided out of the free energy to obtain the emf, computing the standard cell emf is just a matter of subtracting the reduction potential of the oxidized species from that of the reduced species. The emf under particular experimental conditions is then obtained via the Nernst equation (10.2).

> **You must *never* multiply a reduction potential by a stoichiometric coefficient when calculating the emf of a cell.**

**Example 10.3 Calculating cell emf** A cell with the following cell diagram is constructed at 25 °C:

$$Cr_{(s)}|CrCl_{2(aq)}(1 \text{ mol L}^{-1})\|HCl_{(aq)}(pH = 3.0)|O_{3(g)}(0.01 \text{ bar}), O_{2(g)}(0.2 \text{ bar})|Pt$$

The platinum electrode is inert. From the table of Appendix B, we find the following half-reactions:

$$O_{3(g)} + 2H^+_{(aq)} + 2e^- \rightarrow O_{2(g)} + H_2O_{(l)} \qquad E^\circ = +2.07V$$
$$Cr^{2+}_{(aq)} + 2e^- \rightarrow Cr_{(s)} \qquad E^\circ = -0.91V$$

In order for the overall process to be thermodynamically allowed, the emf must be positive. For these half-reactions, the standard emf will be positive when the second half-reaction runs in reverse, i.e. chromium is oxidized. It is often the case that the signs of the emf and standard emf are the same, so we guess that this cell oxidizes chromium. The overall reaction is therefore probably

$$O_{3(g)} + 2H^+_{(aq)} + Cr_{(s)} \rightarrow O_{2(g)} + H_2O_{(l)} + Cr^{2+}_{(aq)}.$$

The standard cell emf is $E^\circ = 2.07 - (-0.91) = 2.98$ V. The emf developed by this cell at the given activities is computed from the Nernst equation. We're going to assume ideal behavior since the concentration of $Cr^{2+}$ is outside the range where we can use Debye–Hückel theory and we have no other information about the activity coefficients. In this reaction, $v_e = 2$.

$$E = E^\circ - \frac{RT}{v_e F} \ln \frac{(a_{O_2})(a_{Cr^{2+}})}{(a_{O_3})(a_{H^+})^2}$$

$$= 2.98 \text{ V} - \frac{(8.314\,41 \text{ J K}^{-1}\text{mol}^{-1})(298.15 \text{ K})}{(2)(96\,485.342 \text{ C mol}^{-1})} \ln \frac{(0.2)(1)}{(0.01)(10^{-3})^2}$$

$$= 2.76 \text{ V}$$

If we had incorrectly guessed which species was oxidized, we would have gotten the same value for the emf, but with the opposite sign. The sign would have told us that we wrote the reaction backwards. There would be no need to recalculate anything once this fact had been noted.

**Example 10.4 Determining standard reduction potentials from emf measurements**  At 25°C, the cell

$$Pt_{(s)}|H_{2(g)}(1 \text{ bar})|H^+_{(aq)}(pH\ 5)\|Al(NO_3)_{3(aq)}(0.001 \text{ mol L}^{-1})|Al_{(s)}$$

develops an emf of 1.425 V and bubbles are observed to form on the inert platinum electrode. What is the standard reduction potential of the aluminum ion?

We first need to write down the two half-reactions and the overall reaction. If bubbles form on the inert electrode, hydrogen gas is being produced. Therefore

$$2H^+_{(aq)} + 2e^- \to H_{2(g)}.$$

The other half-reaction must be

$$Al_{(s)} \to Al^{3+}_{(aq)} + 3e^-.$$

The two half-reactions don't have the same number of electrons. We can fix this by multiplying the first by 3 and the second by 2. Each reaction then involves $\nu_e = 6$ electrons, and the overall reaction is

$$2Al_{(s)} + 6H^+_{(aq)} \to 2Al^{3+}_{(aq)} + 3H_{2(g)}.$$

We have a measurement of $E$. We can use this to calculate $E^\circ$, given the activities of the species participating in the reaction. For the solution on the left, we have the activity of hydrogen ions directly from the pH.

$$a_{H^+,\text{left}} = 10^{-pH} = 10^{-5}.$$

For the aluminum nitrate solution, we can use Debye–Hückel theory:

$$I_{c,\text{right}} = \frac{1}{2}\left[(3)^2(0.001) + (-1)^2(0.003) \text{ mol L}^{-1}\right] = 0.006 \text{ mol L}^{-1}$$

$$\therefore \ln \gamma_{Al^{3+}} = -(1.107 \times 10^{-10})(+3)^2$$

$$\times \left[(6.939 \times 10^{-10} \text{ C}^2\text{N}^{-1}\text{m}^{-2})(298.15 \text{ K})\right]^{-3/2} \sqrt{0.006 \text{ mol L}^{-1}}$$

$$= -0.820$$

$$\therefore \gamma_{\pm,\text{right}} = \exp(-0.820) = 0.440$$

$$\therefore a_{Al^{3+}} = \gamma_{Al^{3+}}[Al^{3+}]/c^\circ = (0.442)(0.001) = 4.40 \times 10^{-4}$$

$$a_{H_2} = \frac{1 \text{ bar}}{1 \text{ bar}} = 1$$

$$E^\circ = E + \frac{RT}{\nu_e F} \ln Q$$

$$\therefore E^\circ = 1.425 \text{ V} + \frac{(8.314\,472 \text{ J K}^{-1}\text{mol}^{-1})(298.15 \text{ K})}{6(96\,485.342 \text{ C mol}^{-1})} \ln\left(\frac{(4.42 \times 10^{-4})^2(1)^3}{(10^{-5})^6}\right)$$

$$= 1.655 \text{ V}$$

The reaction is written with aluminum being oxidized. If we take $E^\circ_{Al^{3+}}$ to be the standard reduction potential of the aluminum ion and $E^\circ_{H^+}$ to be the standard reduction potential of the hydrogen ion, we have

$$E^\circ = E^\circ_{H^+} - E^\circ_{Al^{3+}}.$$

$E^\circ_{H^+} = 0$ by definition, so $E^\circ = -E^\circ_{Al^{3+}}$, which implies that $E^\circ_{Al^{3+}} = -1.655$ V.

**Example 10.5 Determining solubility products from emf measurements** You may have wondered how very small solubility products are measured. There are two commonly used techniques, both of which are electrochemical in nature: one is to measure the conductivity of a saturated solution, a method that we shall not consider in this book; the other is to measure an emf in a cell involving the salt in question. Consider the cell

$$Ag_{(s)}|Ag^+_{(aq)}, I^-_{(aq)}|AgI_{(s)}, Ag_{(s)}$$

The half-reactions are

$$Ag_{(s)} \rightarrow Ag^+_{(aq)} + e^-, \quad E^\circ = -0.7991 \text{ V};$$
$$\text{and } AgI_{(s)} + e^- \rightarrow Ag_{(s)} + I^-_{(aq)}, \quad E^\circ = -0.1518 \text{ V}.$$

The $E^\circ$ values for the half-reactions can be obtained from emf measurements, as shown in the previous example. The overall reaction for this cell is

$$AgI_{(s)} \rightarrow Ag^+_{(aq)} + I^-_{(aq)}$$

with a standard cell emf of $-0.9509$ V. This reaction is exactly that for which the equilibrium constant is the $K_{sp}$ of AgI. Therefore

$$\Delta_r G^\circ_m = -v_e F E^\circ = -(1)(96\,485.342\,\text{C mol}^{-1})(-0.9509\,\text{V})$$
$$= 91.75 \text{ kJ mol}^{-1}$$
$$\therefore K_{sp} = \exp\left(\frac{-91.75 \times 10^3 \text{ J mol}^{-1}}{(8.314\,472\,\text{J K}^{-1}\text{mol}^{-1})(298.15\,\text{K})}\right) = 8.44 \times 10^{-17}$$

---

**Exercise group 10.2**

(1) The cell

$$Pt_{(s)}|H_{2(g)}|H^+_{(aq)}\|V^{2+}_{(aq)}|V_{(s)}$$

is prepared using a $0.001$ mol L$^{-1}$ solution of vanadium (II) nitrate in the right half-cell, and a pH 5 nitric acid solution in the left half-cell. At 298 K under a hydrogen gas pressure of 1 atm, the cell develops an emf of 0.97 V and bubbles are observed to form on the inert platinum electrode. What is the standard reduction potential of the vanadium (II) ion? Use Debye–Hückel theory to estimate the activity coefficients of the ions.

(2) Calculate the equilibrium constant for the reaction

$$Hg_{2(aq)}^{2+} \rightleftharpoons Hg_{(l)} + Hg_{(aq)}^{2+}$$

at $25\,^{\circ}C$. Which of the two aqueous ions will be more abundant under equilibrium conditions? (Metallic mercury is insoluble in water, so it precipitates out.)

(3) (a) Ethanol ($C_2H_5OH$) reacts with the dichromate ion ($Cr_2O_7^{2-}$) in acidic solution to produce ethanal (acetaldehyde, $C_2H_4O$) and the chromium(III) ion. Balance the reaction.

   (b) Calculate the standard reduction potential corresponding to the ethanal/ethanol half-reaction obtained above.

   *Hint*: Start by calculating the standard emf for the electrochemical cell

$$Pt_{(s)}|H_{2(g)}|H_{(aq)}^{+}\,\|C_2H_4O_{(aq)},\,C_2H_5OH_{(aq)},\,H_{(aq)}^{+}|Pt_{(s)}.$$

   (c) Calculate the standard free energy of formation of the aqueous chromium(III) ion.

(4) What emf would be generated by the cell

$$Pt,\,H_2(1\,bar)|HCl(pH = 1.5)\|NaCl(0.004\,mol/L)|Ag,\,AgCl$$

at $25\,^{\circ}C$? Use Debye–Hückel theory to calculate the emf.

(5) Electrochemical cells are sometimes used to measure concentrations. Consider the cell

$$Cu_{(s)}|CuSO_{4(aq)}(1.4 \times 10^{-3}\,mol\,L^{-1})\|NaBr_{(aq)}(8.5 \times 10^{-3}\,mol\,L^{-1}),\,Br_{2(aq)}|Pt_{(s)}.$$

Electrons flow in the external circuit from the copper to the platinum electrode. The equilibrium voltage at $25°C$ is $0.7558\,V$. What is the concentration of bromine in the right half-cell? Use Debye–Hückel theory to calculate the activity coefficients of the ions.

(6) Electrochemical cells can also be used to measure activity coefficients. Consider the cell

$$Cu_{(s)}|CuCl_{2(aq)}(0.20\,mol\,L^{-1})|AgCl_s,\,Ag_{(s)}.$$

Here, we use the convention that the anode is on the left. The emf of this cells is $-0.074\,V$ at $25\,^{\circ}C$. Estimate the mean ionic activity coefficient of copper(II) chloride.

---

## 10.5 Other types of cells

### 10.5.1 Concentration cells

Suppose that we have constructed a cell such as

$$Cu_{(s)}|Cu_{(aq)}^{2+}(c_1)\|Cu_{(aq)}^{2+}(c_2)|Cu_{(s)}.$$

In other words, the only difference between the two half-cells is the concentration of the solutions in which the electrodes bathe. It is understood that an anion is present to balance charge on both sides, and that (e.g.) a salt bridge is present to close the circuit. This is called a **concentration cell**. The overall reaction is

$$Cu^{2+}_{(1)} + Cu_{(2)} \rightarrow Cu^{2+}_{(2)} + Cu_{(1)}.$$

The standard emf of this cell is clearly zero, so

$$E = -\frac{RT}{2F} \ln \frac{a_2}{a_1} \tag{10.3}$$

where $a_1$ is the activity coefficient of the copper ions in the left half-cell, and $a_2$ is the activity coefficient in the right half-cell. Therefore, if $a_2 < a_1$ (usually implying $c_2 < c_1$), a positive emf is generated, i.e. the overall "reaction" is thermodynamically allowed as written.

Concentration cells are among the simplest electrochemical cells that can be devised. They don't generate very large voltages, but they can sometimes be used in analytical chemistry to determine the concentration of an ion in solution given a half-cell containing a known concentration of the same ion. They are also used to measure activity coefficients since some complications that arise from cells made from solutions with different components are minimized or eliminated when the two half-cells contain the same chemicals.

### 10.5.2 Transmembrane potential

Many living cells rely on the establishment of a transmembrane potential for their function. This is an electric potential difference between the inside and outside of the cell. For instance, neurons require a transmembrane potential in order to generate electrical impulses. The transmembrane potential is due to exactly the same effect as concentration cells rely on, namely that a difference in concentration is equivalent to a difference in potential. However, a membrane isn't an electrochemical cell in the normal sense, so the reasoning that leads to the equation for the transmembrane potential is a little different.

In Section 8.5, we discussed the cotransport of sodium and potassium across cell membranes. This cotransport exists to establish a transmembrane potential. It turns out that most cell membranes are not very permeable to sodium or to chloride, the major small anion *in vivo*. These ions contribute little to the transmembrane potential. Cell membranes are, however, fairly permeable to potassium. As a result, the potassium ions that have been pumped out by the cell leak back across the membrane:

$$K^+_{(in)} \rightarrow K^+_{(out)}.$$

The free energy change associated with this process is

$$\Delta G_m = RT \ln \left( \frac{\gamma_{out}[K^+_{(out)}]}{\gamma_{in}[K^+_{(in)}]} \right)$$

where $\gamma_{in}$ and $\gamma_{out}$ are the activity coefficients of the potassium ions on the two sides of the membrane. Using the same reasoning as we used to derive the Nernst equation, we can convert this free energy change into a leakage potential, the main difference being that our charge carriers are now potassium ions of charge number $z = 1$ instead of electrons:

$$E_{leak} = \frac{RT}{F} \ln \left( \frac{\gamma_{out}[K^+_{(out)}]}{\gamma_{in}[K^+_{(in)}]} \right).$$

Recall that $\Delta G_m = G_m(K^+, \text{outside}) - G_m(K^+, \text{inside})$. The leakage potential is similarly defined as $E_{leak} = E_{outside} - E_{inside}$. This will be important in a little while.

The leak leads to a charge imbalance across the membrane since the membrane is impermeable to the counterion (chloride). As a result, leaking stops almost immediately because the charge imbalance establishes an electric potential $\Delta\phi$ which opposes further movement of potassium ions across the membrane. This potential is the transmembrane potential.

The last step in the derivation of our equation for the transmembrane potential is to equate $\Delta\phi$ to $E_{leak}$. We can do this because there are two canceling sign changes. The first one is perhaps obvious: because $\Delta\phi$ opposes $E_{leak}$, the former should be opposite in sign to the latter. However, by convention, $\Delta\phi$ is measured relative to the *outside* of the cell, which is opposite to the way $E_{leak}$ was defined. Thus we have

$$\Delta\phi = \frac{RT}{F} \ln \left( \frac{\gamma_{out}[K^+_{(out)}]}{\gamma_{in}[K^+_{(in)}]} \right).$$

For mammalian muscle cells, we have $[K^+_{(out)}] = 4\,\text{mmol}\,L^{-1}$ and $[K^+_{(in)}] = 155\,\text{mmol}\,L^{-1}$. If we assume that the activity coefficients are similar inside and outside the cell, we can calculate a transmembrane potential at 37°C of $-98$ mV. The negative sign indicates that the potential is lower inside the cell than out. Positively charged particles (like our potassium ions) move spontaneously from areas of high electric potential to areas of lower electric potential so, as expected, the transmembrane potential opposes movement of potassium ions out of the cell, i.e. it opposes the tendency for concentrations to equalize across a permeable membrane. The value we calculated is fairly close to the experimental value of $-85$ mV. The main source of the discrepancy is that the membrane is not completely impermeable to other ions. However, accounting for these factors properly would require a much more sophisticated theory than we have in hand.

### 10.5.3 *Fuel cells*

A fuel cell is essentially just a voltaic cell whose overall reaction looks like combustion. However, the efficiency of a fuel cell is much greater than that of any engine that functions by burning a fuel. The reason for this greater efficiency is interesting, so let's consider it for a moment.

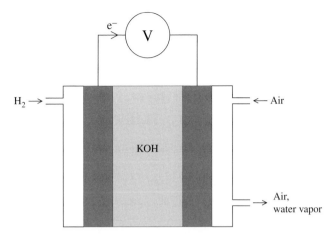

Figure 10.2 A simple hydrogen fuel cell. The gases are admitted to separate chambers, where they are in contact with porous electrodes. These electrodes are both in contact with an electrolyte solution, often aqueous KOH, which mediates the transfer of charge from one electrode to the other. Typically, atmospheric air flows over the oxygen (right-hand) electrode, while the hydrogen can either flow through, with unused hydrogen being recovered and recycled, or, as shown here, be maintained at a constant pressure over its electrode. The flow of air has the additional role of removing excess water, the product of the oxidation of hydrogen, as vapor.

The Second Law essentially tells us that no heat engine can convert heat into work with perfect efficiency (Section 6.1). In other words, at constant pressure, only a portion of the enthalpy can be converted to work, if we, for instance, use this heat to expand a gas and move a piston. In fact, the maximum fraction of the enthalpy that can be converted to work by a heat engine can be calculated using formulas developed in Section 6.8. On the other hand, in an electrochemical cell, we use the free energy rather than the heat. The free energy change is related to the enthalpy change by $\Delta G = \Delta H - T \Delta S$. Usually, $\Delta S$ is not particularly large, so in an isothermal free energy machine such as an electrochemical cell, the maximum work can be similar to $\Delta H$. In fact, if $\Delta S$ is positive, $\Delta G$ can be larger in absolute value than $\Delta H$. (Recall that useful work is performed when $\Delta G$ is negative. This usually means that $\Delta H$ is negative, again because $\Delta S$ is not normally very large.) Thus, a fuel cell should be able to produce more work than a heat engine using the same reaction.

As with all electrochemical cells, the trick is to design a fuel cell in such a way that the reactants don't combine directly. Many common fuels (hydrogen, methane etc.) are gases at normal ambient temperatures, and of course the oxidant, which is usually just atmospheric oxygen, is also a gas. Typically, the fuel and oxidant are admitted into separate chambers where they are placed in contact with appropriate electrodes. A hydroxide solution often provides the medium necessary for ion exchange between the electrodes. Figure 10.2 shows a sketch of a simple hydrogen fuel cell. This fuel cell has the diagram

$$H_{2(g)}|OH^-_{(aq)}|O_{2(g)}.$$

The half-reactions are

$$O_{2(g)} + 2H_2O_{(l)} + 4e^- \rightarrow 4OH^-_{(aq)}, \qquad E^\circ = +0.401 \text{ V}$$

$$H_{2(g)} + 2OH^-_{(aq)} \rightarrow 2H_2O_{(l)} + 2e^-, \qquad E^\circ = +0.8277 \text{ V}$$

and the overall reaction is the familiar

$$2H_{2(g)} + O_{2(g)} \rightarrow 2H_2O_{(l)}$$

with $E^\circ = 1.229$ V and $\nu_e = 4$. The hydroxide solution appearing in the diagram has the role of facilitating the transfer of charge, in this case in the form of hydroxide ions, between one side of the cell and the other.

Suppose that hydrogen gas is admitted at a pressure of 1 atm. Oxygen is typically provided from the air, so the partial pressure of $O_2$ would be 0.21 atm at sea level. The cell emf at 25 °C is therefore

$$E = 1.229 - \frac{(8.314\,472\,\text{J K}^{-1}\text{mol}^{-1})(298.15\,\text{K})}{4(96\,485.342\,\text{C mol}^{-1})} \ln \frac{1}{(1.01325)^2(0.21)} = 1.199 \text{ V}.$$

Just as with chemical cells, several fuel cells can be joined in series to form a battery if higher voltages are required.

Fuel cells have two main advantages over heat engines: first, as mentioned above, they are more efficient, and thus use less fuel to do the same amount of work; second, because there is no combustion, there are no combustion by-products (sulfur oxides etc.), so they produce little if any pollution. The main obstacles to widespread adoption of fuel cell technology at present all revolve around the porous electrodes required to separate the gases from the solution. Typically, polymeric membranes are used. Some of the materials that have been tried so far are expensive. Others aren't very durable. There are also issues with water management since the membranes have to stay wet, but can't be so wet that the electrolyte leaks out.

Since we can get most of the data we need from free energy tables to calculate the emf of a fuel cell, the only number we need from the half-reactions is the ratio of electrons to oxygen. However, this is always 4:1 – look at the two possible oxygen half-reactions in Appendix B – so we don't generally need to work out the half-reactions.

**Example 10.6** Suppose that a fuel cell is powered by methane and oxygen. The overall reaction is

$$CH_{4(g)} + 2O_{2(g)} \rightarrow CO_{2(g)} + 2H_2O_{(l)}.$$

The standard free energy change for this reaction is

$$\Delta_r G^\circ = -817.93 \text{ kJ mol}^{-1}.$$

To convert this to a cell emf, we need to know how many electrons are exchanged in the reaction as written. Four electrons are transferred for every oxygen, and there are two oxygens, so $v_e = 8$. Therefore

$$E^\circ = -\frac{\Delta_r G_m^\circ}{v_e F} = \frac{817.93 \times 10^3 \,\text{J mol}^{-1}}{8(96\,485.342 \,\text{C mol}^{-1})} = 1.0597 \,\text{V}.$$

If the methane pressure is held at 1 atm and the oxygen pressure is held at 0.2 atm, carbon dioxide being allowed to accumulate to a pressure of 0.1 atm, the Nernst equation gives us a maximum output voltage of

$$E = 1.0597 - \frac{(8.314\,472 \,\text{J K}^{-1}\text{mol}^{-1})(298.15 \,\text{K})}{8(96\,485.342 \,\text{C mol}^{-1})} \ln \frac{0.101\,325}{(1.013\,25)(0.202\,65)^2}$$

$$= 1.0568 \,\text{V}$$

at $25\,^\circ$C. Note that this voltage is a maximum because it is the electric potential we would measure under equilibrium (reversible) conditions.

---

## Exercise group 10.3

(1) Methanol is often touted as a potential replacement for fossil fuels because it can be cheaply produced in large quantities by a number of different processes.

*Note*: Attempt this question only if you have covered the material in Section 6.8.

  (a) Suppose that gaseous methanol is burned in an internal combustion engine with upper and lower operating temperatures of 400 and 900 K, respectively. One of the products is water vapor. How much work can be obtained per mole of methanol under these conditions?

  (b) Suppose that liquid methanol is instead oxidized in a fuel cell. The methanol is kept separate from all other reactants and products at all times in the particular cell used. The carbon dioxide produced shows up in a gaseous stream flowing through another part of the cell at a pressure of approximately 0.1 atm. Air also flows through so that the oxygen pressure is always 0.22 atm. How much work can be obtained from this fuel cell at 298 K?

(2) A fuel cell operates on liquid ethanol at $25\,^\circ$C. Oxygen is supplied at a constant pressure of 0.21 atm and carbon dioxide is removed at such a rate that its pressure is an essentially constant 0.1 atm.

  (a) What maximum voltage does this fuel cell generate?

  (b) If the cell is improperly operated, it is also possible for it to generate carbon monoxide instead of carbon dioxide. Assuming that the carbon monoxide pressure is 0.1 atm and that the cell otherwise operates under identical conditions, what is the maximum operating voltage?

(3) (a) Propane is a gas at room temperature. A fuel cell operates on propane at a pressure of 1 atm at $25\,^\circ$C. Assume that propane is an ideal gas and calculate the maximum electrical work available from the fuel cell per cubic meter of propane oxidized if

the oxygen pressure is 0.2 atm and the carbon dioxide pressure is 0.03 atm. *Hint*: Calculate the maximum work per mole first.
(b) Calculate the maximum voltage produced by the fuel cell under the operating conditions described above.

---

### Key ideas and equations

- $\Delta_r G_m = -\nu_e F E$
- $E = E^\circ - \dfrac{RT}{\nu_e F} \ln Q$
- Procedure for balancing redox reactions
- Transmembrane potential: $\Delta\phi = \dfrac{RT}{F} \ln \left( \dfrac{\gamma_{out}[K_{(out)}^+]}{\gamma_{in}[K_{(in)}^+]} \right)$
- In fuel cells, four electrons are transferred for each oxygen used.

---

### Review exercise group 10.4

(1) When cyanide ions ($CN^-$) react with permanganate ($MnO_4^-$) in basic solution, the products are manganese(IV) oxide ($MnO_{2(s)}$) and cyanate ions ($OCN^-$). Balance the reaction.

(2) What emf (voltage) would be generated by the cell

$$\text{Pt, } H_2(1\,\text{bar})|H^+(0.03\,\text{mol kg})^{-1}\|Cl^-(0.004\,\text{mol kg})^{-1}|Ag, AgCl$$

at 25°C, assuming ideal behavior?

(3) Gold is often known as a noble metal because it is highly unreactive under most conditions. For instance, most acids won't dissolve gold. Gold can, however, be dissolved by aqua regia, a mixture of concentrated nitric and hydrochloric acid. When nitric acid reacts with gold, the following reaction occurs:[1]

$$2Au_{(s)} + 3NO_{3(aq)}^- + 9H_{(aq)}^+ \rightarrow 2Au_{(aq)}^{3+} + 3HNO_{2(aq)} + 3H_2O_{(l)}$$

(a) Calculate the equilibrium constant for this reaction at 25°C. From the result of your calculation, argue that this reaction alone won't dissolve much gold.

(b) The hydrochloric acid provides chloride ions which go on to react with the gold (III) ions:

$$Au_{(aq)}^{3+} + 4Cl_{(aq)}^- \rightleftharpoons AuCl_{4(aq)}^-$$

Explain qualitatively why this reaction could help dissolve gold.

(4) Sketch a methane/oxygen fuel cell. What are the half-reactions? Where do these half-reactions occur? In what direction do electrons flow? What is the standard emf of this cell? (Some of these questions can be answered in your diagram.)

---

[1] This is a bit of a cartoon. The reaction generates a range of nitrogen oxide products, including the brown gas $NO_2$.

(5) Nickel–cadmium (nicad) batteries are based on the following half-cell reactions:

$$NiO(OH)_{(s)} + H_2O_{(l)} + e^- \rightarrow Ni(OH)_{2(s)} + OH^-_{(aq)}$$

$$Cd_{(s)} + 2OH^-_{(aq)} \rightarrow Cd(OH)_{2(s)} + 2e^-$$

(a) The emf of a nicad cell is 1.4 V. The standard reduction potential of cadmium (II) hydroxide is $-0.809$ V. What is the standard reduction potential of NiO(OH)?

(b) What is the standard free energy of formation of NiO(OH)?

# Part Three

## Kinetics

# 11

# Basics of chemical kinetics

Thermodynamics tells you what reactions can happen, but not how fast they will happen. In fact, some reactions are so slow that they are never actually observed, even though they are thermodynamically allowed. Other reactions are stunningly fast. In this part of the book, we will discuss the factors that affect rates of reaction, and consider some basic theories that can be used to rationalize them. This chapter introduces a few definitions and ideas that we will need in our study of kinetics.

## 11.1 The business of kinetics

Chemical kinetics is the study of the rates of chemical reactions or, to put it in simpler language, of the speeds of chemical reactions. There are many different questions one can ask about how fast reactions go, and many different approaches one can take to answering these questions. Roughly though, we can break down the field into two major branches:

(1) Phenomenological kinetics: this is the part of kinetics that is concerned with measuring rates of reactions and with the relationship between rates and chemical mechanisms.
(2) Kinetic theory and dynamics: this field focuses on the relationship between rates of reactions and events on a molecular scale. In the best cases, we are able to predict a reaction mechanism and the rates of the reactions that make up that mechanism. More commonly, kinetic theory or dynamics let us relate molecular properties to observed reaction rates. Kinetic theory and dynamics are attempts to answer questions about why the rate of a reaction has a particular value, rather than just taking the rates as quantities determined by experiment, as we do in phenomenological kinetics.

There is of course not a clear line separating these two approaches to kinetics. Nevertheless, this is a useful classification if only because it brings out the fact that we might have different reasons to study kinetics.

Regardless of the approach taken, the central concept in kinetics is the rate of reaction. We express rates as an amount of change divided by the time, or

$$\text{rate} = \frac{\Delta x}{\Delta t}. \tag{11.1}$$

When $\Delta t$ is small, the rate becomes a derivative:

$$\text{rate} = \lim_{\Delta t \to 0} \frac{\Delta x}{\Delta t} = \frac{dx}{dt}.$$

Accordingly, the study of rate processes inevitably leads to the use of calculus.

In general, the rate will have units of $x$, whatever those are, over time. In solution, for example, the rate will often have units of $mol\,L^{-1}s^{-1}$.

## 11.2 Subtleties of the rate concept

The concept of rate of reaction, which seems quite simple on the surface, has a number of subtleties. In particular, the rates of appearance and disappearance of reactants and products in a reaction need not be the same, although they are usually simply related.

Suppose that we are discussing a reaction with the stoichiometry

$$A \to 2B$$

To save myself some writing when I'm analyzing abstract reactions like this one, I usually write $a$ instead of [A] for the concentration of A, and similarly for other reactants and products. In this reaction, the rate $da/dt$ would be a negative number telling us how fast the concentration of A decreases with time due to this reaction, which consumes A. Similarly, $db/dt$ would be the rate at which the concentration of B increases. These rates will normally be related by

$$\frac{db}{dt} = -2\frac{da}{dt}. \tag{11.2}$$

because

(1) B is formed when A is removed so the signs are opposite, and
(2) since 2 units of B are formed for every unit of A consumed, the rate of appearance of B is twice as large (in absolute value) as the rate of removal of A. To see the latter, consider that, by stoichiometry, $\Delta b = -2\Delta a$. Equation (11.2) then follows from the definition of the rate (11.1). The one exception is the case where there are intermediates that accumulate to significant concentrations, in which case they need to be considered in the stoichiometric balance. This is, however, a rare event.

Since each reactant or product can, in principle, have a different rate of formation or removal, the concept of the "rate of reaction" is ill-defined unless we introduce a convention: the rate of reaction is the rate of appearance of a (possibly imaginary) product with a stoichiometric coefficient of 1. In the above example, the rate of reaction is therefore

$$\text{reaction rate} = v = -\frac{da}{dt} = \frac{1}{2}\frac{db}{dt}.$$

Note that the symbol $v$ is usually used for the rate of reaction because many classical authors used the term "reaction velocity" where we would now use "reaction rate."

**Example 11.1 Rate of reaction** For the reaction

$$2A + 3B \rightarrow 4C$$

the rate of reaction is

$$v = -\frac{1}{2}\frac{da}{dt} = -\frac{1}{3}\frac{db}{dt} = \frac{1}{4}\frac{dc}{dt}.$$

Note that the relationship between the rate of reaction and the rates of formation or removal depends on how the reaction is written. If we rewrite the reaction $\frac{1}{2}A + \frac{3}{4}B \rightarrow C$, we obtain, for instance, $v = dc/dt$. Thus, it is important to always show the reaction for which we are giving a reaction rate.

**Example 11.2 Rate of reaction with "excess" molecules** Consider the reaction

$$3A \rightarrow A + B.$$

There is a complication here: A is both a reactant and a product of this reaction. In some sense, the reaction is just

$$2A \rightarrow B.$$

We will see later that there are sometimes sensible reasons to write reactions with "excess" molecules, but for the purpose of determining the relationships between rates, the second "simplified" form, which just shows the stoichiometry of the reaction, is the correct one to use. Thus we have

$$v = -\frac{1}{2}\frac{da}{dt} = \frac{db}{dt}.$$

## 11.3 Kinetics experiments

Before we get too deeply into the subject, it is useful to know that there are two fundamentally different kinds of kinetics experiments in common use:

**Initial rate experiments**: in an initial rate experiment, we measure how fast a reaction proceeds in its initial moments. This often allows us to determine the **rate law** of a reaction, which is the relationship between rate and concentration.
    The initial rate is typically computed using the approximate Formula (11.1).
**Progress curve experiments**: we can also let a reaction go for a while, measuring the concentration as a function of time. This is called a progress curve experiment. Progress curves can also sometimes be used to determine a rate law, but more importantly they allow us to see how the reaction behaves beyond the first moments.

**Example 11.3 Calculating the initial rate of reaction** Suppose that for the reaction $2A \rightarrow B$ at $25\,^\circ C$ with an initial concentration of A of $0.9\,\text{mol}\,L^{-1}$, $0.05\,\text{mol}\,L^{-1}$ of A is converted to product in 3.4 s. Then

$$\frac{da}{dt} \approx \frac{-0.05 \text{ mol L}^{-1}}{3.4 \text{ s}} = -0.015 \text{ mol L}^{-1}\text{s}^{-1}.$$

The rate of reaction is therefore

$$v = -\frac{1}{2}\frac{da}{dt} = 7.4 \times 10^{-3} \text{ mol L}^{-1}\text{s}^{-1}.$$

---

### Exercise group 11.1

(1) Suppose that, for a reaction A → 3B, $1.4 \times 10^{-4}$ mol L$^{-1}$ of product accumulates in the first 3.5 s of the reaction. What is the initial rate?

(2) A set of initial rate experiments was performed for the reaction A → 2P by stopping the reaction after approximately 1 min and measuring the concentration of P (initially zero). In one run, the initial concentration of A was 0.04 mol L$^{-1}$. The reaction was stopped at 64 s, at which time $p$ had reached $1.3 \times 10^{-5}$ mol L$^{-1}$. What was the initial rate?

---

### 11.4 Elementary and complex reactions

Most reactions are **complex**, which is to say that they occur via more than one step. Each step of the reaction is called an **elementary reaction**. An elementary reaction is a process that really occurs, exactly as written. To put it slightly differently, an elementary reaction can't be broken down into simpler chemical reactions. A reaction **mechanism** is the complete sequence of elementary reactions that contribute to the kinetics of a given reaction.

For example, consider the formation of $NO_2$ from NO and $O_2$. The overall reaction is

$$2NO_{(g)} + O_{2(g)} \rightarrow 2NO_{2(g)}.$$

This is, as it turns out, a complex reaction. The elementary reactions making up this mechanism are

$$2NO_{(g)} \rightarrow N_2O_{2(g)},$$
$$N_2O_{2(g)} + O_{2(g)} \rightarrow 2NO_{2(g)}.$$

When we say that $2NO_{(g)} \rightarrow N_2O_{2(g)}$ is elementary, we mean that if we were to watch the reaction happen on a molecular scale, we would actually see two NO molecules collide and form a bond to produce an $N_2O_2$ molecule in one step.

Note that the elementary reactions, when added, give the overall reaction. A mechanism can include reactions that don't contribute to the overall reaction. These might be side reactions or other processes that affect the rate of reaction, but aren't on the pathway from reactants to products. Nevertheless, there will always be some subset of the reactions that adds up to give the overall reaction.

$N_2O_2$ appears in the mechanism, but not in the overall reaction. We call such a species an **intermediate**. Intermediates need not be added to the reaction mixture. They are generated during the reaction, either from the reactants or from other intermediates, and are consumed to make products.

There is no sure way to tell whether a reaction is elementary or complex. However, the following rules are helpful:

(1) Reactions that involve more than two reactant molecules are rarely elementary, especially in the gas phase. The elementary reaction $2NO_{(g)} \rightarrow N_2O_{2(g)}$ requires a collision between two molecules of NO. Collisions between pairs of molecules are frequent events. Three-body collisions, i.e. events in which three molecules arrive at the *same* place at *exactly* the same time, are rare in the gas phase. Elementary steps involving three gaseous reactant molecules are therefore rare. In solution, elementary steps involving three reactant molecules are a little more likely but are still rare. The difference is that once two molecules come into close proximity in solution, the solvent tends to keep them from moving away from each other for a while. There is therefore a little time for a third molecule to arrive. Elementary reactions involving four or more reactant molecules are unknown, either in the gas phase or in solution.

(2) The proposed elementary step should correspond to sensible chemistry; the step should not involve more than a few bond-forming and bond-breaking events. Intermediates formed may be highly reactive species such as radicals or diradicals, but they should not be chemically nonsensical.

**Example 11.4 Elementary reaction or not?** The reaction $F_{2(g)} + 2ClO_{2(g)} \rightarrow 2FClO_{2(g)}$ is almost certainly not elementary because of the number of reactant molecules and the fact that this is a gas-phase reaction.

**Example 11.5 Is combustion of graphite elementary?** The reaction $C_{(s)} + O_{2(g)} \rightarrow CO_{2(g)}$ is not elementary. Too many bonds must be broken and formed for this reaction to be elementary; the reaction must remove a carbon atom from graphite (which requires the breaking of three C–C bonds), break an O–O bond and create two C–O bonds. An additional clue to the non-elementary nature of this reaction is provided by the fact that, in an oxygen-poor environment, this reaction produces carbon monoxide. There are two possible interpretations of this fact: either CO is an intermediate that goes on to react with an oxygen-containing species ($O_2$ itself or another intermediate like an oxygen atom) to make $CO_2$ in an oxygen-rich atmosphere, or the reaction has other intermediates that can make CO in the absence of sufficient oxygen. In the latter case, we would say that the reaction has two **reaction channels**, one of which leads to $CO_2$, and the other to CO. The proportion we get of one or the other product will then depend on reaction conditions. A major thrust of modern research in reaction dynamics is the control of chemical reactions in order to select a particular reaction channel.

**Exercise group 11.2**

(1) Classify each of the following reactions as possibly, probably not or certainly not elementary and explain the basis for your decision.

(a) $H_2O_{2(aq)} + 2Fe^{2+}_{(aq)} + 2H^+_{(aq)} \rightarrow 2H_2O_{(l)} + 2Fe^{3+}_{(aq)}$

(b) $Cl_{(g)} + O_{3(g)} \rightarrow ClO_{(g)} + O_{2(g)}$

(c) $CH_{4(g)} + 2O_{2(g)} \rightarrow CO_{2(g)} + 2H_2O_{(g)}$

## 11.5 The law of mass action

Given a mechanism, we can write down rate equations using the

**Law of mass action**: the rate of an *elementary* reaction is proportional to the product of the *reactant* concentrations.

The proportionality constant is an **elementary rate constant**. The value of the rate constant depends on the reaction conditions (temperature, pH, ionic strength), but not (directly) on the concentrations of reactants or products.

The law of mass action is best studied by example:

**Example 11.6 Application of the law of mass action** Consider the elementary reaction $A + B \xrightarrow{k} C$. The rate of the reaction is $kab$. The rates of change of the concentrations are

$$\frac{da}{dt} = \frac{db}{dt} = -v = -kab,$$

$$\frac{dc}{dt} = v = kab.$$

In the foregoing example, note the placement of the rate constant over the reaction arrow to associate it with the reaction. This notation will become useful when we have several elementary reactions.

The **partial order** of a reaction with respect to a given reactant is the exponent of the corresponding concentration in the rate law. In the preceding example, the (partial) order with respect to A is 1, and the order with respect to B is also 1. The (overall) **order of reaction** is the sum of the partial orders. Thus, we say that the reaction in Example 11.6 is a second-order reaction. According to the law of mass action, the partial and overall orders of an elementary reaction can only be positive whole numbers. In complex reactions, we will see that the orders can be fractions or even negative numbers.

**Example 11.7 2A → B reactions** For the elementary reaction $2A \xrightarrow{k} B$, the rate of the reaction is $ka^2$. According to our definition of rate of reaction, we have

$$\frac{db}{dt} = -\frac{1}{2}\frac{da}{dt} = ka^2,$$

or

$$\frac{da}{dt} = -2ka^2$$

$$\text{and} \quad \frac{db}{dt} = ka^2.$$

You should get used to writing the rate equations directly in the latter form. The logic behind our equation for $da/dt$ is that the reaction removes two molecules of A per elementary step, leading to a coefficient of $-2$. When we start adding more steps, going back to the basic definition of the rate of reaction is actually confusing.

When there is more than one elementary step, the overall rate equation for a given reactant or product is obtained by adding the rates due to the individual steps. For now, we will only deal with reversible elementary reactions, but the principle would be the same for a complex reaction.

**Example 11.8 The law of mass action and reversible reactions** Suppose that the reversible elementary reaction A $\underset{k_-}{\overset{k_+}{\rightleftharpoons}}$ B is a complete mechanism. The rate of the forward step is $k_+a$. The rate of the reverse step is $k_-b$. The forward step consumes A while the reverse step produces it, so

$$\frac{da}{dt} = -k_+a + k_-b.$$

The reverse is true for B so for this compound,

$$\frac{db}{dt} = k_+a - k_-b.$$

Again note the use of rate constants over and under the reversible harpoons to associate rate constants to each direction of the reaction, each of which is to be thought of as a separate elementary reaction.

**Example 11.9 A more complex example of the law of mass action** Suppose that 2A $\underset{k_-}{\overset{k_+}{\rightleftharpoons}}$ B is the complete mechanism of a reaction. The rates of the forward and reverse steps are, respectively, $k_+a^2$ and $k_-b$. The rate of production of B is therefore

$$\frac{db}{dt} = k_+a^2 - k_-b.$$

For each B formed in the forward reaction, two As are consumed. Similarly, for each B consumed in the reverse reaction, two As are produced. Accordingly,

$$\frac{da}{dt} = -2k_+a^2 + 2k_-b.$$

Note where the stoichiometric factors go in this last equation: the forward reaction removes two molecules of A, so we put $-2$ in front of $k_+a^2$. The reverse reaction creates two molecules of A, so the stoichiometric coefficient $+2$ appears in front of $k_-b$.

## Exercise group 11.3

(1) Give the rate of reaction and rate of change of $a$ for each of the following reactions:

    (a) $A + B \xrightarrow{k} 2C$

    (b) $2A \xrightarrow{k} A + B$

    (c) $A \underset{k_{-1}}{\overset{k_1}{\rightleftharpoons}} B + C$

(2) Give the rate equations for all the chemical species in the following mechanism:

$$A + B \underset{k_{-1}}{\overset{k_1}{\rightleftharpoons}} C$$

$$2C \xrightarrow{k_2} P$$

### 11.6 Microscopic reversibility and chemical equilibrium

Chemical mechanisms are subject to an additional constraint, known as the

**Law of microscopic reversibility**: every elementary reaction is reversible. In other words, a complete mechanism should include both forward and reverse versions of every elementary step.

This law is necessary to make kinetics compatible with thermodynamics. In practice, it is often possible to obtain a useful mechanism without including all reverse steps. However, we should be aware that when we leave out a reverse step, we are making an approximation to the true mechanism.

    To understand why the law of microscopic reversibility is necessary, consider the following elementary reaction, which we assume to be irreversible:

$$A \rightarrow B$$

This reaction implies that *all* of the A initially present in the system will be transformed to B. In other words, since the equilibrium constant is $K = b/a$, it implies an infinitely large equilibrium constant. Moreover, since $K = \exp(-\Delta_r G^\circ_m / RT)$, $\Delta_r G^\circ_m$ must be infinitely large and negative. Infinite values of $\Delta_r G^\circ_m$ or, equivalently, of $K$, are not physically possible since A and B must each have a finite free energy. The reaction step must therefore be at least somewhat reversible. In other words, we must have

$$A \underset{k_-}{\overset{k_+}{\rightleftharpoons}} B.$$

For this mechanism,

$$\frac{da}{dt} = -\frac{db}{dt} = -k_+ a + k_- b.$$

Equilibrium is reached when the rates of change of the concentrations are zero. In this case, equilibrium requires

$$k_+ a = k_- b$$

or

$$K = \frac{b}{a} = \frac{k_+}{k_-}.$$

The empirical equilibrium constant (which may have units) can therefore be calculated from the rate constants. Note that the rates of the forward and reverse reactions are not zero at equilibrium. The reactions are still happening, but the forward and reverse reactions occur at equal rates so there is no net change in the concentrations. This is what we mean by **dynamic equilibrium**.

It is worth pausing here to discuss the two different kinds of equilibrium constants we have encountered thus far. In Chapter 8, we encountered thermodynamic equilibrium constants, which have no units. These are constructed as ratios of activities which will in general contain activity coefficients as well as concentrations. On the other hand, the empirical equilibrium constant encountered here is just a ratio of concentrations. If all substances in a system behave ideally, then the two are numerically equal, provided we use units in the empirical equilibrium constant that are consistent with the desired standard state (e.g. $mol\,L^{-1}$ for solutes in the chemists' standard state). On the other hand, if any of the substances in a reaction display non-ideal behavior, then the two will deviate by a factor of the ratio of activity coefficients.

**Example 11.10 Rate and equilibrium constants** Suppose that, for the reaction

$$A \underset{k_-}{\overset{k_+}{\rightleftharpoons}} 2B,$$

the equilibrium constant is $1.8 \times 10^{18}\,mol\,L^{-1}$ and $k_+ = 14\,s^{-1}$. The equilibrium condition is

$$\frac{da}{dt} = -k_+ a + k_- b^2 = 0.$$

$$\therefore K = \frac{b^2}{a} = \frac{k_+}{k_-}.$$

$$\therefore k_- = \frac{k_+}{K} = \frac{14\,s^{-1}}{1.8 \times 10^{18}\,mol\,L^{-1}} = 7.8 \times 10^{-18}\,L\,mol^{-1}s^{-1}.$$

Note that measuring such a small rate constant directly would be impractical.

Although every reaction is *in principle* reversible, some are *effectively* irreversible. The last example shows this: for all but the tiniest of values of $a$, $\frac{da}{dt} \approx -k_+ a$. This suggests that many reactions may have simple rate laws, a topic that we will pick up in the next chapter.

## Exercise group 11.4

(1) For the elementary reaction

$$H_{(g)} + Br_{2(g)} \underset{k_-}{\overset{k_+}{\rightleftharpoons}} HBr_{(g)} + Br_{(g)},$$

$k_+ = 8.47 \times 10^6 \, \text{bar}^{-1}\text{s}^{-1}$ and the equilibrium constant $K = 1.18 \times 10^{31}$ at 298.15 K. What is the value of $k_-$?

## Key ideas and equations

- Rate of reaction: rate of formation of a product with a stoichiometric coefficient of 1.
- Elementary reactions occur as written. Complex reactions can be broken down into several elementary reactions.
- Law of mass action: the rate of an elementary reaction is proportional to the product of the reactant concentrations.
- Order of reaction, rate constant.
- Law of microscopic reversibility.

## Review exercise group 11.5

(1) The rate constant for the elementary reaction

$$I_{(g)} + I_{(g)} + Ar_{(g)} \rightarrow I_{2(g)} + Ar_{(g)}$$

is $5.9 \times 10^3 \, \text{m}^6\text{mol}^{-2}\text{s}^{-1}$ at 293 K. The empirical equilibrium constant for this reaction can be estimated from thermodynamic data to have the value $1.15 \times 10^{20} \, \text{m}^3 \, \text{mol}^{-1}$.
   (a) What is the rate constant for the reverse reaction?
   (b) What is unusual about this reaction?
   (c) What is the order of the reaction with respect to argon? With respect to the iodine atoms?
   (d) What rate of reaction does the law of mass action predict if the partial pressure of iodine atoms is 15 mPa and the partial pressure of argon is 80 kPa?
      *Hint*: Given the units of the rate constant, in what units do you need [Ar] and [I]?
   (e) At the partial pressures given above, roughly how long would it take to use up 1% of the iodine atoms?
(2) Proflavin is a large aromatic ion used to stain DNA. Like many other planar aromatic

species, proflavin dimerizes in water:

$$2P_{(aq)} \underset{k_-}{\overset{k_+}{\rightleftharpoons}} D_{(aq)}$$

where P is a proflavin ion and D is the dimer. The rate constants for this reaction are

$$k_+ = 8 \times 10^8 \, \text{L mol}^{-1}\text{s}^{-1},$$
$$k_- = 2.0 \times 10^6 \, \text{s}^{-1}.$$

(a) Calculate the equilibrium constant for the dimerization reaction.
(b) Write down the rate equations for proflavin and its dimer.
(c) If $[P] = 10^{-4} \, \text{mol L}^{-1}$ and $[D] = 10^{-2} \, \text{mol L}^{-1}$, does the reaction proceed in the forward or reverse direction?

# 12

# Initial rate experiments and simple empirical rate laws

This chapter will introduce you to the analysis of a key class of kinetics experiments, namely initial rate experiments. We will consider only the simplest class of rate laws, those where the rate is proportional to some powers of the concentrations. This is a limited class, but one that comes up surprisingly often in practice. Initial rate experiments are also useful when a reaction has a complex rate law. However, complex rate laws are best understood in the context of a study of reaction mechanisms, so we put off this discussion for now.

## 12.1 Initial rate studies

Rate laws of chemical reactions are often complicated. They can depend on the concentrations of reactants, products, or other chemicals present in the system in almost arbitrary ways. It would be impractical to try to deal at once with all possible complications, so we often try to study reactions under conditions in which we can expect simpler behavior.

Perhaps the most widely used experimental simplification is the **method of initial rates**, which was briefly mentioned in Section 11.3. In this method, we study the reaction only for a very short time after initiation. During this time, the concentrations change very little so that a number of complications arising from changes in the reaction mixture over time are avoided. For instance, if we start with a mixture containing only reactants, then very little product will accumulate during the reaction and we can neglect the reverse reaction whose rate, according to the law of mass action, depends on the product concentrations.

How long a period of time should an initial rate study cover? Unfortunately, there isn't a simple answer to this question. Chemical reactions occur over a vast range of time scales, ranging from geological periods (thousands of years or more) to the microsecond range, or faster. The key is to study the reaction for a time that is short compared to the time over which products accumulate. One useful measure of what is a short and what is a long time in a reaction is the **half-life**. The half-life is the time it takes for the reaction to get half-way to equilibrium. In a nearly irreversible reaction, the half-life would be the time required for the concentration of the limiting reagent to fall to half its initial value. An initial rate study would cover a period of time that is short (usually very short) compared to the half-life. However, the period of the study must be sufficiently long for there to have been a measurable change in some observable property of the reaction mixture, for instance the accumulation of an accurately measurable amount of product. Thus, one wishes to follow a reaction for a short time, but not too short a time. There is no good way to decide a priori what is too long or too short a time. We can, however, check experimentally that we have a reaction time in the right range: increasing or decreasing the reaction time by a small factor shouldn't make much difference to the rate calculated if the reaction time is in an appropriate range for an initial rate study.

## 12.2 Simple empirical rate laws

Suppose that we have collected some initial rate data, having systematically varied, one by one, the concentrations of the reactants and perhaps also of the products. We would typically want to summarize our data with some kind of rate law. Since we will derive this rate law directly from the data, we would call this an **empirical rate law**. It is surprisingly common for the empirical rate law for a reaction to be as simple as

$$v = ka^{\alpha}b^{\beta}c^{\gamma}\dots$$

where $a, b, c \dots$ are concentrations of reactants or products of the reaction and $k, \alpha, \beta, \gamma \dots$ are constants. Given the law of mass action, we can imagine why we might get reaction rates of this form with integer exponents. However, in complex reactions, the partial orders of reaction need not be positive integers and are generally unrelated to the stoichiometric coefficients. Similarly, the empirical rate constant $k$ is not necessarily an elementary rate constant of the mechanism. The partial orders and rate constant are best thought of here as empirical values to be determined by experiment. They are related to the mechanism and to its rate constants and partial orders, but we will have to put off this discussion until later.

Also note that the empirical rate constant and partial orders can depend on reaction conditions (temperature, pH, ionic strength etc.) but that these constants should all be independent of the concentrations of reactants and products, at least over some range of concentrations.

The following examples will show how simple empirical rate laws can be obtained from experimental data, assuming of course that the data admit such a rate law.

**Example 12.1 Determining a simple empirical rate law from initial rates** Consider the following initial rate data for the reaction

$$F_{2(g)} + 2ClO_{2(g)} \rightarrow 2FClO_{2(g)}$$

| Experiment | $[F_2]/\text{mol L}^{-1}$ | $[ClO_2]/\text{mol L}^{-1}$ | $v/10^{-3}\text{mol L}^{-1}\text{s}^{-1}$ |
|---|---|---|---|
| 1 | 0.10 | 0.010 | 1.2 |
| 2 | 0.10 | 0.040 | 4.8 |
| 3 | 0.20 | 0.010 | 2.4 |

Experiments 1 and 3 were carried out under identical conditions except that the concentration of $F_2$ was doubled. The reaction rate also doubled so the rate is proportional to $[F_2]$. In other words, this reaction is of the first order with respect to $F_2$. Analogously, experiments 1 and 2 were carried out under identical conditions except that the concentration of $ClO_2$ is four times higher in experiment 2 than in experiment 1. The rate also goes up by a factor of four so the reaction is also of the first order with respect to $ClO_2$. The data are therefore consistent with the rate law

$$v = k[F_2][ClO_2].$$

This is a second-order reaction (sum of partial orders $= 2$). Note that the order with respect to $ClO_2$ is different from its stoichiometric coefficient. This proves that this reaction is not elementary, otherwise the law of mass action would require a partial order of 2 with respect to $ClO_2$.

Moreover, it is important to understand that the data suggest the above rate law, but fall far short of proving it. The problem is that there is just enough data to calculate the partial orders, but not enough to verify that these orders are consistent across a range of concentrations. We should really have additional data points in which $[ClO_2]$ is varied at fixed $[F_2]$, and vice versa.

We can also calculate the rate constant $k$ from the data:

$$k = \frac{v}{[F_2][ClO_2]}.$$

Any one point will give a value for $k$. For instance, using the first point, we get

$$k = \frac{1.2 \times 10^{-3}\,\text{mol}\,L^{-1}s^{-1}}{(0.10\,\text{mol}\,L^{-1})(0.010\,\text{mol}\,L^{-1})} = 1.2\,L\,\text{mol}^{-1}s^{-1}.$$

In principle, we should repeat this calculation with the other two points and average the results. In this case, you will get exactly the same value in all three cases.

**Example 12.2 Another empirical rate law determined from initial rates** For the reaction

$$2HgCl_{2(aq)} + C_2O_{4(aq)}^{2-} \rightarrow Hg_2Cl_{2(s)} + 2Cl_{(aq)}^- + 2CO_{2(g)},$$

the following initial rate data were obtained:

| Experiment | $[HgCl_2]/mol\,L^{-1}$ | $[C_2O_4^{2-}]/mol\,L^{-1}$ | $v/10^{-7}mol\,L^{-1}s^{-1}$ |
|---|---|---|---|
| 1 | 0.10 | 0.10 | 1.3 |
| 2 | 0.10 | 0.20 | 5.2 |
| 3 | 0.20 | 0.20 | 10 |

Compare experiments 1 and 2: at identical $[HgCl_2]$, doubling $[C_2O_4^{2-}]$ increases the rate by a factor of $\frac{5.2}{1.3} = 4$. The rate therefore increases as the *square* of the $C_2O_4^{2-}$ concentration. The order of the reaction with respect to $C_2O_4^{2-}$ is therefore 2. Similarly, experiments 2 and

3 imply a partial order of 1 with respect to $HgCl_2$. The rate law is therefore

$$v = k[HgCl_2][C_2O_4^{2-}]^2.$$

The rate constant is

$$k = \frac{1.3 \times 10^{-7} \, \text{mol} \, \text{L}^{-1} \text{s}^{-1}}{(0.10 \, \text{mol} \, \text{L}^{-1})(0.10 \, \text{mol} \, \text{L}^{-1})^2} = 1.3 \times 10^{-4} \, \text{L}^2 \text{mol}^{-2} \text{s}^{-1}.$$

**Example 12.3 Partial orders that can't be determined by inspection** Consider the following data for a reaction $A + B \rightarrow P$:

| Experiment | $a/\text{mol} \, \text{L}^{-1}$ | $b/\text{mol} \, \text{L}^{-1}$ | $p/\text{mol} \, \text{L}^{-1}$ | $v/10^{-3} \text{mol} \, \text{L}^{-1} \text{s}^{-1}$ |
|---|---|---|---|---|
| 1 | 0.2 | 0.1 | 0.5 | 1.6 |
| 2 | 0.4 | 0.1 | 0.5 | 2.3 |
| 3 | 0.8 | 0.1 | 0.5 | 3.3 |
| 4 | 0.2 | 0.2 | 0.5 | 1.6 |
| 5 | 0.2 | 0.3 | 0.5 | 1.6 |
| 6 | 0.2 | 0.1 | 0.3 | 2.7 |
| 7 | 0.2 | 0.1 | 0.1 | 8.0 |

Let's start by considering experiments 1, 4 and 5. When the initial concentration of B is changed, there is no effect on the rate. Accordingly the order of the reaction with respect to B is zero.

Experiments 1, 2 and 3 show the variation of the rate with $a$, but in this case, most of us wouldn't be able to figure out the order of reaction by inspection as we have previously done. We start by calculating the ratios of the concentrations and rates:

| Experiments $(i, j)$ | $a_j/a_i$ | $v_j/v_i$ |
|---|---|---|
| 1, 2 | 2 | 1.44 |
| 2, 3 | 2 | 1.43 |

Each time we double the concentration, the rate increases by the same factor. This indicates that the relationship is a simple power law. The question then is to what exponent must we raise 2 to get roughly 1.43? Some of you might be able to see the answer right away, but for the rest of us, here's how we go about this problem. We are looking for a relationship of the form $v \propto a^\alpha$ (holding everything else constant). Therefore, the ratio of rates at different values of $a$ is

$$\frac{v_j}{v_i} = \frac{a_j^\alpha}{a_i^\alpha} = \left(\frac{a_j}{a_i}\right)^\alpha.$$

To find the value of $\alpha$, we take the logarithm (any base) of both sides of this equation

$$\log\left(\frac{v_j}{v_i}\right) = \log\left(\frac{a_j}{a_i}\right)^\alpha = \alpha\log\left(\frac{a_j}{a_i}\right);$$

$$\therefore \alpha = \frac{\log(v_j/v_i)}{\log(a_j/a_i)}.$$

In this case, we have $\alpha \approx \log 1.43/\log 2 = 0.52$. We could report this value as the order of the reaction, however, we will see later that orders that are simple fractions come up fairly frequently in certain types of mechanisms. In this case, the order we calculated is close enough to $\frac{1}{2}$ that, considering the precision of the data, we would be justified in reporting the order as $\alpha = \frac{1}{2}$.

We now turn to experiments 1, 6 and 7 in which the concentration of P was varied. Again, we compute the ratios of concentrations and rates in the different experiments:

| Experiments $(i, j)$ | $p_i/p_j$ | $v_i/v_j$ |
|---|---|---|
| 1, 6 | 1.67 | 0.59 |
| 6, 7 | 3 | 0.34 |

Here, the rate *increases* as the concentration of P *decreases*. The partial order with respect to P is therefore *negative*. We could solve for this exponent as we did for the order of reaction with respect to A, but this one is not quite as difficult. Note that $1/1.67 = 0.60$ and that $1/3 = 0.33$. Accordingly, the order is $-1$. The complete rate law is therefore

$$v = ka^{1/2}p^{-1}.$$

This rate law implies that the product P slows down the reaction, a common phenomenon known as **product inhibition**.

The rate constant can be calculated from $k = vp/a^{1/2}$. If we calculate the rate constant for each experiment and average the values, we get $k = 1.80 \times 10^{-3}\,\mathrm{mol^{3/2}L^{-3/2}s^{-1}}$.

---

### Exercise group 12.1

(1) The rate of a reaction triples when the concentration of a reactant is doubled. What is the order of the reaction with respect to the concentration of this reactant?
(2) For the reaction of ozone with nitrogen monoxide $(O_{3(g)} + NO_{(g)} \rightarrow O_{2(g)} + NO_{2(g)})$ at 340 K, the following initial rate data have been obtained:

| $[O_3]/\mu\mathrm{mol\,L^{-1}}$ | $[NO]/\mu\mathrm{mol\,L^{-1}}$ | $v/\mu\mathrm{mol\,L^{-1}s^{-1}}$ |
|---|---|---|
| 2.1 | 2.1 | 16 |
| 4.2 | 2.1 | 32 |
| 6.3 | 2.1 | 48 |
| 6.3 | 4.2 | 96 |
| 6.3 | 6.3 | 144 |

Determine the empirical rate law and the value of the rate constant.

(3) For the bromination of acetone $(CH_3COCH_{3(aq)} + Br_{2(aq)} \rightarrow CH_3COCH_2Br_{(aq)} + H^+_{(aq)} + Br^-_{(aq)})$, the following initial rate data were obtained:

| Experiment | $[CH_3COCH_3]$ mol L$^{-1}$ | $[Br_2]$ mol L$^{-1}$ | $[H^+]$ mol L$^{-1}$ | $v$ $10^{-4}$ mol L$^{-1}$ s$^{-1}$ |
|---|---|---|---|---|
| 1 | 0.30 | 0.050 | 0.050 | 0.57 |
| 2 | 0.30 | 0.10 | 0.050 | 0.57 |
| 3 | 0.30 | 0.050 | 0.10 | 1.2 |
| 4 | 0.40 | 0.050 | 0.20 | 3.1 |
| 5 | 0.40 | 0.050 | 0.050 | 0.76 |

Determine the rate law (orders and rate constant).

(4) Iodine reacts with ketones (organic molecules containing a non-terminal C=O group) in aqueous solution according to the general reaction scheme

$$I_2 + ketone \rightarrow iodoketone + H^+ + I^-$$

For one particular ketone, the following initial rate data were obtained:

| | $[I_2]$ $10^{-4}$ mol L$^{-1}$ | $[ketone]$ mol L$^{-1}$ | $[H^+]$ $10^{-2}$ mol L$^{-1}$ | $v$ $10^{-4}$ mol L$^{-1}$ s$^{-1}$ |
|---|---|---|---|---|
| 1 | 5.0 | 0.2 | 1.0 | 0.70 |
| 2 | 3.0 | 0.2 | 1.0 | 0.70 |
| 3 | 5.0 | 0.5 | 1.0 | 1.7 |
| 4 | 5.0 | 0.5 | 3.2 | 5.4 |

(a) What is the rate law for this reaction?
(b) Calculate the rate constant.

## 12.3 The van't Hoff method

There exists another method for determining simple empirical rate laws. Suppose that we have a set of initial rate measurements for a reaction that we suspect of having a simple rate law of the form

$$v = ka^n,$$

$a$ being the concentration of a reactant. Taking a logarithm of each side of the equation, we get

$$\ln v = \ln k + n \ln a.$$

A plot of $\ln v$ as a function of $\ln a$ should therefore give a straight line of slope $n$, the order of the reaction, and of intercept $\ln k$. This method was first devised by van't Hoff. The advantage of van't Hoff's method is that it allows us to use more than two data points at a time. This tends to increase our confidence in the results.

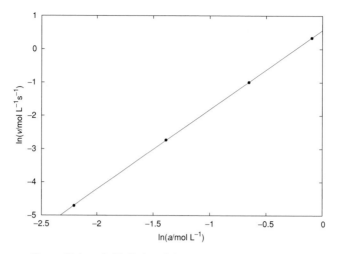

Figure 12.1 van't Hoff plot of the data from Example 12.4.

We will get the slope and intercept of the line by linear regression. In addition, we should *always* plot the data since it is nearly impossible to figure out if the data fit a linear relationship without looking at the graph. The correlation coefficient, which you may have been told to use for such purposes, is actually not that useful, as we will see in Section 13.2.

**Example 12.4 Application of the van't Hoff method**  Consider the following initial rate data for a reaction A → P:

| $a/\mathrm{mol\,L^{-1}}$ | 0.11 | 0.25 | 0.52 | 0.91 |
|---|---|---|---|---|
| $v/\mathrm{mol\,L^{-1}s^{-1}}$ | $9.0 \times 10^{-3}$ | $6.5 \times 10^{-2}$ | 0.37 | 1.4 |

To use van't Hoff's method, we take the logarithms of both quantities:

| $\ln(a/\mathrm{mol\,L^{-1}})$ | $-2.21$ | $-1.39$ | $-0.65$ | $-0.09$ |
|---|---|---|---|---|
| $\ln(v/\mathrm{mol\,L^{-1}s^{-1}})$ | $-4.71$ | $-2.73$ | $-0.99$ | $0.34$ |

The graph of this data set is shown in Figure 12.1. The data plotted in this fashion fit a straight line of slope $n = 2.4$ and intercept $\ln k = 0.56$. Thus, the reaction does obey a simple rate law: $v = ka^{2.4}$. The rate constant is $k = \exp(0.56) = 1.75\,\mathrm{L^{1.4}\,mol^{-1.4}\,s^{-1}}$. The units of the rate constant are obtained by unit analysis from the rate law.

While van't Hoff's method is a reasonably good way to determine the order of a reaction, it must be used with caution. The problem is that log–log plots tend to make things look linear, even when the underlying relationship isn't a simple power law. Thus, while we *can* use a van't Hoff plot to determine whether a reaction obeys a simple rate law, a careful experimentalist will always:

(1) Try to get data for the van't Hoff plot spanning a few orders of magnitude in the reactant concentration. In our last example, the largest and smallest concentrations of A were in a ratio of approximately 9:1. One would normally like this ratio to be more like 100:1 to make sure that a power law is obeyed over a reasonable range of concentrations.
(2) Confirm the order obtained by the van't Hoff methods by other methods. The best methods for confirming the order of reaction are based on integrated rate laws, to be studied in Chapter 13.

Also, while we *can* calculate the rate constant from the van't Hoff method, we should regard the value obtained in this manner as a rough approximation. There are two problems: first, as a general rule, intercepts are more sensitive to experimental error than slopes; second, we need to apply the exponential function to get a rate constant. This tends to magnify errors, especially if we are dealing with large values, which is common in kinetics. Suppose, for instance, that the intercept in a van't Hoff plot was 11.5. This would imply a rate constant of $9.9 \times 10^4$ (in appropriate units). However, there is always some uncertainty in the estimate of the intercept due to the experimental scatter in the points. Suppose that in this case, the intercept is known to $\pm 0.4$ (i.e. to an accuracy of about 3.5%, which is actually very good for such a determination). Then the rate constant might be as small as $\exp(11.4 - 0.4) = 6.0 \times 10^4$ or as large as $\exp(11.4 + 0.4) = 1.3 \times 10^5$. This is quite a large spread of values!

It is also possible to use the van't Hoff method for simple rate laws where the rate depends on more than one concentration. If, for instance,

$$v = ka^n b^m$$

then

$$\ln v = \ln k + n \ln a + m \ln b.$$

If we perform a series of experiments holding $b$ constant, then $n$ is the slope of a plot of $\ln v$ vs. $\ln a$. The intercept of this plot is $i_1 = \ln k + m \ln b_1$, where $b_1$ is the value of $b$ in the first set of experiments. Conversely, if we hold $a$ constant and vary $b$, then the partial order $m$ is the slope of a plot of $\ln v$ vs. $\ln b$. The intercept of the second set of experiments is $i_2 = \ln k + n \ln a_2$, where $a_2$ is the value of $a$ in this set of experiments. Once we know $n$ and $m$, the two intercepts give us two independent estimates of $k$.

When would we use the van't Hoff method rather than the simpler methods of Section 12.2? If you have just two or three observations per concentration varied, then you might as well just compare rates between experiments, as we did in Section 12.2. You should prefer the van't Hoff method when you have four or more data points per concentration.

Whatever method is used, obtaining rate constants from initial rate data is problematic. Sometimes, it's the best we can do, but we must be aware that these methods have limitations. On the other hand, initial-rate methods, including the van't Hoff method, are very useful in making a preliminary determination of the order and approximate size of the rate constant, from which additional experiments can be designed.

## Exercise group 12.2

(1) For the decomposition of dinitrogen pentoxide in tetrachloromethane ($2N_2O_5 \rightarrow 4NO_2 + O_2$), the following initial rate data have been obtained:

| $[N_2O_5]/mol\,L^{-1}$ | 0.92 | 1.23 | 1.79 | 2.00 | 2.21 |
|---|---|---|---|---|---|
| $v/\mu mol\,L^{-1}s^{-1}$ | 9.5 | 12.0 | 19.3 | 21.0 | 22.6 |

Determine the empirical rate law (order of reaction and rate constant).

(2) Succinimide can be made by oxidation of pyrrolidone with tetraphenylporphyrinatoiron (III) chloride (TPPFe) and *tert*-butyl hydroperoxide ($Bu^tOOH$):

pyrrolidone          succinimide

Varying only the concentration of pyrrolidone, the following initial rate data were obtained:[1]

| $[pyrrolidone]/mmol\,L^{-1}$ | 0.5 | 3.0 | 5.0 | 10.0 |
|---|---|---|---|---|
| $v/mmol\,L^{-1}h^{-1}$ | 0.908 | 3.86 | 5.98 | 6.52 |

Does this reaction have a simple rate law with respect to pyrrolidone and if so, what is the order?

(3) The reaction $2HBr_{(g)} + NO_{2(g)} \rightarrow H_2O_{(g)} + NO_{(g)} + Br_{2(g)}$ has been studied at $127\,°C$. A series of initial-rate experiments were performed at a constant $NO_2$ pressure of 10 torr but different HBr pressures:

| $p_{HBr}/torr$ | 4.7 | 11.0 | 15.3 | 20.3 | 26.1 | 34.1 |
|---|---|---|---|---|---|---|
| $v/torr\,min^{-1}$ | 0.98 | 2.35 | 3.16 | 3.96 | 5.56 | 6.77 |

A second series of initial-rate experiments were carried out at a constant HBr pressure of 10 torr while varying the $NO_2$ pressure:

| $p_{NO_2}/torr$ | 5.0 | 10.2 | 15.4 | 20.3 | 25.7 |
|---|---|---|---|---|---|
| $v/torr\,min^{-1}$ | 1.27 | 2.37 | 3.64 | 4.94 | 6.19 |

(a) Show that the reaction obeys a simple rate law and determine the orders with respect to the two reactants.
(b) What is the value of the rate constant?

---

[1] J. Iley et al., J. Chem. Soc., Perkin Trans. **2**, 1299 (2001).

## Key ideas and equations

- Many reactions have simple rate laws of the form $v = ka^{\alpha}b^{\beta} \ldots$
- In general, the partial orders of reaction can have any value, positive or negative, although in practice small whole numbers or fractions with small denominators are most common.
- van't Hoff method: if $v = ka^n$, $\ln v = \ln k + n \ln a$, which allows you to extract $k$ and $n$ by a graphical method.

---

### Review exercise group 12.3

(1) In an initial rate experiment, it is found that increasing the concentration of a reactant by a factor of 10 increases the rate by a factor of 3.2. What is the order of the reaction?

(2) (a) In a particular experiment, the rate of hydrolysis of nitrogen dioxide

$$2NO_{2(g)} + H_2O_{(g)} \rightarrow HNO_{2(g)} + HNO_{3(g)}$$

was found to be $0.26 \, \text{mol L}^{-1}\text{min}^{-1}$. What is the rate of change of the concentration of $NO_{2(g)}$?

(b) The empirical rate law for this reaction is $v = k[H_2O][NO_2]$. Suppose that we repeated the experiment for which data was given in question 2a, but tripled the initial concentration of water and doubled the initial concentration of $NO_2$. What rate of reaction would you predict would be measured?

(3) Consider the following table of initial rate data for a reaction $A \rightarrow 2B$:

| Experiment | $a/\text{mol L}^{-1}$ | $b/\text{mol L}^{-1}$ | $v/\text{kmol L}^{-1}\text{s}^{-1}$ |
|---|---|---|---|
| 1 | 1.5 | 1.3 | 17 |
| 2 | 3.5 | 1.3 | 23 |
| 3 | 9.2 | 1.3 | 31 |
| 4 | 1.5 | 0.55 | 18 |
| 5 | 1.5 | 0.13 | 17 |

Verify that this reaction has a simple rate law. Determine the rate law, the overall order and the rate constant.

---

# 13

## Integrated rate laws

Rate laws, whether obtained from the law of mass action or from empirical rate studies, are in fact **differential equations**. What does this mean? In general, a differential equation is a relationship between some quantities and their derivatives. Our rate laws are relationships between concentrations and their derivatives with respect to time. The solution of a set of rate equations is a set of equations that tell us how the concentrations depend on time.

As we saw in the last chapter, even complex reactions can have simple-looking rate laws. In this chapter, we will learn how to solve simple rate equations, and how to use these solutions to analyze data from experiments in which we measure concentration vs. time.

### 13.1 First-order reactions

Let's start by examining a very simple differential equation, namely that arising from a reaction with a first-order rate law:

$$\frac{dx}{dt} = -kx.$$

In this equation, $x$ represents the concentration of a reactant (hence the negative sign). This equation can be solved by **separation of variables**, which we encountered previously when we were deriving the Beer–Lambert law. In this technique, we treat $dx$ and $dt$ as independent entities. We rearrange the differential equation to isolate terms that depend on $x$ from those that depend on $t$:

$$\frac{dx}{x} = -k\, dt. \tag{13.1}$$

(The constant $-k$ depends neither on $x$ nor on $t$ and so could have gone on either side of the equation.) Next, we integrate both sides from an initial condition up to some arbitrary time. Suppose that at $t = 0$, $x = x_0$. Then

$$\int_{x_0}^{x(t)} \frac{dx'}{x'} = -k \int_0^t dt'.$$

A couple of things are worth noting:

- The primed variables are merely intended to distinguish the integration variables from the "real" variables. If we didn't modify the variables under the integral signs, we would end up with the same variable names inside the integrals as in the limits, which might be confusing.
- Starting the integration from $t = 0$ is purely a matter of convention and convenience. If we knew the concentration at some other time, we could equally well start from there.

From here, it is purely a matter of elementary calculus and algebra:

$$\int_{x_0}^{x(t)} \frac{\mathrm{d}x'}{x'} = -k \int_0^t \mathrm{d}t'$$

$$\therefore \ln x' \Big|_{x_0}^{x(t)} = -k \, t' \Big|_0^t$$

$$\therefore \ln x(t) - \ln x_0 = -k(t - 0) \tag{13.2}$$

$$\therefore \ln \left( \frac{x(t)}{x_0} \right) = -kt$$

$$\therefore \frac{x(t)}{x_0} = \exp(-kt)$$

$$\therefore x(t) = x_0 e^{-kt} \tag{13.3}$$

We now have a solution to the differential equation, i.e. an equation that tells us the value of $x$ at any time $t$. This solution can be verified by substituting it back into the differential equation (13.1); taking a derivative of Equation (13.3), we get

$$\frac{\mathrm{d}x(t)}{\mathrm{d}t} = -kx_0 \exp(-kt) = -kx(t)$$

which is just the original differential equation. The solution is therefore correct.

We usually take it for granted that $x$ depends on $t$, so we can drop the explicit argument. Equation (13.2) can then be written as

$$\ln x = \ln x_0 - kt. \tag{13.4}$$

This form gives us a direct method for measuring rate constants of first-order reactions. Suppose we know that a certain reaction obeys a first-order rate law. If we plot $\ln x$ as a function of $t$, we should get a straight line of slope $-k$.

**Example 13.1 Decomposition of dimethyldiazene** At sufficiently high pressures, the gas-phase decomposition of dimethyldiazene (a.k.a. azomethane)

$$CH_3N = NCH_{3(g)} \rightarrow C_2H_{6(g)} + N_{2(g)}$$

obeys a first-order rate law. The following data were obtained at 300°C:

| $t/s$ | 0 | 100 | 150 | 200 | 250 | 300 |
|---|---|---|---|---|---|---|
| $P_{(CH_3N)_2}/\text{mmHg}$ | 284 | 220 | 193 | 170 | 150 | 132 |

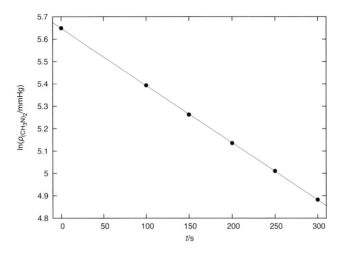

Figure 13.1 Plot of data from Example 13.1.

In Equation (13.4), we can use any quantity that is proportional to the number of moles of reactant for $x$. In particular, the partial pressure $p = nRT/V$ is often a convenient choice for gases:

| $t/s$ | 0 | 100 | 150 | 200 | 250 | 300 |
|---|---|---|---|---|---|---|
| $\ln\left(p_{(CH_3N)_2}/mmHg\right)$ | 5.65 | 5.39 | 5.26 | 5.14 | 5.01 | 4.88 |

The data fit a straight line beautifully (Figure 13.1). If the reaction obeyed a different rate law, this would not normally be the case. We determine the slope by linear regression of the $\ln p_{(CH_3N)_2}$ vs. $t$ data. We find a slope of $-2.55 \times 10^{-3}$ s$^{-1}$, implying a rate constant $k = 2.55 \times 10^{-3}$ s$^{-1}$.

In Equation (13.4), $x$ refers to the concentration of a *reactant*. Depending on the nature of the reactants and products, it may, however, be easier to measure the concentration of a product. It is possible to derive integrated rate laws for these cases, but it's generally a little easier to calculate the concentration or pressure of one of the reactants by stoichiometry before applying Equation (13.4).

### 13.1.1 Half-life

The half-life ($t_{1/2}$) is the time required to get half-way to equilibrium from some given reaction conditions. The half-life is used most often for nearly irreversible reactions, in which case it is the time required for half the reactant to be consumed. This quantity turns out to be particularly useful in studying first-order reactions, although it has other uses.

To compute the half-life of a reaction, we determine the value of $t = t_{1/2}$ at which $x = \frac{1}{2}x_0$ from the integrated rate law.

$$x(t_{1/2}) = x_0 \exp(-kt_{1/2}) = \frac{1}{2}x_0$$

$$\therefore -kt_{1/2} = \ln(1/2) = -\ln 2$$

$$\therefore t_{1/2} = \frac{\ln 2}{k}$$

You can see that the half-life of a first-order reaction is constant. We will soon learn that this is not the case for any other order of reaction, which explains why the half-life is a much more useful statistic for first-order reactions than it is for other orders.

**Example 13.2 Half-life and rate constant** If a first-order reaction has a half-life of 1 h, the rate constant is

$$k = \frac{\ln 2}{t_{1/2}} = \frac{\ln 2}{1\,\text{h}} = 0.69\,\text{h}^{-1} = 1.9 \times 10^{-4}\,\text{s}^{-1}.$$

The half-life of a first-order reaction is *independent* of the initial amount of reactant present. This means that, for a first-order reaction, it takes the same amount of time to eliminate half of what is left of the reactant, no matter how much of it has already reacted. This idea can be given mathematical form by rewriting Equation (13.3) as follows:

$$x(t) = x_0 \exp(-kt)$$

$$= x_0 \exp\left(-\ln 2\frac{t}{t_{1/2}}\right)$$

$$= x_0\,(\exp(\ln 2))^{-t/t_{1/2}}$$

$$= x_0 2^{-t/t_{1/2}}.$$

$$\therefore x(t) = x_0 \left(\frac{1}{2}\right)^{t/t_{1/2}} \tag{13.5}$$

This is just another way to write the integrated rate law for a first-order reaction. If we take $t = t_{1/2}$, we get $x = \frac{1}{2}x_0$. After two half-lives ($t = 2t_{1/2}$), $x = \frac{1}{4}x_0$; after three half-lives, $x = \frac{1}{8}x_0$ and so on.

---

### Exercise group 13.1

(1) The radical cation of 2,5-dihydrofuran (2,5-DHF$^{+\bullet}$) can be generated by irradiation in a CF$_3$CCl$_3$ matrix at 77 K. (Think of the matrix as a solid solvent.) The radical cation decays with first-order kinetics. It can be detected by electron paramagnetic resonance (EPR). The intensity of the EPR signal is proportional to concentration. The following

data were obtained:[1]

| $t$/min | 2.6 | 3.4 | 4.4 | 5.4 | 6.9 | 8.4 | 10.1 |
|---|---|---|---|---|---|---|---|
| EPR intensity | 0.54 | 0.43 | 0.36 | 0.26 | 0.21 | 0.15 | 0.12 |

(a) Calculate the half-life of the radical cation.
(b) How long would it take for the 2,5-DHF$^{+\bullet}$ concentration to fall to 1% of its initial value?

(2) Indium (I) chloride dissolves quickly in dilute acid. The following reaction then occurs:

$$3\text{In}^+_{(aq)} \rightarrow 2\text{In}_{(s)} + \text{In}^{3+}_{(aq)}$$

The reaction is known to obey a first-order rate law. The concentration of indium (I) ions varies with time as follows:

| $t$/s | 0 | 240 | 480 | 720 | 1000 |
|---|---|---|---|---|---|
| $[\text{In}^+]$/mmol L$^{-1}$ | 8.23 | 6.41 | 5.00 | 3.89 | 3.03 |

(a) Confirm that the reaction obeys a first-order rate law.
(b) Obtain the rate constant for the process.
(c) The total volume of the solution is 0.4 L. What mass of indium metal is formed in 15 min? The molar mass of indium is 114.82 g mol$^{-1}$.

(3) Suppose that a certain drug is administered in pill form at a rate of one pill every 6 hours. Each pill, as it is manufactured, contains 18 mg of the drug. The minimum effective dose for an adult of average mass is 12 mg every 6 hours. The drug is somewhat unstable and breaks down in a first-order process with a half-life of 47 days. What is the shelf life of the drug? In other words, how long does it take before each pill contains less than the minimum effective dose?

---

### 13.1.2 Other processes with first-order rate laws

In addition to chemical reactions, there are many processes with first-order rate laws. We examine several examples in this section.

### Radioactive decay

Radioactive decay processes are almost certainly the best-known examples of processes obeying first-order rate laws. The kinetics of radioactive decay is almost always characterized by the half-life.

It is fairly easy to measure the half-life of a short-lived radioisotope; we can measure the rate of radioactive decays using a Geiger counter. Because radioactive decay is a first-order process, the rate of decay is proportional to the number of atoms left. Thus we can apply our usual treatment of first-order data directly to the radioactivity measurements without

---

[1] W. Knolle *et al.*, *J. Chem. Soc., Perkin Trans.* **2**, 2447 (1999).

even needing to know how many atoms of the radioactive substance we have: a plot of $\ln A$ ($A$ for (radio)activity) vs. $t$ should have a straight line of slope $-k$.

Half-lives of radioisotopes can be very long. For very long-lived radioisotopes, the activity of a sample might not change detectably during a person's lifetime. How are these half-lives measured? The key is that the rate of disintegrations, i.e. the radioactivity, is directly measurable using a Geiger counter. We simply count up the number of radioactive particles emitted from a sample in a given amount of time. This allows us to calculate the rate:

$$\text{rate} = A = -\frac{\Delta x}{\Delta t} = \frac{\text{disintegrations}}{\text{time}}.$$

Even if the half-life is long, the number of disintegrations can be large for a macroscopic sample because the number of atoms is large. For instance, the half-life of $^{40}$K is $1.25 \times 10^9$ y, from which we can calculate $k = 5.55 \times 10^{-10}$ y$^{-1}$. The natural abundance of this isotope is $0.0117\%$. One gram of potassium ($0.026$ mol) therefore contains about $3 \times 10^{-6}$ mol of potassium-40 atoms, which corresponds to about $1.8 \times 10^{18}$ atoms. The rate of disintegration is therefore $kx = (5.55 \times 10^{-10}$ y$^{-1})(1.8 \times 10^{18}) = 1 \times 10^9$ y$^{-1} \approx 3 \times 10^6$ day$^{-1}$. It is therefore entirely feasible to measure the rate of disintegration since there is a large number of events in one day, even when the half-life is very large. Given the rate, we simply divide by the number of atoms of the radioactive isotope in the sample to get the rate constant and thence the half-life. The isotopic abundance must be independently measured, but this can be straightforwardly accomplished by mass spectroscopy.

### Radioisotope dating

Living organisms get almost all their carbon from one of two sources, namely from the atmosphere during photosynthesis, or from eating other living organisms. At the very bottom of the food pyramid are photosynthesizing organisms, and since the time scale from carbon fixation to consumption is short (several decades at most), the mixture of carbon isotopes in any living organism is essentially the same as that in the atmosphere during its life. In particular, the ratio of carbon-14 to carbon-12 in living organisms is essentially identical to that in the atmosphere. The atmospheric abundance of carbon-14 is in a steady state. In other words, the rate of production of this isotope is balanced by its rate of decay such that the atmospheric concentration is approximately constant. Carbon-14 is produced from nitrogen-14 in the upper atmosphere by cosmic-ray neutron bombardment:

$$^{14}_{7}\text{N} + ^{1}_{0}\text{n} \rightarrow ^{14}_{6}\text{C} + ^{1}_{1}\text{H}$$

The rate of this reaction depends on the intensity of cosmic rays striking the Earth. This does vary with time, but mostly only by a few percent. There are occasionally periods of unusually low or high cosmic-ray activity,[2] but these only last a few hundred years or less, which won't affect carbon dates by much unless the artifact to be dated happens to have

---

[2] D. Lal *et al.*, *Earth Planet. Sci. Lett.* **234**, 335 (2005).

been made during one of these periods. Obviously, this is potentially a problem, which is why archaeologists and paleontologists generally try to confirm carbon dates using other evidence.

Carbon-14 decays back to nitrogen-14 by spontaneous beta emission:

$$^{14}_{6}C \rightarrow {}^{14}_{7}N + {}^{0}_{-1}\beta.$$

Again, the rates of these two processes balance each other such that the relative amount of carbon-14 in the atmosphere is nearly constant. Recently, another factor has started to affect the relative abundance of carbon-14: due to heavy use of fossil fuels containing "old" carbon depleted of carbon-14, the relative abundance of this isotope has been decreasing over the past hundred years or so. The data given in this section is appropriate to pre-industrial samples. Carbon dating will still be useful to date artifacts of the industrial era, but the calculations will be more complex since they will need to take into account the variation of the ratio of $^{14}C$ to $^{12}C$ due to the burning of fossil fuels. A similar comment can be made about variations in cosmic-ray activity. It is possible to correct for this factor, it just takes a little extra work.

Ignoring all of these higher-order complications, we can use average data to get reason-able estimates of the age of an artifact: for typical samples, the ratio of carbon-14 to total carbon is such that $^{14}C$ atoms disintegrate at a rate of $0.255 \, \text{Bq g}^{-1}$ (0.255 disintegrations per second per gram of carbon). When an organism dies, it ceases to take in carbon. Since carbon-12 and carbon-13 are stable, while carbon-14 decays with a half-life of 5730 y, the ratio of carbon-14 to total carbon decreases as the remains age. Consequently, the specific radioactivity (the number of disintegrations per minute per gram of carbon) also decreases.

**Example 13.3 Radiocarbon dating** Suppose that a wooden tool has a specific radioactivity of $0.195 \, \text{Bq g}^{-1}$. Since this rate is proportional to the concentration of carbon-14 atoms, we have (from Equation 13.5),

$$\frac{t}{t_{1/2}} = \frac{\ln(x_0/x)}{\ln 2} = \frac{\ln\left(\dfrac{0.255 \, \text{Bq g}^{-1}}{0.195 \, \text{Bq g}^{-1}}\right)}{\ln 2} = 0.387.$$

Since the half-life is 5730 y, the wood from which the tool was made is approximately 2220 y old.

Carbon dating is only reliable if:

(1) Enough $^{14}C$ has decayed that the rate of disintegration is significantly different from that of an atmospheric sample, and
(2) There are still enough atoms of $^{14}C$ that a reasonable number of disintegrations occur in a few hours.

Generally speaking, we would want a difference of at least several percent from the atmo-spheric reading before we could reliably use carbon dating. If, for instance, our equipment allows us to resolve differences in disintegration counts of 2%, we could use carbon dating with samples as young as

$$t = t_{1/2} \frac{\ln(1.02)}{\ln(2)} = 160 \, \text{y}.$$

At the other end of the range, we want a few disintegrations per hour so that, over a day, the statistical fluctuations will average out and a reasonably accurate rate of disintegration will be obtained. Very small samples are generally used since one generally wishes to preserve the artifacts being studied. A typical sample size is 0.05 g. If we want one disintegration per hour, the minimum specific activity is

$$A = \frac{1 \, \text{h}^{-1}}{(0.05 \, \text{g})(3600 \, \text{s} \, \text{h}^{-1})} = 0.006 \, \text{Bq} \, \text{g}^{-1}.$$

The age of an artifact with this specific activity would be

$$t = (5730 \, \text{y}) \frac{\ln \left( \dfrac{0.255 \, \text{Bq} \, \text{g}^{-1}}{0.006 \, \text{Bq} \, \text{g}^{-1}} \right)}{\ln 2} = 31\,000 \, \text{y}.$$

This is the approximate reliable range of the carbon dating technique: a few hundred to a few tens of thousands of years.

Other radioisotopes with longer half-lives can be used to date objects too old for carbon dating. For instance, the following nuclear reaction is used to date rocks:

$$^{40}\text{K} + \text{e}^- \rightarrow \, ^{40}\text{Ar}$$

Because it is an unreactive gas, argon is almost exclusively found in the Earth's crust as a result of radioactive decay. The ratio of argon-40 to potassium-40 can thus be used to measure the age of a non-porous rock. Since the half-life of potassium-40 is 1.25 billion years, very ancient rocks can be dated by this technique. In the experiment, a rock sample is vaporized with a high-powered laser. The gas sample is pumped off to a mass spectrometer which measures the ratio of $^{40}\text{Ar}$ to $^{40}\text{K}$. The potassium-40 has been undergoing radioactive decay since the rock solidified, so its concentration in the rock obeys the equation

$$\left[ ^{40}\text{K} \right] = \left[ ^{40}\text{K} \right]_0 \exp(-kt).$$

The amount of argon-40 created by radioactive decay of potassium-40 equals the amount of potassium-40 used, i.e.

$$\left[ ^{40}\text{Ar} \right] = \left[ ^{40}\text{K} \right]_0 - \left[ ^{40}\text{K} \right] = \left[ ^{40}\text{K} \right]_0 - \left[ ^{40}\text{K} \right]_0 \exp(-kt).$$

The ratio of the two isotopes is therefore

$$r = \frac{\left[ ^{40}\text{Ar} \right]}{\left[ ^{40}\text{K} \right]} = \frac{\left[ ^{40}\text{K} \right]_0 - \left[ ^{40}\text{K} \right]_0 \exp(-kt)}{\left[ ^{40}\text{K} \right]_0 \exp(-kt)} = \frac{1}{\exp(-kt)} - 1 = \exp(kt) - 1.$$

Note that the initial amount of potassium-40 drops out of the equation. The ratio $r$ is obtained by mass spectrometry. The rate constant $k$ is calculated from the known half-life.

The only unknown in this equation is $t$:

$$\exp(kt) = r + 1;$$
$$\therefore kt = \ln(r + 1);$$
$$\therefore t = \frac{1}{k} \ln(r + 1);$$

or, in terms of the half-life,

$$t = \frac{t_{1/2}}{\ln 2} \ln(r + 1).$$

## Population growth

Population growth is often described by equations similar to those for first-order chemical processes. The rate law for population growth is often given as

$$\frac{dp}{dt} = kp.$$

In this context, $k$ is called the specific growth rate. In other words, it is the growth rate per unit population. The solution to this differential equation is

$$p = p_0 \exp(kt).$$

$k$ is related to the doubling time $t_2$ of the population by

$$t_2 = \ln 2/k,$$

by analogy to the half-life of a chemical reaction.

**Example 13.4 World population growth** The following table shows estimates of the world's population for the last several decades:

| Year | 1950 | 1960 | 1970 | 1980 | 1990 | 2000 | 2010 |
|------|------|------|------|------|------|------|------|
| Population/$10^9$ | 2.56 | 3.04 | 3.71 | 4.45 | 5.27 | 6.07 | 6.91 |
| $\ln(p/10^9)$ | 0.940 | 1.112 | 1.311 | 1.493 | 1.662 | 1.803 | 1.933 |

A plot of $\ln p$ vs. $t$ is not really linear (Figure 13.2). By fitting only the 1950–1990 data, we see that this apparent curvature is mostly due to the last two points. Up to 1990, the specific growth rate (slope) was $k = 0.0183 \text{ y}^{-1}$, corresponding to a doubling time of 38 y. If this trend had been maintained, the world population would have hit 12 billion sometime in the late 2030s. However, growth has clearly slowed in recent years, and many demographers believe that the world's population will stabilize somewhere around 10 billion. There are several reasons for this slowing in population growth, but by far the largest factor is the one-child policy in China. As a result of this policy, population growth in China has slowed tremendously in recent years, from a peak of 2.3% per year in the 1960s to about 0.6% per year now.

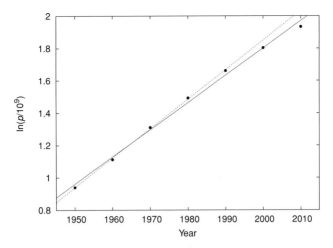

Figure 13.2 Plot of world population data from Example 13.4. The solid line is a fit of the full data set. The dashed line is a fit of the 1950–1990 data.

### Cooling and warming

When a warm object is placed in a cool environment, the object's temperature drops until it reaches the temperature of the surroundings. Conversely, a cool object warms to the temperature of its surroundings. The difference in temperature between the object and its environment obeys a first-order law:

$$\frac{d\Delta T}{dt} = -k\Delta T.$$

In this equation, $k$ is a constant that depends both on the object's and on the environment's thermal properties. This equation is known as **Newton's law**. The solution of this differential equation is just

$$\Delta T = \Delta T_0 \exp(-kt). \tag{13.6}$$

These equations have obvious physical applications, but they are also extensively used in forensic science.

**Example 13.5 Time of death** Detectives arrive at a murder scene at 2:04 a.m. The body is found in a climate-controlled room in which the temperature is 20.0 °C. They record the temperature of the body at 2:08 a.m. and every 20 minutes thereafter for an hour. The body is only removed from the room after the measurements are complete. Their results are as follows:

| $t$ | 2:08 | 2:28 | 2:48 | 3:08 |
|---|---|---|---|---|
| $T/°C$ | 32.2 | 31.1 | 30.0 | 29.1 |

Equation (13.6) implies that if we plot $\ln \Delta T$ vs. $t$, we should get a straight line. It is also convenient to convert the time into hours since midnight rather than leave it as hours and

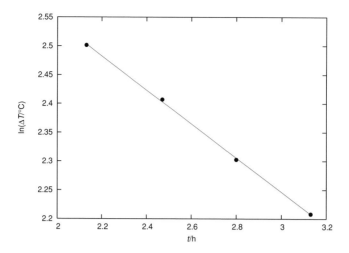

Figure 13.3 Plot of temperature data from Example 13.5.

minutes.

| $t/$h | 2.13 | 2.47 | 2.80 | 3.13 |
|---|---|---|---|---|
| $\Delta T/^\circ$C | 12.2 | 11.1 | 10.0 | 9.1 |
| $\ln(\Delta T/^\circ$C) | 2.501 | 2.407 | 2.303 | 2.208 |

The data fit a straight line quite well (Figure 13.3). The equation of the line of best fit is

$$\ln \Delta T = 3.13 - 0.295t.$$

Since the body's temperature at the time of death would have been approximately $37\,^\circ$C, $\Delta T$ at that moment would have been $17\,^\circ$C. This would mean that the time of death would have been approximately $t = 1.0\,$h or 1:00 a.m.

---

## Exercise group 13.2

(1) $^{90}$Sr has a half-life of 28.1 y. This radioisotope is particularly hazardous because it becomes incorporated into bones and thus remains in the body and is never excreted. If a person accidentally ingests 1 µg of $^{90}$Sr, how much will remain after 40 y?

(2) Over a period of several days, a sample of 1.00 g of osmium displays an average of 47 decays per day. These decays are due to the very long-lived radioisotope $^{186}$Os. The natural abundance of this isotope is 1.58%. What is the half-life? Express your answer in years.

(3) $^{238}$U decays by alpha emission. A natural sample of uranium weighing 0.0532 g produces 651 alpha particles per second. This sample is analyzed by mass spectrometry and found to contain 99.2832% $^{238}$U and to have an average molar mass of 238.029 229 g mol$^{-1}$. What is the half-life of $^{238}$U? Express your answer in years.

(4) The following table gives the population of Finland in the past two centuries:

| Year | 1800 | 1850 | 1900 | 1950 | 2000 | 2010 |
|---|---|---|---|---|---|---|
| Population/$10^3$ | 837 | 1637 | 2656 | 4030 | 5181 | 5365 |

   (a) Doubling times are computed on the assumption that the population is growing exponentially. What is the doubling time for the population of Finland?
   (b) Assuming exponential growth, in what year would you predict that the population will reach six million?
   (c) Is the population really growing exponentially?

(5) All stars contain traces of metals which can be detected by analyzing their spectra. The heavier metals are almost exclusively made during extremely violent events such as supernova explosions. The debris from a supernova is widely scattered and is incorporated into other stellar bodies (stars and planets) during their formation. Some of the very heavy metals are (relatively) stable, while others are not. If we know the ratio in which these heavy metals are initially present and the half-life of an unstable metal, we can estimate the average ages of the supernovas that contributed to the sample. Since young galaxies have many more supernova explosions than older galaxies, this gives a rough idea of the age of the galaxy. The initial ratios are only known from theoretical calculations, so their values are a source of considerable uncertainty which must be taken into account in deriving these estimates.

   $^{232}$Th has a very long half-life and can be taken to be stable for the purpose of these calculations. $^{238}$U has a much shorter half-life of about 4.5 Gy. According to stellar nucleosynthesis models, the initial ratio of $^{238}$U to $^{232}$Th was between 0.56 and 0.79. The currently observed ratio of these two isotopes in star CS31082-001, a star in the Milky Way galaxy, is 0.18. Using these data, determine a range of possible ages for our galaxy.

(6) While cooking a beef roast recently, I recorded the following data: the oven temperature was 175°C; I recorded the internal temperature of the roast at various times (given as hours and minutes p.m.) during the cooking process:

| Time | 4:37 | 4:53 | 5:13 | 5:25 | 5:33 |
|---|---|---|---|---|---|
| $T/°C$ | 6 | 19 | 39 | 51 | 60 |

The experiment stopped at 5:33 p.m. because the internal temperature had reached the medium-rare point. Do these data obey Newton's law for warming?

## 13.2 Reactions of other orders

Reactions with simple rate laws whose order is different than unity can be treated together. The rate equation is

$$\frac{dx}{dt} = -kx^n. \tag{13.7}$$

We solve this equation by separation of variables:

$$\frac{dx}{x^n} = -k\,dt$$

$$\therefore \int_{x_0}^{x} (x')^{-n}\,dx' = -k \int_{0}^{t} dt'$$

$$\therefore \frac{1}{-n+1} (x')^{-n+1}\Big|_{x_0}^{x} = -k\,t'\Big|_{0}^{t}$$

$$\therefore \frac{1}{1-n} \left(x^{1-n} - x_0^{1-n}\right) = -kt$$

$$\text{or} \quad x^{1-n} = x_0^{1-n} + kt(n-1) \tag{13.8}$$

As in our treatment of first-order processes, $x_0$ is the value of $x$ at time zero.

We typically use Equation (13.8) the same way as we do the first-order equation, i.e. by constructing a plot in which the data appear as a straight line. If we think (or know, from other evidence) that a reaction might obey an $n$th order rate law, we plot $x^{1-n}$ vs. $t$. For an $n$th order reaction, this will give a straight line of slope $k(n-1)$ so that the rate constant can be obtained. On the other hand, if we have the wrong rate law (wrong $n$, or data that don't obey a simple rate law at all), our graph will be curved. If the order of the reaction is not known, we can try plots appropriate to different values of $n$ to see if one fits particularly well.

Second-order reactions are particularly important. For $n = 2$, Equation (13.8) specializes to

$$\frac{1}{x} = \frac{1}{x_0} + kt \tag{13.9}$$

so a plot of $1/x$ as a function of $t$ gives a straight line of slope $k$.

**Example 13.6 Determining the order of reaction**  The decomposition of chlorine dioxide to chlorine and oxygen is known to be a complex reaction. The overall reaction is

$$2ClO_{(g)} \rightarrow Cl_{2(g)} + O_{2(g)}.$$

The following data were obtained in one experiment:

| $t$/ms | 0.12 | 0.96 | 2.24 | 3.20 | 4.00 | 5.20 | 6.81 |
|---|---|---|---|---|---|---|---|
| [ClO]/$\mu$mol L$^{-1}$ | 8.49 | 7.10 | 5.79 | 5.20 | 4.77 | 4.17 | 3.61 |

We would like to determine if this reaction obeys a simple rate law, and if so, what the order is. Generally in these cases, we might try the first- and second-order rate laws because these are the two most common alternatives, unless something about the reaction suggests to us that we should try something else. If one of the two plots fits the data well, we would conclude that the reaction is of that order.

If the reaction is of the first order, a plot of ln[ClO] against time should be linear. The data don't fit the first-order rate law very well (Figure 13.4). Real experimental data often have noticeable scatter around a line of best fit. What we're looking for when trying to

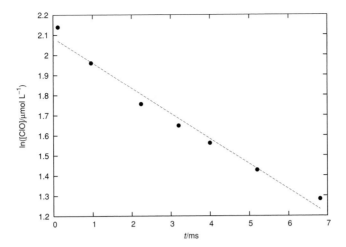

Figure 13.4 First-order plot for Example 13.6.

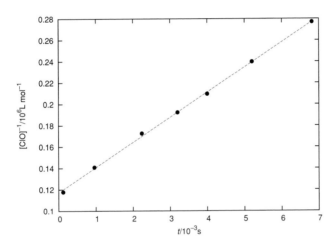

Figure 13.5 Second-order plot for Example 13.6.

decide whether data fit a line or not is whether we can detect curvature in the data set. Plotting the data with the line of best fit as we have done here generally makes it easy to see curvature. In this case, the data clearly curve around the line, so we would conclude that this reaction is *not* a first-order reaction.

We next try the integrated rate law for a second-order reaction. The appropriate plot is one of 1/[ClO] vs. $t$ (Figure 13.5). The points fit the line well, indicating that we have probably found the correct rate law. We have a small problem with units to deal with. The concentrations are given in $\mu mol\,L^{-1}$ and the times in ms. Having SI prefixes in the "denominator" of a unit (e.g. $L\,(\mu mol)^{-1}(ms)^{-1}$) is very bad style. We can do our regression using the raw data and then convert to $L\,mol^{-1}s^{-1}$ at the end, but I find this to be an error-prone process. A better approach is to convert the concentrations to $mol\,L^{-1}$ and the times to s *before* doing the linear regression. (In effect, I do this before I draw the graph, although I may still take out a power of 10 when I label my axes, as I have done here.) If you do this, you will find a rate constant (the slope of the line) of $k = 2.35 \times 10^7\,L\,mol^{-1}s^{-1}$.

There's one little subtlety here. If we just want to use Equation (13.9) to predict [ClO] (or, by stoichiometry, the concentrations of some of the products) at future times, then the rate constant we found above is perfectly adequate. However, if we want the rate constant for this reaction to agree with the definition of rate in Section 11.2, we have to look at the stoichiometry of the reaction, which in this case has a stoichiometric coefficient of 2 for the ClO. By convention, the rate of reaction is therefore

$$v = -\frac{1}{2}\frac{d[ClO]}{dt} = k[ClO]^2$$

or

$$\frac{d[ClO]}{dt} = -2k[ClO]^2.$$

This differs in form from Equation (13.7) by a factor of 2. In other words, the "rate constant" we calculated above is really twice the real rate constant of this reaction. We should therefore report

$$k = \frac{1}{2}\left(2.35 \times 10^7\,L\,mol^{-1}s^{-1}\right) = 1.18 \times 10^7\,L\,mol^{-1}s^{-1}.$$

Note that either answer is acceptable *provided* it's clear which one you're providing.

This method of distinguishing between simple rate laws requires a reasonably long time series. Data that only span one half-life or less will probably not be sufficient to distinguish between rate laws on the basis of the time course of the reaction, particularly if the experimental uncertainties are large, as we will see in the following example.

**Example 13.7 Too little data**   Consider the following data for a reaction $A \rightarrow B$:

| $t$/ms | 0 | 10 | 20 | 30 | 40 | 50 | 60 |
|---|---|---|---|---|---|---|---|
| $a$/mol L$^{-1}$ | 0.50 | 0.46 | 0.43 | 0.39 | 0.36 | 0.34 | 0.31 |

Note that these data cover less than one half-life. Compare the first- and second-order plots shown in Figure 13.6. Neither of these plots is obviously better than the other.

You will note that I decide which plot is better based on visual appearance. Many students who have taken statistics are tempted to use the correlation coefficient. This is indeed very tempting: most scientists like quantitative measures, and the correlation coefficient measures how good a linear relationship we have. Why not use this coefficient? The problem is that the correlation coefficient can't distinguish between scatter, which is expected in experimental measurements, and non-linearity, which indicates that we are using the wrong

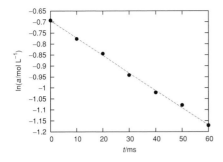
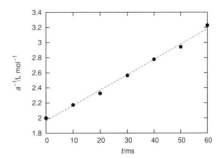

Figure 13.6 Plots of data from Example 13.7

model for the data. The following example demonstrates how the correlation coefficient can fail to distinguish between scatter and curvature.

**Example 13.8 Scatter vs. curvature** Consider the following data set from a kinetics experiment:

| $t/s$ | $x/\mathrm{mol\,L^{-1}}$ |
|---|---|
| 0 | 0.356 |
| 0.12 | 0.260 |
| 0.24 | 0.167 |
| 0.36 | 0.142 |
| 0.48 | 0.103 |
| 0.60 | 0.085 |
| 0.72 | 0.051 |
| 0.84 | 0.040 |
| 0.96 | 0.032 |

Figure 13.7 shows these data plotted on a logarithmic scale, along with the line of best fit. There is some scatter, but overall the data fit a first-order relationship well.

Now consider the following data set:

| $t/s$ | $x/\mathrm{mol\,L^{-1}}$ |
|---|---|
| 0 | 0.320 |
| 0.12 | 0.286 |
| 0.24 | 0.249 |
| 0.36 | 0.213 |
| 0.48 | 0.179 |
| 0.60 | 0.149 |
| 0.72 | 0.122 |
| 0.84 | 0.100 |
| 0.96 | 0.081 |

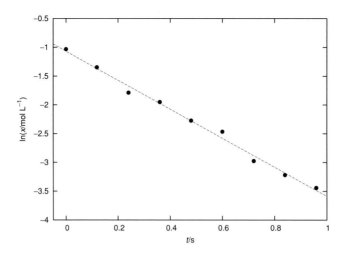

Figure 13.7 First-order plot for the first data set from Example 13.8. The correlation coefficient for this data set is −0.9962.

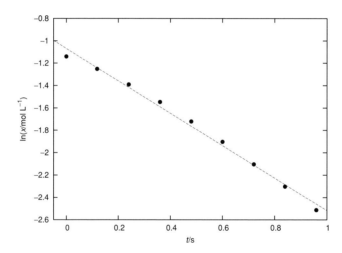

Figure 13.8 First-order plot for the second data set from Example 13.8. The correlation coefficient for this data set is also −0.9962.

The first-order plot for these data is shown in Figure 13.8. This data set clearly shows curvature rather than random scatter. However, the linear correlation coefficient of this data set is *exactly the same* as that of the first data set. Clearly then, correlation coefficients don't tell the whole story. The lesson to learn from this example is that in typical applications in the physical sciences, we should look at the graph of the data, *not* at the correlation coefficient.

## Half-life

The half-life of a reaction of general order is found by setting $x = \frac{1}{2}x_0$ and solving for $t = t_{1/2}$ from the general integrated rate law (13.8):

$$t_{1/2} = \frac{2^{n-1} - 1}{kx_0^{n-1}(n-1)} \tag{13.10}$$

The half-life in general depends on the initial concentration of reactant. First-order reactions therefore stand out as an exceptional case for having constant half-lives. For this reason, half-lives are often used to describe first-order reactions, and rarely used for other orders. In principle though, it is possible to determine the order of a reaction from the dependence of the half-life on initial concentration.

---

### Exercise group 13.3

(1) For the essentially irreversible reaction $2A_{(aq)} \rightarrow B_{(aq)}$, the following data were obtained:

| $t/\text{min}$ | 0 | 10 | 20 | 30 | 40 | $\infty$ |
|---|---|---|---|---|---|---|
| $b/\text{mol L}^{-1}$ | 0 | 0.089 | 0.153 | 0.200 | 0.230 | 0.312 |

Previous experiments showed that the rate did not depend on $b$. The order of the reaction with respect to A is either $\frac{1}{2}$ or 1. Can these data be used to determine the order? If so, report the order and rate constant. If not, demonstrate that the data are inadequate to this purpose.

*Hint*: The amount of B that accumulates after a long time can be used to calculate the initial amount of A.

(2) The reaction of ethyl acetate with hydroxide to form acetate ions and ethanol has a rate law $v = k[\text{AcOEt}][\text{OH}^-]$ with a rate constant of $0.11\,\text{L mol}^{-1}\text{s}^{-1}$. If the initial concentrations are $[\text{AcOEt}] = [\text{OH}^-] = 0.05\,\text{mol L}^{-1}$, what would the concentration of ethanol be after 10 min?

(3) $\pi$–$\pi$ interactions in molecules with several double or triple bonds often result in the formation of loosely bound dimers or, in extreme cases, of stacks of molecules. Butadiene dimerizes in this way in the gas phase, i.e. it undergoes the reaction 2 butadiene $\rightarrow$ dimer. The following data were obtained in a reaction vessel held at 326 °C that originally contained pure butadiene:

| $t/\text{min}$ | $p/\text{torr}$ |
|---|---|
| 0 | 632.0 |
| 20.78 | 556.9 |
| 49.50 | 498.1 |
| 77.57 | 464.8 |
| 103.58 | 442.6 |

Note that the pressure measured is the total pressure, not the pressure of reactant.

(a) Could this reaction be elementary? If so, what would the order of the reaction be?
(b) Are the data consistent with the order of reaction you predicted above? If so, what is the rate constant?

(4) Suppose that a company wants to remove a pollutant X from a solution produced as a byproduct of an industrial process. Their research department finds a reaction $X + A \rightarrow Y + 2Z$ that can convert the pollutant into the harmless substances Y and Z.

(a) Preliminary experiments indicate that the rate of the reaction only depends on the concentration of X and that the order with respect to X is either 2 or 3. To determine the order of the reaction, an experiment was performed in which the concentration of the product Z was followed as a function of time:

| $t$/min | 0 | 3.22 | 6.58 | 10.02 | 13.91 |
|---|---|---|---|---|---|
| $z$/mmol L$^{-1}$ | 0 | 13.0 | 22.6 | 30.0 | 36.4 |

In this experiment, the initial concentration of X was $0.0400 \, \text{mol L}^{-1}$. What is the order of the reaction and rate constant?

(b) The industrial waste produced by this company contains X at a concentration of $0.2 \, \text{mol L}^{-1}$ and is considered safe when this concentration drops below $1 \, \mu\text{mol L}^{-1}$. If the reaction conditions are more or less as given above, how long would the treatment of the industrial waste take? In your opinion, is this an acceptable length of time?

(5) Isoxaflutole (denoted I in the table below) is a herbicide used to halt the growth of grass and broadleaf weeds in corn fields. In a recent study,[3] the concentration of isoxaflutole was measured in the top soil of corn fields after application of this herbicide. In a sandy loam field, the following data were obtained:

| $t$/days | 0 | 14 | 29 | 45 | 70 | 87 | 105 |
|---|---|---|---|---|---|---|---|
| $[I]/\mu\text{g (kgsoil)}^{-1}$ | 99 | 42 | 29 | 21 | 12 | 9 | 7 |

Leaching of chemicals out of soil generally follows first-order kinetics, as do various degradation processes. However, the authors of this study claim that the decay of the isoxaflutole concentration obeys second-order kinetics. Unfortunately, they do not show their analysis. Do these data unambiguously support the authors' claim? If the data are clearly consistent with a particular order of reaction (whether first or second order) give the corresponding rate constant. Otherwise, explain clearly why these data cannot be used to claim either a first- or a second-order relationship.

## 13.3 Two-reactant reactions

### 13.3.1 How to avoid complex integrated rate laws

It is possible to obtain integrated rate laws for more complicated cases, such as reactions with two or more reactants. These equations are more complex than those for simple

---

[3] J. Rouchaud *et al.*, *Arch. Environ. Contam. Toxicol.* **42**, 280 (2002).

reactions, so they tend to be harder to use in practice, and they are of almost no use if we don't already know the partial orders of the reaction. In general, kineticists avoid them.

Kineticists like to keep things as simple as possible, particularly when exploring a new reaction. Ideally, we would study the dependence of the rate on *one* concentration at a time. It turns out that we can do exactly that by using a large excess of one reactant. The concentration of the reactant in large excess will remain roughly constant, so we will then have isolated the kinetics of the other reactant.

Suppose we have a two-reactant reaction, again with a stoichiometry $A + B \rightarrow C$. We assume this reaction has a rate law $v = ka^{\alpha}b^{\beta}$. If we have a large (say, hundredfold) stoichiometric excess of B, then $b = b_1$ is approximately constant, and the rate law reduces to $v = k'a^{\alpha}$, where $k' = kb_1^{\beta}$ is a **pseudo-first-order rate constant**. We treat this reaction under these conditions as if A were the only reactant and determine $k'$ and $\alpha$. Once we know these, we turn around and use a large stoichiometric excess of A ($a = a_2$), so that $v = k''b^{\beta}$, and find $k''$ and $\beta$. Once we know all of these constants, we can calculate the true rate constant $k$ by $k = k'/b_1^{\beta} = k''/a_2^{\alpha}$. The two estimates of $k$ should of course agree within the experimental errors.

In an important variation on this technique, we do experiments at several different concentrations of B, each of which is much larger than the initial concentration of A. We determine the order of the reaction with respect to A as above, with the added caveat that we should get the same order of reaction at each value of $b$. We will, however, get a different value of $k'$ for each concentration of B since $k' = kb^{\beta}$. Then we plot $\ln k'$ vs. $\ln b$ which, if all goes well, should give a straight line of slope $\beta$ and intercept $k$.

### *13.3.2 Integrated rate laws*

Suppose that you do want an integrated rate law for a reaction with two reactants. These cases can also be treated by separation of variables because the reactant concentrations are related by stoichiometry. It's much easier to see how this works through examples than by trying to explain it in general, so here goes:

**Example 13.9** Suppose that the rate of the reaction $A + B \rightarrow C$ is known to be $v = kab$. There are two cases to consider:

(1) If the initial concentrations of A and B are equal, then we know by stoichiometry that $a = b$ throughout the course of the reaction. It follows that $v = ka^2$ and we can use Equation (13.9) to describe the time course of the reaction.
(2) Otherwise, the two concentrations are still related by stoichiometry. Suppose that the initial reaction mixture contains none of the product C. Then $c$ represents both the amount of A reacted and the amount of B reacted. By stoichiometry, it follows that $a = a_0 - c$ and $b = b_0 - c$ where $a_0$ and $b_0$ are the initial concentrations of the two reactants. The rate law becomes

$$v = \frac{dc}{dt} = k(a_0 - c)(b_0 - c).$$

This differential equation can be solved by separation of variables:

$$\int_0^{c(t)} \frac{dc'}{(a_0 - c')(b_0 - c')} = \int_0^t k \, dt'.$$

The integral on the right-hand side is trivial. The integral on the left-hand side can be solved using the table of integrals given in Appendix I.1. The appendix tells us that

$$\int \frac{dx}{(a + bx)(c + ex)} = \frac{1}{ae - bc} \ln\left(\frac{c + ex}{a + bx}\right).$$

Our variable of integration is $c'$ rather than $x$. The other correspondences are more-or-less obvious ($a \leftarrow a_0$, $b \leftarrow -1$ etc.). Our integral is

$$
\begin{aligned}
kt &= \int_0^{c(t)} \frac{dc'}{(a_0 - c')(b_0 - c')} \\
&= \frac{1}{-a_0 - (-b_0)} \ln\left(\frac{b_0 - c'}{a_0 - c'}\right)\Bigg|_0^{c(t)} \\
&= \frac{1}{b_0 - a_0}\left[\ln\left(\frac{b_0 - c(t)}{a_0 - c(t)}\right) - \ln\left(\frac{b_0}{a_0}\right)\right] \\
&= \frac{1}{b_0 - a_0} \ln\left(\frac{a_0(b_0 - c(t))}{b_0(a_0 - c(t))}\right).
\end{aligned}
\tag{13.11}
$$

With a little effort, we can isolate $c(t)$:

$$c(t) = \frac{a_0 b_0 \left[e^{kt(b_0 - a_0)} - 1\right]}{b_0 e^{kt(b_0 - a_0)} - a_0}.$$

If we know the initial concentrations and rate constant, we can calculate the concentration of the product, and therefore, by stoichiometry, of the reactants, at any time $t$.

We can also rearrange Equation (13.11) to a form suitable for extracting the rate constant by a graphical method: recall that $b_0 - c = b$ and $a_0 - c = a$. Thus we have

$$\ln\left(\frac{b}{a}\right) = (b_0 - a_0)kt + \ln\left(\frac{b_0}{a_0}\right).$$
$$\tag{13.12}$$

A graph of $\ln(b/a)$ vs. $t$ should give us a straight line with a slope of $(b_0 - a_0)k$. In almost every instance, we know the initial concentrations of $a$ and $b$, so we can calculate the rate constant from the slope.

## 13.4 Reversible elementary reactions

If a mechanism consists of a single reversible elementary reaction, we can solve the rate equation after the application of a simple stoichiometric argument. As an example, we consider aspartic acid dating, a technique that is sometimes used to date biological materials.

Most living organisms only produce the L stereoisomers of the amino acids. Thus, living tissue mostly contains the L amino acids. However, once an organism dies, it stops replenishing its supply of amino acids and natural processes result in racemization, i.e. in equilibration between the D and L forms. In principle then, we can use the ratio of the D to L amino acids to determine how long it has been since a piece of living tissue died. The amino acid of choice in these studies is aspartic acid.

Interestingly, this technique can also be used to determine the ages of individuals in a population because some types of tissue consist essentially of dead material deposited at birth. Aspartic acid dating has been used to determine the ages of bowhead whales, with some surprising results.[4]

Although the isomerization is not an elementary reaction, both directions behave like simple first-order reactions:

$$L \underset{k_{DL}}{\overset{k_{LD}}{\rightleftharpoons}} D$$

The rate equation for (say) the L form is

$$\frac{d[L]}{dt} = -k_{LD}[L] + k_{DL}[D].$$

Since the total amount of the D and L forms is constant, we have

$$[L] + [D] = [L]_0 + [D]_0 \tag{13.13}$$

where $[L]_0$ is the initial concentration of the L amino acid and $[D]_0$ is the initial concentration of the D form. The differential equation to solve is therefore

$$\frac{d[L]}{dt} = -k_{LD}[L] + k_{DL} ([L]_0 + [D]_0 - [L]).$$

Changing the stereochemistry at a carbon center involves essentially identical steps, whether one is going in one direction or the other. Thus the rate constants $k_{LD}$ and $k_{DL}$ are roughly equal. The differential equation therefore simplifies to

$$\frac{d[L]}{dt} = -k[L] + k ([L]_0 + [D]_0 - [L]) = k ([L]_0 + [D]_0 - 2[L])$$

where $k = k_{LD} = k_{DL}$. By separation of variables, we obtain

$$\int_{[l]_0}^{[l]} \frac{d[L]}{2[L] - [L]_0 - [D]_0} = -kt.$$

(Both sides were multiplied by $-1$ to make some of the algebra that follows a little simpler.) This integral can be solved by the simple substitution $u = 2[L] - [L]_0 - [D]_0$. Then $du = 2\,d[L]$ and the integral becomes

$$\frac{1}{2} \int \frac{du}{u} = \frac{1}{2} \ln u = \frac{1}{2} \ln (2[L] - [L]_0 - [D]_0).$$

[4] J. C. George *et al.*, *Can. J. Zool.* **77**, 571 (1999).

If we substitute in the limits and rearrange the result a little, we get

$$\ln \{2[\text{L}] - ([\text{L}]_0 + [\text{D}]_0)\} = -2kt + \ln ([\text{L}]_0 - [\text{D}]_0).$$

Using Equation (13.13), we can simplify the term on the left-hand side:

$$\ln ([\text{L}] - [\text{D}]) = -2kt + \ln ([\text{L}]_0 - [\text{D}]_0).$$

In amino acid dating, it is not the absolute amounts of D and L amino acids that are measured but their ratio, i.e. $[\text{D}]/[\text{L}]$. We now want to rearrange the above equation so that it involves the $[\text{D}]/[\text{L}]$ ratio. From Equation (13.13), we have

$$\ln ([\text{L}] + [\text{D}]) = \ln ([\text{L}]_0 + [\text{D}]_0).$$
$$\therefore \ln ([\text{L}] - [\text{D}]) - \ln ([\text{L}] + [\text{D}]) = -2kt + \ln ([\text{L}]_0 - [\text{D}]_0) - \ln ([\text{L}]_0 + [\text{D}]_0).$$
$$\therefore \ln \left( \frac{[\text{L}] - [\text{D}]}{[\text{L}] + [\text{D}]} \right) = -2kt + \ln \left( \frac{[\text{L}]_0 - [\text{D}]_0}{[\text{L}]_0 + [\text{D}]_0} \right).$$
$$\therefore \ln \left( \frac{1 - \frac{[\text{D}]}{[\text{L}]}}{1 + \frac{[\text{D}]}{[\text{L}]}} \right) = -2kt + \ln \left( \frac{1 - \frac{[\text{D}]_0}{[\text{L}]_0}}{1 + \frac{[\text{D}]_0}{[\text{L}]_0}} \right). \tag{13.14}$$

Now suppose that we have access to a number of samples of known age. A plot of the left-hand side of this equation vs. $t$ should have a slope of $-2k$ and an intercept that is related to $[\text{D}]_0/[\text{L}]_0$. In some whale species, the teeth display growth rings that can be counted, much as tree rings are. Humans are also a good source of data for this calibration since their ages are generally known quite accurately from birth records. From such measurements, the following estimates have been obtained: $k = 1.18 \times 10^{-3} \text{ y}^{-1}$ and $[\text{D}]_0/[\text{L}]_0 = 0.0285$. The latter value is worthy of comment. In principle, in most living organisms $[\text{D}]_0/[\text{L}]_0$ is essentially zero. However, the analytical technique used to measure the $[\text{D}]/[\text{L}]$ ratio involves the acidification of the sample. As you may have learned in organic chemistry, acids catalyze racemization, so that even a sample that initially contained no D amino acids will show a non-zero $[\text{D}]/[\text{L}]$ ratio after analysis. The above value of $[\text{D}]_0/[\text{L}]_0$ corrects for this experimental artifact.

Once we have $k$ and $[\text{D}]_0/[\text{L}]_0$, we can solve Equation (13.14) for $t$, i.e. the age of an individual with a given $[\text{D}]/[\text{L}]$ ratio:

$$t = \frac{1}{2k} \left[ \ln \left( \frac{1 - \frac{[\text{D}]_0}{[\text{L}]_0}}{1 + \frac{[\text{D}]_0}{[\text{L}]_0}} \right) - \ln \left( \frac{1 - \frac{[\text{D}]}{[\text{L}]}}{1 + \frac{[\text{D}]}{[\text{L}]}} \right) \right]. \tag{13.15}$$

The procedure for estimating the age of recently dead whales is as follows: first, the eye is harvested and the lens nucleus (the oldest, central part of the lens) is carefully removed. A biochemical analysis is performed to measure the ratio of D to L aspartic acid. Equation (13.15) can then be used to determine the age of the whale. For example, one whale in the study was found to have a D to L ratio of 0.1850. We can now calculate from Equation (13.15) that this whale was 134 years old. Another bowhead whale in the sample may have been over 200 years old! If this result is correct, bowhead whales would be the longest-living mammals, and perhaps even the longest-living vertebrates on our planet.

This method has several significant sources of uncertainty. As we will see later, rate constants increase with temperature. The estimate of $k$ used in this study is based on an average of data from humans and fin whales. The average body temperature of humans is $37\,^{\circ}\text{C}$ while the body temperature of fin whales is approximately $36\,^{\circ}\text{C}$. The bowhead is an Arctic species. Its eyes are in almost constant contact with freezing water so that the temperature of its eye lens is probably somewhat lower than either of the above. Accordingly, the appropriate rate constant for aspartic acid racemization in bowhead whale eye lenses is probably smaller than the estimate used. This would mean that bowheads would be *older* than calculated. Other sources of error, mostly relating to bacterial contamination, would cause the ages calculated by the aspartic acid method to be overestimates of the true ages of the whales. It is impossible to tell without considerable further study which source of error is the most important. However, one other line of evidence supports the hypothesis that bowheads are extraordinarily long-lived: it is not uncommon to find harpoon heads deeply buried in the flesh of recently deceased bowheads. These harpoon heads would have been implanted into the whales as a result of unsuccessful hunts. Anthropologists have examined these harpoons. Based on the style and materials used, some of the recovered harpoon heads may be over 100 years old. A whale that escaped a hunt was probably a mature individual at the time, so some of these whales must be very old indeed.

---

### Exercise group 13.4

(1) Suppose that the reaction $2A + 3B \rightarrow P$ has a rate law $v = kab$. Write down an integral that relates the concentration of P to time if $a(0) = a_0$, $b(0) = b_0$ and $p(0) = 0$. Do not solve the integral.

(2) Consider the reaction

$$2H_{2(g)} + 2NO_{(g)} \rightarrow 2H_2O_{(g)} + N_{2(g)}$$

(a) The empirical rate law for the reaction is $v = k[H_2][NO]^2$. Derive an integrated rate law for this reaction with general initial conditions. Leave your answer in the form $t = $ something.

(b) At $700\,^{\circ}\text{C}$, $k = 0.384\,\text{L}^2\,\text{mol}^{-2}\,\text{s}^{-1}$. If the initial concentrations of hydrogen and of nitrogen monoxide are, respectively, $0.10$ and $0.050\,\text{mol}\,\text{L}^{-1}$, how long would it take to consume 95% of the NO?

---

### 13.5 Fluorescence kinetics

Way back in Section 3.5.1, I mentioned that quantum yields are difficult to measure directly, and that there was a method based on fluorescence to measure the FRET efficiency. We now have enough kinetics to look at these methods.

Imagine the following experiment: we give a fluorescent sample a short flash of light; we then watch the fluorescent emission over time. After the flash, the energy stored in electronic excitation of a molecule can be lost in the following ways:

- The molecule can fluoresce.
- The molecule can transfer its excitation to another molecule by FRET.
- The molecule can lose its electronic energy in some other way (e.g. collisional deactivation).

These alternatives can be expressed as reactions. The donor will be denoted D, and the acceptor A. Excitation will be represented by an asterisk. The three reactions corresponding to the above three deactivation pathways are, in order,

$$D^* \xrightarrow{k_f} D + h\nu$$
$$D^* + A \xrightarrow{k_{FRET}} D + A^*$$
$$D^* \xrightarrow{k_c} D$$

The last reaction represents all deactivation mechanisms other than fluorescence and FRET.

True second-order processes arise because two molecules have to come together to react. When FRET is used as a spectroscopic ruler, the donor and acceptor are generally part of a single molecular assembly (molecule or complex), so really we're talking about FRET being a first-order process, despite the representation above, which suggests a second-order process.

The fluorescence intensity at any given time is just the rate at which photons are being emitted by fluorescence, i.e. $I_f = k_f[D^*]$. Thus, the fluorescence intensity is proportional to the concentration of excited fluorophores, $[D^*]$. This concentration will decrease with time due to the above processes. The rate equation for $[D^*]$ is

$$\frac{d[D^*]}{dt} = -\left(k_f + k_c + k_{FRET}\right)[D^*].$$

This is a first-order process with effective rate constant

$$k = k_f + k_c + k_{FRET}.$$

The solution to the first-order equation is, of course

$$[D^*] = [D^*]_0 \exp(-kt).$$

By convention, spectroscopists write the exponential term in the form $\exp(-t/\tau)$, where $\tau$ is the **fluorescence lifetime**. By comparison, you can see that $k = \tau^{-1}$, or

$$\tau_{DA}^{-1} = k_f + k_c + k_{FRET}$$

where we have added the subscript "DA" to emphasize that this is the lifetime of the donor in the presence of the acceptor. If we measure the lifetime in the absence of the acceptor, then we would have

$$\tau_D^{-1} = k_f + k_c.$$

The FRET efficiency is the fraction of the photons absorbed by the donor that are transferred to the acceptor. For a single molecule, the probability that the excitation decays by any particular process is proportional to its rate constant, so in particular, the probability that a molecule transfers its excitation by FRET is proportional to $k_{FRET}$. Since the excitation has to eventually be lost by one of the three processes considered here, then $k_{FRET}/k$ is the probability that the excitation will be lost by FRET or, averaged over a large number of molecules, the efficiency:

$$\eta_{FRET} = \frac{k_{FRET}}{k_f + k_c + k_{FRET}} = \frac{\tau_{DA}^{-1} - \tau_D^{-1}}{\tau_{DA}^{-1}} = 1 - \frac{\tau_D^{-1}}{\tau_{DA}^{-1}} = 1 - \frac{\tau_{DA}}{\tau_D}.$$

The FRET efficiency can therefore be expressed in terms of the fluorescence lifetimes measured with and without the acceptor. The lifetimes are easy to measure: since fluorescence decay is a first-order process, we just plot the logarithm of the fluorescence intensity vs. time. The slope of this graph is $k$, and the lifetime is the inverse of this value. Fluorescence lifetimes are generally of the order of milliseconds, so the flash that starts off the experiment and the subsequent intensity measurements must be coordinated electronically, but this is still much easier than photon counting, which is what we need to do to get quantum yields.

**Example 13.10** Cameleon proteins are genetically engineered proteins that consist of a donor and an acceptor fused to either side of a protein that undergoes a conformational change on calcium binding. The change in fluorescence on calcium binding can therefore be used as a calcium indicator. YC3.60 uses enhanced cyan fluorescent protein (ECFP) as the donor, and Venus, an enhanced yellow fluorescent protein, as the acceptor. For this donor–acceptor pair, $R_0 = 4.90$ nm. The fluorescence lifetime of the donor ECFP alone is 2.71 ns. The fluorescence lifetime of YC3.60 in the absence of calcium is 1.33 ns, while the lifetime in the presence of calcium is 0.095 ns.[5] These data allow us to calculate the distance between ECFP and Venus in the presence and absence of calcium.

Without calcium, we have

$$\eta_{FRET} = 1 - \frac{\tau_{DA}}{\tau_D} = 1 - \frac{1.33 \text{ ns}}{2.71 \text{ ns}} = 0.509.$$

[5] J. W. Borst *et al.*, *Biophys. J.* **95**, 5399 (2008).

Using Equation (3.9), we have

$$\left(\frac{R}{R_0}\right)^6 = \frac{1}{\eta_{FRET}} - 1 = \frac{1}{0.509} - 1 = 0.96.$$

$$\therefore \frac{R}{R_0} = (0.96)^{1/6} = 0.99.$$

$$\therefore R = 0.99(4.90\,\text{nm}) = 4.9\,\text{nm}.$$

If we repeat these calculations for YC3.60 in the presence of calcium, we get $R = 2.8\,\text{nm}$. The interpretation of these numbers is that YC3.60 collapses to a compact conformation on calcium binding, with the two ends of the protein close together.

---

### Exercise group 13.5

(1) Polyglutamine repeats in proteins are thought to facilitate amyloid plaque formation, which in turn is thought to be important in several neurological diseases. Digambaranath and coworkers have studied the structure of polyglutamine using FRET.[6] Specifically, they studied a polyglutamine they called $DQ_{16}$ which had an $N$-terminal $o$-aminobenz-amide (a fluorescent donor group), and another called $DQ_{16}A$ which was identical to $DQ_{16}$ except that it had an acceptor group, 3-nitrotyrosine, at its $C$-terminal. The $R_0$ value for this donor–acceptor pair is 24.3 Å in acidic solution.

(a) Following a light pulse, the fluorescence intensity of $DQ_{16}$ decayed as follows:

| $t/\text{ns}$ | 5.05 | 9.00 | 13.18 | 17.57 |
|---|---|---|---|---|
| $I$ | 0.827 | 0.551 | 0.323 | 0.218 |

What is the fluorescence lifetime of $DQ_{16}$?

(b) A similar set of measurements for $DQ_{16}A$ gave the following results:

| $t/\text{ns}$ | 4.79 | 9.00 | 13.18 | 17.57 |
|---|---|---|---|---|
| $I$ | 0.696 | 0.320 | 0.150 | 0.071 |

What is the fluorescence lifetime of $DQ_{16}$ A?

(c) What is the distance between the $N$- and $C$-terminals of this peptide?

---

### Key ideas and equations

- Separation of variables
- For first-order reactions
  - $x = x_0 \exp(-kt)$ or $\ln x = \ln x_0 - kt$
  - $t_{1/2} = \ln 2 / k$
- For reactions of order $n \neq 1$, $x^{1-n} = x_0^{1-n} + kt(n-1)$
- $\eta_{FRET} = 1 - \tau_{DA}/\tau_D$

[6] J. L. Digambaranath et al., *Proteins* **79**, 1427 (2011).

## Review exercise group 13.6

(1) Suppose that someone told you that a second-order reaction has a half-life of 30 s at 25 °C. Is this a meaningful statement? Explain briefly.

(2) As we have seen, exponential growth or decay, corresponding to a first-order rate law, are commonly observed in a variety of contexts. Consider the following atmospheric $CO_2$ data:[7]

| Year | 1744 | 1847 | 1943 | 1962 | 1980 | 2000 |
|---|---|---|---|---|---|---|
| $[CO_2]$/ppm | 276.8 | 286.8 | 307.9 | 317.62 | 336.98 | 367.01 |

Is the atmospheric $CO_2$ concentration growing exponentially? If so, what is the doubling time? If not, is the growth faster or slower than exponential?

Aside: A very deep ice core taken at the Russian Vostok research station in East Antarctica provides data for the last 400 000 years. The atmospheric $CO_2$ levels during this period have fluctuated quite a bit, but have never been higher (until the last few years) than 300 ppm. A single reading of 298.7 ppm was obtained for a sample corresponding to 323 000 years before the present. All the other readings are below 290 ppm. Recent $CO_2$ levels are therefore higher than *any* in the recent geological history of the planet, which is one of the major sources of concern about the rapid increase in atmospheric carbon dioxide concentrations in our times: we are rapidly getting away from a range of concentrations where historical trends provide any clue as to what will happen next.

(3) When ion channels are opened in a cell membrane (e.g. the membranes of neurons), the flow of ions creates a voltage difference between the two sides of the membrane. This voltage is not established instantaneously but obeys the rate law

$$\frac{d(\Delta V)}{dt} = -\frac{\Delta V}{RC}$$

where $\Delta V$ is the difference between the transmembrane voltage and its equilibrium value,

$$\Delta V = V - V_{eq},$$

$R$ is the membrane electrical resistance and $C$ is its capacitance. Resistance is measured in ohms ($\Omega$), and capacitance is measured in farads (F). The units of resistance and capacitance combine as follows: $1\,\Omega \times 1\,F = 1\,s$.

(a) Derive an integrated rate law for the voltage. Write your final answer in the form $V =$ something. Show all your steps.

---

[7] The first three points are from the Antarctic ice core data of Neftel *et al.* (1994), available on the web at http://cdiac.esd.ornl.gov/trends/co2/siple.htm. The last three points are annual averages of direct atmospheric measurements taken since 1957 at the South Pole and summarized by Keeling and Whorf (2002) at http://cdiac.esd.ornl.gov/trends/co2/sio-spl.htm.

(b) A membrane has a capacitance of 15 pF and a resistance of 10 MΩ. If the initial transmembrane voltage is 0 mV and the equilibrium voltage is 100 mV, how long would it take for the membrane voltage to reach 99 mV?

(4) The reaction $2NO_{(g)} + Cl_{2(g)} \rightarrow 2NOCl_{(g)}$ is thought to be elementary. (Yes, it's one of those supposedly rare termolecular gas-phase elementary reactions. There's nothing a physical chemist likes better than a good exception, so there's a disproportionate amount of information available on termolecular reactions.) Suppose that there is initially no NOCl present, and that the initial concentrations of NO and of $Cl_2$ have no particular relationship. Work out the integrated rate law for this reaction, right up to the step of taking the integral. *Do not evaluate the integral.* In other words, leave your answer in the form $t = \int_a^b f(z)\,dz$.

(5) We normally write simple integrated rate laws (e.g. the first-order equation) in terms of the concentration of a reactant. This is not the only possibility. Suppose that a reaction has the stoichiometry $A \rightarrow 2B$ and has a second-order rate law, $v = ka^2$. Derive an integrated rate law for the concentration of the product, $b$, assuming that $b(0) = 0$.

(6) Fungal pellets are roughly spherical masses that sometimes form when fungi are grown in a liquid medium. The growth of the fungal population results in an increase in pellet mass which has empirically been found to obey the following equation:

$$m^{1/3} = m_0^{1/3} + kt$$

(a) What is the kinetic order of this growth process?
(b) Derive an equation for the doubling time.

(7) In question 1 of Exercise group 13.1, we discussed the first-order decay of the 2,5-dihydrofuran radical cation ($2,5\text{-DHF}^{+\bullet}$) in a $CF_3CCl_3$ matrix. First-order decay of radicals can result in one of two ways:

(a) A unimolecular reaction occurs in which the radical rearranges itself. This is the case for $2,5\text{-DHF}^{+\bullet}$ which rearranges to the radical cation $2,4\text{-DHF}^{+\bullet}$ which is stable up to 140 K.
(b) The radical reacts with the matrix. The resulting process is really a pseudo-first-order reaction.

These two possibilities are difficult to distinguish by kinetic methods alone.

Radicals can also react by recombination, i.e. by a reaction of the type $R^\bullet + R^\bullet \rightarrow R_2$, or by other similar bimolecular reactions that pair up the odd electrons and produce non-radical products.

The decay of 2,3-dihydrofuran radicals ($DHF^\bullet$) was studied by electron paramagnetic resonance (EPR) in an F-113 matrix at 140 K. (F-113 is the freon $CFCl_2CF_2Cl$.) Recall that the intensity of an EPR signal is proportional to the concentration of the corresponding radical species. The results were as follows:[8]

---

[8]  W. Knolle *et al., J. Chem. Soc., Perkin Trans.* **2**, 2447 (1999).

| $t$/min | 1.85 | 3.70 | 6.17 | 12.35 | 16.67 | 25.31 | 37.04 |
|---|---|---|---|---|---|---|---|
| EPR intensity | 0.71 | 0.56 | 0.43 | 0.30 | 0.25 | 0.18 | 0.13 |

Based on these data, what is the likeliest mechanism for the removal of DHF radicals under the experimental conditions described above?

(8) Silicon atoms enter the upper atmosphere either from interplanetary dust or from the evaporation of meteoroids as they enter the atmosphere. These silicon atoms can then react either with oxygen or with ozone to form silicon oxides. The following is one of the elementary reactions that can occur:

$$Si_{(g)} + O_{2(g)} \xrightarrow{k} SiO_{(g)} + O_{(g)}$$

(a) Give the mass-action rate law for this elementary reaction.

(b) This reaction has been studied under conditions in which oxygen was present in great excess. The following pseudo-first-order rate constants were measured at different oxygen concentrations at 190 K:[9]

| $[O_2]/10^{-6} \, mol \, L^{-1}$ | 0.052 | 0.284 | 0.879 | 1.12 | 1.63 | 2.152 |
|---|---|---|---|---|---|---|
| $k'/10^5 \, s^{-1}$ | 0.090 | 0.322 | 0.798 | 1.01 | 1.59 | 1.91 |

(i) Explain why a graph of $k'$ vs. $[O_2]$ will give the second-order rate constant $k$. What should the intercept of this graph be?

(ii) Determine $k$, and discuss whether the value of the intercept you calculated is in reasonable agreement with your theoretical prediction.

(iii) For the experiment with an oxygen concentration of $1.12 \times 10^{-6} \, mol \, L^{-1}$ at 190 K, how long would it take for 90% of the silicon atoms to react?

[9] J. C. Gómez Martín, M. A. Blitz and J. M. C. Plane, *Phys. Chem. Chem. Phys.* **11**, 671 (2009).

# 14

# Complex reactions

I have mentioned several times in the past few chapters that complex reactions can have either simple or complex rate laws. We will now see how complex reactions can give rise to complex rate laws, but also how these rate laws often simplify to simple forms.

You may notice as you read this chapter that it seems to focus a lot on gas-phase chemistry, and that there are few biological examples and problems. This is a deliberate choice. First of all, gas-phase chemistry is of interest to biologists since the atmosphere ties us all together. My second reason for focusing on gas-phase chemistry here has to do with pedagogy: examples of complex mechanisms deriving from gas-phase chemistry are often simpler than those arising in biochemistry. We will get a good grounding in the basic methods for treating complex mechanisms in this chapter before moving on to enzyme kinetics in the next chapter.

## 14.1 Two-step mechanisms

In order to get our feet wet with complex reactions, we will try to resolve a puzzle: on the surface, unimolecular gas-phase reactions, i.e. reactions with the stoichiometry $A \rightarrow$ product(s), should be the simplest reactions around. However, these reactions always have complex rate laws: at low pressures, they display second-order kinetics, while at high pressures, they follow first-order kinetics with respect to the reactant. (At intermediate pressures, the rate can't be described by either a first- or a second-order rate law.) There's a fairly straightforward explanation for this behavior, first proposed by Lindemann in 1922: a stable molecule doesn't just decompose or isomerize on its own. Decomposition reactions, for instance, require bonds to be broken. A molecule therefore has to be energized somehow to undergo reaction. In the gas phase, molecules can gain energy through collisions. Suppose that A is a reactant molecule and P a product, and that X is an energized version of A. Lindemann's hypothesis is equivalent to the following mechanism:

$$A + A \underset{k_{-1}}{\overset{k_1}{\rightleftharpoons}} A + X$$

$$X \overset{k_2}{\rightarrow} P$$

In the first step, two molecules of A collide, and one of them acquires enough energy to react. The reactive intermediate X then goes on to form the product(s). We show the product-forming step as being irreversible because these reactions are often very strongly thermodynamically favored such that reversion to the reactant is unlikely. It then doesn't matter whether P is a single product (as shown here) or a set of breakdown products, since it won't appear in the rate law.

In Section 11.5, the rule for writing rate laws for complex mechanisms was given. Basically, the rate of change of a given concentration is the sum of the rates of change due to each elementary reaction in which that species appears. This is our first example of the application of this rule to a complex mechanism, so make sure you understand where each term in the following rate equations came from:

$$\frac{da}{dt} = -k_1 a^2 + k_{-1}ax \tag{14.1a}$$

$$\frac{dx}{dt} = k_1 a^2 - k_{-1}ax - k_2 x \tag{14.1b}$$

These are coupled differential equations, meaning that we have two equations in two dependent variables, each of which appears in both equations. Unlike the simple cases studied in the last section, these equations cannot be solved analytically. There exists no procedure analogous to separation of variables for solving coupled differential equations.

We don't need to solve the rate equations to address Lindemann's puzzle. Note that this puzzle was stated in terms of the rate of reaction. All we need to do is to be able to arrive at an equation for the rate of reaction. If you look back at the mechanism, you should be able to convince yourself that the overall reaction is just A → P. From the definition of the rate of reaction and using the law of mass action, we would therefore write

$$v = \frac{dp}{dt} = k_2 x.$$

Note that this rate involves the concentration of the intermediate X. Intermediate concentrations are often difficult to measure. Furthermore, the observations we are trying to explain are phrased in terms of the reactant pressure $a$. We should therefore try to find some expression that relates $x$ to $a$.

Recall that the intermediate X is an energized version of the reactant A. It should be highly reactive, since it is carrying excess energy. Indeed, this is the reason why it eventually breaks down into products, whereas A doesn't do so directly. That being the case, in the initial stages of the reaction, we expect X to build up, but we will soon reach a situation where it breaks down as fast as it can be produced by A + A collisions. We then say that $x$ has reached a **steady state**. In mathematical form, we expect that

$$dx/dt \approx 0 \tag{14.2}$$

or, using Equation (14.1b),

$$k_1 a^2 - k_{-1}ax - k_2 x \approx 0.$$

We wanted to eliminate $x$ from the rate equation. We can solve this last equation for $x$:

$$x \approx \frac{k_1 a^2}{k_{-1} a + k_2}.$$

The rate of reaction would then be given by

$$v = k_2 x \approx \frac{k_1 k_2 a^2}{k_{-1} a + k_2}. \tag{14.3}$$

This procedure for obtaining a rate law is called the **steady-state approximation**. Specifically, the steady-state approximation involves the assumption that, after an initial transient, the concentrations of any intermediates change slowly, which we express mathematically by a condition such as Equation (14.2). This approximation will usually be valid when the intermediates are highly reactive so that they do not accumulate to high concentrations.

We are now ready to resolve Lindemann's puzzle. The denominator of Equation (14.3) is a sum of terms, one of which depends on $a$. If the pressure of A is sufficiently small, then $k_{-1} a$ eventually becomes much smaller than $k_2$ and can be neglected. This gives

$$v \approx \frac{k_1 k_2 a^2}{k_2} = k_1 a^2.$$

This agrees with the observation that a second-order rate law is observed at low pressures. Note in fact that the rate of reaction under these conditions is just the rate of the first step, $A + A \rightarrow A + X$. At sufficiently low pressures, we expect this second-order process, which requires collisons between molecules of A, to be slow. A reaction can't proceed faster than its slowest step, so this first step is said to be **rate limiting** under these conditions.

Now consider what happens if $a$ is sufficiently large that $k_{-1} a \gg k_2$:

$$v \approx \frac{k_1 k_2 a^2}{k_{-1} a} = \frac{k_1 k_2}{k_{-1}} a.$$

This is the first-order behavior observed at high pressures. Lindemann's mechanism thus explains both extremes of the rate law for gas-phase unimolecular reactions.

Note the following:

- In general, the Lindemann mechanism has a complex rate law that is different in form from the simple rate laws we discussed in Chapter 12.
- Even in limits in which we get simple rate laws, the observed rate constant may not correspond to an elementary rate constant. For the Lindemann mechanism at high pressures, the observed first-order rate constant would be related to the elementary rate constants by $k = k_1 k_2 / k_{-1}$.

### 14.1.1 The oxidation of NO to $NO_2$

As a second example, consider the oxidation of NO to $NO_2$,

$$2NO_{(g)} + O_{2(g)} \rightarrow 2NO_{2(g)}$$

The mechanism of this reaction is

$$2NO_{(g)} \underset{k_{-1}}{\overset{k_1}{\rightleftharpoons}} N_2O_{2(g)};$$

$$N_2O_{2(g)} + O_{2(g)} \overset{k_2}{\rightarrow} 2NO_{2(g)}.$$

$$v = \frac{1}{2}\frac{d[NO_2]}{dt} = k_2[N_2O_2][O_2]. \tag{14.4}$$

You should take a moment to verify Equation (14.4), and in particular to see how the factor of $\frac{1}{2}$, which comes from the overall stoichiometry, cancels a factor of 2 from the product-forming step.

We are faced with the same problem as in the Lindemann mechanism, namely that the rate law depends on the concentration of the intermediate $N_2O_2$. Once again, because it's hard to do experiments in which we control the concentration of intermediates, we would like to obtain a rate law that depends only on reactant concentrations. We could apply the steady-state approximation to the intermediate $N_2O_2$, but in this case we have an extra piece of experimental information: it is known that the equilibrium $2NO \rightleftharpoons N_2O_2$ is fast. This isn't something we could have guessed, but sometimes this kind of information is available from studies of other, related reactions. Given this extra piece of information, we might reason that the dimerization of NO can be treated as being in equilibrium, with the second reaction in the mechanism acting only as a small disturbance to the equilibrium. This reasoning leads to the relationship

$$k_1[NO]^2 \approx k_{-1}[N_2O_2],$$

from which we obtain

$$[N_2O_2] \approx \frac{k_1}{k_{-1}}[NO]^2.$$

The rate of reaction is then

$$v = \frac{k_1 k_2}{k_{-1}}[NO]^2[O_2].$$

This procedure is called the **equilibrium approximation**.

What if we had used the steady-state approximation instead? The steady-state condition reads

$$\frac{d[N_2O_2]}{dt} = k_1[NO]^2 - k_{-1}[N_2O_2] - k_2[N_2O_2][O_2] \approx 0.$$

Solving for $[N_2O_2]$ and substituting into Equation (14.4), we get

$$v = \frac{k_1 k_2[NO]^2[O_2]}{k_{-1} + k_2[O_2]}.$$

If $k_2[O_2] \ll k_{-1}$, which is what we mean when we say that the product-forming step is slow compared to the equilibrium, we recover the equilibrium approximation rate. In other words, the steady-state approximation includes the equilibrium approximation as a special case, so if the equilibrium approximation is correct, then so is the steady-state approximation.

The reverse is not generally true, on the other hand. This observation turns out to hold for many simple mechanisms.

### 14.1.2 Choosing between the equilibrium and steady-state approximations

The following rules should help you decide which approximation to use in a given case:

- The steady-state approximation (SSA) is usually applied to highly reactive intermediates.
- The equilibrium approximation (EA) is used when a reversible elementary reaction is expected to be much faster than other steps in the mechanism.
- The SSA is more general, so if there is any doubt at all about the validity of the EA, use the SSA. On the other hand, the EA is generally easier to apply, so it should be used when it is valid.

## 14.2 Chain reactions

In order to sharpen our skills in using the classical steady-state and equilibrium approximations, we will now study an example of an important class of reactions known as chain reactions. We will also see how these approximations can be applied when we have more than one intermediate to eliminate from a rate law.

The gas-phase reaction of hydrogen with bromine to form hydrogen bromide displays surprisingly complex kinetics, as first shown by Bodenstein and Lind in 1906. For one thing, the reaction rate depends on a fractional power of the bromine concentration. For another, the reaction is noticeably inhibited by its product. Bodenstein and Lind showed that the experimental rate data are well fit by the equation

$$v = \frac{k_a[H_2][Br_2]^{1/2}}{k_b + [HBr]/[Br_2]}.$$

This peculiar rate law went unexplained for over a decade. Then, in 1919–1920, J. A. Christiansen, Karl Ferdinand Herzfeld and Michael Polanyi, working independently, all proposed the following mechanism for the hydrogen-bromine reaction:

$$Br_2 \xrightarrow{k_1} 2Br \qquad\qquad\qquad \text{(CHP1)}$$

$$Br + H_2 \xrightarrow{k_2} HBr + H \qquad\qquad \text{(CHP2)}$$

$$H + Br_2 \xrightarrow{k_3} HBr + Br \qquad\qquad \text{(CHP3)}$$

$$2Br \xrightarrow{k_{-1}} Br_2 \qquad\qquad\qquad \text{(CHP4)}$$

$$H + HBr \xrightarrow{k_{-2}} H_2 + Br \qquad\qquad \text{(CHP5)}$$

This is called a **chain reaction**; Steps CHP2 and CHP3 generate each other's radical reactants so that, once the reaction gets going, it is self-sustaining. These two steps are called **propagation** reactions. Chain propagation reactions make at least as many radicals as they consume. The first step is the **initiation** step or, in other words, the step that starts off the reaction. The elementary reaction CHP4 is the **termination** or **chain-breaking** step. It eliminates any excess bromine atoms once the reaction starts to slow down. Every chain reaction has initiation, propagation and termination steps. Reaction CHP5 is not responsible for product formation. In fact, it is the reverse of step CHP2. It is an **inhibition** step: its overall effect is to slow (inhibit) the reaction by consuming the product. This is the reaction that will account for the inverse relationship on the HBr concentration in the empirical rate law. Note that our mechanism does not include the reverse of reaction CHP3. This is because the elementary process $HBr + Br \rightarrow H + Br_2$ has a very low rate, due in part to the extraordinarily small bond dissociation energy of the bromine molecule, which makes this reaction very endothermic. We will see later that a key factor in determining the sizes of rate constants is the activation energy, the amount of energy the reactants must gain in order to react. The larger the activation energy is, the smaller the corresponding rate constant. For an endothermic reaction, the minimum activation energy is $\Delta_r H$, so reactions with large, positive enthalpies of reaction tend to be slow. The small bond dissociation energy of bromine is, incidentally, the same factor that makes reaction CHP1 a significant process at reasonable temperatures. It is also worth noting that reaction CHP1 is actually a complex reaction obeying a variation of the Lindemann mechanism. Specifically, a $Br_2$ molecule can be energized by collision with any other molecule in the reaction vessel. We can treat CHP1 as a pseudo-first-order reaction because the stoichiometry of the reaction, $Br_2 + H_2 \rightarrow 2HBr$, means that the total number of molecules in the gas phase, and thus the total pressure, is constant. However, if we systematically varied the pressure, we would find that $k_1$ actually depends on pressure. A mechanism is a model, and not necessarily a catalog of everything we know about a reaction. When we build a model of a complex reaction, we often leave out reactions (in this case, collisional activation of $Br_2$) that are not directly relevant to the phenomenon we are trying to explain (different rate laws in different pressure ranges) in order not to distract ourselves from the central issues. This is what we have done here. Similar comments could be made for the reverse reaction, CHP4, which will require a collision with another molecule to carry away the bond energy before the bromine molecule made in this process can dissociate again. We can also neglect steps like $H + Br \rightarrow HBr$, because the concentrations of the two radicals should remain quite low due to their reactivities. Collisions between rare species are much less likely than collisions between rare species and abundant molecules. Of the possible radical recombination steps, the Br+Br step turns out to have the highest rate by far, and thus to be the only significant reaction of this type.

Since free radicals are highly reactive, it should be reasonable to apply the steady-state approximation to their concentrations:

$$\frac{d[Br]}{dt} = 2k_1[Br_2] - k_2[Br][H_2] + k_3[H][Br_2] - 2k_{-1}[Br]^2 + k_{-2}[H][HBr] \approx 0;$$

$$\frac{d[H]}{dt} = k_2[Br][H_2] - k_3[H][Br_2] - k_{-2}[H][HBr] \approx 0.$$

We treat these equations as two equations in the two unknowns [H] and [Br]. If we first add the two equations, we get the relationship

$$2k_1[\mathrm{Br_2}] - 2k_{-1}[\mathrm{Br}]^2 \approx 0$$

or

$$[\mathrm{Br}] \approx \sqrt{k_1[\mathrm{Br_2}]/k_{-1}}.$$

(Adding steady-state relationships in order to get simpler equations to solve is a common trick.) Substituting this relationship into the second steady-state equation, we get

$$[\mathrm{H}] \approx \frac{k_1^{1/2}k_2[\mathrm{Br_2}]^{1/2}[\mathrm{H_2}]}{k_{-1}^{1/2}(k_3[\mathrm{Br_2}] + k_{-2}[\mathrm{HBr}])}.$$

The overall reaction is $\mathrm{H_2} + \mathrm{Br_2} \rightarrow 2\mathrm{HBr}$ so the reaction rate is

$$v = \frac{1}{2}\frac{d[\mathrm{HBr}]}{dt} = \frac{1}{2}\left(k_2[\mathrm{Br}][\mathrm{H_2}] - k_{-2}[\mathrm{H}][\mathrm{HBr}] + k_3[\mathrm{H}][\mathrm{Br_2}]\right).$$

We can substitute the steady-state concentrations of H and Br into this equation and then spend a good deal of time simplifying terms. However, there is a slightly easier way to go about this chore. The steady-state condition for H can be rearranged to the form

$$k_2[\mathrm{Br}][\mathrm{H_2}] - k_{-2}[\mathrm{H}][\mathrm{HBr}] \approx k_3[\mathrm{H}][\mathrm{Br_2}].$$

We use this equation to simplify the reaction rate, and get

$$v \approx k_3[\mathrm{H}][\mathrm{Br_2}] = \frac{k_1^{1/2}k_2k_3[\mathrm{Br_2}]^{3/2}[\mathrm{H_2}]}{k_{-1}^{1/2}(k_3[\mathrm{Br_2}] + k_{-2}[\mathrm{HBr}])} = \frac{\dfrac{k_1^{1/2}k_2k_3}{k_{-1}^{1/2}k_{-2}}[\mathrm{Br_2}]^{1/2}[\mathrm{H_2}]}{k_3/k_{-2} + [\mathrm{HBr}]/[\mathrm{Br_2}]}.$$

This is just the Bodenstein–Lind rate law, except that the empirical constants $k_a$ and $k_b$ are now expressed in terms of elementary rate constants.

The dependence of the rate on the square root of the concentration of one of the reactants is typical of reactions in which radical intermediates appear. The reason is simple: the initiation step generally involves the splitting of a reactant into two radicals. This generates a quadratic relationship between the reactant and radical concentrations which, when solved, gives us a square root. Note, however, that mechanisms involving radicals don't always give rate laws containing square-root terms, since it sometimes occurs that these terms end up being squared away (literally) at some point during the derivation.

## 14.3 Deriving rate laws of complex mechanisms

Deriving rate laws is one of those exercises that strikes fear in the hearts of many students of chemistry and biochemistry. Although there may occasionally be opportunities for creativity, the truth is that deriving a rate law for a complex mechanism can largely be reduced to an algorithm. If you always do things in the order indicated below, you can get these derivations right every time, give or take the odd algebraic error. Here are the steps you should go through:

(1) Identify any intermediates in the mechanism.
(2) If you don't have the overall reaction, figure out what it is.
(3) Write down an expression for $v$ based on the overall reaction using the law of mass action and the definition of the rate of reaction.
(4) You will need to apply *one* approximation for *every* intermediate in the mechanism, be it a steady-state or an equilibrium approximation. Start with the reversible steps. Ask yourself if any of the information you have would lead you to think that the equilibrium approximation would be valid for that step. If so, apply the EA. Add one SSA equation for each intermediate not already appearing in an EA equation.
(5) The intermediate concentrations are your unknowns, and you should now have as many equations as intermediates. You want to solve your equations for the unknown intermediate concentrations. Before you start, look at your expression for $v$. In order to avoid back-substitutions (i.e. extra work), you will want to solve for the intermediate concentration(s) appearing in $v$ *last*. Keeping this in mind, solve for the intermediate concentrations. Look for opportunities to simplify the work by adding equations before you go too far.
(6) Substitute the intermediate concentration(s) into $v$. You're done!

---

## Exercise group 14.1

(1) The $N_2O_5$-catalyzed decomposition of ozone proceeds by the following mechanism:

$$N_2O_{5(g)} \underset{k_{-1}}{\overset{k_1}{\rightleftharpoons}} NO_{2(g)} + NO_{3(g)}$$

$$NO_{2(g)} + O_{3(g)} \overset{k_3}{\rightarrow} NO_{3(g)} + O_{2(g)}$$

$$2NO_{3(g)} \overset{k_4}{\rightarrow} 2NO_{2(g)} + O_{2(g)}$$

(a) Write down the overall reaction.
   *Hint*: The nitrogen oxides are catalytic species which do not appear in the overall process. You can ignore the first reaction which simply serves to create the true catalyst in this reaction, namely nitrogen dioxide.
(b) The first reaction is known to equilibrate rapidly relative to the rate at which the other reactions proceed. Using this information, and assuming that the nitrogen trioxide concentration reaches a steady state, derive a rate law for this reaction involving only the ozone and dinitrogen pentoxide concentrations.
(c) Is the rate law simple? If so, what are the orders of reaction with respect to $N_2O_5$ and $O_3$?
(d) Do you get the same rate law if you simply use the SSA for both nitrogen dioxide and nitrogen trioxide?

(2) Here is a variation on the Lindemann mechanism:

$$2A \underset{k_{-1}}{\overset{k_1}{\rightleftharpoons}} B$$

$$B \overset{k_2}{\rightarrow} A + P$$

(a) What is the overall reaction?
(b) Assume that the intermediate is highly unstable. Derive an approximate rate law for this mechanism.
(c) What overall order of reaction does your rate law predict?

(3) Consider the following mechanism:

$$W + W \underset{k_{-1}}{\overset{k_1}{\rightleftharpoons}} W + 2X$$

$$X + Y \overset{k_2}{\rightarrow} Z$$

(a) What is the overall reaction? Identify the reactants, products and intermediates in the mechanism.
(b) Assuming that W is a normal molecule with fully satisfied valencies, what kind of chemical species is X?
(c) Suppose that the first step reaches equilibrium rapidly. Derive an approximate rate law valid in this case.
(d) Under what condition(s) could you apply the steady-state approximation? Derive an approximate rate law using this approximation.

(4) Two mechanisms have been proposed for the reaction $CO_{(g)} + Cl_{2(g)} \rightarrow COCl_{2(g)}$:

$$Cl_{2(g)} \underset{k_{-1}}{\overset{k_1}{\rightleftharpoons}} 2Cl_{(g)}$$

$$Cl_{(g)} + Cl_{2(g)} \underset{k_{-2}}{\overset{k_2}{\rightleftharpoons}} Cl_{3(g)} \qquad (14.5)$$

$$Cl_{3(g)} + CO_{(g)} \overset{k_3}{\rightarrow} COCl_{2(g)} + Cl_{(g)}$$

$$Cl_{2(g)} \underset{k_{-1}}{\overset{k_1}{\rightleftharpoons}} 2Cl_{(g)}$$

$$Cl_{(g)} + CO_{(g)} \underset{k_{-2}}{\overset{k_2}{\rightleftharpoons}} COCl_{(g)} \qquad (14.6)$$

$$COCl_{(g)} + Cl_{2(g)} \overset{k_3}{\rightarrow} COCl_{2(g)} + Cl_{(g)}$$

(a) Identify the radical intermediates appearing in these mechanisms.
(b) Identify the initiation, chain propagation and termination steps in both mechanisms. Are there any elementary reaction in either mechanism that don't fall into one of these categories?
(c) Derive rate laws for both mechanisms involving only concentrations of stable molecules. Could these mechanisms be distinguished experimentally? Explain.

(5) The experimental rate law for the reaction $2NO_{(g)} + Br_{2(g)} \rightarrow 2NOBr_{(g)}$ is $v = k[NO]^2[Br_2]$. Show that this rate law is consistent with the mechanism

$$NO_{(g)} + Br_{2(g)} \underset{k_{-1}}{\overset{k_1}{\rightleftharpoons}} NOBr_{2(g)}$$

$$NOBr_{2(g)} + NO_{(g)} \overset{k_2}{\rightarrow} 2NOBr_{(g)}$$

Give an equation for the experimental rate constant $k$ in terms of the elementary rate constants $k_1$, $k_{-1}$ and $k_2$. Briefly discuss, based on your derivation of the rate equation, which elementary processes are fast and which are slow in this reaction.

(6) Although elementary trimolecular reactions are rare, reactions with third-order rate laws are not. Consider the mechanism

$$A + A \underset{k_{-1}}{\overset{k_1}{\rightleftharpoons}} B$$

$$B + C \overset{k_2}{\rightarrow} P$$

Under what conditions will this reaction display third-order kinetics?

## 14.4 Equilibria in complex reactions

Back in Section 11.6, we discussed the relationship between equilibrium and kinetics for single elementary reactions. The principle is the same for complex reactions as it is for elementary reactions, namely that the rates of change of the concentrations are zero at equilibrium. We have more equations to solve, however. We are assisted in this task by the **principle of detailed balance**, a consequence of the laws of mass action and of microscopic reversibility: at equilibrium, every elementary reaction is itself in equilibrium.

**Example 14.1 Detailed balance in action** Consider the following mechanism:

$$A \underset{k_{-1}}{\overset{k_1}{\rightleftharpoons}} 2B$$

$$B \underset{k_{-2}}{\overset{k_2}{\rightleftharpoons}} C$$

The overall reaction is

$$A \rightleftharpoons 2C$$

with empirical equilibrium constant

$$K = \frac{c^2}{a}.$$

According to the principle of detailed balance, at equilibrium

$$k_1 a = k_{-1} b^2$$

$$\text{and} \quad k_2 b = k_{-2} c.$$

These relationships can be rearranged to the form

$$K_1 = \frac{k_1}{k_{-1}} = \frac{b^2}{a};$$

$$K_2 = \frac{k_2}{k_{-2}} = \frac{c}{b}.$$

If we square the second of these equations and multiply by the first, the concentration of the intermediate cancels out and we get

$$K = K_1 K_2^2 = \frac{k_1 k_2^2}{k_{-1} k_{-2}^2} = \frac{c^2}{a}.$$

Relationships such as those derived in the previous example between the equilibrium and rate constants mean that the rate constants are not all independent; one of them is fixed by the thermodynamics of the overall reaction. This has important implications in mechanisms containing cycles, as we shall see in the following example.

**Example 14.2 Cyclic reaction mechanisms** Reaction cycles are common in biochemical reaction networks. Consider the following cycle:

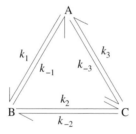

The equilibrium constants for the three steps are

$$K_1 = \frac{b}{a} = \frac{k_1}{k_{-1}};$$

$$K_2 = \frac{c}{b} = \frac{k_2}{k_{-2}};$$

$$K_3 = \frac{a}{c} = \frac{k_3}{k_{-3}}.$$

If we multiply the three equilibrium constants together, we get

$$K_1 K_2 K_3 = 1 = \frac{k_1 k_2 k_3}{k_{-1} k_{-2} k_{-3}}.$$

Therefore, if we know any five of the rate constants, the sixth is automatically fixed.

**Exercise group 14.2**

(1) Consider the mechanism

$$A \underset{k_{-1}}{\overset{k_1}{\rightleftharpoons}} 2B + C$$

$$B \underset{k_{-2}}{\overset{k_2}{\rightleftharpoons}} C$$

(a) What is the overall reaction?
(b) If $k_1 = 1.2 \, \text{s}^{-1}$, $k_{-1} = 3.4 \times 10^8 \, \text{L}^2 \, \text{mol}^{-2}\text{s}^{-1}$, $k_2 = 2800 \, \text{s}^{-1}$ and $k_{-2} = 2.5 \, \text{s}^{-1}$, what is the equilibrium constant for the overall reaction?

**Key ideas and equations**

- Useful rate laws typically do not depend on the concentrations of intermediates.
- The steady-state approximation is applied to highly reactive intermediates.
- The equilibrium approximation is applied to fast reversible reactions.
- Principle of detailed balance: at equilibrium, each elementary reaction is in equilibrium.

**Review exercise group 14.3**

(1) The oxidation of indole by $HSO_5^-$ in acidic solution is thought to proceed by the following mechanism:[1]

$$\text{indole} + HSO_5^- \underset{k_{-1}}{\overset{k_1}{\rightleftharpoons}} C \overset{k_2}{\rightarrow} \text{oxindole} + HSO_4^-$$

(a) Derive a rate law for this mechanism.
(b) Experimentally, it is found that the rate depends on the concentrations as follows:

$$v = k[\text{indole}][HSO_5^-].$$

How is the experimental rate constant $k$ related to the elementary rate constants?
(c) At 293 K in a 20:80 (by volume) mixture of acetonitrile and water, $k = 4.0 \times 10^{-3} \, \text{L} \, \text{mol}^{-1}\text{s}^{-1}$. If we start the reaction with $[\text{indole}] = [HSO_5^-] = 0.030 \, \text{mol} \, \text{L}^{-1}$, how long would it take before 90% of the indole was oxidized under these reaction conditions?
(2) Peracetic acid (PAA) is the following compound:

$$CH_3 - \overset{\overset{\displaystyle O}{\|}}{C} - O - OH$$

---

[1] S. Meenakshisundaram and N. Sarathi, *Int. J. Chem. Kinet.* **39**, 46 (2007).

Recently, Zhao and coworkers have proposed the following mechanism for the decomposition of peracetic acid in aqueous solution at elevated temperatures:[2]

- First, PAA is protonated on the carbonyl oxygen, creating a carbocation:

$$PAA + H^+ \underset{k_{-1}}{\overset{k_1}{\rightleftharpoons}} PAAH^+$$

- The carbocation forms an intermediate addition product with another PAA:

$$PAAH^+ + PAA \overset{k_2}{\rightarrow} X + H^+$$

- The intermediate decomposes to acetic acid (symbolized AH), acetate ($A^-$), a proton and oxygen:

$$X \overset{k_3}{\rightarrow} AH + A^- + H^+ + O_2$$

- Finally, the acetate reprotonates:

$$A^- + H^+ \overset{k_4}{\rightarrow} AH$$

(a) Determine the overall reaction.
(b) Protonation and deprotonation steps are extremely fast in aqueous solution. Using this information and any reasonable additional assumptions you find necessary, determine the rate law for this reaction.
(c) What is the order of the reaction with respect to PAA?
(d) What would you predict the effect of raising the pH on the reaction rate to be?

---

[2] X. Zhao *et al.*, *J. Mol. Catal. A: Chem.* **284**, 58 (2008). I have changed their numbering of the steps for convenience, and expanded one of their steps.

# 15

# Enzyme kinetics

Enzymes are biological catalysts. Recall that a catalyst is a substance that accelerates a chemical reaction without itself being consumed by the reaction. Enzyme catalysis holds a special place in the history of chemistry. The law of mass action was enunciated in the nineteenth century in a thermodynamic context by Guldberg and Waage. By the early twentieth century, the law of mass action was supported by a growing mass of both thermodynamic and kinetic evidence. Nevertheless, some reactions had complex rate laws which were not obviously compatible with mass action. Enzyme catalysis posed such a difficulty. In fact, there was some debate as to whether enzyme molecules even had to come in contact with the reactant molecules to accelerate the reaction. In 1902, Victor Henri showed, with some help from Max Bodenstein, that enzyme kinetics is in fact compatible with the law of mass action. The complex rate law obeyed by enzyme-catalyzed reactions was simply a manifestation of a complex mechanism. Although debates with respect to the mode of action of enzymes persisted for some time, it was at least clear from this point on that enzyme catalysis was compatible with the then-emerging theory of chemical kinetics. This is the vein we will pick up in this chapter. Our study of enzyme kinetics will particularly emphasize the derivation of rate laws and the analysis of data from initial rate experiments.

## 15.1 Properties of enzymes

While enzymes are biological molecules, much of what we will learn about enzymes in this chapter is equally true for inorganic catalysts. However, enzymes do have some remarkable features that set them apart from inorganic catalysts:

- Enzymes are remarkably selective. Most enzymes catalyze a very specific reaction. The rate enhancement is often sensitive to even minor structural changes in the reactants (generally called **substrates** in enzymology). Moreover, enzymes are almost always stereospecific. In other words, if a compound has two stereoisomers, a given enzyme will generally only accelerate the reaction with one of the stereoisomers. Inorganic catalysts are generally much less specific and are almost never sensitive to the stereochemistry of the reactants.

- The rate enhancements obtained with an enzyme are often spectacular. The rate of an enzyme-catalyzed reaction is typically greater than the rate of the corresponding uncatalyzed reaction by a factor of $10^6$ to $10^{12}$. Inorganic catalysts seldom achieve such large rate enhancements.
- Enzymes are much more sensitive to temperature and pH than typical inorganic catalysts. Most enzymes rapidly lose their activity if the temperature is changed by more than 10 or 20 °C, or if the pH is changed by more than two or so units. Many enzymes are *completely* inactive outside their optimal temperature and pH operating ranges. In some cases, the enzyme does not regain its activity even if the temperature and pH are returned to their optimal values. Some inorganic catalysts lose their activity, and sometimes do so irreversibly, as the temperature and pH are changed, but this loss of activity is rarely seen over such narrow ranges of conditions as is the case for enzymes.

What sets enzymes apart from other catalysts? A large part of the answer is their complexity. Enzymes are proteins, assemblies of proteins or, sometimes, complexes of proteins and RNA with molar masses of tens or hundreds of thousands of grams per mole. This complexity makes high selectivity and large rate enhancements possible, but makes them more fragile than most inorganic catalysts (metal surfaces and particles, transition-metal complexes etc.). From the point of view of kinetics, however, this complexity is largely irrelevant, except insofar as it limits the experimental conditions under which enzymes can be studied.

## 15.2 The Michaelis–Menten mechanism

The mechanism by which enzymes accelerate reactions was uncovered in stages. Perhaps the first clear indication of the mechanism of enzyme action was obtained by an elegant experiment carried out in 1880 by Wurtz with the enzyme papain. Papain is a proteolytic enzyme, i.e. one that breaks down proteins. In case you're curious, the name of this enzyme comes from its source, the papaya fruit. Papaya is often used as a meat tenderizer precisely because it contains papain which helps break down protein, the main biochemical constituent of the connective tissue that sometimes makes meat tough. Wurtz was particularly interested in the action of papain on fibrin, one of the proteinaceous components of blood clots. Fibrin is insoluble in water, but its breakdown products (small peptides) do dissolve. Wurtz took a sample of fibrin and soaked it in a papain solution for a while, long enough to see that the fibrin was dissolving, but not long enough to destroy all of it. The remaining fibrin was filtered out and carefully washed over and over again. When the washed fibrin was placed in pure water, two things happened: first, the fibrin started to dissolve; second, if he added additional fibrin, this too dissolved. Both of these observations indicate that the enzyme had remained attached to the first batch of fibrin through several thorough washings. Wurtz had shown that the enzyme forms a stable complex with its substrate. From this experimental demonstration, Wurtz hypothesized that this complex was catalytically significant. Specifically, his hypothesis was that an enzyme acted on its substrate while the two were complexed together, and that the products were subsequently released from the enzyme, the latter having suffered no chemical transformation in the process.

In modern notation, Wurtz's hypothesis is equivalent to the following mechanism:

$$E + S \underset{k_{-1}}{\overset{k_1}{\rightleftharpoons}} C \overset{k_{-2}}{\longrightarrow} E + P \tag{MM}$$

where E is the enzyme, S is the substrate (reactant), C is an enzyme–substrate complex and P is the product. This mechanism is called the **Michaelis–Menten mechanism**, although Wurtz clearly thought of it first, an example of Stigler's law that no scientific discovery is ever named for its original discoverer.

The details of how an enzyme achieves catalysis is a subject worthy of its own course. It is, however, useful to have some basic ideas of the nature of the enzyme–substrate complex. As mentioned earlier, an enzyme is a large protein. Some of the amino acids that make up a protein are hydrophobic. As a result, a protein in aqueous solution will fold up in such a way as to "hide" the hydrophobic amino acids from the solvent. Folding does not in general lead to a unique structure, but typically there will be just a few dominant structures in solution, and sometimes just one. In the case of enzymes, at least one of the dominant structures will have a cleft in which the substrate can fit. The shape of this cleft is usually such that very few molecules other than the intended substrate fit properly, which accounts for the great specificity of enzymes. This cleft is called the **active site** of the enzyme, because that is where catalysis happens. Various chemical groups will be positioned within the active site to facilitate the reaction. The precise positioning of these chemical groups accounts for the very high catalytic efficiency of enzymes. This description of enzyme function is sometimes referred to as the **lock-and-key theory**, originally developed by Emil Fischer. There are a couple of subtleties not properly captured by the simplest version of lock-and-key theory. We will return to these issues in Chapter 17. However, lock-and-key theory is sufficient to develop the theory of enzyme kinetics.

Let's now work out the rate equations. If you look back at the mechanism, you will note that the enzyme is never destroyed. It forms a complex C with the substrate, but this complex either dissociates back into enzyme and substrate or into enzyme and product. The total amount of enzyme is therefore a constant:

$$e + c = e_0. \tag{15.1}$$

We can use this equation to eliminate $e$:

$$e = e_0 - c.$$

Note that we don't want to eliminate $c$ yet since $v = k_{-2}c$, so that we will eventually need to find an expression for $c$ involving only reactant concentrations. Furthermore, since we assume that product formation is irreversible, $p$ doesn't appear in the rate equations. We can therefore write just two rate equations for this mechanism:

$$\frac{ds}{dt} = -k_1 es + k_{-1}c = -k_1 s(e_0 - c) + k_{-1}c, \tag{15.2a}$$

$$\frac{dc}{dt} = k_1 es - k_{-1}c - k_{-2}c = k_1 s(e_0 - c) - (k_{-1} + k_{-2})c. \tag{15.2b}$$

We now develop the classical equilibrium and steady-state approximations for this mechanism. We start with the simpler equilibrium approximation, the approach taken by Henri in his 1902 paper. Suppose that equilibrium is established rapidly between the enzyme–substrate complex and the free species. Then

$$k_1 es = k_1 s(e_0 - c) \approx k_{-1}c$$

or

$$c \approx \frac{k_1 e_0 s}{k_1 s + k_{-1}} = \frac{e_0 s}{s + k_{-1}/k_1}.$$

$k_{-1}/k_1$ is the equilibrium constant for the *dissociation* of the enzyme–substrate complex into enzyme and substrate (the reverse of reaction 1). If we let

$$K_E = k_{-1}/k_1, \tag{15.3}$$

we get

$$c \approx \frac{e_0 s}{s + K_E}.$$

If we now use this approximate equilibrium value of $c$ in our expression for the rate of reaction, we get

$$v \approx \frac{k_{-2} e_0 s}{s + K_E} = \frac{v_{max} s}{s + K_E}$$

where $v_{max} = k_{-2} e_0$.

We will come back to this equation in a minute, but first let's look at the steady-state approximation, first applied to this mechanism by Briggs and Haldane. Suppose that the enzyme–substrate complex is in a steady state, i.e. that $k_{-2}$ is large so that C reacts rapidly after its production. Then

$$\frac{dc}{dt} = k_1 s(e_0 - c) - (k_{-1} + k_{-2})c \approx 0.$$

$$\therefore c \approx \frac{k_1 e_0 s}{k_1 s + k_{-1} + k_{-2}} = \frac{e_0 s}{s + K_S}. \tag{15.4}$$

$$\therefore v \approx \frac{k_{-2} e_0 s}{s + K_S} = \frac{v_{max} s}{s + K_S}.$$

In these equations,

$$K_S = (k_{-1} + k_{-2})/k_1 \tag{15.5}$$

and $v_{max}$ is defined as in Henri's equilibrium approach.

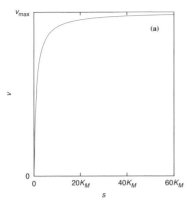
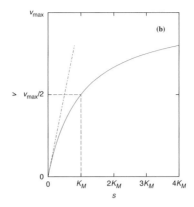

Figure 15.1 Properties of the Michaelis–Menten rate law (15.6). (a) The rate tends to $v_{max}$ at very large substrate concentration $s$. At sufficiently large $s$, $v$ is essentially constant and the reaction displays zero-order kinetics with respect to the substrate. (b) The Michaelis constant $K_M$ is the value of $s$ at which $v = \frac{1}{2}v_{max}$. The dotted line has a slope of $v_{max}/K_M$. Note that the Michaelis–Menten curve is tangent to this line at small $s$, i.e. the reaction there displays first-order kinetics with an apparent rate constant of $v_{max}/K_M$.

The two methods for obtaining a simplified rate law yield equations of the same form, namely

$$v \approx \frac{v_{max}s}{s + K_M}.$$ (15.6)

$K_M$ is best thought of as an experimentally measured quantity called the **Michaelis constant**. To the extent that an enzyme-catalyzed reaction obeys this rate law, it will not be possible to tell which interpretation we should give to $K_M$ ($K_E$ or $K_S$). In either interpretation, $K_M$ can be thought of as a **dissociation constant**, i.e. a ratio of the rate constant for dissociation of an enzyme–substrate complex to the rate constant for formation of the complex. If $K_M$ is very large, the substrate dissociates easily from the enzyme. If $K_M$ is small, the substrate is strongly bound. For two enzyme–substrate pairs that have the same $v_{max}$, the pair with the lowest $K_M$ has the higher rate of reaction at similar substrate concentrations.

The Michaelis–Menten rate law (15.6) describes a hyperbolic relationship of the rate to the substrate concentration. In particular:

(1) At large substrate concentrations ($s \gg K_M$), $v$ becomes asymptotic to $v_{max}$ (Figure 15.1(a)), which is the maximum rate of reaction, as its name suggests. We say that the enzyme is **saturated**. Literally, this means that almost all of the enzyme exists in the intermediate complex (C) form. The enzyme is transforming substrate into product as fast as it can and the reaction displays zero-order kinetics with respect to the substrate. The only way to raise the rate of the reaction under these conditions is to increase the amount of enzyme.

The proper use of conservation relations like (15.1) is essential in deriving rate laws. In this mechanism, for instance, the conservation of enzyme is directly responsible for the saturation phenomenon observed at large $s$. You are therefore strongly encouraged to look for conservation relations *before* deriving rate equations.

(2) At small substrate concentrations ($s \ll K_M$), $v \approx v_{max}s/K_M$, i.e. the reaction behaves as if it had a simple first-order rate law (Figure 15.1(b)) with rate constant $v_{max}/K_M$.

(3) Half the maximal rate is reached when $s = K_M$ (Figure 15.1(b)). This is sometimes used to get a crude estimate of $K_M$ and is a useful check on other methods for computing the value of this constant.

(4) In theory, if we know the concentration of enzyme ($e_0$) and $v_{max}$, we can calculate $k_{-2}$ by $k_{-2} = v_{max}/e_0$. In practice, many enzymes have more complex mechanisms than the simple Michaelis–Menten mechanism (additional steps, multiple substrates or products etc.). Moreover, we may not know the exact mechanism at the time of a kinetic study. We thus define the **turnover number** of an enzyme by $k_{cat} = v_{max}/e_0$. The turnover number, which has units of inverse time, is the maximum number of molecules of substrate that a single molecule of enzyme can convert to product per unit time. For the simple Michaelis–Menten mechanism, $k_{cat} = k_{-2}$. In other mechanisms, $k_{cat}$ may have a more complicated relationship to the rate constants.

There is an additional complication we often face in enzyme kinetic studies: some enzymes are hard to purify. It is not unusual for an enzyme to lose all activity during purification procedures. In these cases, we often study an enzyme in an unpurified, or only partially purified, cell extract. We would then have, at best, a crude estimate of the enzyme concentration, and we would therefore not calculate $k_{cat}$. Rather, we would use $v_{max}$ as a relative measure of enzyme concentration since $v_{max}$ is proportional to $e_0$. In a variation on this idea, people often report the **specific activity**, which is defined as the (maximum) rate of substrate conversion (often in $\mu$mol min$^{-1}$) by a given preparation divided by the total protein mass in the sample. (I put maximum in parentheses because specific activity is sometimes used at less than saturating substrate concentrations, although the results under such conditions are often difficult to reproduce.) The units of the specific activity are therefore mol min$^{-1}$g$^{-1}$ (or equivalent). The specific activity is a convenient way to check if, for example, a given enzyme preparation has degraded during storage, or if a sample I made up today is as good as the one I made yesterday. If we multiply the specific activity by the protein concentration in g L$^{-1}$, we get $v_{max}$. If the specific activity was calculated per gram of the particular enzyme under study rather than per gram of total protein, then the specific activity multiplied by the molar mass of the enzyme gives $k_{cat}$.

### 15.2.1 Determining $v_{max}$ and $K_M$

The experimental parameters $v_{max}$ and $K_M$ can be determined in a number of ways, some better than others. It is first important to note that enzymologists generally use initial rate methods. This avoids complications related to the reverse reaction. It also avoids difficulties related to the fact that enzymes are often not very stable once taken out of their original cellular environment so that long experiments to be analyzed with an integrated rate law are not always practical.

The classic method for treating initial rate data from an enzyme-catalyzed reaction is called the **Lineweaver–Burk** or **double-reciprocal** plot. The method is simple: take the reciprocal of both sides of the Michaelis–Menten rate law (Equation 15.6):

$$\frac{1}{v} = \frac{s + K_M}{v_{max} s};$$

$$\therefore \frac{1}{v} = \frac{1}{v_{max}} + \frac{K_M}{v_{max}} \frac{1}{s}. \tag{15.7}$$

A plot of $1/v$ as a function of $1/s$ should therefore give a straight line of slope $K_M/v_{max}$ and intercept $1/v_{max}$.

**Example 15.1 Lineweaver–Burk plot**   Consider the following initial rate data for an enzyme-catalyzed reaction:

| $s/\text{mmol L}^{-1}$ | 2.0 | 4.0 | 8.0 | 12.0 | 16.0 | 20.0 |
|---|---|---|---|---|---|---|
| $v/\mu\text{mol L}^{-1}\text{s}^{-1}$ | 130 | 200 | 290 | 330 | 360 | 380 |

To use the Lineweaver–Burk method, we compute $1/s$ and $1/v$ and plot these quantities:

| $s^{-1}/\text{L mol}^{-1}$ | 500 | 250 | 125 | 83.3 | 62.5 | 50.0 |
|---|---|---|---|---|---|---|
| $v^{-1}/\text{L s mol}^{-1}$ | 7692 | 5000 | 3448 | 3030 | 2778 | 2632 |

Note that I got rid of the SI prefixes before taking the inverses to avoid having odd units for the slope. This isn't necessary, but I find that it saves me from making errors handling the units during the subsequent calculations. The plot is shown in Figure 15.2. The slope is 11.3 s and the intercept is $2.08 \times 10^3$ L s mol$^{-1}$. From the intercept, we find $v_{max} = 480 \, \mu\text{mol L}^{-1}\text{s}^{-1}$. The slope is $K_M/v_{max}$ so $K_M = (11.3 \text{ s})(480 \, \mu\text{mol L}^{-1}\text{s}^{-1}) = 5.43 \text{ mmol L}^{-1}$.

> **The Lineweaver–Burk plot is the worst possible way to obtain $v_{max}$ and $K_M$ values, short of just eyeballing the data.**

One of the problems can clearly be seen in Figure 15.2: data obtained at larger values of $s$ are bunched up together at small values of $1/s$, and small-$s$ data end up widely spaced and far from the other points. This gives the one or two points at the very smallest values of $s$ a large statistical moment. Statistical moment is analogous to the moment of a force in physics; think of a mechanical force applied to a long lever at a point far from the fulcrum. The idea is the same here: when the points are not evenly spaced, the sum of squared errors (which is what linear regression minimizes) is typically smallest when the line passes near

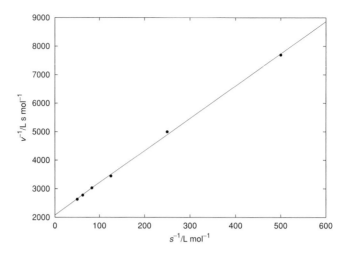

Figure 15.2 Lineweaver–Burk plot of the data of Example 15.1.

points that are isolated from the others, even if the other points suggest a different trend. If the point at smallest $s$ happens to have been badly measured, the slope and intercept returned by fitting a Lineweaver–Burk plot could be very wrong. Having one or two points dominate the fitting procedure defeats the purpose of measuring points at several different values of $s$. This problem could be overcome by choosing substrate concentrations that are equally spaced on a $1/s$ axis, but then we would have very few points at large $s$, and those points are necessary to accurately estimate $v_{max}$. It's a no-win situation.

There's another problem with Lineweaver–Burk plots. Let's say that we have a data set like the one in the previous example, which has points at both small and large values of $s$. This results in the bunching up of a lot of our data in the lower left-hand corner of the plot, as noted above. This tends to exaggerate our impression that the data fit a straight line, since only a few points are out of the lower left-hand mass. My own observations of published Lineweaver–Burk plots over a number of years, as well as a systematic study by Hill, Waight and Bardsley,[1] suggest that true Michaelis–Menten behavior is far from universal, and may in fact be the exception rather than the rule. Lineweaver–Burk plots often make non-Michaelian behavior less obvious, which is problematic, particularly since non-Michaelian behavior of an enzyme is sometimes physiologically relevant, or may be a sign of problems with an experimental protocol.

Despite the serious shortcomings of Lineweaver–Burk plots, these are still in very common use. Accordingly, it's a good idea to learn to interpret them, but not a particularly good idea to participate in propagating their use.

There are other ways to transform the Michaelis–Menten law into a linear form. The Eadie–Hofstee transformation starts from the Lineweaver–Burk equation. If we multiply

---

[1]  C. M. Hill, R. D. Waight and W. G. Bardsley, *Mol. Cell. Biochem.* **15**, 173 (1977).

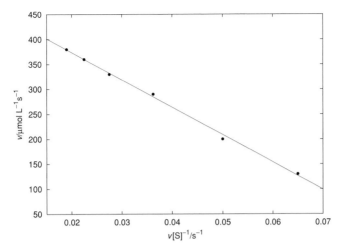

Figure 15.3 Eadie–Hofstee plot of the data presented in Example 15.1.

both sides of Equation (15.7) by $vv_{max}$ and rearrange, we get

$$v = v_{max} - K_M \frac{v}{s}. \tag{15.8}$$

A plot of $v$ as a function of $v/s$ should give a straight line of slope $-K_M$ and intercept $v_{max}$.

**Example 15.2 Eadie–Hofstee plot** We return to the data of Example 15.1. In the Eadie–Hofstee plot, the data take the form

| $v[S]^{-1}/s^{-1}$ | 0.0650 | 0.0500 | 0.0363 | 0.0275 | 0.0225 | 0.0190 |
|---|---|---|---|---|---|---|
| $v/\mu mol\,L^{-1}s^{-1}$ | 130 | 200 | 290 | 330 | 360 | 380 |

(Note that I have reverted to the square bracket notation in $v/[S]$ to avoid confusion with the units in my table headings and graph axis labels.) The graph is shown in Figure 15.3. As illustrated in this graph, the Eadie–Hofstee plot stretches the data at small $s$ (large $v/s$) less than does the Lineweaver–Burk plot. This allows us to use a reasonable spread of values of $s$ to get accurate estimates of both $v_{max}$ and $K_M$. It will also allow us to see non-Michaelian behavior, when it arises, more clearly than would be the case in a Lineweaver–Burk plot. Moreover, note that the parameters of the hyperbola are trivially related to the slope and intercept of the Eadie–Hofstee plot. Unlike the Lineweaver–Burk plot then, there is no need to do any arithmetic with the slope and intercept to get $v_{max}$ and $K_M$. Given all these advantages, it is frankly amazing that anyone still uses Lineweaver–Burk plots!

In any event, for this particular data set, we find $v_{max} = 484\,\mu mol\,L^{-1}s^{-1}$ and $K_M = -(\text{slope}) = 5.50\,mmol\,L^{-1}$. These values are slightly different than those obtained using the Lineweaver–Burk plot. The differences are attributable to the statistical difficulties previously mentioned with respect to bunching up of the large-$s$ data in a Lineweaver–Burk plot. In other cases, the two methods can give results that differ substantially. To be clear: the Eadie–Hofstee results are more reliable.

Although the Eadie–Hofstee plot is much better behaved than the Lineweaver–Burk plot, the best way to get the Michaelis–Menten parameters is to do a non-linear regression using a computer program. Non-linear regression directly fits the parameters of an expression such as the Michaelis–Menten equation to your data. As an added bonus, non-linear regression routines almost always return a standard error in addition to the value of the parameter itself. If we run the data of Examples 15.1 and 15.2 through a non-linear regression program, we get $v_{max} = 484 \pm 5 \, \mu\text{mol} \, L^{-1} s^{-1}$ and $K_M = 5.55 \pm 0.16 \, \text{mmol} \, L^{-1}$. These are our best estimates of these parameters. You will note that the Eadie–Hofstee results are much closer to these best estimates than the Lineweaver–Burk values.

If we should be using non-linear regression, why do we bother with Eadie–Hofstee plots? There are two good reasons and one not-so-good one. Let's start with the not-so-good reason: I don't know of any hand-held calculators that can do a non-linear regression, so an Eadie–Hofstee plot is something you can actually do in an exam, unlike a non-linear regression. Now for some better reasons: as previously mentioned, not every enzyme follows the Michaelis–Menten mechanism. If you don't know ahead of time if your enzyme is Michaelian, you would like to check this graphically, and that is one of the things that the Eadie–Hofstee method lets you do. The other reason has to do with enzyme inhibitors, which we will come to in the next section. There are different kinds of inhibition, and they look different from each other in Eadie–Hofstee plots. Figuring out what kind of inhibition you have from a graph is much easier than trying to do the same thing by looking at numbers spit out by a non-linear regression program.

---

## Exercise group 15.1

(1) Myosin is an enzyme that catalyzes the hydrolysis of ATP, harnessing the free energy of this reaction for muscle contraction. The initial rates of this reaction are as follows:

| $[\text{ATP}]/\mu\text{mol} \, L^{-1}$ | $v/\mu\text{mol} \, L^{-1} s^{-1}$ |
|:---:|:---:|
| 75 | 0.067 |
| 125 | 0.095 |
| 200 | 0.119 |
| 325 | 0.149 |
| 625 | 0.185 |
| 1550 | 0.191 |
| 3200 | 0.195 |

Calculate $v_{max}$ and $K_M$ for this enzyme–substrate system. Comment on the fit of the data to the Michaelis–Menten rate law.

(2) Consider the following initial rate data for an enzyme:

| $s/\text{mmol L}^{-1}$ | $v/\mu\text{mol L}^{-1}\text{s}^{-1}$ |
|---|---|
| 2.0 | 130 |
| 4.0 | 200 |
| 8.0 | 290 |
| 12.0 | 330 |
| 16.0 | 360 |
| 20.0 | 380 |

(a) Given that the concentration of enzyme is $2.0\,\text{g L}^{-1}$ and that its molar mass is approximately $50\,\text{kg mol}^{-1}$, and assuming ordinary Michaelis–Menten kinetics, calculate $K_M$ and $k_{-2}$.

(b) The sum $k_{-1} + k_{-2}$ appearing in the expression for $K_M$ derived from the steady-state approximation ($K_S$) can't be smaller than $k_{-2}$. From this observation, compute a lower bound for the value of $k_1$.

(3) Chlorotriazines are widely used as herbicides. Their principal mode of action is to inhibit photosynthesis. They do not degrade readily in the environment so that they tend to accumulate in soils and in water. They are not very toxic to mammals, but there are some questions about their long-term effects on the endocrine system. Studies of the *in vivo* fates of the compounds are therefore of considerable interest. Simazine (2-chloro-4,6-bis(ethylamino)-1,3,5-triazine) is metabolized, chiefly in the liver, to 2-chloro-4-ethylamino-6-amino-1,3,5-triazine:

simazine

Hanioka *et al.*[2] have studied the kinetics of this reaction catalyzed by a rat liver extract containing 400 µg of protein in a total volume of 1 mL. Their data are shown below:

| $[\text{simazine}]/\mu\text{mol L}^{-1}$ | 13 | 25 | 48 | 93 | 194 |
|---|---|---|---|---|---|
| $v/\text{nmol L}^{-1}\text{s}^{-1}$ | 0.26 | 0.39 | 0.53 | 0.56 | 0.62 |

Determine the Michaelis constant and the specific activity.

(4) Enzyme catalysis is not the only biochemical process that obeys the Michaelis–Menten rate law. Transport of nutrients and other chemicals across cell membranes also frequently obeys the Michaelis–Menten equation. In the simplest case, the transport

---

[2] N. Hanioka *et al.*, *Toxicol. Appl. Pharmacol.* **156**, 195 (1999).

mechanism can be represented by

$$T + N_{(out)} \rightleftharpoons C \rightarrow T + N_{(in)}$$

where N is a nutrient and T is a transporter (a protein whose role is to facilitate the movement of N across the membrane). Note that, give or take a change in notation, this is just the Michaelis–Menten mechanism except that the "reactant" and "product" are chemically identical, but located on opposite sides of a membrane. The following data have been obtained for the rate transport of sucrose into a strain of *Bacillus*:[3]

| [sucrose]/$\mu$mol L$^{-1}$ | 6.3 | 17.1 | 23.3 | 34.7 | 48.3 |
|---|---|---|---|---|---|
| $v$/$\mu$mol min$^{-1}$g$^{-1}$ | 0.151 | 0.296 | 0.389 | 0.488 | 0.540 |

*Note*: The rates are given as a specific activity relative to the total mass of cellular protein. The division by the amount of protein is meant to normalize the results to the amount of transporter present in the cells studied. The assumption is that the amount of transporter is a constant fraction of the total amount of protein in the cells.

(a) Do these data obey Michaelis–Menten kinetics? Discuss briefly.

(b) Regardless of your answer to question 4a, determine the best-fit values of $v_{max}$ and $K_M$.

## 15.3 Enzyme inhibition

### 15.3.1 Competitive inhibition

Suppose that, in addition to the substrate, an enzyme is presented with a second species whose binding to the enzyme prevents the binding of the substrate, i.e.

$$E + S \underset{k_{-1}}{\overset{k_1}{\rightleftharpoons}} C \overset{k_{-2}}{\longrightarrow} E + P$$

$$E + I \underset{k_{-3}}{\overset{k_3}{\rightleftharpoons}} H \qquad\qquad\text{(CI)}$$

I is a **competitive inhibitor** of the reaction of S catalyzed by E. Competitive inhibitors are almost always molecules that fit in the active site of an enzyme, thus blocking out the substrate.

When trying to derive rate equations for a complex mechanism like this one, it's good to remind ourselves of what we want to do before we proceed: the rate of reaction (rate of formation of product) is $v = dp/dt = k_{-2}c$. C is an intermediate, so we don't want its concentration appearing in the rate equations. A similar comment could be made about H; it's neither a reactant nor a product, so it shouldn't appear in the rate equation we obtain in the end. We will use the equilibrium and/or steady-state approximations to eliminate these two concentrations. It's also important in deriving rate equations to take proper account

---

[3]  C. J. Peddie *et al.*, *Extremophiles* **4**, 291 (2000).

of any conservation relations that arise, as we saw in our treatment of the Michaelis–Menten mechanism. These should be used to eliminate concentrations other than those of intermediate complexes like C and H, since we will use the equilibrium or steady-state approximations to get rid of these.

There are two useful conservation relations in this mechanism. First of all, the total amount of enzyme in its three possible forms (E, C and H) is constant:

$$e + c + h = e_0,$$

where $e_0$ is the total amount of enzyme. Moreover, the total amount of inhibitor is a constant:

$$i + h = i_0.$$

We can use these conservation relations to eliminate $e$ and $i$:

$$i = i_0 - h, \tag{15.9}$$
$$e = e_0 - c - h.$$

If we try to use the steady-state or equilibrium approximation along with these conservation relations, we eventually get a quadratic equation. There is nothing wrong with that of course, and the resulting steady-state rate should be valid across a range of experimental conditions, so arguably it is a very important equation to have at our disposal. However, this equation doesn't lead to graphical methods and, as we saw above, graphical methods can be very useful, even though we will usually use a non-linear regression program to fit the parameters.

If we want a graphical method, we will have to use the fact that, in a laboratory experiment, we have a lot of freedom to choose the concentrations. Suppose that we use a lot more inhibitor than enzyme, i.e. that $i_0 \gg e_0$. Then, since $h < e_0$, we can conclude that $h \ll i_0$. Equation (15.9) then becomes simply $i \approx i_0$.

At this point, we have to decide whether we are going to use the steady-state or equilibrium approximation. In enzyme kinetics, it rarely makes a difference to the form of the equations we will eventually obtain. That being the case, I usually pick the equilibrium approximation since it cuts down a little on the algebra we need to do. We apply the equilibrium approximation to the two reversible steps:

$$k_1 es = k_1 s(e_0 - c - h) \approx k_{-1} c; \tag{15.10}$$
$$k_3 ei \approx k_3 i_0 (e_0 - c - h) \approx k_{-3} h. \tag{15.11}$$

In order to reduce the number of symbols we are carrying, and in accord with common practice in enzyme kinetics, we divide Equation (15.10) by $k_1$, and Equation (15.11) by $k_3$. We get

$$s(e_0 - c - h) \approx K_S c, \tag{15.12}$$
$$i_0 (e_0 - c - h) \approx K_1 h, \tag{15.13}$$

where we have defined the dissociation constants $K_S = k_{-1}/k_1$ and $K_I = k_{-3}/k_3$. Note that $k_{-1}/k_1$ had previously been called $K_E$. I use $K_S$ here to emphasize that this is the Michaelis constant of the substrate, and not to denote any particular approximation. I'm holding back the symbol $K_M$ for another use.

We want to solve these two equations for $c$ and $h$, the concentrations of the two enzyme complexes. Recall that $v = k_{-2}c$ so that we want to solve for $c$ *last*. We can start by solving Equation (15.12) for $h$. (It doesn't actually matter which equation you solve for $h$, only that you start by solving for $h$.) We get

$$h \approx e_0 - c - K_S c/s.$$

Now substitute this into Equation (15.13), and isolate $c$. After a few rearrangements, we get

$$c \approx \frac{K_I e_0 s}{K_I s + K_S(K_I + i_0)}.$$

$$\therefore v = k_{-2}c \approx \frac{K_I k_{-2} e_0 s}{K_I s + K_S(K_I + i_0)}.$$

When we derive rate laws in enzyme kinetics, we usually try to put them in the Michaelis–Menten form. This allows us to use Eadie–Hofstee plots to extract the parameters, among other advantages. In this case, the equation we have obtained is almost in the form of Equation (15.6), except that there is a factor of $K_I$ multiplying $s$ in the denominator of the expression. If we now divide the top and bottom of our expression by $K_I$, we can put the equation into the Michaelis–Menten form:

$$v \approx \frac{v_{max} s}{s + K_S \left(1 + \frac{i_0}{K_I}\right)} \tag{15.14}$$

where, as in the Michaelis–Menten mechanism, $v_{max} = k_{-2}e_0$. It is worth emphasizing that this equation is *only* valid if the inhibitor concentration is very much larger than the enzyme concentration.

Suppose that we have performed a series of experiments at different values of $i_0$, keeping all other conditions identical. Suppose also that each of our experiments satisfies $i_0 \gg e_0$, except perhaps for one set of measurements performed without the inhibitor. Note that Equation (15.14) looks like Equation (15.6), except that $K_M = K_S(1 + i_0/K_I)$. In other words, if we create an Eadie–Hofstee plot of our data, the points collected for a given value of $i_0$ would appear as a straight line. The lines corresponding to the different values of $i_0$ would have a common intercept because competitive inhibition doesn't affect $v_{max}$. However, the absolute value of the slope, which is just $K_M$, would increase with increasing inhibitor concentration. If we calculate the apparent $K_M$ for each $i_0$ and replot $K_M$ as a function of $i_0$, we should get a straight line of intercept $K_S$ and slope $K_S/K_I$.

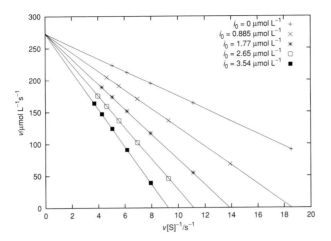

Figure 15.4 Eadie–Hofstee plot of the enzyme inhibition data from Example 15.3.

**Example 15.3 Treatment of competitive inhibition data** Consider the following initial rate data from an enzyme inhibition study:

$i_0 = 0$

| $s/10^{-6}\,\mathrm{mol\,L^{-1}}$ | 4.91 | 14.72 | 24.54 | 34.36 | 44.17 |
|---|---|---|---|---|---|
| $v/10^{-6}\,\mathrm{mol\,L^{-1}s^{-1}}$ | 91.0 | 163.9 | 195.1 | 212.4 | 223.5 |

$i_0 = 0.885\,\mathrm{\mu mol\,L^{-1}}$

| $s/10^{-6}\,\mathrm{mol\,L^{-1}}$ | 4.91 | 14.72 | 24.54 | 34.36 | 44.17 |
|---|---|---|---|---|---|
| $v/10^{-6}\,\mathrm{mol\,L^{-1}s^{-1}}$ | 68.3 | 136.6 | 170.7 | 191.2 | 204.8 |

$i_0 = 1.77\,\mathrm{\mu mol\,L^{-1}}$

| $s/10^{-6}\,\mathrm{mol\,L^{-1}}$ | 4.91 | 14.72 | 24.54 | 34.36 | 44.17 |
|---|---|---|---|---|---|
| $v/10^{-6}\,\mathrm{mol\,L^{1}s^{-1}}$ | 54.6 | 117.0 | 151.7 | 173.8 | 189.1 |

$i_0 = 2.65\,\mathrm{\mu mol\,L^{-1}}$

| $s/10^{-6}\,\mathrm{mol\,L^{-1}}$ | 4.91 | 14.72 | 24.54 | 34.36 | 44.17 |
|---|---|---|---|---|---|
| $v/10^{-6}\,\mathrm{mol\,L^{1}s^{-1}}$ | 45.5 | 102.4 | 136.6 | 159.3 | 175.6 |

$i_0 = 3.54\,\mathrm{\mu mol\,L^{-1}}$

| $s/10^{-6}\,\mathrm{mol\,L^{-1}}$ | 4.91 | 14.72 | 24.54 | 34.36 | 44.17 |
|---|---|---|---|---|---|
| $v/10^{-6}\,\mathrm{mol\,L^{1}s^{-1}}$ | 39.0 | 91.0 | 124.1 | 147.1 | 163.9 |

Our first task is to obtain $K_M$ and $v_{max}$ for each set of experiments. We use an Eadie–Hofstee plot since that is the better of the two commonly used graphical methods. As we can see from Figure 15.4, the slope $(-K_M)$ changes with $i_0$, but the intercept $(v_{max})$ is

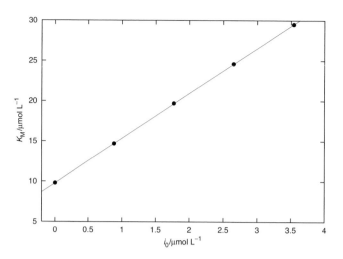

Figure 15.5 Secondary plot of the enzyme inhibition data from Example 15.3.

constant. This is consistent with competitive inhibition. The parameters obtained by fitting are as follows:

| $i_0/\mu\mathrm{mol\,L^{-1}}$ | $K_M/\mu\mathrm{mol\,L^{-1}}$ | $v_{max}/\mu\mathrm{mol\,L^{-1}\,s^{-1}}$ |
|---|---|---|
| 0 | 9.8 | 273 |
| 0.88 | 14.7 | 273 |
| 1.77 | 19.7 | 273 |
| 2.65 | 24.6 | 274 |
| 3.54 | 29.5 | 273 |

$v_{max}$ is therefore $273\,\mu\mathrm{mol\,L^{-1}s^{-1}}$. If we plot $K_M$ against $i_0$, we get a straight line (Figure 15.5). The slope of this line is 5.57 and the intercept is $9.79\,\mu\mathrm{mol\,L^{-1}}$. The intercept is $K_S$ while the slope is the ratio of $K_S$ to $K_I$. It follows that $K_I = 1.76\,\mu\mathrm{mol\,L^{-1}}$.

### 15.3.2 Uncompetitive inhibition

In uncompetitive inhibition, the inhibitor does not bind to the free enzyme, but does bind to and inactivate the enzyme–substrate complex:

$$\mathrm{E + S} \underset{k_{-1}}{\overset{k_1}{\rightleftharpoons}} \mathrm{C} \overset{k_{-2}}{\longrightarrow} \mathrm{E + P}$$

$$\mathrm{C + I} \underset{k_{-3}}{\overset{k_3}{\rightleftharpoons}} \mathrm{H} \tag{UI}$$

Uncompetitive inhibition usually involves binding of the inhibitor to a regulatory site rather than to the active site. The binding of the substrate to the active site modifies the conformation (shape) of the enzyme in such a way that it becomes possible for the inhibitor to bind to the regulatory site. If an inhibitor does bind to this site, it also alters the conformation of the enzyme, but this time in such a way as to make catalysis impossible. This is an example of what biochemists call an allosteric interaction, an interaction between different sites that alters the activity of the enzyme.

We are again seeking to express the rate as a function of $s$. There are two conservation relations, namely conservation of enzyme and of inhibitor:

$$e_0 = e + c + h,$$
$$i_0 = i + h.$$

We can use these equations to eliminate $e$ and $i$:

$$e = e_0 - c - h,$$
$$i = i_0 - h.$$

If we carried on from here without making any further approximations, we would get a quadratic equation. As before, we're going to focus on laboratory experiments where we can select reaction conditions such that $i_0 \gg e_0$, which implies that $i \approx i_0$. We apply the EA for the two reversible steps of the reaction. We obtain the equations

$$k_1 s(e_0 - c - h) \approx k_{-1} c,$$
$$k_3 i_0 c \approx k_{-3} h.$$

Defining $K_S = k_{-1}/k_1$ and $K_I = k_{-3}/k_3$, we get

$$s(e_0 - c - h) \approx K_S c, \tag{15.15}$$
$$i_0 c \approx K_I h. \tag{15.16}$$

The latter of these two equations gives us $h \approx i_0 c/K_I$. If we now substitute this value of $h$ into Equation (15.15) and rearrange, we get

$$c = \frac{e_0 s}{s\left(1 + \frac{i_0}{K_I}\right) + K_S}.$$

$$\therefore v = k_{-2} c = \frac{k_{-2} e_0 s}{s\left(1 + \frac{i_0}{K_I}\right) + K_S}. \tag{15.17}$$

As with the other mechanisms we have studied in this chapter, the rate equation is a hyperbola. We can always rewrite such rate equations in the Michaelis–Menten form (Equation 15.6). In this case, we can do this by dividing the top and bottom of Equation (15.17) by $(1 + i_0/K_I)$:

$$v = \frac{\dfrac{k_{-2} e_0}{1 + i_0/K_I} s}{s + \dfrac{K_S}{1 + i_0/K_I}}$$

which is now in the Michaelis–Menten form with

$$v_{max} = \frac{v^0_{max}}{1 + i_0/K_I}$$

$$\text{and } K_M = \frac{K_S}{1 + i_0/K_I},$$

where $v^0_{max} = k_{-2}e_0$ is the maximum velocity for the uninhibited reaction.

In uncompetitive inhibition, *both* $v_{max}$ and $K_M$ decrease with $i_0$. Now note that the intercept of an Eadie–Hofstee plot on the $v/s$ axis (i.e. the intercept with the line $v = 0$) is, from Equation (15.8), $v/s = v_{max}/K_M$. For uncompetitive inhibition, this intercept evaluates to $v/s = v^0_{max}/K_S$, so lines corresponding to different values of $i_0$ should all meet at this point on the $v/s$ axis.

The dependence of $v_{max}$ and of $K_M$ on $i_0$ is hyperbolic. (The hyperbola is oriented differently than a Michaelis–Menten hyperbola, but it's still a hyperbola.) To extract $K_S$, $K_I$ and $v_{max}$ from experimental data, we have to rearrange these equations by taking their reciprocals:

$$\frac{1}{v_{max}} = \frac{1}{v^0_{max}} + \frac{i_0}{K_I v^0_{max}}$$

$$\text{and } \frac{1}{K_M} = \frac{1}{K_S} + \frac{i_0}{K_I K_S}.$$

If we plot $1/v_{max}$ vs. $i_0$, provided $i_0 \gg e_0$ in all our experiments, we should get a straight line with an intercept of $1/v^0_{max}$ and a slope of $1/K_I v^0_{max}$. By combining the slope and intercept of this plot, we can obtain estimates of $v^0_{max}$ and of $K_I$. Similarly, a plot of $1/K_M$ vs. $i_0$ will allow us to estimate $K_S$ and, again, $K_I$. The two estimates of $K_I$ should of course agree.

**Example 15.4 Treatment of uncompetitive inhibition data** The enzyme cytochrome b558 catalyzes the transfer of an electron from NADPH to oxygen in immune system cells, generating superoxide ($O_2^{-\bullet}$) which is then converted by other enzymes to hydrogen peroxide. The hydrogen peroxide is used to kill micro-organisms. The synthetic peptide L418 has been claimed to be an uncompetitive inhibitor of this enzyme.[4] The following initial rate data were obtained:

| $[L418]_0/\mu mol\,L^{-1}$ | $v/nmol\,min^{-1}$ | | | | |
| --- | --- | --- | --- | --- | --- |
| 0 | 2.94 | 2.70 | 2.44 | 1.82 | 1.19 |
| 40 | 1.59 | 1.47 | 1.38 | 1.10 | 0.81 |
| 80 | 0.95 | 0.88 | 0.85 | 0.74 | 0.56 |
| $[NADPH]/pmol\,L^{-1}$ | 234 | 132 | 97 | 48 | 24 |

[4] T. Tsuchiya *et al.*, *Biochem. Biophys. Res. Commun.* **257**, 124 (1999).

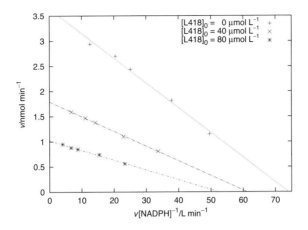

Figure 15.6 Eadie–Hofstee plot of the data from Example 15.4.

In Eadie–Hofstee form, the data become

$[L418]_0 = 0$

| $v[\mathrm{NADPH}]^{-1}/\mathrm{L\,min}^{-1}$ | 12.56 | 20.45 | 25.15 | 37.92 | 49.58 |
|---|---|---|---|---|---|
| $v/\mathrm{nmol\,min}^{-1}$ | 2.94 | 2.70 | 2.44 | 1.82 | 1.19 |

$[L418]_0 = 40\,\mu\mathrm{mol\,L}^{-1}$

| $v[\mathrm{NADPH}]^{-1}/\mathrm{L\,min}^{-1}$ | 6.79 | 11.14 | 14.23 | 22.92 | 33.75 |
|---|---|---|---|---|---|
| $v/\mathrm{nmol\,min}^{-1}$ | 1.59 | 1.47 | 1.38 | 1.10 | 0.81 |

$[L418]_0 = 80\,\mu\mathrm{mol\,L}^{-1}$

| $v[\mathrm{NADPH}]^{-1}/\mathrm{L\,min}^{-1}$ | 4.06 | 6.67 | 8.76 | 15.42 | 23.33 |
|---|---|---|---|---|---|
| $v/\mathrm{nmol\,min}^{-1}$ | 0.95 | 0.88 | 0.85 | 0.74 | 0.56 |

The Eadie–Hofstee graph is shown in Figure 15.6. Incidentally, these are superb experimental results. The Eadie–Hofstee graph shows that the intercept and absolute value of the slope both decrease with increasing inhibitor concentration, which would be consistent with uncompetitive inhibition. However, the $v/[\mathrm{NADPH}]$ intercepts are different. This could be due to one of two things:

(1) These data do not correspond to uncompetitive inhibition, but to some other type of inhibition. There are, in fact, a great many inhibition mechanisms, some of which are only subtly different when plotted in Eadie–Hofstee form. It's entirely possible that the data arise from a type of inhibition other than the two we have studied here. In particular, the fact that there is a systematic decrease in the intercept with increasing concentration of L418 suggests that this is the case.

Table 15.1 *Kinetic parameters for the inhibition of cytochrome b558 by L418.*

| $[L418]_0/\mu mol\,L^{-1}$ | $K_M/pmol\,L^{-1}$ | $v_{max}/nmol\,min^{-1}$ | $(v_{max}/K_M)/L\,min^{-1}$ |
|---|---|---|---|
| 0 | 49 | 3.64 | 74 |
| 40 | 29 | 1.79 | 62 |
| 80 | 20 | 1.02 | 51 |

(2) We have just three lines, so it's possible that random error has just accidentally caused the three intercepts to look like they are drifting systematically. This is always a possibility, but I would argue that the relatively good fits to the Eadie–Hofstee lines make it unlikely that there were large enough errors to account for the differences in the intercepts.

Given the appearance of the Eadie–Hofstee graph, I would tend to conclude that this is not an example of uncompetitive inhibition. In their paper, Tsuchiya and coworkers wrote that they decided that they did have uncompetitive inhibition based on their Lineweaver–Burk plots for which "all lines obtained with different concentration of L418 have identical slopes of Km/Vmax." You will recognize this criterion for uncompetitive inhibition as being equivalent to a common $v/[\text{NADPH}]$ intercept of $v^0_{max}/K_S$ in an Eadie–Hofstee plot, despite the slight differences in notation. Their Lineweaver–Burk lines do look like they have similar slopes. Unfortunately, they do not give the values of these slopes, so it's difficult to judge how well they agree.

Suppose that we were to accept that this is a case of uncompetitive inhibition. We could then work out the kinetic parameters as described above. Table 15.1 gives the values of $v_{max}$ and $K_M$ for each inhibitor concentration, as well as the $v/[\text{NADPH}]$ intercept. We see from this table that the spread of values of the intercept is not great. It is possible that Tsuchiya and coworkers computed a similar spread of values for the slopes of their Lineweaver–Burk plots and concluded that these were in reasonable agreement with each other. Be that as it may, we can replot these data in the form $1/v_{max}$ and $1/K_M$ vs. $[L418]_0$ (Figure 15.7). The slope and intercept of the $1/K_M$ plot are, respectively, $3.70 \times 10^{14}\,L^2\,mol^{-2}$ and $2.02 \times 10^{10}\,L\,mol^{-1}$. The intercept is $1/K_S$ so we have $K_S = 50\,pmol\,L^{-1}$. The intercept divided by the slope gives $K_I = 54\,\mu mol\,L^{-1}$. Similarly, the $1/v_{max}$ plot has a slope and intercept of $8.82 \times 10^{12}\,min\,L\,mol^{-2}$ and $2.52 \times 10^8\,min\,mol^{-1}$, respectively, from which we can calculate $v^0_{max} = 4.0\,nmol\,min^{-1}$ and $K_I = 29\,\mu mol\,L^{-1}$. The two estimates of $K_I$ disagree somewhat. Whether this is due to random errors or to this not being a case of uncompetitive inhibition is a matter of judgment. If, again, we assume that this is random error, we would report the following final data:

$$v^0_{max} = 4.0\,nmol\,min^{-1}$$
$$K_S = 50\,pmol\,L^{-1}$$
$$K_I = 42 \pm 13\,\mu mol\,L^{-1}$$

Table 15.2 *Relationship between the Michaelis–Menten parameters $v_{max}$ and $K_M$ and the inhibitor concentration in two types of enzyme inhibition studied at large inhibitor concentrations.*

| Type | $v_{max}$ ($v$ intercept) | $K_M$ | $v_{max}/K_M$ ($v/s$ intercept) |
|---|---|---|---|
| competitive | $k_{-2}e_0$ | $K_S(1 + i_0/K_I)$ | varies |
| uncompetitive | $\dfrac{1}{v_{max}} = \dfrac{1}{v_{max}^0} + \dfrac{i_0}{K_I v_{max}^0}$ | $\dfrac{1}{K_M} = \dfrac{1}{K_S} + \dfrac{i_0}{K_I K_S}$ | $v_{max}^0/K_S$ |

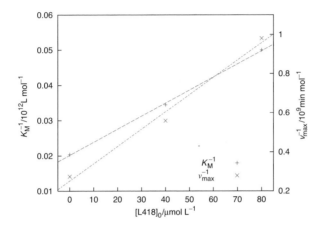

Figure 15.7 Secondary plots of the uncompetitive inhibition data from Example 15.4. Note the large amount of scatter from the line for the $1/v_{max}$ data.

The value of $K_I$ was obtained by averaging the two estimates, while the uncertainty is half the difference between them. Interestingly, Tsuchiya and coworkers give a very different value for $K_I$, namely $11\ \mu mol\,L^{-1}$. The difference is difficult to explain since they do not report any of their other kinetic parameters.

### 15.3.3 Determining the type of inhibition

Assume that we have carried out a set of inhibition experiments with $i_0 \gg e_0$. In this case, the data from a set of rate measurements at a particular value of $i_0$ will generally plot out as a hyperbola. Each type of inhibitory interaction has a different effect on $K_M$ and $v_{max}$. The two examples studied in this chapter are summarized in Table 15.2. There are many other ways in which inhibitors can act on an enzyme. Other types of inhibition would lead to different patterns in Eadie–Hofstee plots. Reference books on enzyme kinetics, such as those listed at the end of this chapter, provide rate equations for several more examples of inhibition mechanisms than it is possible to include in this book.

## Exercise group 15.2

(1) In a series of enzyme inhibition experiments, the following data were obtained:

| $[I]_0/\text{mmol L}^{-1}$ | 0 | 1.5 | 3.0 | 4.5 |
|---|---|---|---|---|
| $v_{max}/\mu\text{mol L}^{-1}\text{s}^{-1}$ | 10.0 | 11.2 | 10.3 | 10.4 |
| $K_M/\text{mmol L}^{-1}$ | 4.0 | 6.6 | 9.0 | 11.8 |

    (a) What is the inhibition mechanism? Explain briefly.
    (b) Calculate the kinetic parameters ($K_I$ etc.) of this reaction.

(2) Consider the following enzyme inhibition data:

$[I]_0 = 0$

| $s/\mu\text{mol L}^{-1}$ | 41.7 | 125 | 208 | 292 | 375 |
|---|---|---|---|---|---|
| $v/\mu\text{mol L}^{-1}\text{s}^{-1}$ | 248 | 447 | 532 | 580 | 610 |

$[I]_0 = 9.02 \text{ mmol/L}$

| $s/\mu\text{mol L}^{-1}$ | 41.7 | 125 | 208 | 292 | 375 |
|---|---|---|---|---|---|
| $v/\mu\text{mol L}^{-1}\text{s}^{-1}$ | 186 | 373 | 466 | 522 | 559 |

$[I]_0 = 18.0 \text{ mmol/L}$

| $s/\mu\text{mol L}^{-1}$ | 41.7 | 125 | 208 | 292 | 375 |
|---|---|---|---|---|---|
| $v/\mu\text{mol L}^{-1}\text{s}^{-1}$ | 149 | 319 | 414 | 474 | 516 |

$[I]_0 = 27.1 \text{ mmol/L}$

| $s/\mu\text{mol L}^{-1}$ | 41.7 | 125 | 208 | 292 | 375 |
|---|---|---|---|---|---|
| $v/\mu\text{mol L}^{-1}\text{s}^{-1}$ | 124 | 279 | 373 | 435 | 479 |

Determine the type of inhibition and calculate the kinetic parameters.

(3) Many of the most commonly used antibiotics are members of a class of biochemical compounds called $\beta$-lactams. Antibiotic resistance in bacteria is often associated with the expression of a $\beta$-lactamase, an enzyme that hydrolyzes $\beta$-lactams. One possible method for combating antibiotic resistance is to include a $\beta$-lactamase inhibitor with the antibiotic. This would reduce the effectiveness of the lactamase and thus allow the antibiotic to do its job. As a step in this direction, Yang and Crowder have studied the inhibition of the hydrolysis of nitrocefin by a lactamase from *Stenotrophomonas maltophilia* by *N*-benzylacetyl-D-alanylthioacetic acid (BATA).[5] Nitrocefin (represented by N in the table below) is a $\beta$-lactam that is not used as an antibiotic, but has spectroscopic properties that make it particularly suitable for enzyme assays. Based on

---

[5] K.W. Yang and M.W. Crowder, *Arch. Biochem. Biophys.* **368**, 1 (1999). Thanks to Michael Crowder for making his raw data available.

a Lineweaver–Burk plot, Yang and Crowder concluded that BATA is a competitive inhibitor. Their results are as follows:

| [N]/$\mu$mol L$^{-1}$ | [BATA]/$\mu$mol L$^{-1}$ | | | |
| --- | --- | --- | --- | --- |
| | 0 | 1.9 | 3.8 | 7.6 |
| | $v$/$\mu$mol L$^{-1}$s$^{-1}$ | | | |
| 71.9 | 0.303 | 0.255 | 0.222 | 0.164 |
| 54.1 | 0.313 | 0.256 | 0.204 | 0.149 |
| 36.0 | 0.278 | 0.208 | 0.179 | 0.125 |
| 22.2 | 0.251 | 0.175 | 0.135 | 0.093 |
| 7.2 | 0.160 | 0.089 | 0.061 | 0.035 |

Do you agree that BATA is a competitive inhibitor? Explain your reasoning in detail.

(4) In bacteria, phosphoglucose isomerase catalyzes the transformation of fructose-6-phosphate into glucose-6-phosphate (or vice versa), allowing the cell to select between two different fermentative pathways according to the prevailing conditions. This selection requires a metabolic switch, i.e. a means of turning the activity of phosphoglucose isomerase on or off. This can be accomplished if appropriate metabolites (intermediates in biochemical pathways) inhibit the enzyme. Richter and coworkers have studied the inhibition of phosphoglucose isomerase from *Oenococcus oeni* by metabolites of the two fermentative pathways.[6] In particular, for the conversion of fructose-6-phosphate (F6P) to glucose-6-phosphate in the presence of varying concentrations of erythrose-4-phosphate (E4P), the following data were obtained:

| [F6P]/mmol L$^{-1}$ | $v$/$\mu$mol g$^{-1}$min$^{-1}$ | | | | |
| --- | --- | --- | --- | --- | --- |
| 0.2 | 132 | 74 | 75 | 30 | 18 |
| 0.5 | 283 | 205 | 183 | 97 | 43 |
| 1 | 392 | 257 | | 155 | 92 |
| 1.5 | 471 | 398 | 325 | 243 | 132 |
| 2 | 525 | 434 | 373 | 296 | 169 |
| 5 | 560 | 551 | 435 | 374 | 250 |
| 7 | 580 | 581 | 396 | 380 | 280 |
| 10 | 640 | 591 | 448 | 439 | 336 |
| [E4P]/$\mu$mol L$^{-1}$ | 0 | 1.5 | 3 | 4.5 | 6 |

These data were generated from a cell extract rather than a purified enzyme, so the rates were normalized by the dry weight of the preparation, hence the units given above.

Determine whether these data exhibit competitive or uncompetitive kinetics, or neither. If the type is competitive or uncompetitive, determine the values of $K_S$, $K_I$ and $v_{max}$ (or $v_{max}^0$). Otherwise, explain exactly how you reached your negative conclusion.

---

[6]  H. Richter, A. A. De Graaf, I. Hamann and G. Unden, *Arch. Microbiol.* **180**, 465 (2003). I would like to thank Professor Unden for making his raw data available.

### 15.4  Deriving rate laws for enzyme mechanisms

The algorithm for deriving rate laws for enzyme mechanisms or, in general, for catalyzed reactions, is similar to the system presented in Section 14.3, but there are a few extra steps. I won't repeat the entire algorithm here. Rather, I will highlight the differences:

- Instead of intermediates, look for enzyme complexes.
- Write down an enzyme conservation relation before you try to apply any approximations. Also write down any other conservation relations you notice. Conservation of inhibitors is a common condition. In reactions that pass a chemical group around (e.g. a phosphate group), conservation of this chemical group (called a "conserved moiety" in the enzyme literature) is another potentially useful relationship.
- If you have any information about some concentrations being large or small compared to each other, use this immediately to simplify the conservation relations.
- Use the enzyme conservation relation to eliminate the concentration of free enzyme in all equations where it appears. Similarly use any other conservation relationships to eliminate additional concentrations.
- Unless you have a special reason for doing otherwise, apply the equilibrium approximation for any reversible complexation steps that appear in the mechanism. If necessary, round out the set of approximations with steady-state approximations for some of the enzyme complexes.

### Key ideas and equations

- For the Michaelis–Menten mechanism, $v = \dfrac{v_{max}s}{s + K_M}$
  - $k_{cat} = v_{max}/e_0$
  - $K_M$ is a dissociation constant for the enzyme-substrate complex.
- Eadie–Hofstee plot: $v = v_{max} - K_M\dfrac{v}{s}$
- Table 15.2.

### Suggested reading

There are a number of useful books on enzyme kinetics. Here are a few I particularly like:

A. Cornish-Bowden, *Principles of Enzyme Kinetics*; Butterworths: London, 1976.

  Cornish-Bowden's books – he has written several more – present the ideas of enzyme kinetics in an extremely clear manner.

K. J. Laidler and P. S. Bunting, *The Chemical Kinetics of Enzyme Action*, 2nd edn; Clarendon: Oxford, 1973.

This book is a wonderful summary of ideas, methods and results in enzyme kinetics, with many references to the literature.

A. Fersht, *Enzyme Structure and Mechanism*, 2nd edn.; Freeman: New York, 1985.

This is a much wider ranging book than Laidler and Bunting's, but it still manages to provide considerable depth and is absolutely beautifully written.

I. H. Segel, *Enzyme Kinetics*; Wiley: New York, 1975.

If you ever want a rate law for an enzyme mechanism and don't want to derive it yourself, this is the first place I would look. This book is absolutely encyclopedic when it comes to mechanisms and rate laws.

As you may have gathered, I have a soft spot for the history of enzymology. If you're interested in this subject, here are a couple of books I would recommend:

J. S. Fruton, *Molecules and Life*; Wiley: New York, 1972.

This book contains a series of essays on the history of biochemistry.

H. C. Friedmann, *Enzymes*; Hutchinson Ross: Stroudsburg, PA, 1981.

This book contains excerpts and translations of most of the key papers in the history of enzymology, along with extensive, insightful notes by the editor.

---

## Review exercise group 15.3

(1) Micro-organisms are increasingly being used to degrade pollutants. Accordingly, the kinetics of these processes are of considerable interest. Because biochemical degradation pathways often have a rate-limiting enzyme-catalyzed step, it is often (but not always) the case that biodegradation obeys simple Michaelis–Menten kinetics. Consider the following initial rate data obtained for the degradation of phenol by *Pseudomonas putida* immobilized on calcium alginate beads:[7]

| [phenol]/mmol $L^{-1}$ | 1.01 | 2.67 | 5.29 | 7.94 | 10.4 |
|---|---|---|---|---|---|
| $v$/µmol $L^{-1}h^{-1}$ | 3.1 | 6.4 | 10 | 12 | 14 |

(a) Does this reaction follow Michaelis–Menten kinetics? Explain briefly.
(b) Regardless of your answer to part (a), provide estimates of the Michaelis constant and maximum reaction rate.

(2) DNA polymerase $\alpha$ (pol $\alpha$) catalyzes the first step in the replication of DNA. Pol $\alpha$ is inhibited by replication factor C (RF-C). (RF-C is a DNA-binding protein that plays a role in helping other enzymes involved in DNA synthesis (e.g. pol $\delta$) attach to the DNA. The complete story is too complex to explain in any detail here.) In this case,

---

[7] K. Bandhyopadhyay *et al.*, *Biochem. Eng. J.* **8**, 179 (2001).

instead of having two small molecules competing for the active site of an enzyme, we have two enzymes competing for a single substrate, namely DNA.

(a) A crude cartoon of the mechanism for this competitive process would be

$$\text{pol } \alpha + \text{DNA} \underset{k_{-1}}{\overset{k_1}{\rightleftharpoons}} C_P \overset{k_{-2}}{\longrightarrow} \text{pol } \alpha + P + \text{DNA}$$

$$\text{RF-C} + \text{DNA} \underset{k_{-3}}{\overset{k_3}{\rightleftharpoons}} C_R$$

Derive a rate law for this mechanism valid when the pol $\alpha$ concentration is much smaller than the DNA concentration, and the latter is in turn much smaller than the RF-C concentration. Show that this rate law is of identical form to that for ordinary competitive inhibition.

*Hint*: Start by writing down three conservation relations, then use the conditions given above to simplify these relations. After that, the derivation follows the normal path.

(b) The following initial rate data have been obtained in the presence of 0.2 mu of pol $\alpha$:[8]

| [RF-C]/mu | [DNA]/g L$^{-1}$ | | | |
|---|---|---|---|---|
|  | 0.025 | 0.049 | 0.121 | 0.241 |
|  | $v$/fmol min$^{-1}$ | | | |
| 0 | 85 | 159 | 246 | 328 |
| 0.2 | 47 | 90 | 167 | 246 |

A unit (u) of enzyme is an empirical unit used to measure the amount of enzyme in a preparation based on its catalytic activity (as described on page 292). The definition of a unit varies somewhat from study to study. Here, a unit was taken to be the amount of RF-C that allows polymerase $\delta$ to polymerize 1 nmol of DNA bases in 60 minutes. This definition of a unit doesn't actually matter all that much here. Just take it as a measure of concentration.

Show that these data are consistent with competitive inhibition, and calculate $K_S$, $K_I$ and the specific activity.

*Note*: Since you will only have two points, it won't be necessary to draw a graph to extract $K_S$ and $K_I$.

(3) Show that the following mechanism is kinetically indistinguishable from (leads to the same rate law as) the ordinary Michaelis–Menten mechanism:

$$E + S \underset{k_{-1}}{\overset{k_1}{\rightleftharpoons}} C_1 \underset{k_{-2}}{\overset{k_2}{\rightleftharpoons}} C_2 \overset{k_3}{\rightarrow} E + P$$

Provide equations for $K_M$ and $v_{max}$ appropriate to this mechanism.

[8]  Maga *et al.*, *J. Mol. Biol.* **295**, 791 (2000).

(4) When determining rate laws for complex mechanisms, we generally assume that the final product-forming step is irreversible. This is appropriate when data from initial-rate experiments are to be treated, but violates the law of microscopic reversibility. Moreover, in some cases we *do* want to know how the rate depends on the product concentration.

Derive a rate law for the fully reversible Michaelis–Menten mechanism

$$E + S \underset{k_{-1}}{\overset{k_1}{\rightleftharpoons}} C \underset{k_2}{\overset{k_{-2}}{\rightleftharpoons}} E + P$$

under the assumption that the enzyme concentration is small compared to the substrate concentration. Simplify your answer, but don't attempt to introduce Michaelis constants or a maximum velocity into your equation.

*Hints*: Think carefully about the conservation relations. The EA can't be applied here because there are two elementary reactions that might both equilibrate rapidly. The final result will be in a form similar (but not identical) to the Michaelis–Menten equation.

# 16

# Techniques for studying fast reactions

Many elementary chemical reactions occur on extremely short time scales (down to femtoseconds). Special experimental and mathematical methods are required to study fast processes. Here, we will focus on reactions in solution. Some of the methods described here are also adaptable to gas-phase reactions, but there are also some gas-phase methods that are quite different from anything we would use in solution. Also note that this chapter does not provide a comprehensive survey of methods, even for reactions in solution. The intention is to discuss a few particularly important methods, which hopefully will give you a flavor of the various approaches available.

## 16.1 Flow methods

### 16.1.1 The continuous flow method

The basic idea behind the continuous flow method is very simple: the reactants are supplied at a constant rate into a tube where the reaction takes place. At constant flow velocity, positions in the tube correspond to specific times since mixing, the relationship between time and the distance from the mixing chamber $L$ being $t = L/v$. By varying the flow speed or moving a detector along the length of the tube, we can therefore study the course of a fast reaction at our leisure.

Figure 16.1 shows a diagram of a typical solution-phase continuous-flow apparatus. The mixing chamber must be so designed that the two reactants mix quickly. The chamber then corresponds to $t = 0$ (freshly mixed reactants). We also need to operate at high flow rates because we want some turbulence in the tube. The alternative is laminar flow; in a laminar flow, the flow velocity varies with the distance from the tube wall, higher flow velocities being achieved near the center than at the wall (Figure 16.2). In the direction transverse to the flow direction, only diffusion operates, so mixing in this direction is very slow. The result is that molecules close to the center of the tube have traveled from the mixing chamber for a shorter period of time than molecules closer to the walls. Turbulent liquid flows, on the other hand, are well mixed in the radial direction (the direction perpendicular to the tube wall) so that the average speeds of the molecules are all the same, regardless of whether they started close to the wall or not when they entered the tube. Turbulent flows

314

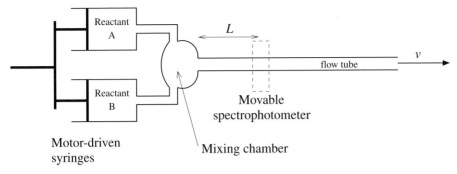

Figure 16.1 Diagram of a solution-phase continuous-flow apparatus. The two reactant solutions are injected together by a motorized syringe assembly which ensures a constant flow velocity ($v$) through the tube. The geometry of the mixing chamber must be such that the two reactants are thoroughly and rapidly mixed. (Note that the relative size of the mixing chamber is exaggerated in this diagram. It would normally have a very small volume relative to the flow tube. Also, mixing chambers come in a variety of shapes, ranging from a simple T, with the reactants approaching from opposite directions, to elaborately machined units whose precise specifications are either patented or treated as trade secrets by their manufacturers.) A movable detector (commonly a spectrophotometer) measures a property of the reactive mixture that can be related to the concentration of a reactant or product. The distance of the detector from the mixing chamber ($L$) can be related to the time since the two reactants were mixed by $t = L/v$. After leaving the tube, the solution simply goes to a waste container.

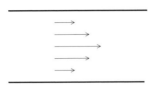

Figure 16.2 Laminar flow. The heavy lines represent the walls of a tube through which a fluid flows. The vectors represent the velocity of the fluid at different distances from the walls. In laminar flow, the velocity is highest at the center and approaches zero at the walls of the tube. This is to be avoided in a kinetics experiment since it implies that in any given slice transverse to the tube, there are parcels of fluid that have been mixed at different times in the past.

have large Reynolds numbers, where the Reynolds number $\mathcal{R}$ is given by

$$\mathcal{R} = \frac{vd\rho}{\eta}.$$

In this equation, $v$ is the flow speed, $d$ is the diameter of the tube, $\rho$ is the density and $\eta$ is the viscosity of the solution. (Viscosity measures how "thick" a liquid is.) There is, as with all such things, no sharp cut-off between large and small Reynolds numbers. However, as a rule of thumb, people generally try to design their instruments so that $\mathcal{R}$ is at least 2000. If we use a fixed detector and vary the flow velocity, the requirement that $\mathcal{R} \gtrsim 2000$ will limit the range of reaction times we can study. The movable detector experiment is more commonly used. In this type of experiment, it will generally be necessary to move the detector at least 1 cm between readings. Consider, for instance, a spectrophotometer.

The light beam used by the spectrophotometer might be a few millimeters wide, or perhaps as little as 1 mm for a laser source. We want each of our measurements to correspond to a distinct time, so we have to move the detector at least the width of the beam to make sure that our sample volumes don't overlap. A distance of about 1 cm satisfies this criterion while being easily measured.

Having taken all of this information in, we can examine the consequences of these various ideas for the design of a continuous flow apparatus. Suppose that we want to study a reaction that is complete in about 15 ms. We will be moving the detector 1 cm at a time. Say we want to get samples corresponding to every millisecond of the reaction. This would mean that the tube would have to be at least 15 cm long. If it takes 15 ms for the fluid to travel 15 cm, the flow speed must be $v = L/t = 10 \, \mathrm{m \, s^{-1}}$. The density and viscosity of the solution will both be close to those of water (assuming that this is the solvent). Water has a density of about $1000 \, \mathrm{kg \, m^{-3}}$ and a viscosity of about 1 mPa s near room temperature. If the Reynolds number is to be greater than 2000, we must therefore have

$$d > \frac{2000\eta}{v\rho} = \frac{2000(1 \times 10^{-3} \, \mathrm{Pa \, s})}{(10 \, \mathrm{m/s})(1000 \, \mathrm{kg \, m^{-3}})} = 2 \times 10^{-4} \, \mathrm{m} \equiv 0.2 \, \mathrm{mm}$$

which shouldn't pose much of a problem. There are some tradeoffs here: if the diameter of the tube is too small, the detector won't be working with a very large sample and may not be able to pick up enough signal to give reliable results. It is sometimes possible to make up for this by letting the detector average its readings over a longer period of time, but this takes more solution. On the other hand, if we make the tube too large, we'll need to put enormous amounts of solution through the tube.

Suppose that we decide on a 1 mm tube and that our detector can take a reading in 1 s. We can stop pushing fluid through the tube while we reposition the detector, so we don't need to waste our reactant solutions in between readings. The cross-sectional area of the tube is

$$A = \pi r^2 = \pi (0.5 \times 10^{-3} \, \mathrm{m})^2 = 7.9 \times 10^{-7} \, \mathrm{m^2}.$$

The volume of solution that passes through a particular point in the tube every second (the flow rate) is therefore

$$f = Av = (7.9 \times 10^{-7} \, \mathrm{m^2})(10 \, \mathrm{m \, s^{-1}}) = 7.9 \times 10^{-6} \, \mathrm{m^3 \, s^{-1}} = 8 \, \mathrm{mL \, s^{-1}}.$$

Each reading will therefore require at least 8 mL of the mixed solution. There are startup issues to consider as well (e.g. flushing out any old solution before taking a reading), so say we really need about twice that much solution per reading, or 15 mL. Fifteen readings would then take 225 mL of solution. This probably doesn't sound like much, but if the reactants are expensive or difficult to make, this could be a problem.

There is another design challenge to be met. In order to push a fluid through a 1 mm tube at a speed of $10 \, \mathrm{m \, s^{-1}}$, a large pressure has to be applied. The exact pressure depends on a number of details, but would typically be a few atmospheres. These are hardly huge pressures, but they are high enough that some care must be taken in building the apparatus if it is to hold together during the experiment.

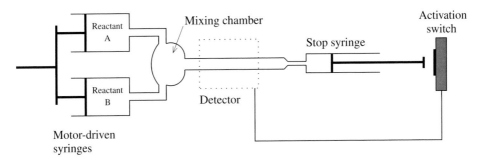

Figure 16.3 Diagram of a stopped-flow apparatus. As in the continuous flow method, reactants are rapidly injected into a mixing chamber (again, exaggerated in size in this diagram, and possibly of a very different shape than shown here). From there, they enter a tube that leads to a third syringe. As this syringe fills, it pushes out a plunger which eventually activates the detector and turns off the drive motors.

### 16.1.2 The stopped-flow method

What do we do if one or both of our reactant solutions are difficult to make in large quantities? In this case, we want to rapidly mix the reactant solutions and then rapidly start acquiring data without shooting our reactive mixture out of the apparatus. Both of these can be achieved using a variation on the experiment described in the last section called the stopped-flow method. A sketch of the apparatus is shown in Figure 16.3. In this case, instead of being discarded, the reactive mixture fills a third syringe. This causes the plunger to move out. It eventually hits a switch that stops the drive motors and activates the detector. The detector in this case is not mobile, and takes a series of readings over time.

The stopped-flow technique relies on:

(1) rapid filling of the mixing chamber, tube and syringe, and
(2) a detector that can very rapidly measure concentrations or, more precisely, an observable proportional to concentration, such as the absorbance at a particular wavelength.

The dead time (the time it takes to fill the sample tube and start recording data) can be as short as one or two milliseconds. Equipment that achieves such short mixing times is usually highly miniaturized, with reaction volumes of a few tens of microliters. Because the detector doesn't have to move, a huge variety of detection methods can be used, ranging from more-or-less straightforward spectroscopic methods to more exotic techniques like X-ray and neutron scattering.

## 16.2 Flash photolysis

Flow methods, as we have seen above, can get us experimental information on a millisecond time scale. Unfortunately, many reactions are much faster than this. One of the most productive avenues for studying fast reactions was developed by Norrish and Porter in the

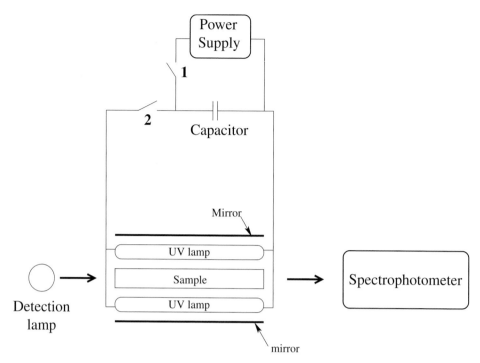

Figure 16.4 Traditional flash photolysis apparatus. The switches labeled **1** and **2** are closed in turn to charge the capacitor and to discharge it through the ultraviolet (UV) lamps. The mirrors ensure that most of the light generated passes through the sample tube. The spectrometer then either collects the transmitted light intensity at a fixed wavelength as a function of time, or for detection of intermediates, takes a complete spectrum as a function of wavelength.

1940s. Suppose we have a reaction in which the reactants can coexist, at least for a time, without reacting. Further suppose that we can trigger this reaction by a pulse of light. This second condition is true of a great many reactions, since the absorption of light can cause electronic excitation and bond breakage, typically forming radicals.

The traditional flash photolysis apparatus is shown in Figure 16.4. The capacitor is charged by closing switch **1**. Once this is done, switch **1** is opened and switch **2** is closed. The charge accumulated in the capacitor discharges through the ultraviolet (UV) lamps, producing a flash of white light lasting 20 μs or so. Recording of data can therefore start within several tens of microseconds. This is sufficiently fast to observe many chemical intermediates using straightforward spectroscopic techniques. Accordingly, this technique not only allows us to study the kinetics of a reaction, but also to identify intermediates. For the latter purpose, the detection lamp would usually also be operated in a short pulse. In a modern version of this experiment, the transmitted light would be sent through a grating to a diode array detector. Complete spectra could be taken at different time points, giving a complete profile of the reaction both in terms of the time evolution, and in terms of the chemical species generated.

Most laboratories carrying out flash photolysis experiments now use lasers instead of flash lamps. Lasers have several advantages:

(1) Instead of white light from a flash lamp, which contains a range of frequencies, lasers produce a beam with a very narrow range of frequencies. This gives the experimentalist more control over the bonds to be photolyzed and over the energy distribution of the photolysis products than is possible with flash lamps. Modern lasers are often tunable (i.e. their emission wavelength can be adjusted) so that different energy ranges can be explored.

(2) Laser flashes can be very fast, allowing faster reactions to be studied than in conventional flash photolysis. Picosecond flashes are now routinely produced.

(3) Using conventional flash lamps, there is a tradeoff between the light energy produced and the duration of the flash: nanosecond flashes are possible, but only microjoules of energy are produced in these flashes. At the other end of the scale, we can generate flashes that release several hundred joules of energy, but they last several microseconds. Furthermore, the energy from a flash lamp is delivered over a large area. A typical flash tube might be up to 1 m long. The result is that any given part of the reactive mixture generally receives only a small amount of energy, particularly if we want a fast flash. On the other hand, high flash energies focused on a very small area are easily obtained with lasers. This allows the generation of very high concentrations of intermediates and thus facilitates their detection.

Lasers though have one important disadvantage: it's hard to make a wide laser beam. Accordingly, a laser flash is delivered to a very small volume. In some experiments, that doesn't matter. In others, this means that we are only making radicals in one part of the reaction vessel, which will make diffusion of radicals an important process in addition to the reaction kinetics we are trying to study. Contrast that to the traditional flash apparatus (Figure 16.4), which is designed to deliver the light evenly to the entire reaction vessel.

What do you do if the reaction you want to study isn't one that can be triggered by light? In a clever and relatively recent variation on the flash photolysis experiment, chemists have developed methods for releasing reactants from chemical "cages" using a light flash. One example is the release of ATP by photolysis of compound **1** in Figure 16.5. Compound **1** is not hydrolyzed by enzymes that hydrolyze ATP. In fact, it generally doesn't even bind in the active site because of the nitrobenzyl group at the "business end" of ATP. If you want to study an enzyme that hydrolyzes ATP, you can therefore mix the enzyme and compound **1**, then release ATP by photolysis and watch the reaction from there.

Norrish and Porter shared the 1967 Nobel Prize in Chemistry for their development of flash photolysis with Manfred Eigen, who developed another approach to studying fast reactions which we will study shortly. More recently, the 1999 Nobel Prize was awarded to Ahmed Zewail for developing methods for studying transition states (Section 17.1) using femtosecond laser pulses. While Zewail's methods are quite different from ordinary flash photolysis, they are undoubtedly the intellectual descendants of Norrish and Porter's work.

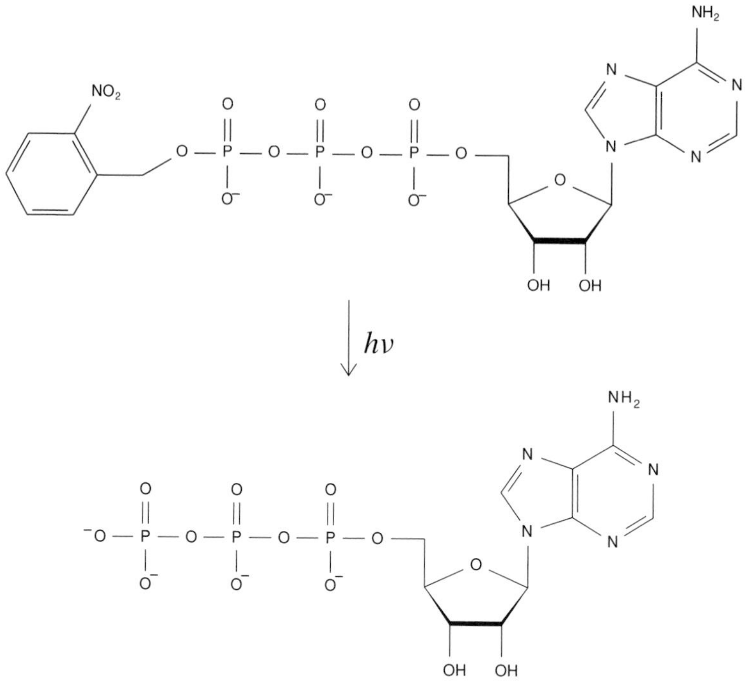

Figure 16.5 Release of ATP (**2**) from compound **1** by photolysis.

## 16.3 Relaxation methods

Relaxation methods were originally developed by Manfred Eigen. Many variations have been used, but the basic idea is always the same: take a reaction at equilibrium, change the equilibrium constant by making a rapid change in the reaction conditions, and watch it go to the new equilibrium (Figure 16.6). As in flash photolysis, if the disturbance is sufficiently rapidly introduced, the very early moments of a reaction can be studied, and then it's just a matter of being able to collect data fast enough.

### 16.3.1 The temperature-jump method

In the temperature-jump method, we are relying on a simple principle: the equilibrium constant depends on the temperature, so if we can make a rapid change in the temperature, the system will find itself out of equilibrium. We can then watch it evolve to the new equilibrium point.

There are several methods for creating a temperature ($T$) jump. Perhaps the simplest is to use Joule heating. Joule heating is the increase in temperature that occurs when a current passes through a resistor. In this case, the solution acts as the resistor. This technique is mostly used with aqueous solutions so the rest of the discussion assumes that water is the

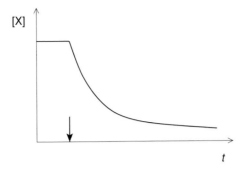

Figure 16.6 Typical course of a relaxation experiment. The system starts out at equilibrium. At a particular point in time, indicated by the arrow, the reaction conditions are rapidly changed. The system then relaxes to its new equilibrium. We observe the system during this relaxation phase, after the sudden change in conditions.

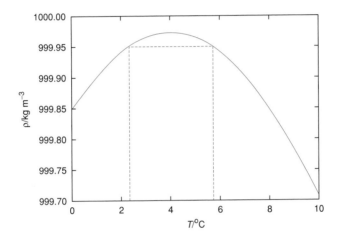

Figure 16.7 Density of liquid water as a function of temperature. Because water has a density maximum near $4\,°C$, it is possible to choose initial and final temperatures straddling $4\,°C$ that correspond to the same density. For instance, the density is $999.95\ kg\ m^{-3}$ at both 2.34 and $5.73\,°C$.

solvent. A large capacitor is charged, and then discharged through a sample cell containing at most a few milliliters of solution. The discharge is typically done in $10^{-7}$ to $10^{-6}$ s and leads to a temperature rise of 7–8 °C. There are a number of complications to worry about. First of all, the density of water varies with temperature (Figure 16.7) so that this large a temperature rise can cause shock waves as the water rapidly expands. To minimize this problem, the experiment is carried out at temperatures near 4 °C. It is then possible to choose initial and final temperatures at which the solution has approximately the same density.

Another issue in $T$-jump experiments is thermal equilibration between the sample cell and the solvent. The walls of the sample cell are usually made of an insulating material and so are not subject to Joule heating. As a result, heat is transferred from the solvent to the cooler walls of the cell. This results in convection currents which begin roughly 1 s after the discharge. These convection currents cause optical turbulence, which tends to make spectrophotometer readings fluctuate, so the experiment must be over before these currents start. This means that Joule heating can be used to study reactions on time scales ranging from microseconds to one or two tenths of a second.

Another option is to use an intense laser beam to heat the sample. By this method, called optical heating, temperature jumps of up to $10\,°C$ can be obtained in a few nanoseconds. Optical heating is therefore suitable for measuring rates of reactions on time scales ranging from nanoseconds up. The main problem here is that the sample volume must be small, since it must be completely contained in the laser beam to obtain uniform heating. Very sensitive detectors will be necessary in these experiments.

However the temperature jump is achieved, the result is the same. The variation of the equilibrium constant with temperature is given by Equation (8.11). Suppose that a reaction has an enthalpy change of $100\,kJ\,mol^{-1}$. Then, for a temperature jump of $8\,°C$ with an average $T$ of $4\,°C$ (to minimize the shock waves due to expansion), we get

$$\ln\left(\frac{K_2}{K_1}\right) = \frac{\Delta_r H_m^\circ}{R}\left(\frac{1}{T_1} - \frac{1}{T_2}\right)$$

$$= \frac{100 \times 10^3\,J\,mol^{-1}}{8.314\,472\,J\,K^{-1}mol^{-1}}\left(\frac{1}{273.15} - \frac{1}{281.15\,K}\right) = 1.25.$$

$$\therefore \frac{K_2}{K_1} = \exp(1.25) \approx 3.5.$$

If $\Delta_r H_m^\circ$ were negative, we would predict a decrease of the equilibrium constant by a similar factor. Larger enthalpy changes or larger temperature jumps would create larger changes in the equilibrium constant, and therefore larger displacements from the new equilibrium.

### 16.3.2 The pressure-jump method

Although equilibrium constants are not very sensitive to pressure, it is possible to change the equilibrium constant of a system by making very large pressure changes. This is the basis of the pressure ($p$)-jump method.

Again, there are many methods for obtaining a pressure jump, some of which are illustrated in Figure 16.8. In one technique [Figure 16.8(a)], a valve is rapidly opened to a tank of pressurized gas. This is a fairly slow method, taking several hundredths of a second, but large pressure jumps of over 100 atm can be achieved.

In a variation on this technique, a valve is opened between the sample cell and a much larger liquid reservoir [Figure 16.8(b)]. Pressure changes are transmitted very rapidly where two liquids are in contact. The pressure of the liquid reservoir determines the pressure of the sample cell. Mixing between the two is minimized because the surface of contact inside the open valve is very small. On the time scale of this experiment, we can therefore neglect effects associated with dilution of the sample by the solvent in the liquid reservoir. Pressure

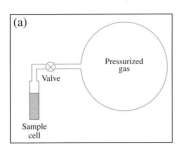
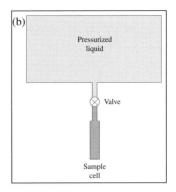
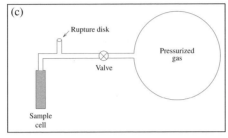

Figure 16.8 Methods for creating a pressure jump. (a) A valve is opened to a pressurized gas cylinder. (b) A valve is opened to a pressurized liquid ballast. Note that the sample cell in this case is filled right up to the valve such that the pressure is transmitted through the liquid. (c) The sample cell is pressurized prior to the start of the experiment. After closing the valve to the pressurized gas reservoir, a rapid downward pressure jump is obtained by breaking a rupture disk.

changes (increases or decreases) of a few hundred atmospheres can be achieved in under 5 ms. Because the pressure can be varied up or down, the effect of pressure on the reaction can be fully studied by this method.

Another simple method involves pressurizing the system to be studied, then rapidly decreasing the pressure (to ambient pressure) by breaking a rupture disk [Figure 16.8(c)]. The pressure change in this case is much faster, taking about 60 μs. One advantage of this method, in addition to the very fast change in pressure, is that the readings are taken near 1 bar rather than at an elevated pressure. This makes the data more directly comparable to data from other experiments near 1 bar (thermochemical measurements, classical kinetics experiments etc.) than would be the case for high-pressure data. However, slightly smaller pressure jumps are obtained (50–60 atm), which in turn means that smaller changes in the equilibrium constant occur.

The pressure-jump method involves one unfortunate complication: when we rapidly change the pressure on a sample, the temperature changes. The temperature increases during pressurization and decreases in a depressurization. This temperature change is usually at most a few tenths of a degree, but this is enough to cause difficulties with some measurement methods. Of course, if this is a problem for $p$-jump experiments, it's an even larger problem for $T$ jumps.

We now derive an equation for the variation of the equilibrium constant with pressure. Recall Equation (7.3), which tells us how $G$ varies with $p$ and $T$. At constant temperature, this simplifies to $dG = V\,dp$. For a reaction, this equation would apply to both reactants and products. Taking the difference between the change in free energy of the reactants and products under standard conditions, we get $d(\Delta_r G_m^\circ) = \Delta_r V_m^\circ\,dp$, where $\Delta_r V_m^\circ$ is the change in the molar volume under standard conditions. Now using the relationship between the standard free energy change and the equilibrium constant, Equation (8.2), we get

$$d(-RT\ln K) = \Delta_r V_m^\circ\,dp,$$

$$\therefore d(\ln K) = -\frac{\Delta_r V_m^\circ}{RT}\,dp$$

since we are holding $T$ constant. If we now integrate this equation assuming that the change in the molar volume of reaction is more or less constant over the pressure range of interest, we get

$$\Delta\ln K =\approx -\frac{\Delta_r V_m^\circ}{RT}\Delta p.$$

Note that $\Delta\ln K = \ln K_2 - \ln K_1 = \ln(K_2/K_1)$, so this last equation becomes

$$\ln\left(\frac{K_2}{K_1}\right) \approx -\frac{\Delta_r V_m^\circ}{RT}\Delta p.$$

For reactions in solution, $\Delta_r V_m^\circ$ is very small, so it takes a large pressure change to have an appreciable effect on $K$. The effect is greatest in charge neutralization reactions (e.g. $A^+ + B^- \to C$) where $\Delta_r V_m^\circ$ might be as large as $-10\,\text{cm}^3\,\text{mol}^{-1}$. Suppose that we carry out an experiment at $25\,°C$ with a pressure jump $\Delta p = -100$ atm for a reaction with $\Delta_r V_m^\circ = -10\,\text{cm}^3\,\text{mol}^{-1}$. This gives $\ln(K_2/K_1) = -0.04$, or $K_2/K_1 = 0.96$, i.e. a decrease in the equilibrium constant of about 4%. This is a small change, but enough to lead to observable changes in the reactant concentrations as they approach the new equilibrium.

### 16.3.3 Analysis of relaxation experiments

Neither the temperature-jump nor the pressure-jump methods cause particularly large changes in the equilibrium constant. Accordingly, the reaction always starts out close to the new equilibrium after the jump. We can exploit this in our analysis of the experimental results. We proceed by example.

#### Reversible first-order process

Suppose that we want to study the elementary reaction

$$A \underset{k_-}{\overset{k_+}{\rightleftharpoons}} B$$

by a relaxation technique. The experiment follows what happens *after* the disturbance so only the values of the parameters (rate and equilibrium constants) after the jump are relevant. The temperature or pressure jump only serves to create slightly out-of-equilibrium conditions in the system.

Let the new equilibrium point be $(a_{eq}, b_{eq})$. Since the jump leaves us close to the new equilibrium, we can write

$$a = a_{eq} + \delta a$$

where $\delta a$ is a small displacement. We similarly define $\delta b$ by $b = b_{eq} + \delta b$. By stoichiometry, $\delta b = -\delta a$ so

$$b = b_{eq} - \delta a.$$

According to the law of mass action,

$$\frac{da}{dt} = -k_+ a + k_- b = -k_+(a_{eq} + \delta a) + k_-(b_{eq} - \delta a) = -k_+ a_{eq} + k_- b_{eq} - \delta a(k_+ + k_-).$$

However, $-k_+ a_{eq} + k_- b_{eq} = 0$, because $da/dt = 0$ at equilibrium. Moreover,

$$\frac{da}{dt} = \frac{d}{dt}(a_{eq} + \delta a) = \frac{d(\delta a)}{dt}$$

because $a_{eq}$ is a constant. (Recall that we are only interested in what happens *after* the jump, when the temperature and pressure have stopped changing.) Therefore

$$\frac{d(\delta a)}{dt} = -\delta a(k_+ + k_-).$$

This is just a first-order rate equation with solution

$$\delta a = \delta a_0 \exp(-[k_+ + k_-]t).$$

By convention, we write the solution in terms of a relaxation time $\tau$ such that

$$\delta a = \delta a_0 \exp(-t/\tau).$$

In this case, $\tau = (k_+ + k_-)^{-1}$. If we measure the relaxation time (by the usual graphical method for first-order rate processes), we then know the sum of the forward and reverse rate constants. Once the reaction reaches equilibrium, we can measure the equilibrium constant:

$$K = b_{eq}/a_{eq} = k_+/k_-.$$

This measurement gives us the ratio of the two rate constants. The equilibrium constant and relaxation time together will let us determine the two rate constants.

**Example 16.1** Transfer RNA can exist in two different conformations in solution:

$$A \underset{k_-}{\overset{k_+}{\rightleftharpoons}} B$$

The equilibrium constant for this reaction at 28 °C is 10. The relaxation time measured after a 3 °C temperature jump from 25 to 28 °C is 3 ms.

The data give us the following two equations:

$$k_+/k_- = 10$$

$$\text{and} \quad (k_+ + k_-)^{-1} = 3 \,\text{ms}.$$

Solving these equations, we get $k_- = 30 \,\text{s}^{-1}$ and $k_+ = 303 \,\text{s}^{-1}$. Note that since all the data correspond to 28 °C (the temperature after the jump), these rate constants also correspond to this temperature.

### Bimolecular reactions with a single product

Now consider reactions of the type

$$A + B \underset{k_-}{\overset{k_+}{\rightleftharpoons}} C.$$

Note that this class of reactions includes, for instance, binding equilibria of enzymes with non-reactive substrate analogs (e.g. competitive inhibitors). Suppose that, after the jump,

$$a = a_{eq} + \delta a$$

where $\delta a$ is small. By stoichiometry then,

$$b = b_{eq} + \delta a$$

$$\text{and} \quad c = c_{eq} - \delta a.$$

The rate equation for $a$ is

$$\frac{da}{dt} = \frac{d(\delta a)}{dt} = -k_+ ab + k_- c.$$

$$\therefore \frac{d(\delta a)}{dt} = -k_+(a_{eq} + \delta a)(b_{eq} + \delta a) + k_-(c_{eq} - \delta a)$$

$$= \left[ -k_+ a_{eq} b_{eq} + k_- c_{eq} \right] - \delta a(k_+ a_{eq} + k_+ b_{eq} + k_-) - k_+(\delta a)^2.$$

At equilibrium, $da/dt = 0$ so the quantity in square brackets is zero by definition. Also, since $\delta a$ is supposed to be small, $(\delta a)^2$ should be very small so we neglect it. It follows that

$$\frac{d(\delta a)}{dt} \approx -\delta a(k_+ a_{eq} + k_+ b_{eq} + k_-).$$

The quantity in parentheses is a constant so this is the differential equation of a first-order process. If we define the relaxation time

$$\tau = \left( k_+ a_{eq} + k_+ b_{eq} + k_- \right)^{-1},$$

the differential equation becomes

$$\frac{d(\delta a)}{dt} \approx -\frac{1}{\tau} \delta a.$$

The solution of this differential equation is

$$\delta a = \delta a_0 \exp(-t/\tau).$$

If we measure the relaxation time and equilibrium concentrations, we can recover both $k_+$ and $k_-$. First, the equilibrium concentrations can be used to determine the equilibrium constant:

$$K = \frac{c_{eq}}{a_{eq}b_{eq}} = \frac{k_+}{k_-}.$$

There is one important note here: the equilibrium constant that appears here is the phenomenological equilibrium constant (with units of $L\,mol^{-1}$) and *not* the dimensionless thermodynamic equilibrium constant. This will be an important consideration in the example to be discussed below.

We can rearrange the expression for the relaxation time slightly:

$$\tau^{-1} = k_- \left[ K \left( a_{eq} + b_{eq} \right) + 1 \right].$$

Thus, if we have measured $\tau$ and the equilibrium concentrations, from which we can also calculate $K$, we can recover $k_-$. Given $k_-$ and $K$, we can calculate $k_+$.

**Example 16.2** Perhaps the most famous application of relaxation methods has been to the measurement of the rate constants for the autoionization of water:

$$H_2O_{(l)} \underset{k_a}{\overset{k_d}{\rightleftharpoons}} H^+_{(aq)} + OH^-_{(aq)}$$

The relaxation time is 37 µs at 25 °C. This is a reaction of the same stoichiometry as studied in this section, except that it is written backwards:

$$H^+_{(aq)} + OH^-_{(aq)} \underset{k_d}{\overset{k_a}{\rightleftharpoons}} H_2O_{(l)}.$$

The equilibrium concentrations are, at 25 °C, $[H^+]_{eq} = [OH^-]_{eq} = 1.0 \times 10^{-7}\,mol\,L^{-1}$ and $[H_2O] = 55.33\,mol\,L^{-1}$. Note that we need the mole density of water because kinetics requires the use of phenomenological equilibrium constants:

$$K = \frac{[H_2O]}{[H^+][OH^-]} = 5.5 \times 10^{15}\,L\,mol^{-1}.$$

$k_d$ can be calculated from the relaxation time by

$$\begin{aligned}
k_d^{-1} &= \tau \left[ K \left( [H^+]_{eq} + [OH^-]_{(eq)} \right) + 1 \right] \\
&= (37 \times 10^{-6}\,s) \left[ (5.5 \times 10^{15}\,L\,mol^{-1})(1.0 \times 10^{-7} + 1.0 \times 10^{-7}\,mol\,L^{-1}) + 1 \right] \\
&= 4.1 \times 10^4\,s.
\end{aligned}$$

$$\therefore k_d = 2.4 \times 10^{-5}\,s^{-1}.$$

Since $K = k_a/k_d$,

$$k_a = k_d K = 1.35 \times 10^{11}\,L\,mol^{-1}s^{-1}.$$

**Exercise group 16.1**

(1) The phenomenological equilibrium constant of the reaction

$$CH_3COOH_{(aq)} \underset{k_a}{\overset{k_d}{\rightleftharpoons}} CH_3COO^-_{(aq)} + H^+_{(aq)}$$

has the value $1.8 \times 10^{-5}$ mol L$^{-1}$ at 25 °C. The relaxation time measured at this temperature in a 0.100 mol L$^{-1}$ solution of acetic acid is 8.3 ns. What are the values of the two rate constants?

*Hint*: You will need the equilibrium concentrations of acetate and of protons. These can be calculated from the equilibrium condition.

(2) The *trp* repressor (R) is a dimeric protein formed by the assembly of two identical subunits in solution:

$$2S \underset{k_d}{\overset{k_a}{\rightleftharpoons}} R$$

This reaction has recently been studied by the pressure-jump method.[1]
   (a) Derive an equation for the relaxation time.
   (b) In the presence of 2.5 mol L$^{-1}$ of the denaturant guanidine hydrochloride (GuHCl) the value of $\Delta_r V_m^\circ$ for the reaction is $-162$ mL mol$^{-1}$. (Without a denaturant, the *trp* repressor is too stable for its association and dissociation kinetics to be studied.) The apparatus used created downward pressure jumps of up to 150 bar at 21 °C. By roughly what factor would the equilibrium constant increase or decrease after these pressure jumps?

(3) The DNA stain proflavin dimerizes in solution:

$$2P \underset{k_d}{\overset{k_a}{\rightleftharpoons}} P_2$$

This is of course the same kind of reaction as in the previous problem, so you can reuse some of the results derived there.
   (a) In the experiments, relaxation times were measured at several different proflavin concentrations. Define the total proflavin concentration $p_t = [P] + 2[P_2]$. Relate $p_t$ to $[P]_{eq}$, the equilibrium concentration of P. Then obtain an equation relating the relaxation time $\tau$ to $p_t$ and propose a graphical method for extracting the rate constants of this reaction.
   (b) The following data have been obtained:

| $\tau/10^{-7}$ s | 3.2 | 1.8 | 1.4 | 1.2 |
|---|---|---|---|---|
| $p_t/10^{-3}$ mol L$^{-1}$ | 0.5 | 2.0 | 3.5 | 5.0 |

   Determine the values of $k_a$ and $k_d$.
   (c) Calculate $\Delta_r G_m^\circ$ for the dimerization of proflavin.

---

[1]  G. Desai et al., *J. Mol. Biol.* **288**, 461 (1999).

## Key ideas and equations

- Mixing is the major limiting factor in studying fast reactions. There are two possible strategies to deal with this: rapid mixing (flow methods) or avoiding mixing altogether (flash photolysis, relaxation methods). Methods that avoid mixing can be used to study much faster reactions than rapid mixing methods.

- For a $T$ jump, $\ln\left(\dfrac{K_2}{K_1}\right) = \dfrac{\Delta_r H_m^\circ}{R}\left(\dfrac{1}{T_1} - \dfrac{1}{T_2}\right)$

- For a $p$ jump, $\ln\left(\dfrac{K_2}{K_1}\right) \approx -\dfrac{\Delta_r V_m^\circ}{RT}\Delta p$

| Reaction | $\tau$ |
|---|---|
| $A \underset{k_-}{\overset{k_+}{\rightleftharpoons}} B$ | $(k_+ + k_-)^{-1}$ |
| $A + B \underset{k_-}{\overset{k_+}{\rightleftharpoons}} C$ | $\left\{k_-\left[K(a_{eq} + b_{eq}) + 1\right]\right\}^{-1}$ |

## Suggested reading

The *Methods in Enzymology* series has always taken a broad view of what might be useful to an enzymologist. Consequently, these volumes are a tremendous resource for anyone working in the molecular sciences. Volume XVI is dedicated to fast reactions. Although it's a little old – it was published in 1969 – the basic ideas are all there so that it's a great starting point for someone wanting to become acquainted with the classical methods for studying reactions on micro- to millisecond time scales.

If you're interested in caged compounds for use in kinetics studies, I would recommend the following review papers:

Günter Mayer and Alexander Heckel, *Angew. Chem., Int. Ed.* **45**, 4900–4921 (2006).
Hsien-Ming Lee, Daniel R. Larson and David S. Lawrence, *ACS Chem. Biol.* **4**, 409–427 (2009).

# 17

# Factors that affect the rate constant

The rate constant is called that because, ideally, it is constant with respect to the concentrations of chemical species in the system. However, it does depend on the reaction conditions. Most obviously, rate constants depend on the temperature. They also depend on the chemical environment in which the reaction occurs. For instance, some reactions can be carried out in different solvents. Provided the two solvents are reasonably similar, the mechanism is likely to be the same. However, the rate constants are generally different. Moreover, it is only true to a first approximation that rate constants do not depend on concentration. Changing the concentrations of species in a system can change the intermolecular forces acting on the participants in a reaction and thus the rate constants. These effects are particularly pronounced in electrolyte solutions, where the concentrations of ions have a profound effect both on the organization of the solvent and on the distribution of solutes.

## 17.1 A simple picture of elementary reactions

Let's consider a specific elementary reaction, namely

$$Br_{(g)} + H_{2(g)} \rightleftharpoons HBr_{(g)} + H_{(g)}$$

We have thermodynamic data for all the species in this reaction in Appendix A.1. Since $\Delta_r \nu_{gas} = 0$, we can calculate that $\Delta_r U_m^{\circ} = \Delta_r H_m^{\circ} = 69.84 \, \text{kJ mol}^{-1}$. (See Section 5.9 for the connection between $\Delta_r U_m^{\circ}$ and $\Delta_r H_m^{\circ}$.) This means that the products are higher in energy than the reactants by $69.84 \, \text{kJ mol}^{-1}$.

Let's now try to construct a picture of this reaction in our minds. Two things have to happen: an H–Br bond must be formed and an H–H bond must be broken. Since this reaction is elementary, both of these things happen more or less at the same time. (That's what we mean by an elementary reaction.) Since we are now going to talk about reactions on a molecular scale, however, we must be clear about what "at the same time" really means. Roughly speaking, it means that the H–Br bond is formed while the H–H bond is lengthening and preparing to "break." The whole event takes about the same time as a bond vibration (about $10^{-13}$ s). Knowing that the bond forming and breaking events occur more or less simultaneously still leaves some room for different possibilities. For instance, the two bond lengths could change at similar rates. Then, if we could look at the atoms

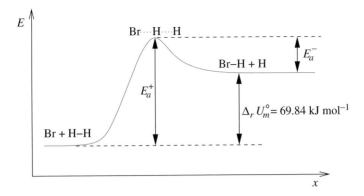

Figure 17.1 Sketch of the potential energy profile of the elementary reaction $Br + H_2 \rightleftharpoons HBr + H$. In the forward direction, the reactants must gain an energy $E_a^+$ to cross the activation barrier. The reverse reaction has an activation energy of $E_a^-$.

during the $10^{-13}$ s or so that it takes to transform the reactants into products, we would see something like this:

$$Br \qquad H-H \qquad \Longrightarrow \qquad Br\cdots H\cdots H \qquad \Longrightarrow \qquad Br-H \qquad H$$

We start with a Br atom and an $H_2$ molecule. The central structure represents a species with very weak H–Br and H–H bonds. The H–Br bond strengthens, forming a stable molecule, and the H–H bond breaks, leaving us with the products of this elementary reaction. Note that the central structure must represent a very high-energy species relative to both reactants and products since the H–H bond has been substantially weakened (which takes energy), but the H–Br bond has not completely formed yet either. If this mental picture of the reaction is correct, we can think of (for instance) the difference between the H–H and H–Br distances as a **reaction coordinate**, i.e. a variable that locates where we are along the path from reactants to products. The reaction coordinate $x = d_{HH} - d_{HBr}$ increases as the reaction proceeds. We can sketch the potential energy along our reaction coordinate (Figure 17.1). The high-energy species between the reactant and products is called the **activated complex** or **transition state**. The amount of energy needed to get to the transition state is called the **activation energy** and is normally denoted by the symbol $E_a$. On the above diagram, we show both the activation energy for the forward ($E_a^+$) and reverse ($E_a^-$) elementary reactions. Note that the activation energies and $\Delta_r U_m^\circ$ are related, as you can see from the figure:

$$\Delta_r U_m^\circ = E_a^+ - E_a^-$$

The above picture is not the only one we can imagine for this reaction. Instead of a "loose" transition state, we might have a "tight" one in which the atoms are all at more-or-less normal bonding distances from each other:

$$Br \qquad H-H \qquad \Longrightarrow \qquad Br-H-H \qquad \Longrightarrow \qquad Br-H \qquad H$$

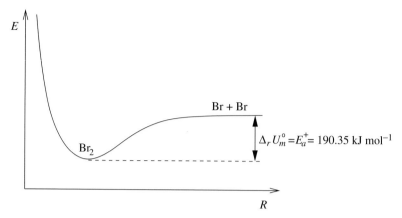

Figure 17.2 Sketch of the potential energy profile of the elementary reaction $Br_2 \rightleftharpoons 2Br$. The reaction coordinate here is the bond length, $R$. The radical recombination reaction has an activation energy of zero. Accordingly, the activation energy for dissociation equals the change in energy for the reaction.

In this scenario, the bromine forms a more-or-less normal bond to hydrogen, using its non-bonding electrons, and *then* the other hydrogen atom leaves. We can guess that the transition state in this picture will also have a high energy since it requires a hydrogen atom to form more than one bond, which in turn requires valence shell expansion, i.e. the participation of 2s and perhaps 2p electrons. (Note that I'm only listing possibilities here for illustrative purposes. In this case, because the energy required to expand the valence shell of hydrogen is so large, a tight transition state is unlikely.) The energy diagram for this pathway will therefore look much like the one for the scenario with a loose transition state.

Either way, we conclude that the reaction must pass through a high-energy transition state. The reactants will have to surmount an **activation barrier** (an energy hill, as pictured above) in order to proceed to products.

Not every reaction involves an activation energy. Consider the elementary reaction

$$Br_{2(g)} \rightleftharpoons 2Br_{(g)}$$

The reverse reaction, recombination of the radicals, is all downhill. The energy diagram for radical formation/recombination reactions is sketched in Figure 17.2. The activation energy for recombination is zero. The activation energy for dissociation is just $\Delta_r U_m^\circ$.

It must be emphasized before we move on that the first case (the existence of a transition state of higher energy than either the reactants or products) is by far the more common of the two cases illustrated here.

Finally, it is useful to distinguish clearly between the transition state and a chemical intermediate. A chemical intermediate is a stable chemical species which can, at least in principle, be isolated. The lifetime of a chemical intermediate is much longer than the time scale of molecular vibrations. On the other hand, a transition state is not, by any

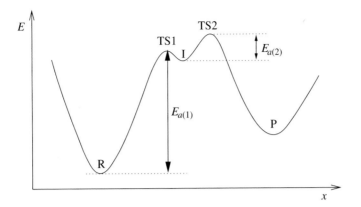

Figure 17.3 Potential energy profile for a reaction with a high-energy intermediate. Only the activation energies for the forward reaction R → P are shown. There are two independent activation energies and two transition states (TS1 and TS2) in this system associated with the two elementary reactions R → I and I → P.

definition whatsoever, stable and cannot, even in theory, be isolated. Transition states can only be studied by ultra-fast spectroscopic methods, their lifetimes being of the order $10^{-13}$ s.

We can picture the difference between intermediates and transition states by reference to an energy profile for a simple isomerization reaction (Figure 17.3). The intermediate (I) occupies a "well" in the potential energy. Intermediates, once formed, need to gain enough energy to surmount an activation barrier to react. These barriers may be relatively low (as shown in the figure) or may be quite high. The transition states, on the other hand, sit at energy maxima and typically do not survive a single molecular vibration.

If you compare Figures 17.1 and 17.3, you may wonder why, in one case, the reactant and product regions are shown as flat regions of the potential energy landscape, while in the other, they are shown as wells. The answer is that the reaction coordinate $x$ can represent very different quantities for different reactions, and that the potential energy profile therefore looks different as we move away from the barrier. For the Br + H$_2$ reaction, I suggested that $x = d_{HH} - d_{HBr}$ might be a good reaction coordinate. Large positive values of this coordinate correspond to an H atom and an HBr molecule that are far apart. At large distances, the intermolecular forces tend to zero, so the potential energy flattens out. Similar comments could be made about large negative values of $x$, where the Br atom and H$_2$ molecule are far apart. On the other hand, for an isomerization reaction like the one sketched in Figure 17.3, the reaction coordinate is an internal coordinate of the molecule (distance between some pair of atoms, bond angle etc.). There is a limit to the values that an internal coordinate can adopt without breaking bonds, so typically in these cases the graph of potential energy vs. internal coordinate rises, sometimes very sharply, at extreme values of $x$. The reactant, product and intermediate (if any) then sit at the bottoms of their respective potential energy wells.

## 17.2 The Arrhenius equation

In general, molecules will have to surmount an activation (energy) barrier to react. The molecules in a sample of matter don't all have the same energy. Some have more, some have less. Moreover, they are constantly exchanging energy with each other through collisions. We would like to know how many reactant molecules have at least $\epsilon_a$ units of energy relative to their ground state (the lowest energy available). (Note that I'm using $\epsilon$ here for energies on a per molecule basis to distinguish these from molar energies.) Only these molecules can react to form products according to the simple picture developed in the last section. Our question can be answered using the Boltzmann equation (Section 3.2). The probability that a given molecule has an energy of $\epsilon$ is given by Equation (3.1), which we rewrite slightly here:

$$P(\epsilon) = g(\epsilon)\exp(-\epsilon/k_{\mathrm{B}}T)/q, \tag{17.1}$$

where $g(\epsilon)$ is the degeneracy of energy level $\epsilon$, i.e. the number of different quantum states of the molecule with the same energy, and $q$ is the molecular partition function. The molecular partition function is a proportionality constant that only depends on the temperature. In the derivation that follows, we will treat the energy as a continuous variable. In other words, we will apply a classical (non-quantum) treatment. The equation we will obtain is known to only be approximately correct, partly because of quantum effects, but also because it is based on a rough kind of reasoning. Assuming classical behavior is, as it turns out, no worse than the other approximations on which this equation is based.

If any energy is allowed, then we have to reinterpret Equation (17.1). Instead of the degeneracy, $g(\epsilon)$ becomes the density of states, which is the number of states with energy between $\epsilon$ and $\epsilon + d\epsilon$. In such a case, we cannot talk of the probability that a molecule has an energy of exactly $\epsilon$. This probability is zero because there are an infinite number of allowed energies in any finite interval or, to put it another way, the energy of the system is bound to differ in some decimal place from any number we care to write down. In the case where the energy varies continuously then, we should think of the right-hand side of Equation (17.1) as a probability density for energy $\epsilon$, i.e. a probability per unit energy. More specifically, if $\bar{\epsilon}$ is a possible outcome of an energy measurement for a molecule, we have

$$P(\epsilon \le \bar{\epsilon} \le \epsilon + d\epsilon) = \frac{g(\epsilon)\exp(-\epsilon/k_{\mathrm{B}}T)}{q}\, d\epsilon.$$

If we add up these probabilities for all the little intervals of width $d\epsilon$, which is equivalent to integrating the probability density, we should get 1 since the molecule has to have some value or other of the energy:

$$1 = \int_0^\infty \frac{g(\epsilon)\exp(-\epsilon/k_{\mathrm{B}}T)}{q}\, d\epsilon = \frac{1}{q}\int_0^\infty g(\epsilon)\exp(-\epsilon/k_{\mathrm{B}}T)\, d\epsilon.$$

$$\therefore q = \int_0^\infty g(\epsilon)\exp(-\epsilon/k_{\mathrm{B}}T)\, d\epsilon.$$

We want to know the probability that a molecule has an energy in excess of $\epsilon_a$, the activation energy. This probability would be the sum of the probabilities that the energy is in an interval of width $d\epsilon$ for all values of $\epsilon > \epsilon_a$:

$$P(\epsilon > \epsilon_a) = \int_{\epsilon_a}^{\infty} \frac{g(\epsilon)\exp(-\epsilon/k_B T)}{q}\, d\epsilon = \frac{\int_{\epsilon_a}^{\infty} g(\epsilon)\exp(-\epsilon/k_B T)\, d\epsilon}{\int_0^{\infty} g(\epsilon)\exp(-\epsilon/k_B T)\, d\epsilon}.$$

If we can't say anything about the density of states, we're stuck here. We now make a radical approximation, namely that there is one constant density of states below the activation energy, which we will call $g_b$, and a different, but also constant, density of states above the barrier, $g_a$. Mathematically, we would write $g(\epsilon) = g_b$ for $\epsilon < \epsilon_a$ and $g(\epsilon) = g_a$ for $\epsilon > \epsilon_a$, with $g_a$ and $g_b$ constant. Then

$$P(\epsilon > \epsilon_a) = \frac{g_a \int_{\epsilon_a}^{\infty} \exp(-\epsilon/k_B T)\, d\epsilon}{g_b \int_0^{\epsilon_a} \exp(-\epsilon/k_B T)\, d\epsilon + g_a \int_{\epsilon_a}^{\infty} \exp(-\epsilon/k_B T)\, d\epsilon}.$$

Usually, the second term in the denominator of this expression will be much smaller than the first because the transition state will be high in energy and thus a low-probability state. We would therefore not make a large error by neglecting this term. For the same reason, we would not make a large error by extending the range of integration in the remaining term in the denominator to infinity:

$$\begin{aligned} P(\epsilon > \epsilon_a) &= \frac{g_a \int_{\epsilon_a}^{\infty} \exp(-\epsilon/k_B T)\, d\epsilon}{g_b \int_0^{\infty} \exp(-\epsilon/k_B T)\, d\epsilon} \\ &= \frac{-g_a\, k_B T\, \exp(-\epsilon/k_B T)|_{\epsilon_a}^{\infty}}{-g_b\, k_B T\, \exp(-\epsilon/k_B T)|_0^{\infty}} \\ &= \frac{g_a}{g_b}\exp(-\epsilon_a/k_B T) = \frac{g_a}{g_b}\exp(-E_a/RT). \end{aligned}$$

In the last equality, we have used the fact that Boltzmann's constant is just the ideal gas constant on a per molecule basis to transform the equation from one involving the activation energy per molecule, $\epsilon_a$, to one involving the activation energy per mole, $E_a$. We conclude that the probability that a molecule has an energy above $\epsilon_a$, or $E_a$ on a molar basis, is proportional to $\exp(-\epsilon_a/k_B T)$ or, equivalently, to $\exp(-E_a/RT)$. We then suppose that, once the reactants have come together, the probability of a reaction occurring is proportional to the probability that the reactants have enough energy. If this is the case, then the rate constant should contain a term proportional to the Boltzmann factor derived above, and we write

$$k = A\exp(-\epsilon_a/k_B T) = A\exp(-E_a/RT). \tag{17.2}$$

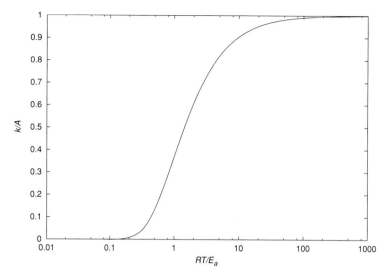

Figure 17.4 General shape of the Arrhenius function. Since $R$ is a constant and $E_a$ is fixed for a given elementary reaction, the quantity plotted as the abscissa is simply proportional to the temperature. Similarly, the ordinate is proportional to the rate constant since $A$ is a constant for a given reaction. Note the logarithmic scale on the temperature axis.

This is the **Arrhenius equation**. The parameter $A$ is rather boringly called the **pre-exponential factor**, and is taken to be constant at this level of theory.

The Arrhenius function has a straightforward shape, shown in Figure 17.4. At very low temperatures, the rate constant decreases to zero because none of the molecules then have any excess energy. As the temperature rises, the rate constant increases, reaching a limiting value of $A$ at very high temperatures. At ultra-high temperatures, all molecules have enough energy to get over the activation barrier such that this barrier no longer matters. Thus, $A$ is the value the rate constant would have if the energy barrier between reactants and products weren't relevant. To put it another way, the pre-exponential factor contains all the factors other than simple energetics that determine the rate of reaction, such as the relative orientations of the two molecules in a bimolecular process. The crossover between low- and high-temperature behavior occurs when $RT$ is of a similar order of magnitude to $E_a$.

We can recover the activation energy and $A$ by studying the dependence of a rate constant on temperature. To do this, we transform the Arrhenius equation to a linear form. If we take a natural logarithm of both sides of the equation, we get

$$\ln k = \ln A - \frac{E_a}{R} \frac{1}{T}. \tag{17.3}$$

Thus, a plot of $\ln k$ against $1/T$ should yield a straight line of slope $-E_a/R$ and intercept $\ln A$.

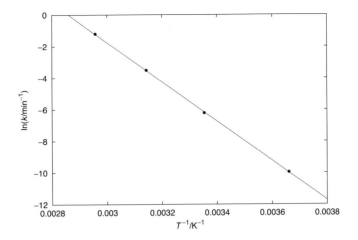

Figure 17.5 Arrhenius plot of the data from Example 17.1.

**Example 17.1 Extraction of Arrhenius parameters** The decomposition of dinitrogen pentoxide follows a simple first-order rate law at moderate pressures. The rate constant is found to vary with temperature as follows:

| $T/°C$ | 0 | 25 | 45 | 65 |
|---|---|---|---|---|
| $k/\text{min}^{-1}$ | $4.7 \times 10^{-5}$ | $2.0 \times 10^{-3}$ | $3.0 \times 10^{-2}$ | $3.0 \times 10^{-1}$ |

To obtain the Arrhenius parameters, we must evaluate $\ln k$ and $1/T$. Note that $T$ must be in kelvins *before* we take its reciprocal.

| $T^{-1}/\text{K}^{-1}$ | $3.661 \times 10^{-3}$ | $3.354 \times 10^{-3}$ | $3.143 \times 10^{-3}$ | $2.957 \times 10^{-3}$ |
|---|---|---|---|---|
| $\ln(k/\text{min}^{-1})$ | $-9.965$ | $-6.215$ | $-3.507$ | $-1.204$ |

The Arrhenius plot is shown in Figure 17.5. The slope and intercept are, respectively, $-1.25 \times 10^4$ K and 35.68. The Arrhenius parameters are therefore

$$E_a = -R(-1.25 \times 10^4 \, \text{K}) = 104 \, \text{kJ mol}^{-1},$$
$$A = \exp(35.68) = 3.1 \times 10^{15} \, \text{min}^{-1}.$$

Note that the activation energy and $A$ will always be positive. Also note that the units of $A$ are the same as those of the original rate constant.

The experimental activation energy found by fitting data to the Arrhenius equation is not exactly equal to the difference in energy between the bottom of the reactant energy well and the top of the energy barrier at which we find the transition state due to a couple of quantum mechanical effects we encountered in Chapters 2 and 3. First, zero-point energy prevents a molecule from having no vibrational energy at all. Accordingly, a molecule's total energy is always somewhat higher than the lowest corresponding point on the energy profile; reactants never quite sit at the bottoms of their valleys and transition states cannot be formed at exactly the energy of the top of the activation barrier. The size of this effect is different at the top and bottom of the energy barrier, so the two zero-point energies do not offset each other. The net effect can either be an increase or a decrease in the activation energy by (at most) a few kJ mol$^{-1}$, which might be significant in reactions with low barriers. Additionally, tunneling allows molecules to pass right through the barrier rather than going over it. This tends to lower the experimentally observed activation energy. Additionally, at very low temperatures where tunneling may become similar in importance to over-the-barrier trajectories, tunneling may cause a deviation from linearity in Arrhenius plots.

---

## Exercise group 17.1

(1) The following second-order rate constants were obtained for the reaction between 2-tolyl isocyanate and 1-butanol:

| $T/°C$ | 0.0 | 7.0 | 15.0 | 25.0 |
|---|---|---|---|---|
| $k/\text{L mol}^{-1}\text{s}^{-1}$ | $4.04 \times 10^{-5}$ | $7.72 \times 10^{-5}$ | $1.29 \times 10^{-4}$ | $2.50 \times 10^{-4}$ |

Calculate the activation energy and pre-exponential factor for this reaction.

(2) Aromatic compounds can be nitrated in strong acids according to the general reaction scheme

$$A + HNO_3 \rightarrow ANO_2 + H_2O.$$

The empirical rate law for these reactions is

$$v = k[A][HNO_3].$$

The rate constant for the nitration of *p*-chloronitrobenzene in 83.70% sulfuric acid has been measured at a few temperatures:[1]

| $T/°C$ | 40 | 60 | 75 |
|---|---|---|---|
| $k/\text{L mol}^{-1}\text{s}^{-1}$ | $6.9 \times 10^{-4}$ | $3.6 \times 10^{-3}$ | 0.43 |

Determine the pre-exponential factor and activation energy for this reaction.

(3) The reaction $2\text{Fe(CN)}_{6(aq)}^{3-} + 2\text{S}_2\text{O}_{3(aq)}^{2-} \rightarrow 2\text{Fe(CN)}_{6(aq)}^{4-} + \text{S}_4\text{O}_{6(aq)}^{2-}$ displays first-order kinetics. Measurements of the rate constant at different temperatures are given below:[2]

| $T/°C$ | 20 | 30 | 40 | 50 | 60 | 70 |
|---|---|---|---|---|---|---|
| $k/10^{-4}\text{min}^{-1}$ | 1.52 | 2.13 | 3.28 | 5.29 | 9.31 | 14.06 |

(a) Calculate the activation energy and pre-exponential factor for this reaction.
(b) Predict the value of the rate constant at $0\,°C$.

---

[1] N. C. Marziano *et al.*, *J. Chem. Soc., Perkin Trans.* **2**, 1973 (1998).    [2] Y. Li *et al.*, *J. Phys. Chem. B* **104**, 10956 (2000).

(4) In living cells, molecular motors transport other molecules from place to place by "walking" along microtubules. The speed at which these motors move along microtubules can be measured. For a molecular motor known as Eg5, the following data have been obtained:[3]

| $T/°C$ | 13 | 17 | 24 | 26 | 30 |
|---|---|---|---|---|---|
| speed/$\mu m\,s^{-1}$ | 0.21 | 0.26 | 0.31 | 0.33 | 0.39 |

(a) Many processes with an underlying chemical basis obey the Arrhenius equation. Is this true of the speed of this molecular motor?

(b) Regardless of your answer to part a, determine an activation energy for this process.

(c) Predict the temperature at which the speed would be $0.5\ \mu m\,s^{-1}$. Report your answer in degrees Celsius.

### 17.3 Transition-state theory

Transition-state theory, sometimes also called absolute-rate theory, takes a thermodynamic approach to the description of the transition state. The starting point is a simple one: we imagine that there is a pseudo-equilibrium established between the reactants and the transition state. In order words, we imagine that the elementary reaction

$$R \xrightarrow{k} P \tag{17.4}$$

is in fact made up of two steps, namely

$$R \underset{}{\overset{K^{\ddagger}}{\rightleftharpoons}} TS \xrightarrow{k^{\ddagger}} P.$$

This "mechanism" must not be taken too seriously. The transition state TS is not a stable species at all, so there is no sense in which it can be thought of as being in equilibrium with anything. However, if the activation barrier is high so that the transition state is rarely reached, this is a reasonable description of the effective dynamics.

If we do accept the above description for the formation of the transition state, we find, assuming ideal behavior,

$$[TS] = K^{\ddagger}[R] \tag{17.5}$$

so that the rate of formation of the product due to this elementary reaction is

$$\frac{d[P]}{dt} = k^{\ddagger}[TS] = k^{\ddagger}K^{\ddagger}[R].$$

Since the rate of the elementary reaction (17.4) should be $k[R]$, the elementary rate constant must be

$$k = k^{\ddagger}K^{\ddagger}. \tag{17.6}$$

[3] I. M.-T. C. Crevel *et al.*, *J. Mol. Biol.* **273**, 160 (1997).

The pseudo-equilibrium constant $K^{\ddagger}$ can be related to the standard free energy change for reaching the transition state from the reactants by the familiar formula

$$K^{\ddagger} = \exp(-\Delta^{\ddagger}G_m/RT). \tag{17.7}$$

We also need a value for $k^{\ddagger}$. This is the "rate constant" for the disintegration of the transition state to product. This process is closely related to a bond vibration as it involves the stretching of a bond that is about to break. Thus, $k^{\ddagger}$ can be thought of as a frequency for the stretching of a bond associated with the reaction coordinate. This frequency is given by a statistical mechanical treatment, which we omit to avoid getting too far from our main line of thought, as

$$k^{\ddagger} = \frac{k_B T}{h}. \tag{17.8}$$

If we now combine Equations (17.6), (17.7) and (17.8), we get

$$k = \frac{k_B T}{h} \exp(-\Delta^{\ddagger}G_m/RT). \tag{17.9}$$

Conceptually, this is a very pivotal equation in transition-state theory. The only part of this equation that depends on the reaction is $\Delta^{\ddagger}G_m$. For two reactions of the same order, the parameter that controls whether one reaction is fast and the other slow at a given temperature is just their values of $\Delta^{\ddagger}G_m$. Arguably, rather than the energy profiles we considered in the previous sections of this chapter, we should be focusing on free-energy profiles. We will return to this idea shortly.

The free energy of activation can be written in terms of an enthalpy and entropy of activation by

$$\Delta^{\ddagger}G_m = \Delta^{\ddagger}H_m - T\Delta^{\ddagger}S_m. \tag{17.10}$$

Written in terms of the enthalpy and entropy, our expression for the rate constant thus becomes

$$k = \frac{k_B T}{h} \exp(\Delta^{\ddagger}S_m/R) \exp(-\Delta^{\ddagger}H_m/RT). \tag{17.11}$$

If we have measurements of $k$ vs. $T$, we can get the activation parameters by a graphical method. First, rearrange Equation (17.11) to the form

$$\frac{kh}{k_B T} = \exp(\Delta^{\ddagger}S_m/R) \exp(-\Delta^{\ddagger}H_m/RT).$$

Then, take a logarithm of both sides:

$$\ln\left(\frac{kh}{k_B T}\right) = \frac{\Delta^{\ddagger}S_m}{R} - \frac{\Delta^{\ddagger}H_m}{R}\frac{1}{T}. \tag{17.12}$$

If we plot $\ln(kh/k_B T)$ vs. $T^{-1}$, we should get a straight line with a slope of $-\Delta^{\ddagger}H_m/R$, and an intercept of $\Delta^{\ddagger}S_m/R$. We can therefore get the activation parameters from the slope and intercept of this graph, which is called an **Eyring plot**.

Strictly speaking, Equation (17.12) only applies to first-order reactions. The reason is that we wrote Equation (17.5) for a first-order reaction in a homogeneous phase (i.e.

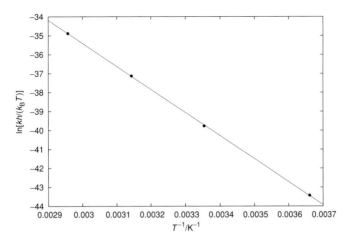

Figure 17.6 Eyring plot from Example 17.2.

with all reactants or products in the same phase). For other types of reactions, we would have ended up with some extra factors involving the standard concentration or pressure. The derivation of the equation is exactly the same, except that we end up with these factors of the standard concentration or pressure that exactly cancel off the corresponding parts of the units of the rate constant. Since these factors are generally chosen to have numeric values of 1, we can still use Equation (17.12) provided we use units for the rate constant that are consistent with our thermodynamic standard state. Specifically, for reactions in solution the rate constants should be in units derived from the $mol \, L^{-1}$ (e.g. $L \, mol^{-1} s^{-1}$ for a second-order reaction), while in the gas phase we should have rate constants in units derived from the bar (e.g. $bar^{-1} s^{-1}$ for a second-order reaction). Also note that the units of time in the rate constant must be seconds in order to agree with the units of $h$.

**Example 17.2 Eyring plot** We will generate an Eyring plot to obtain the activation parameters for the reaction of Example 17.1. We need to convert the values of the rate constant to units of $s^{-1}$ before we calculate $\ln(kh/k_B T)$.

| $T^{-1}/10^{-3} K^{-1}$ | 3.661 | 3.354 | 3.143 | 2.957 |
|---|---|---|---|---|
| $k/s^{-1}$ | $7.8 \times 10^{-7}$ | $3.3 \times 10^{-5}$ | $5.0 \times 10^{-4}$ | $5.0 \times 10^{-3}$ |
| $\ln(kh/k_B T)$ | $-43.43$ | $-39.77$ | $-37.12$ | $-34.88$ |

The Eyring plot is shown in Figure 17.6. The slope of the graph is $-1.22 \times 10^4 \, K$, from which we calculate $\Delta^{\ddagger} H_m = -R(\text{slope}) = 101 \, kJ \, mol^{-1}$. The intercept is 1.11, leading to an entropy of activation of $\Delta^{\ddagger} S_m = R(\text{intercept}) = 9.25 \, J \, K^{-1} mol^{-1}$.

The reaction we have analyzed here is not an elementary reaction. It is a gas-phase unimolecular reaction, so it proceeds via the Lindemann mechanism. We will discuss the interpretation of the activation parameters for complex reactions in Section 17.3.1.

If we have previously calculated the Arrhenius parameters, we can also calculate the enthalpy and entropy of activation from $E_a$ and $A$. To determine the relationship between the two sets of parameters, we need to look at the temperature dependence of the rate constant, since that is how the Arrhenius parameters are determined. Let's start with the logarithmic form of the Arrhenius equation (Equation 17.3). Taking a derivative with respect to $T$, we get

$$\frac{d}{dT} \ln k = \frac{E_a}{RT^2},\tag{17.13}$$

assuming that $A$ and $E_a$ don't depend on temperature. Similarly, we will assume that $\Delta^{\ddagger}S$ varies only negligibly with temperature, as we often did in our study of thermodynamics. Equation (17.11) then gives

$$\ln k = \ln\left(\frac{k_B T}{h}\right) + \frac{\Delta^{\ddagger}S_m}{R} - \frac{\Delta^{\ddagger}H_m}{RT};$$

$$\therefore \frac{d}{dT} \ln k = \frac{1}{T} - \frac{1}{RT}\frac{d}{dT}\Delta^{\ddagger}H_m + \frac{\Delta^{\ddagger}H_m}{RT^2}.$$

Applying Equation (5.15) (page 91) to the middle term of the last equation, we get $\Delta^{\ddagger}H_m = \Delta^{\ddagger}U_m + RT\Delta^{\ddagger}\nu_{gas}$, where $\Delta^{\ddagger}\nu_{gas}$ is the change in the number of molecules of gas required to reach the transition state per reactive event. *This term is zero for reactions in solution*, so that we will only need to consider this effect for reactions in the gas phase. Assuming that $\Delta^{\ddagger}U_m$ depends only weakly on $T$, we get

$$\frac{d}{dT} \ln k = \frac{1}{T} - \frac{\Delta^{\ddagger}\nu_{gas}}{T} + \frac{\Delta^{\ddagger}H_m}{RT^2}.$$

If we now compare this equation to Equation (17.13), which gives the dependence of $\ln k$ on $T$ for the Arrhenius equation, we find

$$\Delta^{\ddagger}H_m = E_a + RT(\Delta^{\ddagger}\nu_{gas} - 1).\tag{17.14}$$

This equation tells us how to calculate the enthalpy of activation from the experimental activation energy. Conversely, it can be used to calculate activation energies from theoretical computations of the enthalpy of activation.

**Example 17.3 Enthalpy of activation in solution** For the elementary reaction $A_{(aq)} + B_{(aq)} \rightarrow C_{(aq)}$, the activation energy is $100\,kJ\,mol^{-1}$. What is the enthalpy of activation at $25\,^{\circ}C$?

To solve this problem, we must first realize that since the reaction proceeds in solution, $\Delta^{\ddagger}\nu_{gas} = 0$. Thus

$$\Delta^{\ddagger}H = 100\,kJ\,mol^{-1} - (8.314472 \times 10^{-3}\,kJ\,K^{-1}mol^{-1})(298.15\,K) = 98\,kJ\,mol^{-1}$$

**Example 17.4 Enthalpy of activation in the gas phase** For the elementary reaction $A_{(g)} + B_{(g)} \rightarrow C_{(g)}$, the activation energy is $100\,kJ\,mol^{-1}$. What is the enthalpy of activation at $25\,^{\circ}C$?

Since this is a gas-phase reaction, we need to consider the term in $\Delta^{\ddagger} v_{gas}$. Since this elementary reaction combines two reactant molecules into one product molecule, the transition state is itself almost certainly a single species that combines A and B. Accordingly, $\Delta^{\ddagger} v_{gas} = -1$ and

$$\Delta^{\ddagger}H = 100 \text{ kJ mol}^{-1} - 2(8.314\,472 \times 10^{-3} \text{ kJ K}^{-1}\text{mol}^{-1})(298.15 \text{ K}) = 95 \text{ kJ mol}^{-1}.$$

If we combine Equations (17.11) and (17.14), we obtain our final expression for the transition-state theory rate constant:

$$k = \frac{k_B T}{h} \exp(\Delta^{\ddagger} S_m / R) \exp(1 - \Delta^{\ddagger} v_{gas}) \exp(-E_a / RT). \tag{17.15}$$

If we compare this equation to the Arrhenius form, we find that

$$A = \frac{k_B T}{h} \exp(\Delta^{\ddagger} S_m / R) \exp(1 - \Delta^{\ddagger} v_{gas}). \tag{17.16}$$

Transition-state theory therefore predicts that $A$ is not exactly constant, but depends linearly on $T$. To see this dependence, however, one needs to perform experiments over a very wide range of temperatures, since the exponential dependence of the rates on $T$ through the activation energy term is much stronger.

If one has measured $A$, one can then calculate $\Delta^{\ddagger} S_m$:

$$\Delta^{\ddagger} S_m = R \left[ \ln \left( \frac{hA}{k_B T} \right) - 1 + \Delta^{\ddagger} v_{gas} \right]. \tag{17.17}$$

This equation is correct as written only for a first-order elementary reaction. The issue is exactly the same as for the Eyring equation, and the solution is also exactly the same: for reactions in solution, the rate constant should be expressed in units derived from the $\text{mol L}^{-1}$, and for reactions in the gas phase, the rate constant should be in units derived from the bar.

**Example 17.5 Entropy of activation in solution** For the elementary reaction $A_{(aq)} + B_{(aq)} \rightarrow C_{(aq)}$, the pre-exponential factor $A$ is $1.4 \times 10^5 \text{ L mol}^{-1}\text{s}^{-1}$. Calculate the entropy of activation at 25 °C.

Since this reaction occurs in solution, $\Delta^{\ddagger} v_{gas} = 0$. Therefore

$$\Delta^{\ddagger}S = (8.314\,472 \text{ J K}^{-1}\text{mol}^{-1})$$
$$\times \left[ \ln \left( \frac{(6.626\,0688 \times 10^{-34} \text{ J Hz}^{-1})(1.4 \times 10^5 \text{ L mol}^{-1}\text{s}^{-1})}{(1.380\,6503 \times 10^{-23} \text{ J K}^{-1})(298.15 \text{ K})} \right) - 1 \right]$$
$$= -155 \text{ J K}^{-1}\text{mol}^{-1}.$$

As described above, the units are troublesome in these calculations. The pre-exponential factor was expressed in units of $\text{L mol}^{-1}\text{s}^{-1}$, which:

(1) is in units compatible with $\text{mol L}^{-1}$, and
(2) uses the SI unit of time (s).

All the other quantities in the equation being in SI units, the result will come out in the appropriate SI unit for entropy, the $J K^{-1} mol^{-1}$, measured relative to the chemists' standard state.

**Example 17.6 Entropy of activation in the gas phase** For the elementary reaction $A_{(g)} + B_{(g)} \rightarrow C_{(g)}$, the pre-exponential factor $A$ is $1.4 \times 10^5 L mol^{-1}s^{-1}$. Calculate the entropy of activation at $25\,°C$.

As in Example 17.4, $\Delta^{\ddagger}\nu_{gas} = -1$. Since this is a gas-phase reaction, the units of $A$ should be $bar^{-1}s^{-1}$, which implies a conversion from the value given to us. We can work out the conversion factor using the ideal gas law: $pV = nRT$ so that $n/V = p/RT$. Since 1 bar is $100\,000$ Pa, 1 bar at $25\,°C$ is equivalent to $n/V = 40.3\,mol\,m^{-3} = 0.0403\,mol\,L^{-1}$. In other words, the conversion factor is $0.0403\,mol\,L^{-1}bar^{-1}$. The pre-exponential factor becomes $A = 5.6 \times 10^3\,bar^{-1}s^{-1}$.

We are now in a position to calculate the entropy of activation:

$$\Delta^{\ddagger}S = (8.314\,472\,J\,K^{-1}mol^{-1})$$
$$\times \left[ \ln \left( \frac{(6.626\,0688 \times 10^{-34}\,J\,Hz^{-1})(5.6 \times 10^3\,bar^{-1}s^{-1})}{(1.380\,6503 \times 10^{-23}\,J\,K^{-1})(298.15\,K)} \right) - 1 - 1 \right]$$
$$= -190\,J\,K^{-1}mol^{-1}.$$

Recall (from Section 6.5) that entropy measures the number of microscopic states compatible with a given macroscopic state. The entropy will therefore decrease, as in the above two examples, when a system loses degrees of freedom. In the foregoing cases, where two reactant molecules form a single product, the transition state involves a bringing together of the reactants and thus should have fewer degrees of freedom than the reactants. In other cases, it is much less clear what the structure of the transition state should be. Calculating the entropy of activation tells us whether the transition state has fewer or more microstates than the reactants. This in turn tells us something about the nature of the transition state.

One useful rule of thumb you might want to remember is that forming a transition state complex from two reactants typically results in an entropy of activation of about $-100\,J\,K^{-1}mol^{-1}$ (give or take 10 or $20\,J\,K^{-1}mol^{-1}$). If, for a bimolecular reaction, you notice that $\Delta^{\ddagger}S$ is much less negative than that, then you know that some other factor is tending to offset some of the usual entropy decrease.

**Example 17.7** For the elementary reaction

$$H_{(g)} + Br_{2(g)} \rightarrow HBr_{(g)} + Br_{(g)}$$

the activation energy and pre-exponential factor are, respectively, $15.5\,kJ\,mol^{-1}$ and $1.09 \times 10^{11}\,L\,mol^{-1}s^{-1}$. Our objective is to calculate the enthalpy and entropy of activation at $800\,°C$.

To calculate the enthalpy and entropy of activation, we need to guess $\Delta^{\ddagger}\nu_{gas}$. In this case, it seems quite likely that the transition state is a single complex, perhaps something like $H\cdots Br\cdots Br$. Thus $\Delta^{\ddagger}\nu_{gas} = -1$. We can confirm that this is a sensible choice by calculating the entropy of activation *first*: we start by converting the pre-exponential factor to $\text{bar}^{-1}\text{s}^{-1}$. At $800\,^\circ\text{C}$ (1073 K), the conversion factor from $\text{mol}\,\text{L}^{-1}$ to bar is $0.0112\,\text{mol}\,\text{L}^{-1}\text{bar}^{-1}$ so $A = 1.22 \times 10^9\,\text{bar}^{-1}\text{s}^{-1}$. The rest of the calculation is straightforward:

$$\Delta^\ddagger S = (8.314\,472\,\text{J}\,\text{K}^{-1}\text{mol}^{-1})$$
$$\times \left[ \ln \left( \frac{(6.626 \times 10^{-34}\,\text{J}\,\text{Hz}^{-1})(1.22 \times 10^9\,\text{bar}^{-1}\text{s}^{-1})}{(1.381 \times 10^{-23}\,\text{J/K})(1073\,\text{K})} \right) - 2 \right]$$
$$= -98\,\text{J}\,\text{K}^{-1}\text{mol}^{-1}.$$

The negative entropy of activation confirms that the transition state is probably a single molecule. If we had instead guessed that this reaction had a very loose transition state in which there were still two distinct molecules (a case for which $\Delta^\ddagger \nu_{\text{gas}} = 0$), we would have calculated $\Delta^\ddagger S = -79\,\text{J}\,\text{K}^{-1}\text{mol}^{-1}$, a significantly negative value which is not compatible with the assumption of a loose transition state.

We can now calculate the enthalpy of activation:

$$\Delta^\ddagger H = E_a - 2RT = -2.3\,\text{kJ}\,\text{mol}^{-1}$$

We have two methods for getting the activation parameters: Eyring plots, and Equations (17.14) and (17.17). The two will give slightly different answers. Why? Equations (17.14) and (17.17) allow you to calculate the activation parameters at a particular temperature, while Eyring plots give you average values over the temperature range covered by a data set. Both conceptually and in practice, these are different quantities. Neither procedure is better nor more accurate than the other, they just give you slightly different results. Moreover, both the Arrhenius equation and the version of transition-state theory presented here are approximate theories. It is therefore not reasonable to expect precise values from these theories. That does not make them any less useful; the entropy of activation, in particular, is one of the simpler pieces of data to obtain that provides some insight into the nature of the transition state.

---

### Exercise group 17.2

(1) In the Diels–Alder reaction of ethene with 1,3-butadiene to form cyclohexene in the gas phase, the activation energy is $115\,\text{kJ}\,\text{mol}^{-1}$. The following thermochemical data for the reactants and products are available:

| Species | $\Delta_f U^\circ/\text{kJ}\,\text{mol}^{-1}$ |
|---|---|
| $C_2H_{4(g)}$ (ethene) | 54.95 |
| $C_4H_{6(g)}$ (1,3-butadiene) | 115.4 |
| $C_6H_{10(g)}$ (cyclohexene) | 14.24 |

(a) Sketch the energy profile of the reaction. Clearly label $\Delta_r U_m^\circ$ as well as the activation energies for the forward and reverse reactions.

(b) Calculate the activation energy for the reverse reaction (cyclohexene → ethene + butadiene).

(c) Calculate the enthalpies of activation for the forward and reverse reactions at 298 K.

(2) For the reaction of aniline with ethyl propiolate in solution in the solvent DMSO,

the following second-order rate constants have been obtained:[4]

| $T/K$ | 318 | 323 | 328 | 333 | 338 |
|---|---|---|---|---|---|
| $k/10^{-3} \text{L mol}^{-1}\text{s}^{-1}$ | 1.44 | 2.01 | 2.71 | 3.73 | 4.87 |

Calculate the enthalpy and entropy of activation at $25\,^\circ$C.

(3) For first-order reactions in solution near room temperature, at what approximate value of $A$ would we have $\Delta^\ddagger S = 0$? On a molecular level, what is the significance of such a value of the entropy of activation?

(4) Methyl radicals react with ethane to form methane and ethyl radicals:

$$CH_{3(g)} + C_2H_{6(g)} \rightarrow CH_{4(g)} + C_2H_{5(g)}$$

The pre-exponential factor for this elementary reaction is $2.00 \times 10^8 \text{ L mol}^{-1}\text{s}^{-1}$. What is the entropy of activation at 298.15 K? What does the value of the entropy of activation of this reaction tell you about the transition state?

(5) Kim and coworkers have studied the first-order thermal isomerization of para-substituted benzyl isocyanides:[5]

The measured rate constants are as follows for the reaction in benzene:
$$k/10^{-5}\text{s}^{-1} \text{ for different substituents X}$$

| $T$ (°C) | −H | −CH₃ | −OCH₃ | −Cl | −NO₂ |
|---|---|---|---|---|---|
| 170 | 2.01 | 2.61 | 3.46 | 1.95 | 1.47 |
| 190 | 9.94 | 13.0 | 16.9 | 9.54 | 7.26 |
| 210 | 45.2 | 59.4 | 78.6 | 44.3 | 33.4 |
| 230 | 228 | 295 | 396 | 226 | 166 |

[4] D. Nori-Shargh *et al.*, *Int. J. Chem. Kinet.* **38**, 144 (2006).     [5] S. S. Kim *et al.*, *J. Org. Chem.* **63**, 1185 (1998).

(a) Calculate the activation energy and pre-exponential factor for each of the reactions.

(b) The reaction is apparently elementary and proceeds via a triangular transition state, i.e. the cyanide group "rotates" during the reaction. Calculate the enthalpy and entropy of activation for each reaction at $200\,^{\circ}$C. Comment on what the results tell us about the effects of the substituents on the transition state.

(6) Fluorinated organic compounds are currently replacing chlorofluorocarbons as refrigerants and anesthetics. Little is known of their atmospheric chemistry and, in particular, of their reactivity toward common stratospheric species such as hydroxide radicals. Consider the elementary reaction

$$OH_{(g)} + CF_3CH_2OCHF_{2(g)} \rightarrow products.$$

The products are unspecified because the reaction can produce a number of different products depending on where the OH radical attacks, the energies of reactant and product, and so on. In other words, this reaction really stands for a whole family of elementary reactions. However, for the calculations performed in this question, this can be treated as a simple elementary reaction. The following data have been obtained for this reaction:[6]

| $T/K$ | 292 | 298 | 347 | 402 |
|---|---|---|---|---|
| $k/10^7 \mathrm{L\,mol^{-1}s^{-1}}$ | 1.3 | 1.0 | 2.5 | 6.6 |

Calculate the enthalpy, entropy and free energy of activation at $25\,^{\circ}$C. Discuss briefly what the value of the entropy at activation tells you about the transition state of this reaction.

### 17.3.1 Complex reactions and transition-state theory

Azomethane ($CH_3N_2CH_3$) decomposes to methyl radicals and nitrogen at elevated temperatures in the gas phase by a first-order process. Suppose that you were given the following data, which relate the rate constant to temperature, and were asked to calculate the entropy of activation:

| $T/K$ | 523 | 541 | 560 | 576 | 593 |
|---|---|---|---|---|---|
| $k/\mathrm{s^{-1}}$ | $1.8 \times 10^{-6}$ | $1.5 \times 10^{-5}$ | $6.0 \times 10^{-5}$ | $1.6 \times 10^{-4}$ | $9.5 \times 10^{-4}$ |

Using an Eyring plot, we find $\Delta^{\ddagger} H_m = 215 \mathrm{\,kJ\,mol^{-1}}$ and $\Delta^{\ddagger} S_m = 53 \mathrm{\,J\,K^{-1}mol^{-1}}$. The overall reaction is

$$CH_3N_2CH_{3(g)} \rightarrow 2\,CH_{3(g)} + N_{2(g)}$$

[6] S. D. Beach *et al.*, *Phys. Chem. Chem. Phys.* **3**, 3064 (2001).

This is *not*, however, an elementary reaction. It is a gas-phase unimolecular reaction. These proceed by the Lindemann mechanism (Section 14.1):

$$2\,CH_3N_2CH_{3(g)} \underset{k_{-1}}{\overset{k_1}{\rightleftharpoons}} CH_3N_2CH_{3(g)} + \{CH_3N_2CH_{3(g)}\}^*$$

$$\{CH_3N_2CH_{3(g)}\}^* \overset{k_2}{\rightarrow} 2\,CH_{3(g)} + N_{2(g)}$$

According to the theory we have seen so far, transition-state theory should only be applied to elementary reactions. What happens if we have a complex reaction, as in this case?

We know from our previous study of the Lindemann mechanism that we get first-order kinetics in this mechanism at high pressures. The rate law in this case is

$$v = \frac{k_1 k_2}{k_{-1}}[CH_3N_2CH_{3(g)}].$$

The observed first-order rate constant is therefore

$$k = \frac{k_1 k_2}{k_{-1}}.$$

If we recognize that $K_1 = k_1/k_{-1}$ is the equilibrium constant of the *first* step, we get

$$k = K_1 k_2. \tag{17.18}$$

$k_2$ is the elementary rate constant for step 2, so we can apply the transition-state theory Equation (17.15) to this rate constant:

$$k_2 = \frac{k_B T}{h}\,\exp(\Delta^{\ddagger}S^{(2)}/R)\exp(1 - \Delta^{\ddagger}v^{(2)}_{gas})\exp(-E_a^{(2)}/RT) \tag{17.19}$$

where I have used a superscript (2) to indicate properties of reaction 2.

$K_1$, on the other hand, is the equilibrium constant for step 1 of the reaction. We therefore have

$$K_1 = \exp\left(\frac{-\Delta_r G^{\circ(1)}}{RT}\right) = \exp\left(\frac{-\Delta_r H^{\circ(1)}}{RT} + \frac{\Delta_r S^{\circ(1)}}{R}\right) \tag{17.20}$$

where the superscript (1) indicates properties of reaction 1.

If we now substitute Equations (17.19) and (17.20) into Equation (17.18), we get

$$k = \frac{k_B T}{h}\,\exp(1 - \Delta^{\ddagger}v^{(2)}_{gas})\exp\left(\frac{\Delta^{\ddagger}S^{(2)} + \Delta_r S^{\circ(1)}}{R}\right)\exp\left(-\frac{E_a^{(2)} + \Delta_r H^{\circ(1)}}{RT}\right).$$

This looks like Equation (17.15), except that the entropy of activation and activation energy have been replaced by

$$\Delta^{\ddagger}S \leftarrow \Delta^{\ddagger}S^{(2)} + \Delta_r S^{\circ(1)} \equiv \Delta^{\ddagger}S^{(app)} \tag{17.21}$$

$$\text{and} \quad E_a \leftarrow E_a^{(2)} + \Delta_r H^{\circ(1)} \equiv E_a^{(app)}. \tag{17.22}$$

The activation parameters calculated above therefore correspond to these apparent parameters. Incidentally, you will note that I calculated these parameters using an Eyring plot rather than using Equations (17.14) and (17.17). While we can usually choose either of

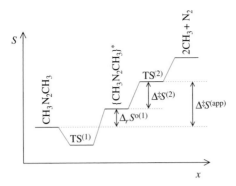

Figure 17.7 Entropy diagram for the gas-phase decomposition of azomethane.

these methods, in the case of complex reactions, it is difficult to decide what one should use for $\Delta^{\ddagger}\nu_{\text{gas}}$, so Eyring plots are to be preferred.

We still need to figure out what these apparent activation parameters mean. Let's focus particularly on the entropies. The story would be similar for the enthalpies. The two terms appearing in Equation (17.21) can be rewritten as differences in entropy:

$$\Delta_r S^{\circ(1)} = S^{\circ}\left(\{CH_3N_2CH_{3(g)}\}^*\right) - S^{\circ}\left(CH_3N_2CH_{3(g)}\right)$$
$$\text{and}\quad \Delta^{\ddagger} S^{(2)} = S^{\circ}(\{TS^{(2)}\}) - S^{\circ}\left(\{CH_3N_2CH_{3(g)}\}^*\right).$$

The apparent entropy of activation is therefore

$$\Delta^{\ddagger} S^{(\text{app})} = S^{\circ}(\{TS^{(2)}\}) - S^{\circ}\left(CH_3N_2CH_{3(g)}\right).$$

This isn't a completely crazy quantity to look at. It's not the entropy of activation for any particular elementary reaction, but it is the entropy gained going from the reactant to the second transition state of the reaction.

Let's put the entropies on a graph to see what this means. Have a look at Figure 17.7. The entropy (per molecule) of the first transition state is probably lower than that of a single azomethane molecule because the first elementary reaction involves the coming together of two molecules. The entropy of the energized azomethane molecule should be higher than that of an ordinary molecule because increasing the energy of a molecule generally makes more microstates available. The entropy of the second transition state should be higher still since it should involve a loosening of chemical bonds, leading to the eventual dissociation. The final, dissociated state (three separate molecules in the gas phase) should have the highest entropy of all.

We can now see that the entropy of activation we calculate in this problem is the change in entropy in going from the reactant azomethane to the second transition state. This total "entropy of activation" involves a sum of the entropy change in reaction 1 (collisional energization of the reactant) and of the entropy of activation of reaction 2. Both are positive contributions to the entropy change. Given that $\Delta^{\ddagger} S^{(\text{app})} = 53\,\text{J K}^{-1}\text{mol}^{-1}$, we see that the changes in going from the reactant to the second transition state, which represents a version of the reactant that has first been energized, and then shifted the appropriate amount of energy into the reaction coordinate, involve an increase in the number of microstates, probably associated with a loosening of some of the bonds.

The last paragraph contains an implicit idea which we should bring out into the open: it's not good enough for a molecule to have enough energy to react. If it were, then unimolecular gas-phase reactions would occur in one step; as soon as a molecule had enough energy to react, it would go to product. In fact, this doesn't happen because the energy has to end up in the right part of the molecule. It's no good for a bond that has nothing to do with the reaction to hold lots of vibrational energy. After the activating collision, the energy gained by the molecule redistributes itself randomly. It's only if that energy finds its way into the reaction coordinate before it is lost that the reaction can actually occur. The energy profiles we have been using as thinking tools in this chapter are thus one-dimensional simplifications of a much more complicated situation.

### 17.3.2 Enzymes and transition-state theory

How exactly does an enzyme or, in general, any catalyst, speed up a reaction? The simplest answer to this question can be obtained by looking at Equation (17.9): enzymes must somehow be decreasing the free energy of activation. Given Equation (17.10), enzymes might be doing either or both of the following:

- Decreasing the enthalpy of activation
- Increasing the entropy of activation

Enzymes use both of these tricks to accelerate reactions. That decreasing the enthalpy of activation is helpful is no great surprise: the enthalpy of activation and the activation energy are almost the same thing (Equation 17.14), so this is just saying that lowering the energy of the transition state speeds up the reaction. We could have come to a similar conclusion from Arrhenius theory, without all the embellishments of transition-state theory.

The idea that increasing the entropy of activation can speed up a reaction takes a little more thought. First, we have to remember that enzyme catalysis takes at least two steps: binding of the substrate(s), and reaction catalysis per se. If the reaction is limited by reaction catalysis (which is often, but not always, the case), then we should focus our attention on $k_{cat}$ ($k_{-2}$ in the Michaelis–Menten mechanism; Section 15.2). One of the key things that an enzyme does is to hold the substrate(s) in its active site in the appropriate position for the reaction to occur. This has the effect of reducing the number of microstates of the reactants (the substrates) relative to the number of microstates they would have if they were free-floating in solution. Since this reduces the entropy of the reactants, all other things being equal, it will increase the entropy of activation, which is a difference between the entropy of the transition state and the entropy of the reactants.

Whether it's through entropic or enthalpic effects, reducing the free energy of the transition state means that we are stabilizing this state with respect to the reactants. This leads us to one of the great insights that transition-state theory has to offer to enzyme kinetics.

Naïvely, we might think that enzymes should be good at binding their substrates, but in fact, what enzymes are really good at is binding their transition states. From this insight flows another one: if you want to design a competitive inhibitor for an enzyme, you should look for compounds that are similar in shape and charge distribution to the transition state, and not to the substrates. This idea is of huge practical importance since many drugs and pesticides are competitive inhibitors. For example, $\beta$-lactams, a group of antibiotics that includes penicillin, are competitive inhibitors of bacterial enzymes that are involved in cell wall synthesis. Some chemotherapeutic agents are also competitive inhibitors, such as methotrexate, which inhibits dihydrofolate reductase, an enzyme necessary for the production of nucleotides. Since cancer cells divide rapidly, inhibiting nucleotide production, which is necessary for DNA replication, tends to be more toxic for these cells than for non-dividing cells. Knowing that our inhibitors should be transition-state analogs has opened up whole new vistas in drug design.

## 17.4 Ionic strength

Our understanding of the effect of ionic strength on elementary rate processes derives from transition-state theory. It turns out that there is little effect of ionic strength in unimolecular elementary reactions, for reasons that will become clear from the derivation below. We therefore consider the bimolecular elementary process

$$A + B \xrightarrow{k} P.$$

As before, we assume that this reaction can be decomposed into accession to the transition state

$$A + B \underset{}{\overset{K^{\ddagger}}{\rightleftharpoons}} TS$$

which is assumed to be in pseudo-equilibrium, and decomposition of the transition state to product

$$TS \xrightarrow{k^{\ddagger}} P.$$

The rate of the reaction depends on the *concentration* of the transition state:

$$v = k^{\ddagger}[TS].$$

However, if the first part of this process can be described as an equilibrium, we should have

$$K^{\ddagger} = \frac{a_{TS}}{a_A a_B} = \frac{\gamma_{TS}[TS]c^{\circ}}{\gamma_A[A]\gamma_B[B]} = \frac{[TS]c^{\circ}}{[A][B]} \frac{\gamma_{TS}}{\gamma_A \gamma_B}.$$

$$\therefore [TS] = K^{\ddagger} \frac{[A][B]}{c^{\circ}} \frac{\gamma_A \gamma_B}{\gamma_{TS}}.$$

The reaction rate is therefore

$$v = k^{\ddagger} K^{\ddagger} \frac{[A][B]}{c^{\circ}} \frac{\gamma_A \gamma_B}{\gamma_{TS}}.$$

If we compare this equation to the mass-action rate for our elementary step ($v = k[\text{A}][\text{B}]$), we find that

$$k = \frac{k^{\ddagger} K^{\ddagger}}{c^{\circ}} \frac{\gamma_{\text{A}} \gamma_{\text{B}}}{\gamma_{\text{TS}}}.$$

If the ionic strength is zero, then $\gamma_i = 1$ for all species. The quantity $k_0 = k^{\ddagger} K^{\ddagger}/c^{\circ}$ is therefore the limiting value of the rate constant at zero ionic strength. This relationship should be compared to the simple transition-state theory expression (17.6). The extra factor of $1/c^{\circ}$ is related to the fact that in our development of transition-state theory we assumed a first-order reaction, whereas in this case we assumed a second-order reaction. These issues are closely connected to those raised in Section 17.3 with respect to the calculation of transition-state entropies.

If we substitute $k_0$ into our expression for $k$ and take a logarithm of both sides of the equation, we get

$$\ln k = \ln k_0 + \ln \gamma_{\text{A}} + \ln \gamma_{\text{B}} - \ln \gamma_{\text{TS}}.$$

We use Equation (9.6) to approximate the activity coefficients. After collecting terms, we get

$$\ln k = \ln k_0 - 1.107 \times 10^{-10} (\varepsilon T)^{-3/2} \sqrt{I_c} \left( Z_{\text{A}}^2 + Z_{\text{B}}^2 - Z_{\text{TS}}^2 \right).$$

The transition state is formed by combining A and B. By conservation of charge, the charge of the transition state is therefore $Z_{\text{TS}} = Z_{\text{A}} + Z_{\text{B}}$. The term in parentheses is thus

$$Z_{\text{A}}^2 + Z_{\text{B}}^2 - Z_{\text{TS}}^2 = Z_{\text{A}}^2 + Z_{\text{B}}^2 - (Z_{\text{A}} + Z_{\text{B}})^2 = -2 Z_{\text{A}} Z_{\text{B}}.$$

Consequently, the rate constant is related to the ionic strength by

$$\ln k = \ln k_0 + 2.214 \times 10^{-10} Z_{\text{A}} Z_{\text{B}} (\varepsilon T)^{-3/2} \sqrt{I_c}. \tag{17.23}$$

This equation, called the **Brønsted–Bjerrum equation**, tells us that if we plot $\ln k$ against $\sqrt{I_c}$ (at fixed $T$), we should get a straight line, at least if the ionic strength is not too high. This relationship between $I_c$ and $k$ has been confirmed in a number of experimental studies. It highlights the importance of ionic strength as a variable to control, on an equal footing with temperature and $p$H. We have to remember that Debye–Hückel theory, from which Equation (17.23) is derived, only applies to ions in solution. The activity coefficients of other solutes will also depend on the ionic strength, since the Coulomb force affects solvent organization and can also polarize other solutes. We just can't predict how the ionic strength will affect rate constants from the Brønsted–Bjerrum equation for reactions that involve at least one uncharged reactant. It is therefore critical to control the ionic strength in kinetic studies, regardless of the natures of the reactants and products.

**Example 17.8 Use of the Brønsted–Bjerrum equation** If the rate constant for an elementary second-order reaction $\text{A}^{2+} + \text{B}^- \rightarrow \text{AB}^+$ is $1.0 \times 10^{-4} \, \text{L mol}^{-1} \text{s}^{-1}$ in an aqueous solution at $25\,^{\circ}\text{C}$ with an ionic strength of $0.003 \, \text{mol L}^{-1}$, what would the value of the rate constant be if the ionic strength were increased to $0.01 \, \text{mol L}^{-1}$?

The charges here are $Z_A = 2$ and $Z_B = -1$. We calculate the permittivity of water from the relative permittivity, and get $\varepsilon = \varepsilon_r \varepsilon_0 = 6.939 \times 10^{-10} \, C^2 N^{-1} m^{-2}$. From the data at $I_c = 0.003 \, mol \, L^{-1}$, we can calculate

$$\ln k_0 = \ln k - \frac{2.214 \times 10^{-10} Z_A Z_B}{(\varepsilon T)^{3/2}} \sqrt{I_c}$$

$$= \ln \left(1.0 \times 10^{-4}\right) - \frac{2.214 \times 10^{-10}(2)(-1)}{\left[(6.939 \times 10^{-10})(298.15)\right]^{3/2}} \sqrt{0.003}$$

$$= -8.953.$$

Note that I don't keep track of my units in these calculations because the constant in the Brønsted–Bjerrum equation, just like the constant in the Debye–Hückel limiting law, has awkward units that cancel with the units of the other quantities in the equation. The rate constant at $I_c = 0.01 \, mol \, L^{-1}$ is therefore calculated by

$$\ln k = \ln k_0 + \frac{2.214 \times 10^{-10} Z_A Z_B}{(\varepsilon T)^{3/2}} \sqrt{I_c}$$

$$= -8.953 + \frac{2.214 \times 10^{-10}(2)(-1)}{\left[(6.939 \times 10^{-10})(298.15)\right]^{3/2}} \sqrt{0.01}$$

$$= -9.423.$$

$$\therefore k = \exp(-9.423) = 8.1 \times 10^{-5} \, L \, mol^{-1} s^{-1}.$$

The effect is certainly not negligible, amounting to approximately 20% of the value of the rate constant in this example.

**Example 17.9 Brønsted-Bjerrum plot** The reaction

$$BrCH_2COO^-_{(aq)} + S_2O_{3(aq)}^{2-} \rightarrow S_2O_3CH_2COO^{2-}_{(aq)} + Br^-_{(aq)}$$

was studied at 25 °C by mixing equimolar solutions of the sodium salts of the two reactants. If we define $c = [BrCH_2COONa_{(aq)}] = [Na_2S_2O_{3(aq)}]$, the following data were obtained:

| $c$/mmol L$^{-1}$ | 0.5 | 0.7 | 1.0 | 1.4 | 2.0 |
|---|---|---|---|---|---|
| $k$/L mol$^{-1}$min$^{-1}$ | 0.298 | 0.309 | 0.324 | 0.343 | 0.366 |

Are these data consistent with the Brønsted–Bjerrum equation?

A plot of $\ln k$ vs. $\sqrt{I_c}$ should give us a straight line. We need to calculate the ionic strengths of these solutions. The ionic strength is defined by Equation (9.5). In this case, we get

$$I_c = \frac{1}{2} \left\{ (-1)^2 [BrCH_2COO^-_{(aq)}] + (-2)^2 [S_2O_{3(aq)}^{2-}] + (1)^2 [Na^+] \right\}.$$

We can simplify this expression a little for these experiments given that the salts were present in equimolar amounts: $[BrCH_2COO^-_{(aq)}] = [S_2O_{3(aq)}^{2-}] = c$ and $[Na^+] = 3c$ (two

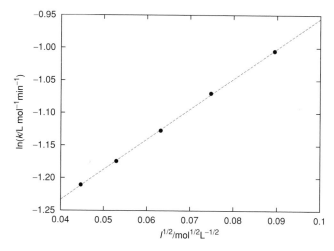

Figure 17.8 Dependence of the rate constant on the ionic strength for the data of Example 17.9.

from the $Na_2S_2O_3$ and one more from the $BrCH_2COONa$). The ionic strength is therefore $I_c = 4c$. Our data become

| $I_c/\text{mmol L}^{-1}$ | 2.00 | 2.80 | 4.00 | 5.60 | 8.00 |
|---|---|---|---|---|---|
| $\sqrt{I_c}/\text{mol}^{1/2}\text{L}^{-1/2}$ | 0.0447 | 0.0529 | 0.0632 | 0.0748 | 0.0894 |
| $k/\text{L mol}^{-1}\text{min}^{-1}$ | 0.298 | 0.309 | 0.324 | 0.343 | 0.366 |
| $\ln(k/\text{L mol}^{-1}\text{min}^{-1})$ | $-1.2107$ | $-1.1744$ | $-1.1270$ | $-1.0700$ | $-1.0051$ |

The resulting graph is shown in Figure 17.8. As predicted by Brønsted-Bjerrum theory, the plot is linear. The intercept, which extrapolates the data to zero ionic strength, is $\ln k_0 = -1.4186$ so that $k_0 = 0.242\,\text{L mol}^{-1}\text{min}^{-1}$. The slope is 4.633. According to the theory, it should be

$$\text{slope} = \frac{2.214 \times 10^{-10} Z_{CH_2COONa^-} Z_{S_2O_3^{2-}}}{(\varepsilon T)^{3/2}}$$

$$= \frac{2.214 \times 10^{-10}(-2)(-1)}{\left[(6.939 \times 10^{-10})\,(298.15)\right]^{3/2}}$$

$$= 4.706.$$

The agreement between theory and data is excellent.

---

## Exercise group 17.3

(1) Fluorescence quenching is a process in which a molecule transfers energy that would otherwise be released as fluorescence to another species in solution, called a quencher:

$$F^* + Q \xrightarrow{k_Q} F + Q^*$$

The rate constants for quenching of the fluorescence of a charged polymer (AP90) by $Tl^+$ in water with potassium nitrate used to vary the ionic strength have been measured.[7] The results are as follows:

| $I_c/\text{mmol L}^{-1}$ | 0.46 | 2.11 | 3.90 | 7.20 | 13.67 |
|---|---|---|---|---|---|
| $k_Q/10^{12}\text{L mol}^{-1}\text{s}^{-1}$ | 2.26 | 0.85 | 0.56 | 0.34 | 0.22 |

Are these data consistent with the Brønsted–Bjerrum equation? If so, what is the charge of the fluorescent polymer?

*Note*: The paper cited above does not give the temperature at which the experiments were performed. It seems likely, given the equipment used, that the temperature was uncontrolled. Since spectroscopic equipment is often a bit warmer than room temperature, assume that the experiments were carried out at $25\,^\circ$C.

(2) Explain why ionic strength would have little effect on most first-order elementary processes.

(3) Show that the Brønsted–Bjerrum theory is consistent with the equilibrium theory of Chapter 9. Specifically, show that the dependence of the rate constant on properties of the transition state cancels out when we relate the equilibrium constant to the rate constants. If it helps, consider a specific reaction.

## Key ideas and equations

- You should make yourself thoroughly familiar with "transition-state thinking": energy profiles, interpretation of activation parameters etc.
- $\Delta_r U^\circ_m = E_a^+ - E_a^-$
- $k = A \exp(-E_a/RT)$
- $k = \dfrac{k_B T}{h} \exp(-\Delta^\ddagger G_m/RT)$
- $\ln\left(\dfrac{kh}{k_B T}\right) = \dfrac{\Delta^\ddagger S_m}{R} - \dfrac{\Delta^\ddagger H_m}{R}\dfrac{1}{T}$
- $\Delta^\ddagger H_m = E_a + RT(\Delta^\ddagger v_{gas} - 1)$
- $\Delta^\ddagger S_m = R\left[\ln\left(\dfrac{hA}{k_B T}\right) - 1 + \Delta^\ddagger v_{gas}\right]$
- $\ln k = \ln k_0 + 2.214 \times 10^{-10} Z_A Z_B\,(\varepsilon T)^{-3/2}\,\sqrt{I_c}$.

## Review exercise group 17.4

(1) What information does each of the following observations give you?
  (a) The rate constant of an elementary bimolecular reaction increases as the ionic strength is increased.

[7] M. E. Morrison *et al.*, *J. Phys. Chem.* **100**, 15187 (1996).

(b) The equilibrium constant of a reaction increases as the temperature is increased.

(2) The enthalpies of formation of ethane ($C_2H_6$) and of methyl radicals ($CH_3$) are, respectively, $-83.85$ and $145.69$ kJ mol$^{-1}$ at $298.15$ K. Calculate the activation energy for the dissociation of ethane into methyl radicals.

*Hint*: Sketch the expected energy profile for this reaction.

(3) (a) A textbook gives the following initial rate data for the reaction $OCl^-_{(aq)} + I^-_{(aq)} \rightarrow OI^-_{(aq)} + Cl^-_{(aq)}$:

| Experiment | $[OCl^-]$ mol L$^{-1}$ | $[I^-]$ mol L$^{-1}$ | $[OH^-]$ mol L$^{-1}$ | $v$ $10^{-4}$ mol L$^{-1}$s$^{-1}$ |
|---|---|---|---|---|
| 1 | 0.0040 | 0.0020 | 1.00 | 4.8 |
| 2 | 0.0020 | 0.0040 | 1.00 | 5.0 |
| 3 | 0.0020 | 0.0020 | 1.00 | 2.4 |
| 4 | 0.0020 | 0.0020 | 0.50 | 4.6 |
| 5 | 0.0020 | 0.0020 | 0.25 | 9.4 |

Determine the rate law and the value of the rate constant.

(b) There's something suspicious about these data. What is it?

(4) In aqueous solution at high temperature and high acid concentrations, formic acid breaks down by decarbonylation:

$$HCOOH_{(aq)} \rightarrow CO_{(g)} + H_2O_{(l)}$$

(a) The rate of this reaction is known to depend on both the formic acid and hydrogen ion concentrations. The following initial rate data have been obtained by Yasaka and coworkers at $200\,^\circ$C:[8]

| Experiment | $[HCOOH]$/mol kg$^{-1}$ | $[H^+]$/mol kg$^{-1}$ | $v/10^{-5}$mol kg$^{-1}$ s$^{-1}$ |
|---|---|---|---|
| 1 | 1.0 | 0.062 | 1.1 |
| 2 | 1.0 | 0.062 | 1.3 |
| 3 | 1.0 | 0.26 | 5.4 |
| 4 | 0.16 | 0.10 | 0.28 |
| 5 | 0.66 | 0.10 | 1.2 |
| 6 | 1.6 | 0.10 | 2.7 |

Determine the rate law and rate constant for this reaction.

*Note*: Pick the whole-number partial orders that best describe these data for your final answer. Because there is considerable experimental scatter in the measurements, you must use *all* the data to calculate the rate constant.

---

[8] Y. Yasaka *et al.*, *J. Phys. Chem. A* **110**, 11082 (2006). The data given here are a subset of the data obtained by these authors.

(b) The rate constant was measured at different temperatures. The following data were obtained:

| $T/°C$ | 170 | 210 | 240 | 280 |
|---|---|---|---|---|
| $k/\text{kg mol}^{-1}\text{s}^{-1}$ | $1.5 \times 10^{-5}$ | $5.1 \times 10^{-4}$ | $3.7 \times 10^{-3}$ | $6.3 \times 10^{-2}$ |

Calculate the Arrhenius parameters for this reaction.

(5) Some chemical processes appear to have negative activation energies. This is of course not possible for an elementary reaction and so must be indicative of a complex reaction. Consider the following two-step mechanism for the conversion of a reactant R to a product P:

$$R \underset{k_{-1}}{\overset{k_1}{\rightleftharpoons}} X \overset{k_2}{\to} P$$

(a) Show that, if the second step is slow, this reaction displays simple first-order kinetics.

(b) Determine the conditions under which the apparent first-order rate constant would display a negative activation energy.

(6) The reaction $A_{(aq)}^{2+} + B_{(aq)} \to C_{(aq)}^{2+}$ has a rate constant of $2.9 \times 10^3 \, \text{L mol}^{-1}\text{s}^{-1}$ at a temperature of 295 K and an ionic strength of $1.6 \times 10^{-4} \, \text{mol L}^{-1}$. The permittivity of water at this temperature is $7 \times 10^{-10} \, \text{C}^2\text{J}^{-1}\text{m}^{-1}$. What rate constant does Brønsted–Bjerrum theory predict at an ionic strength of $2.6 \times 10^{-3} \, \text{mol L}^{-1}$? Is this reasonable? Discuss briefly.

(7) The decomposition of the ethoxy radical ($CH_3CH_2O$) has been studied in the gas phase at very low ethoxy radical pressures in the presence of helium.[9] The mechanism is a variation on the Lindemann mechanism:

$$CH_3CH_2O_{(g)} + He_{(g)} \underset{k_{-1}}{\overset{k_1}{\rightleftharpoons}} He_{(g)} + CH_3CH_2O_{(g)}^*$$

$$CH_3CH_2O_{(g)}^* \overset{k_2}{\to} CH_2O_{(g)} + CH_{3(g)}$$

where $CH_3CH_2O^*$ is an energized radical.

(a) Obtain an approximate rate law giving the dependence of the rate on the concentrations of reactants.

(b) Show that at low pressures of helium, this rate law reduces to $v = k_1[\text{He}][CH_3CH_2O] = k_1'[CH_3CH_2O]$, where $k_1'$ is a pseudo-first-order rate constant.

(c) Measurements of the pseudo-first-order rate constant for the reaction at (known) low helium pressures therefore allow us to recover $k_1$. It was found that $k_1$ varies with temperature according to the Arrhenius law with a pre-exponential factor of $2.0 \times 10^{13} \, \text{L mol}^{-1}\text{s}^{-1}$. Calculate the entropy of activation for the corresponding elementary reaction at 170 °C, a typical temperature at which this reaction is studied.

---

[9] F. Caralp *et al.*, *Phys. Chem. Chem. Phys.* **1**, 2935 (1999).

(8) Reactions of the hydroperoxy radical ($HO_2$) with hydrocarbons are important in combustion processes. The rate constant has been measured as a function of temperature for the elementary reaction

$$HO_{2(g)} + c\text{-}C_5H_{10(g)} \rightarrow H_2O_{2(g)} + c\text{-}C_5H_{9(g)}$$

($c\text{-}C_5H_{10(g)}$ is cyclopentane.) The following data were obtained:[10]

| $T/K$ | 673 | 713 | 753 | 783 |
|---|---|---|---|---|
| $k/10^4 \text{L mol}^{-1}\text{s}^{-1}$ | 2.95 | 5.55 | 12.3 | 18.4 |

(a) Calculate the pre-exponential factor and activation energy.
(b) Calculate the entropy of activation at 298 K. What does the value of the entropy of activation tell us about the transition state?

(9) The inorganic complex $CoRu(CO)_5(\mu\text{-bma})(\mu\text{-PPh}_2)$ loses a carbonyl group to form $CoRu(CO)_4(\mu\text{-bma})(\mu\text{-PPh}_2)$ in solution at moderate temperatures. (bma is 2,3-bis-(diphenylphosphino)maleic anhydride.)

(a) The following first-order rate constants have been measured for this process:[11]

| $T/^{\circ}C$ | 77.2 | 83.0 | 88.0 | 93.2 | 98.1 |
|---|---|---|---|---|---|
| $k/10^{-4}\text{s}^{-1}$ | 2.37 | 7.3 | 10.1 | 20.4 | 30.3 |

Calculate the activation energy and pre-exponential factor.

(b) In reactions in which a ligand is lost, there are two possibilities:
   (i) The ligand is lost "early," i.e. reaching the transition state mostly involves the loosening of the metal-ligand bond.
   (ii) The ligand is lost "late," i.e. reaching the transition state involves an internal rearrangement but no significant weakening of the metal-ligand bond.
   Calculate the entropy of activation at 80 °C and comment on whether you think the carbonyl group leaves early or late in this reaction.

(10) Arrhenius theory assumes that the pre-exponential factor and activation energy are both independent of temperature. However, more advanced theoretical approaches to kinetics typically predict that these quantities are temperature dependent. Using transition-state theory, calculate the rate constant for a first-order solution-phase reaction with $\Delta^{\ddagger}S^{\circ} = 80\,\text{J K}^{-1}\text{mol}^{-1}$ and $\Delta^{\ddagger}H^{\circ} = 100\,\text{kJ mol}^{-1}$ at 300, 350, 400, 450 and 500 K. Produce an Arrhenius plot using your rate constants. Is this graph straight? What activation energy and pre-exponential factor do you get from your graph? How do these values compare to those you would calculate directly from $\Delta^{\ddagger}S^{\circ}$ and $\Delta^{\ddagger}H^{\circ}$ at, say, 300 K? What do you conclude about the temperature dependence of the pre-exponential factor and activation energy predicted by transition-state theory?

[10] S. M. Handford-Styring and R. W. Walker, *Phys. Chem. Chem. Phys.* **4**, 620 (2002).
[11] S. G. Bott *et al.*, *Inorg. Chem.* **39**, 6051 (2000).

# 18

# Diffusion and reactions in solution

A bimolecular elementary reaction requires that the two reactants meet. In solution, molecules are constantly bumping into other molecules, with the net effect that they wander randomly through the solution, which we call diffusion. In this chapter, we will study diffusion and its effect on chemical reactions, specifically focusing on the rate at which molecules meet each other. This will give us an upper bound on the reaction rate, which is useful for a number of purposes.

## 18.1 Diffusion

A molecule in solution is constantly surrounded by other molecules. All the molecules are constantly bumping into their neighbors. They move around the solution by sliding between their neighbors, much as you might slide between people to make your way around a crowded room. This process is a form of **diffusion**, the random motion of molecules in space.

The central quantity in describing diffusion is the **flux**, which is the rate at which molecules cross an imaginary surface in space per unit area of this surface. The flux therefore has units of number per unit time per unit area (e.g. $\mathrm{mol\,s^{-1}m^{-2}}$). We're going to focus on diffusion in one dimension, where it's a little easier to picture what is going on. Imagine that we have a tube of cross-sectional area $A$. The tube is narrow enough that we can treat the concentration across the tube as being constant. We imagine counting the *net* number of molecules that pass a certain cross-section of the tube – this is the imaginary surface mentioned above – from left to right. Some molecules will go from left to right, and some from right to left. We subtract the number of molecules that cross our imaginary surface in the leftward direction from those traveling in the rightward direction. This net rightward traffic, divided by the duration of our counting experiment and by the cross-sectional area of the tube, is the flux. If there are more molecules traveling from left to right than in the reverse direction, then the flux is positive, otherwise it's negative.

In general, the flux is a vector, usually denoted **J**. (You can think of a little cube, and count the molecules that go out of each of the faces.) We can denote the components of **J** by $J_x$, $J_y$ and $J_z$ or, if there's only one component of significance in a given problem, as

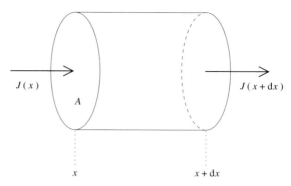

Figure 18.1 Fluxes entering and leaving a small volume of cross-sectional area $A$ and length $dx$.

in our narrow tube, we can simply write $J$. For this one-dimensional case, we will assume that the tube lies along the $x$ axis.

We can connect the flux to the concentration profile through **Fick's first law**, which is valid in its simplest form if we can neglect solute–solute interactions. Since the molecules are moving randomly through the solution, the probability that molecules will move from a region of higher concentration to a region of lower concentration is larger than the probability of the reverse process, simply because there are more molecules in the high-concentration region that could, randomly, find themselves wandering over to the region of low concentration. This suggests that the sign of the flux should be opposite to the sign of the concentration gradient: if the concentration decreases from left to right (negative gradient), then the flux should point in that direction, and vice versa. Fick's first law assumes that these two quantities are in fact directly proportional:

$$J = -D\frac{\partial c}{\partial x}. \tag{18.1}$$

Fick's first law has been found to be an excellent approximation to the relationship between the flux and the concentration gradient in a wide variety of systems. The proportionality constant $D$ is the **diffusion coefficient**. You can think of it as a rate constant for diffusion: large values of $D$ correspond to fast diffusion. The diffusion coefficient depends both on what is diffusing and on the medium in which it is diffusing, as we will see. If you work out the units for the diffusion coefficient, you should find that it has units of area per unit time (e.g. $m^2\,s^{-1}$).

Suppose that we now want to know how the concentration changes at a given point in space as a result of diffusion. Look at Figure 18.1. We consider a small volume of length $dx$ and cross-sectional area $A$. We allow for the possibility that the flux might be different at different points in space, and write $J(x)$ for the flux at position $x$. From the definition of the flux, the number of moles of the diffusible substance entering our volume at $x$ per unit time is $AJ(x)$. The number of moles leaving through the imaginary surface at $x + dx$ per unit time is $AJ(x + dx)$. The net rate of change of the number of moles in this small volume is therefore $AJ(x) - AJ(x + dx)$. Since concentration is number over volume, the

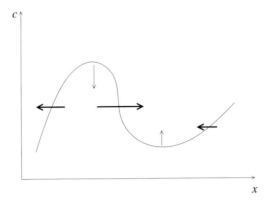

Figure 18.2 In diffusion, the flux vector (heavy arrows) points in the opposite direction to the concentration gradient, i.e. "downhill." The magnitude of the flux is proportional to the gradient, so it is larger where the gradient is steeper, symbolized here by the lengths of the flux vectors. This results in a lowering of high points (where $\partial^2 c/\partial x^2 < 0$) in the concentration profile and in a raising of low points ($\partial^2 c/\partial x^2 > 0$), as indicated by the light arrows.

net rate of change of the concentration in this volume is therefore

$$\frac{\partial c}{\partial t} = \frac{AJ(x) - AJ(x + dx)}{A\,dx} = -\frac{J(x + dx) - J(x)}{dx}.$$

However, by the definition of the derivative, if $dx$ is small, the fraction on the right-hand side is just the derivative of $J$ with respect to $x$:

$$\frac{\partial c}{\partial t} = -\frac{\partial J}{\partial x}. \tag{18.2}$$

This last equation is very general: it tells us how the concentration of a substance changes due to any transport process with flux $J$.

We can now combine Equations (18.1) and (18.2). The result is **Fick's second law**, also known as the **diffusion equation**:

$$\frac{\partial c}{\partial t} = -\frac{\partial}{\partial x}\left(-D\frac{\partial c}{\partial x}\right);$$

$$\therefore \frac{\partial c}{\partial t} = D\frac{\partial^2 c}{\partial x^2}. \tag{18.3}$$

Fick's second law says that the rate of change of the concentration of a substance at a particular point in space is proportional to the second derivative of the concentration with respect to $x$. The second derivative is positive for concave concentration profiles, so diffusion "fills in" points where the concentration is low. Conversely, the second derivative is positive for convex concentration profiles, so diffusion "flattens" high points in the concentration profile. Figure 18.2 illustrates these observations. The overall tendency is for diffusion to homogenize a system.

## 18.2 A peculiar theory for diffusion coefficients in solution

We noted above that the diffusion coefficient depends on what is diffusing and on the medium. We now want to relate the diffusion coefficient to molecular properties. The approach we will take will blend macroscopic theory with a microscopic picture in a most peculiar way, yielding a surprisingly useful theory for diffusion coefficients.

Note again that diffusion causes molecules to move from regions of high concentration to regions of lower concentration. Let's imagine that we can follow one molecule in solution through a small part of its journey. We start by calculating the change in free energy when a single molecule moves from a region of concentration $c$ to one of concentration $c + dc$. This can be calculated from the general relationship between free energy and concentration, Equation (7.7), once we realize that $\Delta G° = 0$ since no chemical reaction is occurring, and assuming that the solute behaves ideally:

$$dG = k_B T \ln \left( \frac{c + dc}{c} \right).$$

(Compare Equation (10.3) which gives the voltage developed due to a concentration difference. Also note the use of $k_B$ instead of $R$ since we are interested in a single molecule.) Since free energy is the negative of the maximum work that can be done by the system, we have

$$dw = -k_B T \ln \left( \frac{c + dc}{c} \right) = -k_B T \ln \left( 1 + \frac{dc}{c} \right).$$

Here, $dw$ is the work that we could extract from a molecule moving down the concentration gradient. Using a Taylor series, we can show that $\ln(1 + x) \approx x$ for small values of $x$. If you don't know about Taylor series, consider that $\ln(1 + 0) = 0$, and that the slope of $\ln(1 + x)$ is $(1 + x)^{-1}$, which tends to 1 as $x \to 0$, so $y = \ln(1 + x)$ passes through the origin with a slope of 1, i.e. it behaves like $f(x) = x$ for small values of $x$. Applying this approximation to the work, we have

$$dw \approx -k_B T \frac{dc}{c}.$$

Moreover, $dw = F\,dx$, so we can think of this work as being the result of a virtual force

$$F_v \approx -\frac{k_B T}{c} \frac{\partial c}{\partial x}.$$

This virtual force is usually known as the **driving force for diffusion**. A molecule subjected to a constant force just keeps accelerating. There has to be an opposing force to limit the increase in speed. As the molecule pushes through the solvent, it experiences a frictional force. For objects of normal size at low speeds, the frictional force is roughly proportional to the speed $v$:

$$F_f = -fv$$

where $f$ is a shape- and medium-dependent constant known as the **frictional coefficient**. A steady diffusive speed is reached when the driving and frictional forces are equal and opposite, i.e. $F_f = -F_v$, which gives

$$fv = -\frac{k_B T}{c}\frac{\partial c}{\partial x},$$

which in turn can be rearranged to

$$cv = -\frac{k_B T}{f}\frac{\partial c}{\partial x}.$$

Now imagine a small surface of area $A$ transverse to the velocity vector. How many moles of material pass through this surface per unit time? Each time unit, the diffusible substance moves $v$ distance units. The volume of material that passes through our surface per unit time is therefore $vA$. The corresponding number of moles is therefore $cvA$. The flux is the number of moles per unit time per unit area, so $J = cvA/A = cv$. Therefore

$$J = -\frac{k_B T}{f}\frac{\partial c}{\partial x}.$$

Comparing this equation to Fick's first law, Equation (18.1), we get

$$D = \frac{k_B T}{f}. \tag{18.4}$$

If we look back at the derivation of this equation, we see some real peculiarities: we started with an equation that comes from thermodynamics, which is a theory that applies to large numbers of molecules, and applied it to a single molecule. We obtained from that an expression for a force that drives diffusion, although individual molecules experience no such force: they just bounce around between solvent molecules. We then brought in the idea of friction, which is appropriate to large objects, but not obviously applicable to molecules. The equation we obtained for the diffusion coefficient (18.4) reflects these oddities: it depends on the temperature, a thermodynamic property of a system which is only meaningful for molecules in an average sense, and on a frictional coefficient which has no obvious molecular interpretation.

If that was all there was to this theory, then it wouldn't be very useful. However, if we stick with the idea that a molecule experiences ordinary friction as it moves through the fluid, we can use expressions obtained in fluid mechanics for the frictional coefficients. The simplest case is that of a spherical solute. For a sphere moving through a fluid, Stokes showed that

$$f = 6\pi r\eta$$

where $r$ is the radius of the sphere (solute) and $\eta$ is the viscosity of the solvent. Viscosity measures the resistance of a fluid to shear forces, i.e. to sliding of layers of fluid relative to each other. Combining Stokes's law with Equation (18.4), we get

$$D = \frac{k_B T}{6\pi r \eta},\tag{18.5}$$

an equation known as the **Stokes–Einstein relationship**. If anything, Equation (18.5) is even more peculiar than (18.4) since it involves the viscosity, a property with no obvious relevance at the molecular level.

Equation (18.5) can be used in one of two ways: the viscosity of a solvent is available from independent experiments. Thus, if we know roughly how large a molecule is, we can use Equation (18.5) to estimate the diffusion coefficient. Conversely, if we know the diffusion coefficient, we can estimate the size of the molecule.

Despite its peculiarities, Stokes–Einstein theory works remarkably well for solutes with a wide range of sizes in a variety of solvents. There are equations for the frictional coefficient for shapes other than a sphere so we are not limited to molecules that are approximately spherical in shape. In fact, the shape dependence of $f$ can be used to determine the solution conformations of biological macromolecules. In the interest of brevity, we will leave this very interesting topic aside.

**Example 18.1 Radius of ribonuclease** The diffusion coefficient of the enzyme ribonuclease in water at $20\,^\circ$C is $1.31 \times 10^{-10}\,\text{m}^2\,\text{s}^{-1}$. The viscosity of water at this temperature is $1.00\,\text{mPa s}$. We can use these data to estimate the radius of a ribonuclease molecule:

$$r = \frac{k_B T}{6\pi \eta D}$$

$$= \frac{(1.380\,6503 \times 10^{-23}\,\text{J K}^{-1})(293.15\,\text{K})}{6\pi (1.00 \times 10^{-3}\,\text{Pa s})(1.31 \times 10^{-10}\,\text{m}^2\,\text{s}^{-1})}$$

$$= 1.64\,\text{nm}$$

---

### Exercise group 18.1

(1) Buckminsterfullerene ($C_{60}$) has a diameter of $10.18\,\text{Å}$. Benzonitrile has a viscosity of $1.24\,\text{mPa s}$ at $25\,^\circ$C. Estimate the diffusion coefficient of $C_{60}$ in benzonitrile, and compare your value to the experimental value[1] of $(4.1 \pm 0.3) \times 10^{-10}\,\text{m}^2\,\text{s}^{-1}$.

---

### 18.3 Bimolecular reactions in solution

We now turn to bimolecular reactions in solution. As we have been discussing, in solution, molecules do not fly around freely because they are in constant contact with the solvent. Rather, we should think of their motion as one in which they slip between solvent molecules from time to time. Since this requires a bit of luck – a molecule has to just happen to hit a

---

[1] Y. Hishida *et al.*, *Anal. Sci.* **22**, 931 (2006).

gap between two solvent molecules to move – once two reactant molecules come in contact, they will tend to stay together for a while. This effect is known as **solvent caging**.

To understand the relative importance of diffusion and reaction, it is useful to decompose the reaction as follows. First, the two reactants (A and B) have to come close enough to each other that reaction is possible. Reactants that have come to this distance of each other are said to have formed an **encounter pair**, denoted {AB}. The reactants can then either move away from each other again, or they can react. We can think of this as a pseudo-mechanism (in the same way as we did in transition-state theory):

$$A + B \underset{k_{-d}}{\overset{k_d}{\rightleftharpoons}} \{AB\} \overset{k_r}{\to} P$$

The rate of reaction is $k_r[\{AB\}]$. Since {AB} is a short-lived configuration, we can apply the steady-state approximation:

$$\frac{d[\{AB\}]}{dt} = k_d[A][B] - (k_{-d} + k_r)[\{AB\}] \approx 0.$$

$$\therefore [\{AB\}] \approx \frac{k_d[A][B]}{k_{-d} + k_r}.$$

$$\therefore \text{rate} \approx \frac{k_d k_r[A][B]}{k_{-d} + k_r}.$$

Since we are really talking about an elementary reaction, the rate should be $k[A][B]$. In other words, the rate constant has the value

$$k = \frac{k_d k_r}{k_{-d} + k_r}. \tag{18.6}$$

If the actual reaction is slow, then $k_r$ is small and the rate constant becomes $k = k_r K_d$ where $K_d = k_d/k_{-d}$ is the "equilibrium constant" for the creation of an encounter pair. In this case, we say that the reaction is **activation controlled**. If, on the other hand, the reaction occurs very easily once the encounter pair has been formed ($k_r$ is large), the rate constant becomes $k = k_d$ and the reaction is said to be **diffusion controlled**. Between these two extremes, we have **diffusion-influenced reactions**, those for which both reaction and diffusion processes are important.

Diffusion-influenced and activation-controlled reactions can only be treated using quantum mechanics, since $k_r$ depends very much on the detailed interaction between the two molecules. On the other hand, we are able to solve the diffusion equation, at least in some simple cases, including one that is directly useful here. Consequently, we are able to calculate rates for diffusion-controlled reactions. In the following, we will again mix ideas from macroscopic and microscopic theory, so watch for that.

For a diffusion-limited reaction, molecules of A and B will react whenever their centers pass within a distance $R_{AB}$ of each other, where $R_{AB} = R_A + R_B$ for non-interacting spherical molecules. $R_{AB}$ is the **reactive distance** for molecules A and B. If the molecules are non-spherical or are subject to strong intermolecular forces, then there will still be a typical distance to which they will need to approach each other to react, but it may be different from the sum of the molecular radii.

We will only sketch the mathematical development from this point. Let $r$ be the distance from a particular molecule of A. Molecules of B should satisfy the three-dimensional analog of the diffusion equation (Equation 18.3). We use a coordinate system centered on the molecule of A we are considering. We can then replace the diffusion coefficient of B in the diffusion equation by the relative diffusion coefficient $D_{AB} = D_A + D_B$, which immediately takes care of the fact that the A molecules are in motion, too.

If molecules of A and B react whenever they pass within $R_{AB}$ of each other, then there won't be any molecules of B within this radius of A. Accordingly, $[B]_{r=R_{AB}} = 0$, where $[B]_r$ denotes the concentration of B at distance $r$ from a molecule of A. On the other hand, far from an A molecule, $[B]_r \rightarrow [B]$, the bulk concentration. It is possible to solve the diffusion equation subject to this pair of conditions. We will only state the result here, since the derivation isn't very enlightening.

$$[B]_r = [B]\left[1 - \frac{R_{AB}}{r} + \frac{R_{AB}}{r}\,\mathrm{erf}\left(\frac{r - R_{AB}}{2\sqrt{D_{AB}t}}\right)\right]$$

where $\mathrm{erf}(x)$ is the error function defined by

$$\mathrm{erf}(x) = \frac{2}{\sqrt{\pi}} \int_0^x \exp(-z^2)\mathrm{d}z. \tag{18.7}$$

If you have never seen the error function before, don't worry. Computer programs (spreadsheets, for instance) often know how to calculate it. Think of it the same way you think of a sine function; you probably don't know how to calculate $\sin x$ by hand, but you really don't have to since your calculator can work it out for you.

The flux towards the molecule of A on which we are concentrating is given by Fick's first law. Here, we calculate $|J|$, i.e. we drop the negative sign in Fick's first law, in order to avoid getting tangled up with our signs later on. We know in this case that the flux is directed toward A, i.e. that $J < 0$, so it's easier conceptually to just keep this in mind and not carry the sign around.

$$|J| = D_{AB}\frac{\partial[B]}{\partial r}$$

$$= D_{AB}[B]\left[\frac{R_{AB}}{r^2} - \frac{R_{AB}}{r^2}\,\mathrm{erf}\left(\frac{r - R_{AB}}{2\sqrt{D_{AB}t}}\right) + \frac{R_{AB}}{r\sqrt{\pi D_{AB}t}}\,\exp\left(\frac{r - R_{AB}}{2\sqrt{D_{AB}t}}\right)\right]. \tag{18.8}$$

This last expression was obtained using the fundamental theorem of calculus and the chain rule.

Recall that the flux is the number of molecules crossing through a surface per unit area per unit time. The surface area of a sphere of radius $R_{AB}$ is $4\pi R_{AB}^2$, so the net rate of arrival of B molecules at the reactive distance is $4\pi R_{AB}^2|J|_{r=R_{AB}}$. This is the rate of reaction per molecule of A. For a diffusion-limited reaction, the first molecule that arrives at this surface reacts. Moreover, the number of molecules of A per unit volume is $[A]LV$, where $V$ is the volume and $L$ is Avogadro's number. The overall rate of change of the number of molecules of B is the negative of the product of the rate of reaction per molecule of A and of the number of molecules of A. Since we usually want the rate of change of the concentration of B, we divide this quantity by $V$ and obtain

$$\frac{d[B]}{dt} = -4\pi R_{AB}^2[A]L|J|_{r=R_{AB}}.$$

We can evaluate $|J|_{r=R_{AB}}$ from Equation (18.8). To do this, note that, from the definition (18.7), $\text{erf}(0) = 0$. After a little algebra, we get the following expression for the rate:

$$\frac{d[B]}{dt} = -4\pi L R_{AB} D_{AB} \left(1 + \frac{R_{AB}}{\sqrt{\pi D_{AB}t}}\right)[A][B].$$

Since the rate of a bimolecular elementary reaction should be $d[B]dt = -k[A][B]$, we have

$$k = k_d = 4\pi L R_{AB} D_{AB} \left(1 + \frac{R_{AB}}{\sqrt{\pi D_{AB}t}}\right).$$

Note that this rate constant depends on time. This makes sense: in the early moments of the reaction, just after mixing, some of the molecules of A will just happen to be very near molecules of B, so the rate will be anomalously large. Note that the time-dependent term becomes small when $\sqrt{\pi D_{AB}t} \gg R_{AB}$, i.e. when

$$t \gg \frac{R_{AB}^2}{\pi D_{AB}}.$$

Typical molecular radii are of the order of 1 nm, and typical diffusion coefficients are of the order of $10^{-10}$ m$^2$ s$^{-1}$ (larger for small molecules), so this time is of the order of a nanosecond. For times much longer than this, we can drop the time-dependent term and we get

$$k_d = 4\pi L(D_A + D_B)R_{AB}$$

where we have replaced $D_{AB}$ by $D_A + D_B$ in the final expression.

In the above derivation, we assumed that the distribution of reactant molecules was more or less random. If the reactants are charged, this will not be true as we saw in our discussion of ionic atmospheres in Section 9.2. A more elaborate treatment gives

$$k_d = 4\pi L(D_A + D_B)R_{AB}\varphi, \tag{18.9}$$

where

$$\varphi = \frac{y}{e^y - 1}$$

with

$$y = \frac{Z_A Z_B e^2}{4\pi \varepsilon k_B T R_{AB}}.$$

The various constants in this last equation were encountered during our development of Debye–Hückel theory: $Z_A$ and $Z_B$ are the charges of the two reactants (in multiples of the elementary charge), $e$ is the elementary charge (in coulombs) and $\varepsilon$ is the permittivity of the solvent. Note that $\varphi = 1$ if either reactant is uncharged.

There are a variety of methods to measure diffusion constants which we will have to leave aside in the interests of brevity. It's also sometimes possible to estimate $R_{AB}$ from structural or other molecular data. From time to time, however, we run into a situation where we have no easy way to estimate $R_{AB}$. One neat way to finesse this problem is to use the Stokes–Einstein relation (Equation 18.5) to estimate $R_A$ and $R_B$ from the corresponding diffusion coefficients:

$$R_{AB} \approx R_A + R_B = \frac{k_B T}{6\pi\eta}\left(\frac{1}{D_A} + \frac{1}{D_B}\right) \tag{18.10}$$

Equation (18.9) can then be used to calculate the diffusional rate constant.

**Example 18.2 Recombination of $H^+$ with $OH^-$** The recombination of $H^+$ with $OH^-$ in water is a typical example of a process that we expect to be diffusion-limited: the reaction is extremely fast. The diffusion constants of the ions are $D_{H^+} = 9.31 \times 10^{-9}$ and $D_{OH^-} = 5.30 \times 10^{-9}$ m$^2$ s$^{-1}$ at 25 °C. We could use Equation (18.10) to calculate the encounter radius, but in this case, because the ions involved are so small, there is a simpler and probably more accurate method for estimating $R_{AB}$; we expect this reaction to occur when the $H^+$ and $OH^-$ ions are separated by a distance equivalent to the OH bond length of water. This would give an encounter radius of approximately $9.58 \times 10^{-11}$ m. The factor $y$ that enters into the computation of $\varphi$ is then

$$y = \frac{(1)(-1)(1.602\,176\times10^{-19}\,\text{C})^2}{4\pi(6.939\times10^{-10}\,\text{C}^2\text{N}^{-1}\text{m}^{-2})(1.380\,6503\times10^{-23}\,\text{J K}^{-1})(298.15\,\text{K})(9.58\times10^{-11}\,\text{m})}$$
$$= -7.465.$$
$$\therefore \varphi = \frac{-7.465}{\exp(-7.449) - 1} = 7.469.$$

In other words, this reaction will be more than seven times faster than it would be if the reactants weren't charged. The diffusional rate constant is

$$k_d = 4\pi(6.022\,142 \times 10^{23}\,\text{mol}^{-1})(9.31 \times 10^{-9} + 5.30 \times 10^{-9}\,\text{m}^2\,\text{s}^{-1})$$
$$\times (9.58 \times 10^{-11}\,\text{m})(7.453)$$
$$= 7.91 \times 10^7\,\text{m}^3\text{mol}^{-1}\text{s}^{-1}$$
$$\equiv 7.91 \times 10^{10}\,\text{L mol}^{-1}\text{s}^{-1}.$$

In Example 16.2, we found that the experimental value of this rate constant (which was called $k_a$ in the previous example) is $1.35 \times 10^{11}$ L mol$^{-1}$s$^{-1}$. This is *larger* than our estimate of the diffusional rate constant. Since the diffusional rate constant is in principle an upper limit for the real rate constant, this suggests that the reaction is indeed diffusion-controlled. We might guess that the problem is our estimate of $R_{AB}$, since that is the most uncertain quantity in our calculations. If we assume that $k_d = 1.35 \times 10^{11}$ L mol$^{-1}$s$^{-1}$ and solve for $R_{AB}$, we get $R_{AB} = 8.1 \times 10^{-10}$ m. This is an *enormous* value for the encounter

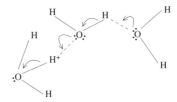

Figure 18.3 Grotthuss mechanism for proton diffusion. The curved arrows show the changes in the organization of the valence electrons, while the dashed lines show the new bonds to be formed.

distance. It suggests that there is some mechanism operating in this reaction that is not accounted for by our simple model of diffusion control. What we're missing is that protons and hydroxide ions in aqueous solution can diffuse by a relay mechanism first proposed by Grotthuss and illustrated in Figure 18.3. By simply rearranging the electrons, the excess charge carried by a proton can move from one water molecule to another. The net effect of the electron motions shown in the figure is that the charge of the proton has moved from the leftmost to the rightmost molecule of water. Since electrons move much faster than nuclei, the overall effect is ultrafast diffusion of protons. As an exercise, you may want to draw an analogous mechanism for the diffusion of hydroxide.

**Example 18.3 Is catalase diffusion controlled?** The diffusion constant of the enzyme catalase is $4.3 \times 10^{-11} \, \text{m}^2 \, \text{s}^{-1}$. This enzyme catalyzes the disproportionation of hydrogen peroxide:

$$2H_2O_2 \rightarrow 2H_2O + O_2$$

The mechanism involves sequential loading of two hydrogen peroxide molecules into the active site:

$$E + H_2O_2 \underset{k_{-1}}{\overset{k_1}{\rightleftharpoons}} C_1$$

$$C_1 + H_2O_2 \underset{k_{-2}}{\overset{k_2}{\rightleftharpoons}} C_2$$

Catalase is one of the more efficient enzymes known. The forward rate constants $k_1$ and $k_2$ have both been measured and are quite high:

$$k_1 = 5 \times 10^6 \, \text{L mol}^{-1} \text{s}^{-1}$$

$$\text{and} \quad k_2 = 1.5 \times 10^7 \, \text{L mol}^{-1} \text{s}^{-1}.$$

We might want to know whether $k_2$ is near the diffusion limit. According to the Stokes–Einstein relation, the diffusion constant mostly depends on the radius of a molecule. Putting a small molecule in the active site of a large enzyme shouldn't make much of a difference to the size, so we will assume that the catalase–peroxide complex $C_1$ has a similar diffusion coefficient to the pure enzyme. The diffusion constants of small molecules in water are typically about $2 \times 10^{-9} \, \text{m}^2 \, \text{s}^{-1}$. All of the data given above correspond to 25 °C. The viscosity of water at this temperature is $8.91 \times 10^{-4}$ Pa s. The encounter radius is therefore approximately

$$R_{AB} = \frac{(1.380\,6503 \times 10^{-23}\,\mathrm{J\,K^{-1}})(298.15\,\mathrm{K})}{6\pi(8.91 \times 10^{-4}\,\mathrm{Pa\,s})} \left( \frac{1}{4.3 \times 10^{-11}} + \frac{1}{2 \times 10^{-9}\,\mathrm{m^2\,s^{-1}}} \right)$$

$$= 5.82 \times 10^{-9}\,\mathrm{m}.$$

Using this value for $R_{AB}$ makes the assumption that the enzyme and peroxide can react if the peroxide meets any part of the surface of catalase. This is of course not true since the peroxide has to find the enzyme's active site. The effect of this assumption is mitigated by the fact that catalase is actually a tetramer, with four active sites on its surface. Still, these active sites represent a small fraction of the surface of the enzyme. We will take a stab at bringing in a correction to account for this factor later.

Carrying on, since the hydrogen peroxide is a neutral molecule, $\varphi = 1$. The diffusional rate constant is therefore

$$k_d = 4\pi(6.022\,142 \times 10^{23}\,\mathrm{mol^{-1}})(4.3 \times 10^{-11} + 2 \times 10^{-9}\,\mathrm{m^2\,s^{-1}})(5.82 \times 10^{-9}\,\mathrm{m})$$

$$= 9.0 \times 10^{7}\,\mathrm{m^3\,mol^{-1}\,s^{-1}}$$

$$\equiv 9.0 \times 10^{10}\,\mathrm{L\,mol^{-1}\,s^{-1}}.$$

This estimate of $k_d$ is quite a bit higher than the experimental rate constants for either of the substrate binding steps, but then again we haven't taken into account the issue that not the entire surface of catalase is reactive. The active site in catalase is accessed through a channel whose mouth is about 15 Å across. If we think of this mouth as defining a disc-shaped opening in the surface of the enzyme, the mouth has an area of

$$A_{\mathrm{mouth}} = \pi r^2 = \pi(7.5 \times 10^{-10}\,\mathrm{m})^2 = 1.8 \times 10^{-18}\,\mathrm{m^2}.$$

There are four active sites, each with its own channel, so the surface area of catalase covered by these openings is $7.1 \times 10^{-18}\,\mathrm{m^2}$.

From the diffusion coefficient of catalase, we calculate a radius of the enzyme of

$$R = \frac{k_B T}{6\pi D\eta} = 5.7 \times 10^{-9}\,\mathrm{m}.$$

The surface area of a sphere of this radius is

$$A_{\mathrm{catalase}} = 4\pi R^2 = 4.1 \times 10^{-16}\,\mathrm{m^2}.$$

The fraction of the total surface of the enzyme accounted for by the channel openings is therefore

$$\frac{A_{\mathrm{openings}}}{A_{\mathrm{catalase}}} = \frac{7.1 \times 10^{-18}\,\mathrm{m^2}}{4.1 \times 10^{-16}\,\mathrm{m^2}} = 0.017.$$

We would only expect binding to occur if hydrogen peroxide finds one of the openings, which we would only expect to happen 1.7% of the time. We should therefore reduce our estimate of the diffusion-limited rate constant by this factor:

$$k_{d,\text{eff}} = 0.017(9.0 \times 10^{10}\,\mathrm{L\,mol^{-1}s^{-1}}) = 1.6 \times 10^{9}\,\mathrm{L\,mol^{-1}s^{-1}}.$$

This is still about 100 times larger than $k_2$. We conclude that the binding of hydrogen peroxide to catalase is diffusion-influenced, but not diffusion-limited.

For reactions between small neutral molecules, we can also estimate $K_d$ and $k_{-d}$ by a simple statistical argument, at least for the case in which intermolecular forces are negligible. The concentration of encounter pairs is given by the concentration of A times the probability that a molecule of B is to be found next to A. Suppose that $\mathcal{N}$ is the average number of molecules (of any kind) in direct contact with a molecule of A. This number is called the **coordination number**. If the intermolecular forces between A and B are about the same as the intermolecular forces between A and the solvent, then we would expect the probability that B would replace a solvent molecule in A's coordination shell to be in rough proportion to the relative abundance of B to the solvent. The probability that any particular solvent molecule coordinated to A is replaced by a molecule of B is then [B]/[S] where [S] is the number of moles of solvent in a liter of solution. Since there are $\mathcal{N}$ sites where this replacement can occur, the probability that any of the solvent molecules has been replaced by B is roughly $\mathcal{N}$ times [B]/[S], assuming [B] $\ll$ [S]. This means that

$$[\{AB\}] \approx [A]\mathcal{N}\frac{[B]}{[S]}.$$

The equilibrium constant $K_d$ is defined by

$$K_d = \frac{[\{AB\}]}{[A][B]},$$

so we have

$$K_d \approx \frac{\mathcal{N}}{[S]}.$$

Once we know $K_d$ and $k_d$, we can calculate $k_{-d}$ by

$$k_{-d} = k_d/K_d.$$

**Example 18.4 Equilibrium constant for formation of encounter pairs** Coordination numbers for small molecules in aqueous solution are typically in the range of 6–8. The mole density of water at $25\,^\circ\mathrm{C}$ is $55.33\,\mathrm{mol\,L^{-1}}$. For typical neutral-molecule reactions in water, $K_d$ would therefore be

$$K_d = 8/(55.33\,\mathrm{mol\,L^{-1}}) = 0.14\,\mathrm{L\,mol^{-1}}$$

Note that the equilibrium constant here is a phenomenological equilibrium constant (with units).

**Example 18.5 Estimate of rate constant for dissociation of encounter pairs** Suppose that, for a neutral-molecule reaction in water at $25\,^\circ\mathrm{C}$, $D_A \approx D_B = 2 \times 10^{-9}\,\mathrm{m^2\,s^{-1}}$ and

quinoid form                                        carbinol form

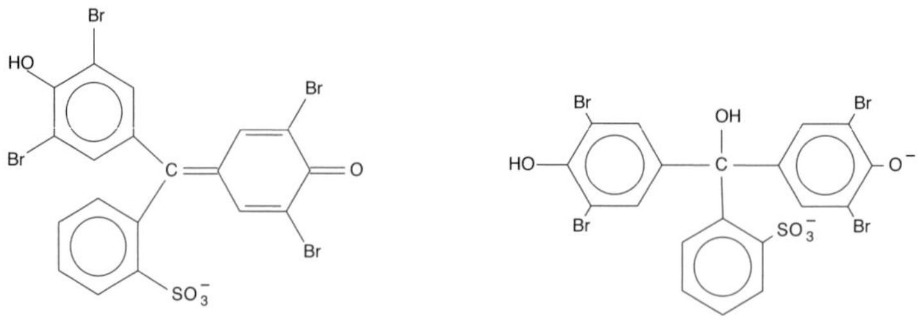

Figure 18.4 The quinoid and carbinol forms of bromphenol blue. The quinoid form is blue while the carbinol is colorless.

the coordination number of both reactants is 8. We can calculate $k_d$ by

$$R_{AB} = \frac{(1.380\,6503 \times 10^{-23}\,\mathrm{J\,K^{-1}})(298.15\,\mathrm{K})}{6\pi(8.91 \times 10^{-4}\,\mathrm{Pa\,s})}\left(\frac{2}{2 \times 10^{-9}\,\mathrm{m^s\,s^{-1}}}\right)$$

$$= 2.45 \times 10^{-10}\,\mathrm{m}.$$

$$\therefore k_d = 4\pi(6.022\,142 \times 10^{23}\,\mathrm{mol^{-1}})(2)(2 \times 10^{-9}\,\mathrm{m^2\,s^{-1}})(2.45 \times 10^{-10}\,\mathrm{m})$$

$$= 7.4 \times 10^{6}\,\mathrm{m^3 mol^{-1}s^{-1}}$$

$$\equiv 7.4 \times 10^{9}\,\mathrm{L\,mol^{-1}s^{-1}}.$$

We calculated $K_d = 0.14\,\mathrm{L\,mol^{-1}}$ in the previous example. Since $K_d = k_d/k_{-d}$, $k_{-d} = 5.3 \times 10^{10}\,\mathrm{s^{-1}}$. This is an extraordinarily large value. Neglecting reaction, it implies that the half-life of the encounter pair is just 13 ps.

**Example 18.6 Estimate of $k_r$** If we can calculate $k_d$ and $k_{-d}$, we can then get $k_r$ from a measured value of $k$ using Equation (18.6). Bromphenol blue (BPB) is a pH indicator that reacts with hydroxide in aqueous solution as follows:

$$\text{quinoid(blue)} + \text{OH}^- \rightarrow \text{carbinol(colorless)}$$

The quinoid and carbinol forms of this indicator are shown in Figure 18.4. At 25 °C, $k = 9.30 \times 10^{-4}\,\mathrm{L\,mol^{-1}s^{-1}}$ for this reaction. We're only going to be able to get ballpark estimates for $k_d$ and $k_{-d}$ since we don't know the coordination number or diffusion coefficient of BPB. Still, it should be enough to let us calculate an order-of-magnitude estimate of $k_r$.

For our two reactants, $Z_A = Z_B = -1$, so $y = 7.449$, which gives $f = 4.338 \times 10^{-3}$. The diffusion coefficient of OH$^-$ is $5.30 \times 10^{-9}\,\mathrm{m^2\,s^{-1}}$. Diffusion coefficients for smallish molecules like BPB are typically around $10^{-9}\,\mathrm{m^2\,s^{-1}}$. As for the reactive distance $R_{AB}$, it might be quite small since the hydroxide has to make a bond with the central carbon. Let's assume this distance is similar to the C–O bond length, which is about 143 pm. We get

$$k_d = 4\pi(6.022\,142 \times 10^{23}\,\text{mol}^{-1})\left[(5.30+1) \times 10^{-9}\,\text{m}^2\,\text{s}^{-1}\right](143 \times 10^{-12}\,\text{m})$$
$$\times (4.338 \times 10^{-3})$$
$$= 3 \times 10^4\,\text{m}^3\,\text{mol}^{-1}\text{s}^{-1}$$
$$\equiv 3 \times 10^7\,\text{L}\,\text{mol}^{-1}\text{s}^{-1}.$$

Since the hydroxide has to react at the central carbon atom, it's the coordination number of this atom that counts, and not the coordination number for the whole molecule. BPB is flat and sterically hindered at this carbon, so the coordination number is likely to be small, perhaps as low as 2 (one above and one below the plane of the central carbon). For the sake of argument, let's use $\mathcal{N} = 2$. This gives $K_d = 4 \times 10^{-2}\,\text{L}\,\text{mol}^{-1}$, from which we can calculate $k_{-d} = k_d/K_d = 8 \times 10^8\,\text{s}^{-1}$. We also have, from Equation (18.6),

$$k_r = \frac{kk_{-d}}{k_d - k} \approx \frac{kk_{-d}}{k_d} = \frac{k}{K_d}$$
$$= \frac{9.30 \times 10^{-4}\,\text{L}\,\text{mol}^{-1}\text{s}^{-1}}{4 \times 10^{-2}\,\text{L}\,\text{mol}^{-1}}$$
$$= 2 \times 10^{-2}\,\text{s}^{-1}.$$

This reaction is strongly activation controlled, as we could also have seen from the relative values of $k$ and $k_d$. The value of $k_r$ we calculated here could now be compared to values obtained from quantum mechanical calculations.

---

### Exercise group 18.2

(1) Diffusion coefficients for large macromolecules such as antibodies are generally of the order of $10^{-11}\,\text{m}^2\,\text{s}^{-1}$. Diffusion coefficients for viruses are much smaller, typically about $10^{-12}\,\text{m}^2\,\text{s}^{-1}$. Assuming that the binding of antibodies to viral particles is diffusion-controlled, estimate the rate constant for this process at $25\,^\circ\text{C}$. The viscosity of water at this temperature is $8.91 \times 10^{-4}\,\text{Pa}\,\text{s}$. Convert your answer to $\text{L}\,\text{mol}^{-1}\text{s}^{-1}$.

(2) Although we usually talk about acid–base chemistry in aqueous solution, we can also study acid–base chemistry in other solvents. Consider the following data for water and methanol at $25\,^\circ\text{C}$.

|  | Water | Methanol |
| --- | --- | --- |
| $\varepsilon_r$ | 78.37 | 32.63 |
| $\eta/\text{mPa}\,\text{s}$ | 0.8904 | 0.547 |

The rate constant for the reaction

$$\text{CH}_3\text{COO}^- + \text{H}^+ \rightarrow \text{CH}_3\text{COOH}$$

has the value $k_{a,w} = 4.5 \times 10^{10}\,\text{L}\,\text{mol}^{-1}\text{s}^{-1}$ in water at $25\,^\circ\text{C}$. Based on the theory of diffusion-limited reactions, what value would you predict for the rate constant in methanol at $25\,^\circ\text{C}$?

*Hints*: Consider the ratio of the rate constants in the two solvents and think about what would be the same or different in the two solvents. When the reactants have opposite charges, the term $e^y$ in the equation for the electrostatic factor $f$ which appears in the diffusion-limited rate constant is very small.

## Key ideas and equations

- $D = k_B T 6\pi r\eta$
- $k = \dfrac{k_d k_r}{k_{-d} + k_r}$
- $k_d = 4\pi L(D_A + D_B)R_{AB}\varphi$ with $\varphi = y/(e^y - 1)$ and $y = \dfrac{Z_A Z_B e^2}{4\pi \varepsilon k_B T R_{AB}}$
- $K_d = k_d/k_{-d} = \mathcal{N}/[S]$

## Review exercise group 18.3

(1) The diffusion coefficient of iodine in $CCl_4$ is $1.5 \times 10^{-9}\ m^2\ s^{-1}$ at 320 K.
  - (a) Making a reasonable assumption, estimate the diffusion coefficient of iodine atoms in $CCl_4$.
  - (b) Iodine atoms can be created in solution by flash photolysis. Recombination of radicals in solution is usually diffusion-limited since the solvent is always present to carry away excess energy liberated in bond formation, and the solvent caging effect tends to keep the radicals together long enough to react, even if they don't react as soon as they first touch. The rate constant at 320 K was found to be $7 \times 10^9\ L\,mol^{-1}s^{-1}$. Assuming that the reaction is diffusion-controlled, calculate the reactive distance. Compare this radius to the bond length of $2.66 \times 10^{-10}\ m$, and comment on the result.

(2) (a) The viscosity of a liquid typically decreases with increasing temperature. This temperature dependence is sometimes described using the Arrhenius-like equation

$$\eta = A_\eta \exp\left(\frac{E_\eta}{RT}\right)$$

Note the lack of a negative sign in the exponential. Using any relevant theories, determine how the activation energy of a diffusion-controlled reaction between neutral molecules is related to the "activation energy" of viscosity.

*Hints*: The size of a solute molecule depends very little on temperature and should be treated as a fixed quantity. To figure out the activation energy, take $d(\ln k)/dT$ both for the Arrhenius equation and for your expression for the rate constant.

  - (b) The viscosity of water is $1.79\ mPa\,s$ at $0\,°C$, and $0.28\ mPa\,s$ at $100\,°C$. What is the value of $E_\eta$ for water?
  - (c) What is the predicted activation energy for a diffusion-controlled reaction at $25\,°C$?

# Appendix A

## Standard thermodynamic properties
## at 298.15 K and 1 bar

The thermodynamic quantities in these tables are appropriate to the molarity-basis (chemists') standard state.

### A.1  Enthalpy, free energy and heat capacity data

| Species | $\dfrac{\Delta_f H^\circ}{\text{kJ mol}^{-1}}$ | $\dfrac{\Delta_f G^\circ}{\text{kJ mol}^{-1}}$ | $\dfrac{C_{p,m}}{\text{J K}^{-1}\text{mol}^{-1}}$ |
|---|---|---|---|
| $Ag^+_{(aq)}$ | 105.79 | 77.11 | |
| $Ag_2O_{(s)}$ | −31.1 | −11.2 | 65.9 |
| $Ag_2S_{(s)}$ | −32.6 | −40.7 | 76.5 |
| $BCl_{3(l)}$ | −427 | −387 | 107.1 |
| $BCl_{3(g)}$ | −403.8 | −388.7 | 62.7 |
| $B_2H_{6(g)}$ | 36 | 86.7 | 56.9 |
| $Ba^{2+}_{(aq)}$ | −537.6 | −560.7 | |
| $BaCO_{3(s)}$ | −1216 | −1138 | 85.4 |
| $Br_{(g)}$ | 111.87 | 82.38 | 20.79 |
| $Br^-_{(aq)}$ | −121.41 | −103.85 | |
| $Br_{2(l)}$ | 0 | 0 | 75.689 |
| $Br_{2(g)}$ | 30.91 | 3.11 | 36.02 |
| $BrF_{3(g)}$ | −255.60 | −229.43 | 66.61 |
| $BrF_{5(g)}$ | −428.8 | −350.6 | 99.62 |
| $C_{(graphite)}$ | 0 | 0 | 8.53 |
| $C_{(diamond)}$ | 1.88 | 2.90 | 6.12 |
| $CH_3COOH_{(l)}$ | −484.5 | −389.9 | 123.43 |
| $CH_3COOH_{(aq)}$ | −485.26 | −396.39 | |
| $CH_3COO^-_{(aq)}$ | −486.01 | −369.31 | |
| $CH_3OH_{(l)}$ | −239.1 | −166.3 | 81.6 |
| $CH_3OH_{(g)}$ | −201.6 | −162.0 | 43.9 |
| $CH_{4(g)}$ | −74.81 | −50.72 | 35.31 |
| $C_2H_{2(g)}$ | 226.73 | 209.20 | 43.9 |
| $C_2H_{4(g)}$ | 52.26 | 68.15 | 43.6 |

(*cont.*)

| Species | $\Delta_f H°$ kJ mol$^{-1}$ | $\Delta_f G°$ kJ mol$^{-1}$ | $C_{p,m}$ J K$^{-1}$mol$^{-1}$ |
|---|---|---|---|
| $C_2H_4O_{(aq)}$ | $-212.34$ | $-139.00$ | |
| $C_2H_5OH_{(aq)}$ | $-276.98$ | $-180.85$ | |
| $C_2H_5OH_{(l)}$ | $-277.69$ | $-174.78$ | 111.5 |
| $C_2H_{6(g)}$ | $-84.68$ | $-32.82$ | 52.6 |
| $C_3H_{8(g)}$ | $-103.85$ | $-23.47$ | 73.6 |
| $C_6H_{6(l)}$ (benzene) | 49.0 | 124.8 | 132 |
| $C_6H_{6(g)}$ (benzene) | 82.9 | 129.8 | 81.6 |
| $C_6H_{12(g)}$ (cyclohexane) | $-123.1$ | 32.0 | 105.3 |
| $C_6H_{12}O_{6(s)}$ ($\alpha$-D-glucose) | $-1274.4$ | $-910.23$ | 218.16 |
| $C_6H_{12}O_{6(aq)}$ ($\alpha$-D-glucose) | $-1263.06$ | $-914.25$ | |
| $C_{12}H_{22}O_{11(aq)}$ ($\alpha$-lactose) | $-2232.37$ | $-1564.36$ | |
| $C_{12}H_{22}O_{11(aq)}$ ($\beta$-lactose) | $-2233.50$ | $-1565.61$ | |
| $C_{12}H_{22}O_{11(aq)}$ (sucrose) | $-2215.85$ | $-1550.89$ | |
| $CO_{(g)}$ | $-110.53$ | $-137.17$ | 29.12 |
| $CO_{2(g)}$ | $-393.51$ | $-394.37$ | 37.1 |
| $CO_{2(aq)}$ | $-413.26$ | $-386.05$ | |
| $CO_{3(aq)}^{2-}$ | $-675.23$ | $-527.90$ | |
| $CaCO_{3(s)}$ | $-1206.9$ | $-1128.8$ | 81.9 |
| $CaO_{(s)}$ | $-634.92$ | $-603.30$ | 42.8 |
| $Cd(OH)_{2(s)}$ | | $-469.8$ | |
| $Cl_{(g)}$ | 121.301 | 105.305 | 21.84 |
| $Cl_{(aq)}^{-}$ | $-167.080$ | $-131.218$ | |
| $ClO_{(g)}$ | 101.22 | 97.48 | 31.54 |
| $Cl_{2(g)}$ | 0 | 0 | 33.91 |
| $Cl_2O_{(g)}$ | 87.88 | 105.10 | 47.50 |
| $Co_{(aq)}^{2+}$ | $-58$ | $-54$ | |
| $Cr_2O_{7(aq)}^{2-}$ | $-1490.3$ | $-1301.1$ | |
| $F_{(aq)}^{-}$ | $-335.35$ | $-281.52$ | |
| $F_{2(g)}$ | 0 | 0 | 31.30 |
| $Fe_{(s)}$ | 0 | 0 | 25.1 |
| $Fe_{(aq)}^{2+}$ | $-90.0$ | $-90.5$ | |
| $FeCl_{2(s)}$ | $-341.6$ | $-302.2$ | 76.7 |
| $FeSO_{4(s)}$ | $-928.4$ | $-820.8$ | 100.6 |
| $HCO_{3(aq)}^{-}$ | $-689.9$ | $-586.8$ | |
| $H_{(g)}$ | 217.998 | 203.276 | 20.79 |
| $H_{2(g)}$ | 0 | 0 | 28.82 |
| $HBr_{(g)}$ | $-36.29$ | $-53.36$ | 29.14 |
| $H_2O_{(l)}$ | $-285.830$ | $-237.140$ | 75.40 |
| $H_2O_{(g)}$ | $-241.826$ | $-228.582$ | 33.58 |
| $H_2S_{(g)}$ | $-20.50$ | $-33.33$ | 34.20 |
| $HPO_{4(aq)}^{2-}$ | $-1299.0$ | $-1096.0$ | |
| $H_2PO_{4(aq)}^{-}$ | $-1302.6$ | $-1137.2$ | |

| Species | $\Delta_f H^\circ$ kJ mol$^{-1}$ | $\Delta_f G^\circ$ kJ mol$^{-1}$ | $C_{p,m}$ J K$^{-1}$mol$^{-1}$ |
|---|---|---|---|
| $Hg_{(l)}$ | 0 | 0 | 27.98 |
| $Hg_{(g)}$ | 61.38 | 31.84 | 20.79 |
| $Hg_2Cl_{2(s)}$ | −265.37 | −210.72 | 101.7 |
| $HgS_{(s)}$ | −58.2 | −50.6 | 48.4 |
| $I^-_{(aq)}$ | −56.78 | −51.72 | |
| $K^+_{(aq)}$ | −252.14 | −282.51 | |
| $KBr_{(s)}$ | −393.8 | −380.1 | 53.6 |
| $Mg^{2+}_{(aq)}$ | −467.0 | −455.4 | |
| $MgCl_2 \cdot 6H_2O_{(s)}$ | −2499.0 | −2114.7 | 315.0 |
| $MnO_{2(s)}$ | −520.0 | −465.2 | 54.1 |
| $Mn_2O_{3(s)}$ | −959 | −881 | 107.7 |
| $N_{2(g)}$ | 0 | 0 | 28.87 |
| $NH_{3(aq)}$ | −80.29 | −26.50 | |
| $NH^+_{4(aq)}$ | −133.26 | −81.19 | |
| $NH_4NO_{3(s)}$ | −365.56 | −183.87 | 84.1 |
| $NO_{2(g)}$ | 33.2 | 51.32 | 37.2 |
| $N_2O_{4(g)}$ | 9.16 | 97.89 | 77.3 |
| $Na_{(s)}$ | 0 | 0 | 28.2 |
| $Na_{(g)}$ | 107.5 | 77.0 | 20.8 |
| $Na^+_{(aq)}$ | −240.34 | −261.95 | |
| $NaCl_{(s)}$ | −411.15 | −384.14 | 49.8 |
| $NaH_{(s)}$ | −56.44 | 36.37 | |
| $NaOH_{(s)}$ | −425.6 | −379.7 | 59.5 |
| $Na_2O_{(s)}$ | −417.98 | −379.18 | 69.09 |
| $Ni^{2+}_{(aq)}$ | −54 | −46 | |
| $NiCl_{2(s)}$ | −305.3 | −259.03 | 71.7 |
| $Ni(OH)_{2(s)}$ | | −444 | |
| $O_{2(g)}$ | 0 | 0 | 29.35 |
| $O_{2(aq)}$ | −12.09 | 16.35 | |
| $O_{3(g)}$ | 142.67 | 163.19 | 39.22 |
| $OH^-_{(aq)}$ | −230.015 | −157.220 | |
| $Ni^{2+}_{(aq)}$ | −54 | −46 | |
| $PCl_{3(g)}$ | −287.0 | −267.8 | 71.8 |
| $PCl_{5(g)}$ | −374.9 | −305.0 | 112.8 |
| $Pb_{(s)}$ | 0 | 0 | 26.4 |
| $Pb_{(g)}$ | 195.2 | 162.2 | 20.79 |
| $Pb^{2+}_{(aq)}$ | 0.92 | −24.24 | |
| $PbBr_{2(s)}$ | −279 | −261.9 | 80.1 |
| $PbCl_{2(s)}$ | −359.4 | −314.2 | |
| $PbI_{2(s)}$ | −175.39 | −173.57 | 77.4 |
| $PbSO_{4(s)}$ | −919.97 | −813.04 | |

(*cont.*)

| Species | $\Delta_f H^\circ$ kJ mol$^{-1}$ | $\Delta_f G^\circ$ kJ mol$^{-1}$ | $C_{p,m}$ J K$^{-1}$mol$^{-1}$ |
|---|---|---|---|
| $S^{2-}_{(aq)}$ | 30 | 79 | |
| $SF_{6(g)}$ | −1220.47 | −1116.44 | 96.88 |
| $SO_{2(g)}$ | −296.81 | −300.09 | 39.8 |
| $SO_{3(g)}$ | −395.7 | −371.1 | 50.67 |
| $SO^{2-}_{4(aq)}$ | −909.34 | −744.00 | |
| $Si_{(s)}$ | 0 | 0 | 20.1 |
| $SrSO_{4(s)}$ | −1453 | −1341 | |
| $UF_{6(g)}$ | −2112.9 | −2029.1 | |
| $U_3O_{8(s)}$ | −3574.8 | −3369.5 | |
| $Zn^{2+}_{(aq)}$ | −153.39 | −111.62 | |
| $ZnF_{2(s)}$ | −764 | −713.4 | 65.6 |
| $ZnO_{(s)}$ | −350.46 | −320.48 | 40.3 |

## A.2 Standard entropies

| Species | $S^\circ$ J K$^{-1}$mol$^{-1}$ |
|---|---|
| $C_{(s)}$ (graphite) | 5.74 |
| $C_2H_5OH_{(l)}$ | 160.67 |
| $C_5H_4N_4O_{2(s)}$ (xanthine) | 161.1 |
| $C_6H_{12}O_{6(aq)}$ ($\alpha$-D-glucose) | 264.01 |
| $CO_{2(g)}$ | 213.785 |
| $CO^{2-}_{3(aq)}$ | −50.0 |
| $Ca_{(s)}$ | 41.6 |
| $CaCl_{2(s)}$ | 104.62 |
| $Cl_{2(g)}$ | 223.081 |
| $H_{2(g)}$ | 130.680 |
| $H_2O_{(l)}$ | 69.95 |
| $N_{2(g)}$ | 191.609 |
| $O_{2(g)}$ | 205.152 |
| $O_{2(aq)}$ | 110.88 |
| $O_{3(g)}$ | 238.92 |
| $U_{(s)}$ | 50.20 |
| $UO^{2+}_{2(aq)}$ | −98.2 |

# Appendix B

## Standard reduction potentials at 298.15 K and 1 bar

| Reduction process | $E^\circ/V$ |
|---|---|
| $O_{3(g)} + 2H^+_{(aq)} + 2e^- \rightarrow O_{2(g)} + H_2O_{(l)}$ | +2.07 |
| $PbO_{2(s)} + 4H^+_{(aq)} + SO^{2-}_{4(aq)} + 2e^- \rightarrow PbSO_{4(s)} + 2H_2O_{(l)}$ | +1.68 |
| $Au^{3+}_{(aq)} + 3e^- \rightarrow Au_{(s)}$ | +1.498 |
| $Cr_2O^{2-}_{7(aq)} + 14H^+_{(aq)} + 6e^- \rightarrow 2Cr^{3+}_{(aq)} + 7\,H_2O_{(l)}$ | +1.33 |
| $O_{2(g)} + 4H^+_{(aq)} + 4e^- \rightarrow 2H_2O_{(l)}$ | +1.229 |
| $Br_{2(aq)} + 2e^- \rightarrow 2Br^-_{(aq)}$ | +1.0873 |
| $NO^-_{3(aq)} + 4H^+_{(aq)} + 3e^- \rightarrow NO_{(g)} + 2H_2O_{(l)}$ | +0.96 |
| $NO^-_{3(aq)} + 3H^+_{(aq)} + 2e^- \rightarrow HNO_{2(aq)} + H_2O_{(l)}$ | +0.934 |
| $Hg^{2+}_{(aq)} + 2e^- \rightarrow Hg_{(l)}$ | +0.851 |
| $Ag^+_{(aq)} + e^- \rightarrow Ag_{(s)}$ | +0.7996 |
| $Hg^{2+}_{2(aq)} + 2e^- \rightarrow 2Hg_{(l)}$ | +0.7973 |
| $O_{2(g)} + 2H_2O_{(l)} + 4e^- \rightarrow 4OH^-_{(aq)},$ | +0.401 |
| $Cu^{2+}_{(aq)} + 2e^- \rightarrow Cu_{(s)}$ | +0.3419 |
| $AgCl_{(s)} + e^- \rightarrow Ag_{(s)} + Cl^-_{(aq)}$ | +0.222 33 |
| $H^+_{(aq)} + e^- \rightarrow \frac{1}{2}H_{2(g)}$ | 0 |
| $AgI_{(s)} + e^- \rightarrow Ag_{(s)} + I^-_{(aq)}$ | −0.1518 |
| $PbSO_{4(s)} + 2e^- \rightarrow Pb_{(s)} + SO^{2-}_{4(aq)}$ | −0.361 |
| $Cd^{2+}_{(aq)} + 2e^- \rightarrow Cd_{(s)}$ | −0.4030 |
| $NiO_{2(s)} + 2H_2O_{(l)} + 2e^- \rightarrow Ni(OH)_{2(s)} + 2OH^-_{(aq)}$ | −0.490 |
| $Zn^{2+}_{(aq)} + 2e^- \rightarrow Zn_{(s)}$ | −0.7618 |
| $Cd(OH)_{2(s)} + 2e^- \rightarrow Cd_{(s)} + 2OH^-_{(aq)}$ | −0.809 |
| $Cr^{2+}_{(aq)} + 2e^- \rightarrow Cr_{(s)}$ | −0.91 |
| $Li^+_{(aq)} + e^- \rightarrow Li_{(s)}$ | −3.0401 |

# Appendix C

## Physical properties of water

**Ice at 0 °C**

Density $= 0.915\,\text{g cm}^{-3}$
Vapor pressure $= 4.579\,\text{torr}$
Heat of fusion $= 333.4\,\text{J g}^{-1} = 6.007\,\text{kJ mol}^{-1}$
Absolute molar entropy $= 41.0\,\text{J K}^{-1}\text{mol}^{-1}$
Specific heat capacity $= 2.113\,\text{J K}^{-1}\text{g}^{-1}$
Molar heat capacity $= 38.07\,\text{J K}^{-1}\text{mol}^{-1}$

**Liquid water**

Specific heat capacity $= 4.184\,\text{J K}^{-1}\text{g}^{-1}$
Molar heat capacity $= 75.37\,\text{J K}^{-1}\text{mol}^{-1}$
*See also next page.*

**Steam at 100 °C**

Density $= 5.880 \times 10^{-4}\,\text{g cm}^{-3}$
Heat of vaporization $= 2257\,\text{J/g} = 40.66\,\text{kJ mol}^{-1}$
Absolute molar entropy at 1 bar $= 196.3\,\text{J K}^{-1}\text{mol}^{-1}$
Molar heat capacity at constant pressure $= 33.76\,\text{J K}^{-1}\text{mol}^{-1}$

Table C.1 *Properties of liquid water as a function of temperature*

| Temperature °C | Density $\text{g cm}^{-3}$ | Mole density $\text{mol L}^{-1}$ | Vapor pressure torr | $\Delta_{vap}h°$ $\text{J g}^{-1}$ | $S°$ $\text{J K}^{-1}\text{mol}^{-1}$ | $\varepsilon_r$ |
|---|---|---|---|---|---|---|
| 0 | 0.9999 | 55.49 | 4.579 | 2501 | 63.2 | 87.85 |
| 4 | 1.0000 | 55.49 | 6.101 | 2492 | 64.3 | 86.26 |
| 20 | 0.9982 | 55.39 | 17.535 | 2454 | 68.6 | 80.18 |
| 25 | 0.9971 | 55.33 | 23.756 | 2443 | 69.95 | 78.37 |
| 40 | 0.9922 | 55.06 | 55.324 | 2408 | 73.7 | 73.18 |
| 60 | 0.9832 | 54.56 | 149.38 | 2362 | 78.3 | 66.78 |
| 80 | 0.9718 | 53.93 | 355.1 | 2316 | 82.7 | 60.95 |
| 100 | 0.9584 | 53.19 | 760.000 | 2269 | 86.9 | 55.63 |

# Appendix D

## The SI system of units

The following is a table of some of the SI units which you will find useful during your study of physical chemistry. This list is by no means exhaustive.

| Observable | SI unit | Abbreviation | Definition |
|---|---|---|---|
| Activity | becquerel | Bq | $1\,\text{Bq} = 1\ \text{event s}^{-1}$ |
| Charge | coulomb | C | |
| Electric potential | volt | V | |
| Energy | joule | J | |
| Force | newton | N | |
| Frequency | hertz | Hz | $1\,\text{Hz} = 1\ \text{cycle s}^{-1}$ |
| Length | meter | m | |
| Mass | kilogram | kg | |
| Number | mole | mol | |
| Power | watt | W | $1\,\text{W} = 1\,\text{J s}^{-1}$ |
| Pressure | pascal | Pa | |
| Temperature | kelvin | K | |
| Time | second | s | |
| Volume | cubic meter | $\text{m}^3$ | |

SI units are often used with prefixes. Here is a list of the more commonly used ones:

| Prefix | Name | Value |
|---|---|---|
| f | femto | $10^{-15}$ |
| p | pico | $10^{-12}$ |
| n | nano | $10^{-9}$ |
| μ | micro | $10^{-6}$ |
| m | milli | $10^{-3}$ |
| c | centi | $10^{-2}$ |
| k | kilo | $10^{3}$ |
| M | mega | $10^{6}$ |
| G | giga | $10^{9}$ |

## D.1 Calculations in SI units

Physical chemistry is a quantitative science. As a result, there will be a lot of calculations to be carried out during your study of the subject. I find that it helps a lot to learn to work with the SI system of units and to try to be consistent about it. There will be times where it's more convenient to use another system, but when in doubt, use SI units. Why? The SI system has one great advantage over most other systems of units: it's consistent. This means that if you use SI units consistently, your calculations will always give answers in appropriate SI units.

Take the ideal gas law. Suppose that you are given the volume, number of moles and temperature of a system. Then you can calculate the pressure by $p = nRT/V$. If you use moles for $n$, express the temperature in kelvins and the volume in m$^3$, and use the SI value of $R$, the answer comes out in the SI units of pressure, namely pascals. There's no messing around with unit analysis and no need to learn several different values of $R$. If you are given data in non-SI units, convert them before doing the calculation. Once you get used to doing things this way, you will probably find that you make fewer errors in handling units.

**Example D.1** Here's a typical first-year ideal gas question, solved in detail here to show you my suggested problem-solving technique. Suppose you are asked to calculate the pressure of 8 mol of argon in a 3 L flask at 870 °C. First, convert all data to SI units using the conversion factors in Appendix E.

$$V = \frac{3\,\text{L}}{1000\,\text{L}\,\text{m}^{-3}} = 0.003\,\text{m}^3$$
$$T = 870 + 273.15\,\text{K} = 1143\,\text{K}$$

Using the SI value of the ideal gas constant (Appendix E), we get

$$p = \frac{(8\,\text{mol})(8.314\,472\,\text{J}\,\text{K}^{-1}\text{mol}^{-1})(1143\,\text{K})}{0.003\,\text{m}^3}$$
$$= 2.5 \times 10^7\,\text{Pa}.$$

How do we know that the answer comes out in pascals? We used SI units consistently in the calculation so the answer comes out in the SI unit of pressure, namely pascals. If we need the answer in some other units, we can then convert it to whatever representation is wanted.

One persistent source of confusion when doing calculations in SI units relates to the unit of mass. It is tempting to think that the gram is the SI unit of mass, but it isn't.

> **The kg is the SI unit of mass.**

When doing calculations in SI units involving mass, it is therefore necessary to use kilograms. In some problems we will encounter in this book, this sometimes means writing molar masses in the unusual units of kg mol$^{-1}$. For now, let's just have a look at a simple and, hopefully, familiar example from your physics course:

**Example D.2** What is the kinetic energy of a 145 g baseball traveling at 145 km h$^{-1}$?

I think that most students who have taken physics would notice that they need to convert the speed to the SI units of m s$^{-1}$ because we need to strip the SI prefix from km h$^{-1}$, and the SI unit of time is the second. What many of us would miss is that we have to convert the mass of the baseball to kg because the latter is the SI unit of mass, not the gram:

$$m = \frac{145\,\text{g}}{1000\,\text{g/kg}} = 0.145\,\text{kg}$$

$$v = \frac{145 \times 10^3\,\text{m}\,\text{h}^{-1}}{3600\,\text{s}\,\text{h}^{-1}} = 40.3\,\text{m}\,\text{s}^{-1}.$$

$$\therefore K = \frac{1}{2}mv^2 = \frac{1}{2}(0.145\,\text{kg})(40.3\,\text{m}\,\text{s}^{-1})^2 = 118\,\text{J}.$$

# Appendix E

## Universal constants and conversion factors

| Constant | Symbol | Value[a] | Units |
|---|---|---|---|
| Avogadro's constant | $L$ | $6.022\,141\,79 \times 10^{23}$ | $mol^{-1}$ |
| Boltzmann's constant | $k_B$ | $1.380\,6503 \times 10^{-23}$ | $J\,K^{-1}$ |
| Electron mass | $m_e$ | $9.109\,3819 \times 10^{-31}$ | kg |
| Elementary charge | $e$ | $1.602\,176\,46 \times 10^{-19}$ | C |
| Faraday's constant | $F$ | $96\,485.342$ | $C\,mol^{-1}$ |
| Ideal gas constant[b] | $R$ | $8.314\,472$ | $J\,K^{-1}mol^{-1}$ |
| Neutron mass | $m_n$ | $1.674\,927\,16 \times 10^{-27}$ | kg |
| Permittivity of the vacuum[c] | $\epsilon_0$ | $8.854\,187\,817 \times 10^{-12}$ | $C^2J^{-1}m^{-1}$ |
| Planck's constant | $h$ | $6.626\,0688 \times 10^{-34}$ | $J\,Hz^{-1}$ |
| | $\hbar$ | $1.054\,571\,68 \times 10^{-34}$ | $J\,s$ |
| Proton mass | $m_p$ | $1.672\,621\,58 \times 10^{-27}$ | kg |
| Speed of light in vacuum[d] | $c$ | $2.997\,924\,58 \times 10^{8}$ | $m\,s^{-1}$ |
| Standard gravity[e] | $g$ | $9.806\,65$ | $m\,s^{-2}$ |

[a] All values in this table are from the NIST web site:
http://physics.nist.gov/cuu/Constants/index.html.
[b] Also known as the "molar gas constant."
[c] Also known as the "electric constant." The value of this constant is fixed in the SI system.
[d] In the SI system of units, the values of $c$ is fixed.
[e] The standard acceleration due to gravity (standard gravity for short) is a value fixed by convention.

*Conversion factors*

**Length**

$1\ \text{Å} = 10^{-10}\ \text{m}$

**Volume**

$1\ cm^3 = 1\ \text{mL}$

$1\ m^3 = 1000\ \text{L}$

**Mass**

| | | | |
|---|---|---|---|
| 1 | amu $=$ | 1 | $g\,mol^{-1}$ |
| 1 | amu $=$ | $1.660\,538\,78 \times 10^{-27}$ | kg |
| 1 | Da $=$ | 1 | $g\,mol^{-1}$ |
| 1 | t $=$ | 1000 | kg |

**Pressure**

| | | | |
|---|---|---|---|
| 1 | atm $=$ | 760 | torr or mmHg |
| 1 | atm $=$ | 101 325 | Pa |
| 1 | bar $=$ | 100 000 | Pa |

**Energy**

| | | | |
|---|---|---|---|
| 1 | cal $=$ | 4.184 | J |
| 1 | eV $=$ | $1.602\,176\,46 \times 10^{-19}$ | J |

**Time**

| | | | |
|---|---|---|---|
| 1 | y $=$ | 365.25 | d |

**Temperature**

To convert degrees Celsius to kelvins, add 273.15.

# Appendix F

## Periodic table of the elements, with molar masses

| 1 | | | | | | | | | | | | | | | | | 18 |
|---|---|---|---|---|---|---|---|---|---|---|---|---|---|---|---|---|---|
| 1 H<br>1.01 | 2 | | | | | | | | | | | 13 | 14 | 15 | 16 | 17 | 2 He<br>4.00 |
| 3 Li<br>6.94 | 4 Be<br>9.01 | | | | | | | | | | | 5 B<br>10.81 | 6 C<br>12.01 | 7 N<br>14.01 | 8 O<br>16.00 | 9 F<br>19.00 | 10 Ne<br>20.18 |
| 11 Na<br>22.99 | 12 Mg<br>24.31 | 3 | 4 | 5 | 6 | 7 | 8 | 9 | 10 | 11 | 12 | 13 Al<br>26.98 | 14 Si<br>28.09 | 15 P<br>30.97 | 16 S<br>32.07 | 17 Cl<br>35.45 | 18 Ar<br>39.95 |
| 19 K<br>39.10 | 20 Ca<br>40.08 | 21 Sc<br>44.96 | 22 Ti<br>47.88 | 23 V<br>50.94 | 24 Cr<br>52.00 | 25 Mn<br>54.94 | 26 Fe<br>55.85 | 27 Co<br>58.93 | 28 Ni<br>58.69 | 29 Cu<br>63.55 | 30 Zn<br>65.41 | 31 Ga<br>69.72 | 32 Ge<br>72.61 | 33 As<br>74.92 | 34 Se<br>78.96 | 35 Br<br>79.90 | 36 Kr<br>83.80 |
| 37 Rb<br>85.47 | 38 Sr<br>87.62 | 39 Y<br>88.91 | 40 Zr<br>91.22 | 41 Nb<br>92.91 | 42 Mo<br>95.94 | 43 Tc | 44 Ru<br>101.07 | 45 Rh<br>102.91 | 46 Pd<br>106.42 | 47 Ag<br>107.87 | 48 Cd<br>112.41 | 49 In<br>114.82 | 50 Sn<br>118.71 | 51 Sb<br>121.76 | 52 Te<br>127.60 | 53 I<br>126.90 | 54 Xe<br>131.29 |
| 55 Cs<br>132.91 | 56 Ba<br>137.33 | 57 La<br>138.91 | 72 Hf<br>178.49 | 73 Ta<br>180.95 | 74 W<br>183.85 | 75 Re<br>186.21 | 76 Os<br>190.2 | 77 Ir<br>192.22 | 78 Pt<br>195.08 | 79 Au<br>196.97 | 80 Hg<br>200.59 | 81 Tl<br>204.38 | 82 Pb<br>207.2 | 83 Bi<br>208.98 | 84 Po | 85 At | 86 Rn |
| 87 Fr | 88 Ra | 89 Ac | 104 Rf | 105 Db | 106 Sg | 107 Bh | 108 Hs | 109 Mt | 110 Ds | 111 Rg | | | | | | | |

| 58 Ce<br>140.12 | 59 Pr<br>140.91 | 60 Nd<br>144.24 | 61 Pm | 62 Sm<br>150.36 | 63 Eu<br>151.97 | 64 Gd<br>157.25 | 65 Tb<br>158.93 | 66 Dy<br>162.50 | 67 Ho<br>164.93 | 68 Er<br>167.26 | 69 Tm<br>168.93 | 70 Yb<br>173.04 | 71 Lu<br>174.97 |
|---|---|---|---|---|---|---|---|---|---|---|---|---|---|
| 90 Th<br>232.04 | 91 Pa<br>231.04 | 92 U<br>238.03 | 93 Np | 94 Pu | 95 Am | 96 Cm | 97 Bk | 98 Cf | 99 Es | 100 Fm | 101 Md | 102 No | 103 Lr |

# Appendix G

## Selected isotopic masses and abundances

| Isotope | Mass/amu | Natural abundance/% |
|---------|----------|---------------------|
| $^{1}$H | 1.007 825 032 | 99.9885 |
| $^{2}$H | 2.014 101 778 | 0.0115 |
| $^{12}$C | 12 | 98.93 |
| $^{13}$C | 13.003 354 837 8 | 1.07 |
| $^{16}$O | 15.994 914 622 | 99.757 |
| $^{17}$O | 16.999 131 70 | 0.038 |
| $^{18}$O | 17.999 161 0 | 0.205 |
| $^{35}$Cl | 34.968 852 71 | 75.76 |
| $^{37}$Cl | 36.965 902 60 | 24.24 |

# Appendix H

## Properties of exponential and logarithmic functions

(1) $a^0 = 1$

(2) $a^x a^y = a^{x+y}$

(3) $a^x / a^y = a^{x-y}$

(4) $a^{-x} = 1/a^x$

(5) $(a^x)^y = a^{xy}$

(6) $a^{1/x} = \sqrt[x]{a}$

(7) $(ab)^x = a^x b^x$

(8) Logarithms are inverse functions of exponentials:
- $\log_a a^x = x$
- $a^{\log_a x} = x$

(9) • log is usually a shorthand notation for $\log_{10}$.[1] I generally prefer to explicitly write down the base, but feel free to use the shorthand if you prefer.

- ln is called the "natural logarithm." ln $x$ means the same thing as $\log_e x$, where $e = 2.718\,28\ldots$ is Napier's number. This very important logarithm comes up in many equations in chemistry. For our purposes, you mostly need to remember that it's just a logarithm that behaves like all other logarithms.

(10) $\log_a 1 = 0$

(11) $\log_a(xy) = \log_a x + \log_a y$

(12) $\log_a(x/y) = \log_a x - \log_a y$

(13) $\log_a(1/x) = -\log_a x$

(14) $\log_a(x^y) = y \log_a x$

(15) Change of base: $\log_a x = \log_a b \log_b x$

---

[1] One of the reasons that I like to explicitly write down the base is that some people, mostly mathematicians, use log for the natural logarithm rather than the base-10 logarithm. Writing down the base avoids any possible confusion.

# Appendix I

## Review of integral calculus

A student needs remarkably little calculus to get by in physical chemistry. Mainly, you will need to be able to take a few derivatives and to do some simple integrals. Hopefully, you remember how to take derivatives. Integration is a little more difficult, but the following two integrals are the only ones you really need to know in this course:

$$\frac{d}{dx} x^n = nx^{n-1} \qquad \Longleftrightarrow \qquad \int x^n \, dx = \frac{x^{n+1}}{n+1} \qquad \text{if } n \neq -1 \qquad (\text{I.1a})$$

$$\frac{d}{dx} \ln x = \frac{1}{x} \qquad \Longleftrightarrow \qquad \int \frac{dx}{x} = \ln x \qquad (\text{I.1b})$$

You will from time to time need to do more complicated integrals, but these can be handled in one of two ways: either they can be reduced to one of the above by a simple substitution, or you should use a table of integrals such as the one given in Section I.1. The latter is a personal opinion: it's good to know all the fancy integration techniques in case you run into an integral that *isn't* in your table of integrals, but most integrals you're likely to run into *have* been done already, and you're much less likely to make errors using a table of integrals than going through the tedious process of computing an integral from scratch.

Let's now look at examples of both procedures:

**Example I.1** Suppose that we want to obtain the integral

$$I = \int \frac{dx}{ax + b}$$

where $a$ and $b$ are constants. This looks a bit like integral (I.1b), except that we have $ax + b$ appearing where we have $x$ alone in Equation (I.1b). When we see a simple linear term replacing a variable in an integral, we can use a simple substitution: let

$$u = ax + b.$$

Then, since $a$ and $b$ are constants,[1]

$$du = a \, dx$$

---

[1] You may recognize this step as a manipulation of differentials, first seen in Chapter 5.

or

$$dx = \frac{1}{a}du.$$

We can now substitute for $x$ and $dx$ in the integral:

$$I = \int \frac{\frac{1}{a}du}{u} = \frac{1}{a}\int \frac{du}{u} = \frac{1}{a}\ln u = \frac{1}{a}\ln(ax+b).$$

**Example I.2**  Suppose that we want to evaluate the integral

$$I = \int \frac{da}{(k_1 a + k_2)(k_3 a + k_4 b_0)}$$

where each of the $k_i$s is a constant, as is $b_0$. This isn't one of the elementary cases, but it looks like integral 5 from Section I.1. In fact, it's exactly the same if we make the substitutions

$$x \leftarrow a \qquad\qquad\qquad a \leftarrow k_2$$
$$b \leftarrow k_1 \qquad\qquad\qquad c \leftarrow k_4 b_0$$
$$e \leftarrow k_3$$

We can substitute these symbols directly into the answer given in the table of integrals:

$$I = \frac{1}{k_2 k_3 - k_1 k_4 b_0} \ln\left(\frac{k_3 a + k_4 b_0}{k_1 a + k_2}\right)$$

Not every integral you will encounter in this course will be quite as simple as these, but you won't need any techniques more advanced than these to solve them.

The only other wrinkle we need to deal with is that we almost always have to compute a *definite* integral rather than an antiderivative. The rule is simple: if $F(x)$ is an antiderivative of $f(x)$, i.e. if $F(x) = \int f(x)\,dx$, then

$$\int_a^b f(x)\,dx = F(x)|_b^a = F(a) - F(b).$$

Note that this is *not* in general equal to $F(a - b)$.

**Example I.3**  Suppose that we want to calculate

$$I = \int_{a_0}^a \frac{da'}{(k_1 a' + k_2)(k_3 a' + k_4 b_0)}.$$

Give or take a very minor change of notation, we have already found the antiderivative in Example I.2. Thus,

$$
\begin{aligned}
I &= \frac{1}{k_2 k_3 - k_1 k_4 b_0} \ln\left(\frac{k_3 a' + k_4 b_0}{k_1 a' + k_2}\right)\Bigg|_{a_0}^{a} \\
&= \frac{1}{k_2 k_3 - k_1 k_4 b_0}\left[\ln\left(\frac{k_3 a + k_4 b_0}{k_1 a + k_2}\right) - \ln\left(\frac{k_3 a_0 + k_4 b_0}{k_1 a_0 + k_2}\right)\right] \\
&= \frac{1}{k_2 k_3 - k_1 k_4 b_0} \ln\left(\frac{(k_3 a + k_4 b_0)(k_1 a_0 + k_2)}{(k_1 a + k_2)(k_3 a_0 + k_4 b_0)}\right).
\end{aligned}
$$

The last equality was obtained using the rule $\ln x - \ln y = \ln(x/y)$. If you're a bit rusty with your logarithms and exponents, you might want to look at Appendix H which summarizes the key properties of these functions.

## I.1 Table of integrals

This is not by any means a complete table of integrals. However, the integrals given here should be sufficient to work through the examples and exercises in this book.

(1) $\displaystyle\int \exp(ax)\,dx = \frac{1}{a}\exp(ax)$

(2) $\displaystyle\int x^n\,dx = \frac{1}{n+1}x^{n+1}$

(3) $\displaystyle\int \frac{dx}{x} = \ln x$

(4) $\displaystyle\int \ln(a/V)\,dV = V\ln(a/V) + V$

(5) $\displaystyle\int \frac{dx}{(a+bx)(c+ex)} = \frac{1}{ae-bc}\ln\left(\frac{c+ex}{a+bx}\right)$

(6) $\displaystyle\int \frac{dx}{x^2(a+bx)} = \frac{b}{a^2}\ln\left(\frac{a+bx}{x}\right) - \frac{1}{ax}$

(7) $\displaystyle\int \frac{dx}{x(a+bx)^2} = \frac{1}{a^2}\ln\left(\frac{x}{a+bx}\right) + \frac{1}{a(a+bx)}$

# Appendix J

## End-of-term review problems

OK, so you've reached the end of term, and you've answered all the problems in the book. What do you do to study for the final exam? Try these problems! They're a mixed bag of problems intended to help you review all the material. They appear in no particular order, so you'll have to think about what these questions relate to and what sections you'll have to review to answer them. The answers are available online at www.cambridge.org/roussel.

(1) If we say, for a particular process on a particular system, that $w = -5\,J$, what does that mean?

(2) Throughout the textbook, when discussing thermodynamics, I avoided the commonly used word "spontaneous," preferring the phrase "thermodynamically allowed." Why?

(3) Explain why the law of microscopic reversibility is required to make kinetics consistent with thermodynamics.

(4) For each of the following statements, indicate whether it is true or false and, **in a few words**, explain your reasoning.

   (a) The dual nature of light is implied by the equation $E = h\nu$.

   (b) The probability that a molecule occupies a certain energy level only depends on the energy (relative to the ground state) and on the temperature.

   (c) A bomb calorimeter measures $\Delta_r H$.

   (d) Exothermic reactions are always thermodynamically allowed.

   (e) According to Debye–Hückel theory, the activity coefficient of a solute depends on its charge, so all uncharged solutes have an activity coefficient of 1.

   (f) The rate of a reaction with overall stoichiometry $A \rightarrow 2B$ is $\frac{1}{2} \frac{d[B]}{dt}$.

   (g) The gas-phase reaction $CH_3N = NCH_3 \rightarrow C_2H_6 + N_2$ is not elementary.

   (h) The activation energy of an elementary reaction can't be negative.

   (i) The encounter pair in the theory of bimolecular reactions in solution is a transition state for the reaction.

   (j) $\Delta_f \bar{H}^\circ = 0$ for an aqueous proton.

   (k) $\Delta S > 0$ for any spontaneous process.

   (l) It is impossible for a system to reach absolute zero.

   (m) The Clausius inequality is equivalent to the Second Law.

   (n) If a reaction is not thermodynamically allowed, we can make it go by adding a catalyst.

   (o) Since the overall reaction is combustion in either case, a fuel cell is no more efficient than a heat engine.

   (p) If, for a certain reaction, $E^\circ > 0$, then that reaction is thermodynamically allowed.

   (q) The reaction $Fe(CO)_5 + OH^- \rightarrow (CO)_4FeH^- + CO_2$ is probably not elementary.

(5) Briefly define or explain each of the following concepts:

   (a) Molar absorption coefficient

   (b) Zero-point energy

  (c) The First Law of Thermodynamics
  (d) Coupled reactions
  (e) Indirect calorimetry
  (f) Microstate
  (g) Third Law of Thermodynamics
  (h) Vapor pressure

(6) Which of the following formulas only apply to photons? Which only apply to ordinary particles? Which apply to both? (a) $E = h\nu$, (b) $p = h/\lambda$, (c) $E = cp$, (d) $E = \frac{1}{2}mv^2$.

(7) What do the following symbols in the Nernst equation represent? Be specific.
  (a) $E$, (b) $E^\circ$, $\nu_e$.

(8) Classify each of the following reactions as possibly, probably not or certainly not elementary. Explain your reasoning briefly.
  (a) $O_{2(g)} + CH_{2(g)} \rightarrow CO_{(g)} + H_2O_{(g)}$
  (b) $OH^-_{(aq)} + NO_{(aq)} \rightarrow NO^-_{2(aq)} + H^+_{(aq)}$

(9) Equilibrium constants are related to rate constants for the underlying elementary processes. Rate constants always increase with increasing temperature, while equilibrium constants can increase or decrease with temperature. Why is this possible?

(10) The Helmholtz free energy can be used to obtain two different pieces of information about a process. What are they?

(11) What is wrong with the following statement?

  $\Delta G$ gives the maximum work which a system can perform at constant temperature and pressure.

(12) Consider the following argument:

  For an endothermic process, $dq > 0$. According to the Clausius inequality, $dS \geq dq/T > 0$, so the entropy change for an endothermic process is always positive. Therefore endothermic processes are always thermodynamically allowed.

  Do you agree with the argument? If so, can we use similar reasoning to say something about the spontaneity of exothermic processes? If not, explain where the argument goes wrong.

(13) Many reactions involve water as a reactant or product. When treating the thermodynamics of such reactions, we generally ignore the activity of the water, i.e. set its value to unity. Why is it often reasonable to do this?

(14) What is an initial rate experiment? Briefly discuss the advantages and disadvantages of initial rate experiments.

(15) Explain why the following statements are true. Consider two energy levels with $E_1 < E_2$. If the two levels have the same degeneracy, then their populations only tend to equalize at very high temperatures. If $g_1 > g_2$, then there will always be more molecules in energy level 1 than in energy level 2. If, on the other hand, $g_1 < g_2$, there will be a temperature at which the two populations are equal.

(16) Antioxidants are molecules that react with strong oxidizing agents generated by metabolism in cells, leaving harmless products. One of the simplest and most common antioxidants is ascorbic acid ($C_6H_8O_6$), also known as vitamin C. Ascorbic acid can react with strong oxidizing agents like hydroxyl radicals (OH), producing water and dehydroascorbic acid ($C_6H_6O_6$) as products. Balance this reaction, assuming that the pH in the cell is 7. Also give the value of $\nu_e$ for this reaction.

(17) In basic solution, $Br_2$ disproportionates into bromate ($BrO_3^-$) and bromide ions. Balance the reaction.

(18) (a) Calculate the lowest two energy levels of an electron in a 1 nm box.
   (b) What are the wavelength, wavenumber and frequency of a photon whose absorption could cause a transition from the ground to the first excited state of this system?
   (c) What is the momentum of the photon?
   (d) The energy of a particle in a box is all kinetic. Calculate the momentum of the particle in each of the two lowest energy levels.
   (e) Is momentum conserved during the transition? Discuss.

(19) The ratio of the populations of two energy levels tends toward 2 at very high temperatures (higher energy population:lower energy population), and the populations are roughly equal at $75\,°C$. What is the spacing between these energy levels? What other information can you extract from these data?

(20) As you know, we often see the pH defined as $-\log_{10}[H^+]$ rather than the more correct definition based on the activity of the hydrogen ion. How large a difference does this make? Using Debye–Hückel theory, calculate the difference between the correct and approximate (concentration-based) definitions of pH for a $0.01\,\mathrm{mol\,L^{-1}}$ HCl solution at $25\,°C$.

(21) The diffusion coefficient of a proton in water is very large: $D = 9.31 \times 10^{-9}\,\mathrm{m^2\,s^{-1}}$ at $25\,°C$. What value does this imply for the radius of this ion in water? Does this value make sense?

(22) Ethanol (molar mass $46.069\,\mathrm{g\,mol^{-1}}$) is a very toxic substance that cells have to keep at low levels. Alcohol dehydrogenase is an enzyme that catalyzes the reaction

$$\mathrm{NAD^{ox} + ethanol \rightarrow NAD^{red} + acetaldehyde + H^+}$$

where $\mathrm{NAD^{ox}}$ and $\mathrm{NAD^{red}}$ are the oxidized and reduced forms of NAD, respectively.

   (a) Ethanol concentrations of $35\,\mathrm{g\,L^{-1}}$ or so are toxic to most strains of *Escherichia coli*. Is the reaction catalyzed by alcohol dehydrogenase thermodynamically allowed at this ethanol concentration at $25\,°C$, pH 5.5 and an ionic strength of $0.25\,\mathrm{mol\,L^{-1}}$ when the concentrations of the other reactants are $[\mathrm{NAD^{ox}}] = 3.7\,\mathrm{mmol\,L^{-1}}$, $[\mathrm{NAD^{red}}] = 0.47\,\mathrm{mmol\,L^{-1}}$ and $[\mathrm{acetaldehyde}] = 1.2\,\mathrm{\mu mol\,L^{-1}}$? Use the following data in the biochemists' standard state at $25\,°C$, pH 7 and ionic strength $0.25\,\mathrm{mol\,L^{-1}}$ to solve this problem:

| Species | $\Delta_f H^{o\prime}$ kJ mol$^{-1}$ | $\Delta_f G^{o\prime}$ kJ mol$^{-1}$ |
|---|---|---|
| Acetaldehyde$_{(aq)}$ | 24.06 | −213.97 |
| Ethanol$_{(aq)}$ | 62.96 | −290.76 |
| NAD$^{ox}_{(aq)}$ | 1059.11 | −10.26 |
| NAD$^{red}_{(aq)}$ | 1120.09 | −41.38 |

   (b) Why don't you need to know anything about the enzyme to answer this question?

(23) Kurt Stern was probably the first person to observe the formation of an enzyme–substrate complex by spectroscopy.[1] He provided an excess of a bad substrate (hydroperoxyethane) to catalase. Catalase binds this substrate rapidly, but catalysis is very inefficient. The result is that the spectrum of the enzyme disappears almost instantly, and is replaced by the spectrum of the enzyme–substrate complex. The spectrum of the free enzyme takes several minutes to reappear. The spectrum of catalase in the visible range consists of three bands, one roughly spanning the range from 490 to 510 nm (maximum at 500 nm), one from 530 to 550 nm (540 nm), and one from 610 to 650 nm (629 nm). The enzyme–substrate complex, on the other hand, has a spectrum consisting of two bands, from 529 to 540 nm (535 nm), and from 564 to 576 nm (570 nm).

   (a) Suppose that you wanted to study this reaction by spectrophotometry. Which would be the most convenient bands of the enzyme and enzyme–substrate complex to use?

[1] K. G. Stern, *Nature* **136**, 335 (1935).

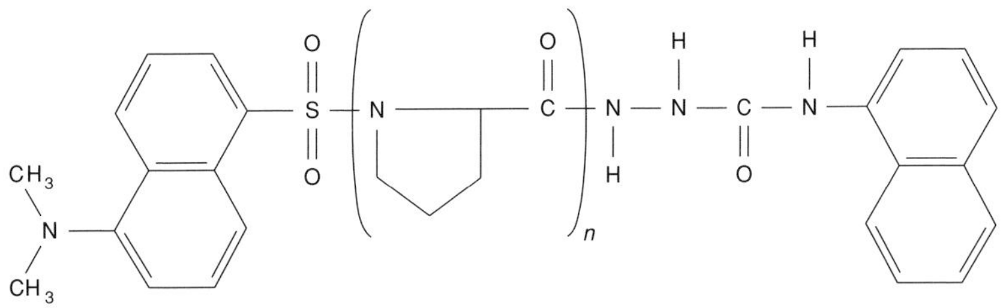

Figure J.1 Compounds synthesized by Stryer and Haugland to test Förster's predicted dependence of the FRET efficiency on $R^6$. The group to the right of the repeated proline units is the donor $\alpha$-naphthyl group. The group to the left is the acceptor dansyl group.

(b) Describe briefly an experiment to determine the molar absorption coefficients and explain how you would calculate these coefficients.

(c) Suppose that you have made a solution of equine liver catalase. The molar absorption coefficient of this enzyme at 623 nm is $8.3 \times 10^4 \, \mathrm{L\,mol^{-1}cm^{-1}}$. Based on the amounts of enzyme and buffer you used, you calculate that the concentration should be $12 \, \mu\mathrm{mol\,L^{-1}}$. When you measure the absorbance of this solution, you get $A_{623} = 0.95$ in a 1 cm cuvette. Is this consistent with the concentration?

(d) Stern observed that the time required for the reappearance of the enzyme's spectrum and disappearance of the enzyme–substrate complex's spectrum was four times longer at $4\,^{\circ}\mathrm{C}$ than at $24\,^{\circ}\mathrm{C}$. What is the activation energy implied by these data?

(24) ATP (adenosine 5′-triphosphate) is the major energy carrier in cells. Energy is mainly released from ATP in the hydrolysis reaction

$$\mathrm{ATP} + \mathrm{H_2O} \rightarrow \mathrm{ADP} + \mathrm{P_i}$$

where ADP is adenosine 5′-diphosphate, and $\mathrm{P_i}$ represents "inorganic phosphate," i.e. one of the protonation states of the phosphate anion. For this reaction, $\Delta_r H_m = -30.88 \, \mathrm{kJ\,mol^{-1}}$. In a turtle hepatocyte (liver cell) with a volume of $10^{-11} \, \mathrm{L}$, $7 \times 10^{-13} \, \mathrm{mol}$ of ATP are consumed per hour. Assume that a cell is mostly water with a density of about $1 \, \mathrm{kg\,L^{-1}}$ and a heat capacity of about $4.2 \, \mathrm{J\,K^{-1}g^{-1}}$. Now suppose that we could put a little insulated jacket around a hepatocyte. Considering only this one reaction, how long would it take for the temperature to rise by $5\,^{\circ}\mathrm{C}$?

(25) Calculate the enthalpy of formation of sodium hydride at $-20\,^{\circ}\mathrm{C}$.

(26) Calculate the base ionization constant of the hydrogen phosphate ion at $80\,^{\circ}\mathrm{C}$, i.e. the equilibrium constant for the transfer of a proton from water to the hydrogen phosphate ion.

(27) When a carbonate solution is acidified, carbon dioxide is produced.

(a) Can we predict the sign of the entropy change in this process without doing any calculations? If so, predict the sign. In either event, explain your reasoning.

(b) Calculate the standard entropy change for the reaction of carbonate ions with hydrogen ions producing carbon dioxide.

(28) Stryer and Haugland tested the dependence of the FRET efficiency on the sixth power of the distance by synthesizing a series of proline polymers with an $\alpha$-naphthyl donor group at one end and a dansyl acceptor group at the other end (Figure J.1).[2] Polyproline forms a relatively rigid helix so that the distance between the ends for a given value of the number of proline monomers, $n$, can be predicted with reasonable accuracy. The following data were obtained:

[2] L. Stryer and R. P. Haugland, *Proc. Natl. Acad. Sci. USA* **58**, 719 (1967).

| $R/\text{Å}$ | 21.2 | 24.5 | 27.8 | 31.0 | 34.4 | 37.0 | 40.5 | 43.1 | 46.0 |
|---|---|---|---|---|---|---|---|---|---|
| $\eta_{FRET}/\%$ | 96.5 | 82.9 | 74.8 | 64.9 | 49.6 | 40.3 | 28.3 | 25.4 | 16.6 |

(a) Rewrite Equation (3.9) in a linearized form, i.e. a form in which the data should fall on a straight line when plotted. Ideally, you would obtain a form in which $R_0$ and the exponent can be recovered from the slope and intercept.

   *Hint*: Start by taking $1/\eta_{FRET}$.

(b) Plot your data according to the theory developed in question 28a. Confirm that the data are consistent with the efficiency–distance relationship, and determine the value of $R_0$.

(29) Enantiomers can often be separated by gas chromatography, using columns packed with a chiral stationary phase. It is also possible to study the kinetics of interconversion between two enantiomers in a gas chromatograph. There are a number of techniques which can be used, ranging from a "stopped-flow" method in which the flow of carrier gas is stopped to allow time for the reaction to occur inside the column, to dynamic methods which require the analysis of chromatogram peak shapes.[3] The chiral stationary phase may affect the kinetics of the reaction due to binding of the material being studied to the stationary phase. However, this is a minor disadvantage to a relatively easy experimental method for studying the interconversion between two enantiomers in the gas phase.

(a) Mydlová has measured the rate constant for the interconversion between the two enantiomers of 1-chloro-2,2-dimethylaziridine by dynamic reaction gas chromatography in a setup with coupled ChiralDex BTA and polydimethylsiloxane columns.[4] The R isomer is illustrated below:

The reaction inverts the configuration at the nitrogen atom. The data given below are averages of two estimates of the rate constant, assuming that $K = 1$ for the interconversion process:

| $T/K$ | 338 | 348 | 358 | 368 |
|---|---|---|---|---|
| $k/10^{-4}\,\text{s}^{-1}$ | 0.8 | 1.8 | 3.2 | 9.4 |

Calculate the activation energy and pre-exponential factor for this interconversion.

(b) Calculate the entropy of activation at 25 °C. What does the value of the entropy of activation tell us about this reaction?

(30) Anikeev, Ermakova and Goto have studied the decomposition reaction of nitromethane in supercritical water.[5] Specifically, they measured the fraction of the initial nitromethane converted to *product* as a function of reaction time at a temperature of 664 K and a pressure of 272 atm:

| $t/\text{s}$ | 0 | 45 | 67 | 134 | 270 | 467 |
|---|---|---|---|---|---|---|
| $X$ | 0 | 0.381 | 0.537 | 0.763 | 0.951 | 0.994 |

Show that the reaction obeys first-order kinetics, and determine the rate constant.

(31) (a) What is the enthalpy of combustion of $\alpha$-D-glucose at 10 °C?

---

[3] J. Krupčik *et al.*, *J. Chromatogr. A* **1186**, 144 (2008).   [4] Data reported in J. Krupčik *et al.* (2008), cited above.
[5] V. I. Anikeev *et al.*, *Kinet. Catal.* **46**, 821 (2005).

(b) If you burned 1.00 g of glucose in a bomb calorimeter near 25 °C, how much heat would be released? The molar mass of glucose is 180.156 g mol$^{-1}$.

(32) Uric acid is a diprotic organic acid formed by the oxidation of purines. Gout is a medical condition in which crystals of uric acid or of sodium hydrogen urate (the sodium salt of the conjugate base of uric acid, also known as monosodium urate) are deposited in a joint, causing pain when the joint is flexed. Under physiological conditions, uric acid mostly exists as its conjugate base, the hydrogen urate ion.

At 25 °C and a physiological ionic strength of 0.15 mol L$^{-1}$, the (first) acid dissociation constant of uric acid is $5.50 \times 10^{-6}$.[6] Calculate the standard free energy of formation of the hydrogen urate ion, given that the standard free energy of formation of aqueous uric acid is $-356.6$ kJ mol$^{-1}$.

(33) The following data were obtained for the gas-phase dimerization of $C_2F_4$ at 300 °C:

| $t$/min | $[C_2F_4]$/mol L$^{-1}$ |
|---|---|
| 0 | 0.0500 |
| 250 | 0.0250 |
| 750 | 0.0125 |
| 1750 | 0.00625 |
| 3750 | 0.00312 |

(a) Show that the data are consistent with this being an elementary reaction and calculate the rate constant.

(b) Suppose that we wanted to collect at least 0.1 mol of the dimer in an experiment starting with 0.25 mol of $C_2F_4$ in a volume of 0.60 L at 300 °C. How long should we let the reaction go?

(c) The rate constant for the reverse reaction is $9.6 \times 10^{-13}$ s$^{-1}$. What is the equilibrium constant?

(34) The mechanism for the reaction

$$Hg_{2(aq)}^{2+} + Tl_{(aq)}^{3+} \rightarrow 2Hg_{(aq)}^{2+} + Tl_{(aq)}^{+}$$

is thought to be

$$Hg_{2(aq)}^{2+} \underset{k_{-1}}{\overset{k_1}{\rightleftharpoons}} Hg_{(aq)}^{2+} + Hg_{(aq)} \quad \text{(fast)}$$

$$Hg_{(aq)} + Tl_{(aq)}^{3+} \overset{k_2}{\rightarrow} Hg_{(aq)}^{2+} + Tl_{(aq)}^{+} \quad \text{(slow)}$$

(a) What rate law would you predict based on this mechanism?

(b) Suppose that we carried out two initial rate experiments where, all other factors being held constant, we doubled the concentration of the mercury(II) ion from one experiment to the next. What effect would this have on the rate?

---

[6] Z. Wang and E. Königsberger, *Thermochim. Acta* **310**, 237 (1998).

(35) For the reaction

$$\text{glycerol} + P_i \rightarrow \text{3-phosphoglycerate} + H_2O$$

in aqueous solution at $25\,°C$, $\Delta_r \bar{G}^{\circ\prime} = 10.0\,\text{kJ/mol}$, where $P_i$ represents "inorganic" phosphate and $\Delta_r \bar{G}^{\circ\prime}$ is the standard free energy change in the biochemists' standard state.

(a) Is this reaction thermodynamically allowed if $[\text{glycerol}] = 1.0\,\text{mmol}\,L^{-1}$, $[P_i] = 18\,\text{mmol}\,L^{-1}$ and $[\text{3-phosphoglycerate}] = 0.35\,\text{mmol}\,L^{-1}$?

(b) Thinking about the definition of the biochemists' standard state, explain why we write $P_i$ instead of (e.g.) $PO_4^{3-}$.

(36) Major thermodynamic reference tables often don't give the standard free energy of formation. Instead, they give the standard enthalpy of formation and standard entropy. The standard enthalpy of formation of $UO_{2(aq)}^{2+}$ is $-1019.0\,\text{kJ}\,\text{mol}^{-1}$. Calculate the standard free energy of formation of $UO_{2(aq)}^{2+}$.

(37) Random numbers have important applications in many fields of computer science, including cryptography. Unfortunately, it is difficult to generate truly random numbers. In cryptography, insufficiently random numbers can result in codes which are easy to break. Some physical devices like Zener diodes do appear to generate true random numbers. Suppose that such a device is used to generate three-bit (bit = binary digit, i.e. a digit which can only take the values 0 or 1) random numbers.

(a) What is the entropy of this random number generator? Give your answer as a multiple of $k_B \ln 2$.

   *Hint*: How many possible values are there for each bit?

(b) What would be the entropy of an $n$-bit random number generator?

(c) Entropy is an extensive property. Explain how your last result is consistent with this statement.

(38) Some humidifiers work by creating a fine mist. Very small droplets of water evaporate very quickly, even at room temperature.

(a) An industrial humidifier vaporizes $5\,\text{kg}$ of water per hour as described above. Assume that the droplets are formed and evaporate at $25\,°C$. How much heat does this remove from the air? Express your final answer in watts.

(b) To put the above number in perspective, suppose that we operate the humidifier in a closed, well-insulated room of dimensions $10 \times 10 \times 4\,\text{m}$. The air pressure in the room is $100\,\text{kPa}$ when the temperature is $18\,°C$. Ignoring all other possible sources of heat and treating the air as an ideal gas, what effect would the operation of the humidifier have on the temperature in the room over the course of an hour? Take the heat capacity of air to be approximately $29\,\text{J}\,K^{-1}\text{mol}^{-1}$.

(39) The reactions of chelated nickel(II) ion complexes with acetonitrile have been studied in the gas phase.[7] In the following reaction, L represents a ligand. The acetonitrile ($CH_3CN$) is a Lewis base which forms a coordinate bond directly to the nickel(II) ion.

$$\text{NiL}^{2+} + CH_3CN \underset{k_d}{\overset{k_a}{\rightleftharpoons}} [\text{NiL}^{2+} \cdot CH_3CN]$$

(a) At $300\,K$ with the ligand illustrated below, the following data were obtained:

$$k_a = 2.4 \times 10^{11}\,\text{L}\,\text{mol}^{-1}\text{s}^{-1}$$
$$K = 8.1 \times 10^{12}\,\text{L}\,\text{mol}^{-1}$$

[7] M. Y. Combariza and R. W. Vachet, *J. Am. Soc. Mass Spectrom.* **15**, 1128 (2004).

($K$ is the empirical equilibrium constant for the reaction.) Calculate $k_d$.

(b) Suppose that we start off with $1.7 \times 10^{-11}$ mol $L^{-1}$ of acetonitrile and $1.0 \times 10^{-11}$ mol $L^{-1}$ of NiL$^{2+}$. What is the initial rate of reaction?

(c) Approximately how long would it take to accumulate $1 \times 10^{-14}$ mol $L^{-1}$ of product under the conditions described above?

(40) There is a great deal of interest in oxidative damage to polymers. This has led to a series of studies on the reactions of polymers with radicals. One such study examined the kinetics of the radical 1,1-diphenyl-2-picrylhydrazyl (DPPH) with polyphenylene (PP).[8]

(a) Typically, polymers react with two equivalents of a radical to yield products without unpaired electrons. The authors of this study proposed the following mechanism:

$$PP + DPPH \underset{k_{-1}}{\overset{k_1}{\rightleftharpoons}} \{PP \cdot DPPH\}$$

$$\{PP \cdot DPPH\} + DPPH \overset{k_2}{\rightarrow} products$$

Develop a rate law for the reaction based on this mechanism. Under what conditions would a simple second-order dependence on the DPPH concentration be observed?

(b) In the presence of an excess of polyphenylene at $65\,^\circ$C, the following data were obtained:

| $t$/min | 2.7 | 14.6 | 26.5 | 38.2 |
|---|---|---|---|---|
| [DPPH]/$10^{-5}$ mol $L^{-1}$ | 8.50 | 5.58 | 4.40 | 3.59 |

Are these data consistent with a second-order reaction? If so, what is the rate constant?

(41) SF$_6$ is inert under most conditions. For example, it won't undergo hydrolysis:

$$SF_{6(g)} + 4H_2O_{(l)} \nrightarrow 6F^-_{(aq)} + SO^{2-}_{4(aq)} + 8H^+_{(aq)}$$

Is the above reaction thermodynamically allowed at $25\,^\circ$C when the pressure of SF$_6$ is 0.2 bar, the concentration of fluoride ions is $1.0 \times 10^{-3}$ mol $L^{-1}$, the concentration of sulfate is 0.43 mol $L^{-1}$ and the pH is 6.0? Does this agree or disagree with the empirical observation that SF$_6$ can't be hydrolyzed? Discuss briefly.

(42) Consider the following simple electrochemical cell:

$$Cd_{(s)}|CdCl_{2(aq)}(0.50\,mol\,L^{-1})\|LiCl_{(aq)}(0.80\,mol\,L^{-1})|Li_{(s)}$$

At $25\,^\circ$C, the mean ionic activity coefficient of CdCl$_2$ at a concentration of 0.50 mol $L^{-1}$ is 0.1006, and the mean ionic activity coefficient of LiCl at 0.80 mol $L^{-1}$ is 0.755. What emf does this cell generate at $25\,^\circ$C, and in what direction do electrons flow?

(43) The designer drug $N$-methyl-benzodioxolyl-butanamine (MBDB) is metabolized by members of the cytochrome P450 family of isozymes (closely related enzymes that catalyze the same reaction(s)). A variety of products are obtained, one of which is 1,2-dihydroxy-4-[2-(methylamino)butyl]benzene (DHMBB). The following initial rate data were obtained for the transformation of the $R$ stereoisomer of MBDB by the P450 isozyme CYP2B6 to DHMBB:[9]

[8] V. A. Vonsyatskii et al., React. Kinet. Catal. Lett. **61**, 280 (1997).
[9] M. R. Meyer et al., Biochem. Pharmacol. **77**, 1725 (2009).

| $[R\text{-MBDB}]/\mu mol\,L^{-1}$ | $v/mol\,min^{-1}(mol\,enzyme)^{-1}$ |
|---|---|
| 21 | 0.65 |
| 55 | 1.17 |
| 100 | 1.94 |
| 132 | 2.01 |
| 258 | 2.73 |
| 452 | 2.73 |
| 585 | 2.98 |
| 845 | 3.02 |

(a) Calculate $v_{max}$ and $K_M$ for this enzyme.
(b) Rates (and therefore $v_{max}$ values) in enzyme kinetics are often reported in units that are not, strictly speaking, units of rate. This particular study is an example. In this case, given the units in which we calculated $v_{max}$, we have actually obtained a different constant than what we would normally call $v_{max}$. Which constant do you have?

(44) In aqueous solution, the acridine homodimer (AcrH)

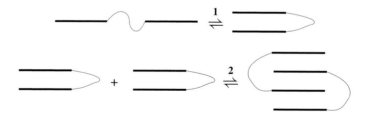

can exist either in an unfolded form or in a folded conformation where the two ring systems stack on top of each other. The folded AcrH molecules can also form aggregates in which the ring systems from different molecules interleave. Schematically, we have

In this diagram, the heavy straight lines represent the ring systems, which we are viewing from the side. Further aggregation reactions are possible involving additional AcrH molecules. However, for the purposes of this question, we will assume that only these two reactions are significant.

(a) Evstigneev and coworkers have measured the equilibrium constant for the folding process (reaction 1) at 298 K and at 308 K.[10] They found $K_{298} = 42$ and $K_{308} = 26$. Calculate
  (i) the standard enthalpy of reaction,
  (ii) the standard free energy of reaction, and
  (iii) the standard entropy of reaction.
  Does the sign of the entropy change make sense? Explain briefly.

[10] M. P. Evstigneev et al., J. Mol. Struct. **784**, 162 (2006).

(b) For the dimerization reaction (reaction 2), Evstigneev and coworkers found $\Delta_r H^\circ = -32\,kJ\,mol^{-1}$. Find the overall enthalpy change for the process which takes unfolded AcrH and creates the two-molecule aggregate.

(c) The equilibrium constant for reaction 2 at 298 K is 540. Suppose that we dissolve $5.0 \times 10^{-4}\,mol\,L^{-1}$ of AcrH in water. At equilibrium at 298 K, what are the concentrations of the unfolded molecule, folded AcrH and the two-molecule aggregate?

*Hint*: You have three unknowns, so you need three equations. The first two equations should be obvious. One way to get a third equation is to consider that the total amount of AcrH in solution is constant.

(45) Volatile organic compounds in the atmosphere can be oxidized by $NO_3$, often yielding undesirable products (e.g. toxic compounds). These reactions often happen in water droplets rather than in the gas phase. Gaillard de Sémainville and coworkers have studied the reactions of $NO_3$ with some ketones, aldehydes and carboxylic acids in aqueous solution.[11] In their experiments, $NO_3$ was generated by laser flash photolysis.

(a) In one set of experiments, peroxodisulfate anions $(S_2O_8^{2-})$ were photolyzed to $SO_4^-$ radical ions using a laser flash at 351 nm. The $SO_4^-$ ions then reacted with nitrate ions to yield sulfate ions and $NO_3$. The reaction of $NO_3$ with either a carboxylic acid or its conjugate base was then followed spectroscopically by recording the absorbance of the solution at a wavelength where $NO_3$ absorbs strongly (632.8 nm).

(i) Briefly explain how the laser flash photolysis experiment works, with specific reference to this reaction system. Make sure to explain what solution(s) must be prepared. Do the reactions have to satisfy any special conditions in order for the measurements described above to give the rate of reaction of $NO_3$ with an organic compound?

(ii) Suppose that the reaction of $SO_4^-$ with nitrate is diffusion-limited. Estimate the rate constant of this reaction at 25 °C, assuming that the diffusion coefficients of $SO_4^-$ and of nitrate are both about $2 \times 10^{-9}\,m^2\,s^{-1}$, and that the radii of the two ions are both about $2 \times 10^{-10}\,m$.

(b) The reaction of $NO_3$ with an organic compound in aqueous solution displays a simple rate law, $v = k[NO_3][R]$, where R stands for the organic reactant. $NO_3$ is a highly reactive radical, so it will typically react with other compounds in solution, yielding an additional "background" decay which obeys first-order kinetics. In these experiments, the organic reactant was present in great excess over the $NO_3$ generated by flash photolysis. Explain why the data from these experiments would be consistent with a first-order rate law, and give an equation for the observed rate constant. Then show that the second-order rate constant for the reaction of $NO_3$ with R is the slope of a graph of the observed first-order rate constant vs. [R].

(c) The following data were obtained for the reaction of excess lactic acid (R) with $NO_3$ at 298 K:

| $[R]/\mu mol\,L^{-1}$ | 100 | 300 | 500 | 700 | 900 |
|---|---|---|---|---|---|
| $k_{obs}/s^{-1}$ | 620 | 911 | 1091 | 1142 | 1535 |

Determine the second-order rate constant for the reaction of lactic acid with $NO_3$.

(d) The following second-order rate constants were recovered for the reaction of lactic acid with $NO_3$:

| $T/K$ | 278 | 288 | 298 | 308 | 318 |
|---|---|---|---|---|---|
| $k/10^6 L\,mol^{-1}s^{-1}$ | 0.74 | 0.84 | 2.06 | 2.75 | 2.59 |

Determine the activation energy and pre-exponential factor of the reaction.

[11] Ph. Gaillard de Sémainville *et al.*, *Phys. Chem. Chem. Phys.* **9**, 958 (2007).

(46) Consider the gas-phase reaction $A + B \rightarrow C$. Sketch a possible entropy profile for this reaction against the reaction coordinate. Label your sketch with the quantities $\Delta_r S$ and $\Delta^{\ddagger} S$.

(47) A classic demonstration involves putting some liquid nitrogen in a balloon, getting rid of as much air as possible, sealing it (usually with a clip), and then watching the balloon expand as the liquid nitrogen evaporates. Typically, we put in a lot of liquid nitrogen and let the balloon explode. However, suppose that we have put just 5 mL of liquid nitrogen (density $0.8086 \, \text{g mL}^{-1}$) in a balloon of sufficient capacity that it doesn't explode.

   (a) The boiling point of liquid nitrogen (i.e. the initial temperature) is 77.34 K. If we let the nitrogen evaporate completely and return to room temperature (20 °C) at a constant external pressure of 1 atm, what is the work done during this process? Is work done on or by the system?

   (b) The enthalpy of vaporization of liquid nitrogen at its boiling point is $5.586 \, \text{kJ mol}^{-1}$. The molar heat capacity of nitrogen gas (in $\text{J K}^{-1}\text{mol}^{-1}$) is given by the equation

$$C_{p,m} = A + BT + CT^2$$

where $A = 28.3$, $B = 2.54 \times 10^{-3}$, $C = 5.4 \times 10^{-5}$ and $T$ is in kelvins. Calculate the heat gained by the nitrogen during this process.

   (c) Calculate the change in internal energy of the nitrogen during this process.

(48) Strontium is used in a number of applications, for instance as a component of alloys in structural or magnetic materials, as a corrosion inhibitor in coatings and in fireworks, where it is often used for its red emission. The most important strontium ore is celestite ($SrSO_4$). Like most ores, celestite is not a pure material, so the strontium needs to be separated from contaminants. Furthermore, the most convenient starting point for industrial applications is strontium carbonate. Strontium carbonate is much less soluble than strontium sulfate, so in principle we can make the former by reacting the latter with (e.g.) a sodium carbonate solution:

$$SrSO_{4(s)} + CO_{3(aq)}^{2-} \rightarrow SrCO_{3(s)} + SO_{4(aq)}^{2-}$$

   (a) Demonstrate how this overall reaction can be broken down into two coupled reactions related to the solubility equilibria of strontium sulfate and of strontium carbonate, and therefore why the relative solubilities of the two compounds determine the direction of equilibrium.

   (b) The solubility products of strontium sulfate and strontium carbonate at 25 °C are, respectively, $3.42 \times 10^{-7}$ and $5.60 \times 10^{-10}$. Calculate the equilibrium constant for the above reaction.

   (c) Calculate the standard free energy of formation of an aqueous strontium ion.

   (d) Using Debye–Hückel theory, calculate the solubility of strontium sulfate in water at 25 °C.

(49) Consider the following table, which gives the results of a set of initial rate experiments for a reaction with the stoichiometry $A + B \rightarrow P$.

| Experiment | $a/\text{mol L}^{-1}$ | $b/10^{-3}\text{mol L}^{-1}$ | $v/10^{-3}\text{mol L}^{-1}\text{s}^{-1}$ |
|---|---|---|---|
| 1 | 0.10 | 1.4 | 0.31 |
| 2 | 0.20 | 1.4 | 1.24 |
| 3 | 0.10 | 2.1 | |

   (a) What is the order with respect to A?

   (b) In a separate set of experiments, it was determined that the order with respect to B is $\frac{1}{2}$. What reaction rate would you predict for experiment 3?

(c) What is the value of the rate constant?

(d) Could this reaction be elementary? Why or why not?

(50) Gallium nitride (GaN) is a solid semiconductor used in light-emitting diodes. Jacob and Rajitha have recently reviewed the available data on the thermodynamic properties of GaN.[12] They found that the standard free energy of formation of GaN was well fit by the equation

$$\Delta_f G^\circ = -131\,530 + 117.4\,T \tag{J.1}$$

over the temperature range 800 to 1400 K, where $\Delta_f G^\circ$ is in $J\,mol^{-1}$ and $T$ is in K.

(a) Over this temperature range, what is the standard enthalpy of formation of GaN?

(b) What does the coefficient of $T$ in Equation (J.1) represent?

(c) At high temperatures, gallium nitride decomposes into its elements. Predict the temperature above which this would happen when GaN is kept in 1 bar of $N_2$.

(d) Above what temperature would GaN decompose into its elements if kept in 20 bar of $N_2$? Does the difference between your answers to this and the previous question agree with Le Chatelier's principle? Explain briefly.

*Note*: Gallium melts at 302.9 K and boils at 2477 K, so it is a liquid over the temperature range for which Equation (J.1) is valid.

(51) In a recent study, the electrochemistry of 1,5,2,4,6,8-dithiatetrazocine (R)

and its radicals was studied by EPR spectroscopy.[13] In one experiment, the radical anion $R^{\bullet-}$ was generated electrochemically, then its decay was followed by monitoring the intensity of the corresponding EPR signal, which is proportional to the concentration of the radical.

(a) Is the formation of the radical anion from the neutral compound an example of oxidation or reduction?

(b) The radical anion can decay by one of at least two mechanisms:

1 $R^{\bullet-} + R^{\bullet-} \rightarrow R + R^{2-}$

2 $R^{\bullet-} + S \rightarrow R + S^{\bullet-}$

where S is the solvent. Explain why these two elementary processes would be expected to lead to different rate laws. What law of chemical kinetics is involved? What simplification(s) arise due to the reaction conditions?

(c) The following data were obtained for the decay of the radical in dichloromethane,[14] starting immediately after the electrolytic current generating the radical had been turned off:

| $t$/s | 14.68 | 27.80 | 40.92 | 54.04 | 67.16 | 80.28 |
|---|---|---|---|---|---|---|
| $I$ | 4.28 | 1.29 | 0.56 | 0.35 | 0.18 | 0.09 |

On the basis of these data, can you determine which of the above two mechanisms is the correct one? If so, also calculate the rate constant. Otherwise, explain what the problem is.

[12] K. T. Jacob and G. Rajitha, *J. Cryst. Growth* **311**, 3806 (2009). [13] R. T. Boeré *et al.*, *Inorg. Chem.* **46**, 5596 (2007). [14] Thanks to Dr. Tracey Roemmele for making the original data available.

(52) Some enzymes have an inactive form that is in equilibrium with the active form, i.e. the mechanism includes, along with the usual Michaelis–Menten reaction steps, a reversible step

$$E \underset{k_{-3}}{\overset{k_3}{\rightleftharpoons}} X$$

where X is the inactive form of the enzyme. Derive a rate law for catalysis of an isomerization reaction S → P by an enzyme with this characteristic. Clearly identify $v_{max}$ and $K_M$.

(53) (a) The enzyme chorismate mutase catalyzes one reaction step in the biosynthesis of phenylalanine and tyrosine in bacteria, fungi and plants. Kast, Asif-Ullah and Hilvert have measured $k_{cat}$ for this enzyme at different temperatures, and obtained the following data:[15]

| $T/K$ | 278 | 288 | 298 | 308 | 318 |
|---|---|---|---|---|---|
| $k_{cat}/s^{-1}$ | 6.54 | 15.7 | 33.8 | 69.9 | 130 |

Calculate the activation energy and pre-exponential factor.

(b) Calculate the entropy of activation at 25 °C. Assuming a simple Michaelis–Menten reaction, what does the sign and/or value of the entropy of activation tell you about the reaction for which $k_{cat}$ is the rate constant?

(54) *Aplysia dactylomela* is a species of sea hare (a type of sea slug) that reproduces very rapidly. *A. dactylomela*'s diet consists of a variety of algae. In one set of lab experiments, *A. dactylomela* was fed a diet of *Cladophora*, a green algae.[16] The animals were weighed before the experiment, then again at the end. They were then dried in an oven. The percentage of the body weight that was water was assumed to be constant throughout the experiment, so the gain in dry weight could be estimated. Any offspring produced were also dried and weighed. Finally, feces were collected and dried. The dried animals, offspring and feces were analyzed by bomb calorimetry to determine their energy content. When dealing with whole animals, some percentage of the material is non-combustible, and shows up as ash after opening the calorimeter. Calculations of metabolic energy are only concerned with the material that can be burned, so the non-combustible ash component, reported as a percentage of the dry mass, must be excluded from the mass of the biological materials in the calculations. The following data were obtained:

|  |  |  |
|---|---|---|
| | Duration of experiment: | 14 days |
| | Energy content of food: | 27.92 kcal |
| Animal | Increase in dry mass: | 2.21 g |
| | Ash: | 33 % |
| | Specific combustion energy: | 4.666 kcal (g ash-free material)$^{-1}$ |
| Offspring | Dry mass: | 0.39 g |
| | Ash: | 31 % |
| | Specific combustion energy: | 4.920 kcal (g ash-free material)$^{-1}$ |
| Feces | Dry mass: | 23.50 g |
| | Ash: | 76 % |
| | Specific combustion energy: | 3.220 kcal (g ash-free material)$^{-1}$ |

How much energy did the animals use per day for metabolism (not counting energy stored in biomass) and physical activity?

[15] P. Kast *et al.*, *Tetrahedron Lett.* **37**, 2691 (1996).
[16] T. H. Carefoot, *J. Exp. Mar. Biol. Ecol.* **5**, 47 (1970).

(55) Many marine organisms contain a substantial amount of calcium carbonate, either as a shell or as skeletal elements. Shells can be removed, but other skeletal elements are often less easy to separate from the soft tissues. Moreover, depending on the diet of the organism, a substantial amount of calcium carbonate can end up in feces. When we carry out bomb calorimetry on marine organisms or their feces, the following reaction occurs:

$$CaCO_{3(s)} \rightarrow CaO_{(s)} + CO_{2(g)}$$

This reaction does not correspond to any biological oxidation so, if we want to use calorimetry to calculate the amount of energy an organism has stored, the amount of calcium carbonate in the sample must be analyzed, and a correction must be made to the raw calorimetric results corresponding to the heat of this reaction. Incidentally, such corrections are included in the data presented in question 54.

(a) The initial and final temperatures in a bomb calorimeter are both close to $25\,^{\circ}C$. Show that this reaction is not thermodynamically allowed at this temperature in the presence of 39 Pa of carbon dioxide (the atmospheric partial pressure of this gas). That being the case, why is this reaction relevant anyway?

(b) Calculate the correction to the measured heat per gram of calcium carbonate in the sample. Explain briefly how you would use this correction. The molar mass of calcium carbonate is $100.09\,g\,mol^{-1}$.

(56) Suppose that we wanted to add chlorine to a molecule and that we thought we could initiate the reaction by breaking the $Cl_2$ bond. The bond enthalpy (the enthalpy change for the reaction $Cl_{2(g)} \rightarrow 2Cl_{(g)}$) is $242\,kJ\,mol^{-1}$ at 298.15 K.

(a) Enthalpy is heat at constant pressure, but photons aren't "heat" in any reasonable sense of the word. It is better to think of them as bundles of energy. Calculate the bond energy for chlorine.

(b) Calculate the frequency of a photon that can break the $Cl_2$ bond. To which spectral region does this frequency correspond?

(57) Hemoglobin (Hb) is the major oxygen carrier in the blood of vertebrates. It is made up of two $\alpha$ and two $\beta$ subunits (i.e. two copies of each of two different kinds of metalloproteins). Each of the four subunits can carry one oxygen molecule. The equilibrium constants for binding of oxygen to human hemoglobin at $21.5\,^{\circ}C$ and pH 7 are as follows:[17]

$$
\begin{array}{ll}
Hb_{(aq)} + O_{2(aq)} \rightleftharpoons HbO_{2(aq)} & K_1 = 9.32 \times 10^3 \\
HbO_{2(aq)} + O_{2(aq)} \rightleftharpoons Hb(O_2)_{2(aq)} & K_2 = 2.10 \times 10^5 \\
Hb(O_2)_{2(aq)} + O_{2(aq)} \rightleftharpoons Hb(O_2)_{3(aq)} & K_3 = 9.07 \times 10^3 \\
Hb(O_2)_{3(aq)} + O_{2(aq)} \rightleftharpoons Hb(O_2)_{4(aq)} & K_4 = 6.60 \times 10^5
\end{array}
$$

(a) Calculate the standard free energy of reaction for the first binding step at $21.5\,^{\circ}C$.

(b) From experiments carried out with purified subunits under conditions where they do not associate, it is known that the equilibrium constants for binding of oxygen to isolated $\alpha$ and $\beta$ subunits are both similar to $K_1$ and $K_3$. If the subunits did not interact with each other, all four equilibrium constants for binding of oxygen to hemoglobin would also be similar to $K_1$ and $K_3$. The larger equilibrium constants in steps 2 and 4 are examples of **cooperativity**, where binding of a ligand (in this case, oxygen) to a multi-subunit protein alters the affinity of the protein for the ligand at additional binding sites.

(i) What is the free energy change for the overall reaction for the first two steps, $Hb_{(aq)} + 2O_{2(aq)} \rightleftharpoons Hb(O_2)_{2(aq)}$? Call this quantity $\Delta_{r(i)}G_m^{\circ}$.

[17] Q. H. Gibson, *J. Biol. Chem.* **245**, 3285 (1970).

(ii) Suppose that there was no cooperativity, i.e. that the equilibrium constant for binding oxygen was the same in step 2 as in step 1. What would the free energy change be for the overall reaction $Hb_{(aq)} + 2O_{2(aq)} \rightleftharpoons Hb(O_2)_{2(aq)}$ in this case? Call this quantity $\Delta_{r(ii)}G_m^\circ$.

(iii) The difference between these two values, i.e. $\Delta_{r(i)}G_m^\circ - \Delta_{r(ii)}G_m^\circ$, is the free energy of cooperativity, i.e. the free energy decrease due to the cooperative binding. Calculate this quantity.

(c) Now consider a solution containing hemoglobin in which there is a large excess of oxygen at a concentration of $300\ \mu mol\ L^{-1}$. Using the equilibrium relationships, show that almost all of the hemoglobin will be in the fully oxygenated form $(Hb(O_2)_{4(aq)})$ under these conditions. *Note:* While you need to do a few calculations, you don't have to be precise about the expected fraction that will be in the fully oxygenated form, provided you explain clearly how the numbers indicate that this form will be strongly dominant.

(d) Would the fully oxygenated form still be overwhelmingly dominant if all the equilibrium constants were similar to $K_1$ and $K_3$, i.e. if there was no cooperativity? Explain briefly.

(58) Barbital (5,5-diethylbarbituric acid, $C_8H_{12}N_2O_3$) is a diprotic acid. Barbital was once sold as a sleeping aid, although it has gone out of use due to the risk of accidental overdoses.

(a) For the dissociation of the first proton, the apparent (i.e. not considering non-ideal effects) $K_a$ in aqueous solution is[18] $1.26 \times 10^{-8}$ at $25\,^\circ C$ and an ionic strength of $0.002\ mol\ L^{-1}$. What is the value of the thermodynamic equilibrium constant? Assume that the undissociated acid behaves ideally.

(b) At an ionic strength of $2.00\ mol\ L^{-1}$, the apparent $K_a$ is $2.75 \times 10^{-8}$. What is the mean ionic activity coefficient at this ionic strength?

(59) Hans-Jürgen Hinz and his group in Münster have been studying the formation of DNA double strands using short (12-base) complementary single-stranded sequences.

(a) In one set of experiments, they studied the binding equilibrium for the following complementary sequences:[19]

5'-TAG GTC AAT ACT-3'

3'-ATC CAG TTA TGA-5'

The following data were obtained:

| $T/^\circ C$ | 10 | 15 | 20 | 25 |
|---|---|---|---|---|
| $K/10^8$ | 3.19 | 2.82 | 2.25 | 1.10 |

Calculate the standard enthalpy, entropy and free energy of reaction.

(b) Discuss the sign and size of the entropy of reaction.

(c) Calculate the equilibrium concentrations of single and double strands at $25\,^\circ C$ in an experiment starting with $4\ \mu mol\ L^{-1}$ of each single strand.

(d) In another set of experiments, they studied the kinetics of double-strand formation by stopped-flow kinetics for the following complementary pair:[20]

5'-ATC CTC AAT ACT-3'

3'-TAG GAG TTA TGA-5'

Briefly explain how a stopped-flow apparatus would be used to study this reaction.

(e) Neglecting the reverse reaction, what rate law would you expect, assuming that double-strand formation is an elementary reaction? How would this rate law simplify if the initial concentrations of the two single strands were equal?

[18] M. E. Krahl, *J. Phys. Chem.* **44**, 449 (1940).     [19] E. Carrillo-Nava et al., *J. Phys. Chem. B*, **114**, 16087 (2010).
[20] E. Carrillo-Nava et al., *Biochemistry*, **47**, 13153 (2008).

(f) When the stopped-flow experiments were carried out at different temperatures in a phosphate buffer at pH 7.0, the following rate constants were obtained:

| $T/°C$ | 10 | 15 | 20 | 25 | 30 | 35 |
|---|---|---|---|---|---|---|
| $k/10^6 \, \text{L mol}^{-1}\text{s}^{-1}$ | 1.092 | 1.340 | 1.591 | 1.864 | 1.951 | 2.269 |

Calculate the activation energy and pre-exponential factor.

(g) Neglecting the reverse reaction, starting with $4 \, \mu\text{mol L}^{-1}$ of each single strand, how long would it take before 80% of the single strands were bound into double strands at $35 \, °C$?

(h) Calculate the enthalpy, entropy and free energy of activation at $25 \, °C$.

(i) Does the sign of the entropy of activation agree with your expectations? Discuss briefly.

(j) Sketch the free energy profile of the reaction, i.e. how the free energy varies going from reactants to products. Label your graph with the values of the free energies calculated in previous parts of this question, assuming that there are only small differences in thermodynamic properties between one pair of sequences and the other.

(60) In the section on enzyme kinetics, we focused our attention on single-substrate reactions. Of course, there are lots of two-substrate reactions out there. The kinetics of multi-substrate reactions is complicated by the question of the order in which the substrates bind. Some enzymes can bind the substrates in either order, while others can only bind the substrates in a particular order. Work out the rate law for the following ordered two-substrate reaction:

$$E + A \underset{k_{-1}}{\overset{k_1}{\rightleftharpoons}} C_1$$

$$C_1 + B \underset{k_{-2}}{\overset{k_2}{\rightleftharpoons}} C_2 \overset{k_{-3}}{\longrightarrow} E + P$$

Try to write your final rate law in a form similar to the Michaelis–Menten form.

# Appendix K

## Answers to exercises

### Exercise group 2.1

(1) 0.61 mol
(2) $p = 7.71 \times 10^{-24}$ kg m s$^{-1}$, $K = 3.14 \times 10^{-22}$ J
(3) $E = 5.81 \times 10^{-19}$ J, $\nu = 8.77 \times 10^{14}$ Hz, $\lambda = 342$ nm, ultraviolet

### Exercise group 2.2

(1) $4.945 \times 10^{-11}$ m
(2) $\lambda = 4 \times 10^{-10}$ m. The question of whether quantum effects are important or not reduces to deciding whether this is a large or a small number. This in turn depends on what kind of experiment we have in mind. If, for instance, we wanted to hold a helium atom in a nanometer-sized atom trap, then quantum effects would be important, since the wavelength is a significant fraction of the size of the trap. If, on the other hand, we wanted to study a helium gas in a large container (several centimeters across, say), then we might conclude that quantum mechanical effects are unimportant, since the wavelength is so much smaller than the size of the container.

### Exercise group 2.3

(1) $2.93 \times 10^{-71}$ J. Small doesn't even begin to describe this number.
(2) $n = 8 \times 10^{7}$, $\Delta E = 5 \times 10^{-29}$ J. Since the gap between energy levels is much, much smaller than the kinetic energy, the energy will behave like a continuous variable so we don't expect quantum effects to be important.

### Exercise group 2.4

(1) 121.5684 nm
(2) (a) $\frac{1}{2}h\nu_0$. (b) $E_{\text{photon}} = h\nu_0\Delta\nu$. (c) $\nu_0 = 3.598 \times 10^{13}$ Hz. (d) 2400 cm$^{-1}$.

### Review exercise group 2.5

(1) There must be two energy levels in the molecule separated by the energy of the photon, i.e. by $hc/\lambda$.
(2) Energy: $1.62 \times 10^{-24}$ J; wavelength: 0.122 m; wavenumber: 8.17 m$^{-1}$
(3) $5.807 \times 10^{-15}$ m. This is a tiny wavelength, smaller than typical atomic dimensions by five orders of magnitude, so quantum effects are not going to be significant in most experiments. However,

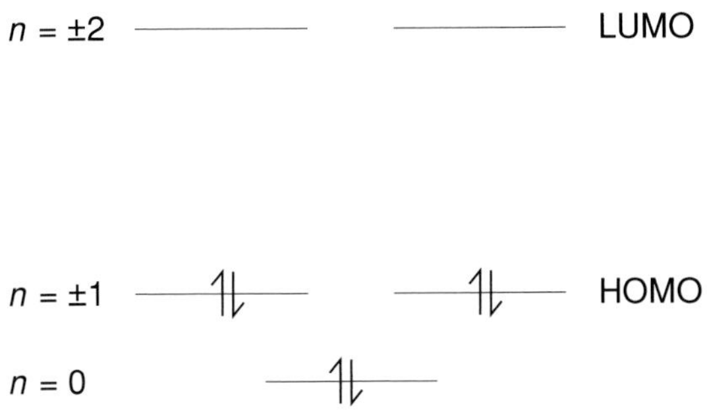

$$n = \pm 2 \quad \underline{\qquad} \qquad \underline{\qquad} \quad \text{LUMO}$$

$$n = \pm 1 \quad \underline{\uparrow\!\downarrow} \qquad \underline{\uparrow\!\downarrow} \quad \text{HOMO}$$

$$n = 0 \qquad \underline{\uparrow\!\downarrow}$$

Figure K.1 Particle-on-a-ring model of benzene, with $\pi$ electrons placed in lowest-energy orbitals.

this wavelength is similar to the size of a nucleus, so quantum effects might be important in experiments where we shoot $\alpha$ particles at nuclei (e.g. Rutherford's gold foil experiment).

(4) (a) $2\pi r = n\lambda$, where $n$ is an integer

    (b) $E_n = \dfrac{n^2 h^2}{8\pi^2 m r^2}$

    (c) $L_z = nh/2\pi$

    (d) See Figure K.1.

    (e) $1.28 \times 10^{-10}$ m

    (f) $1.28 \times 10^{-10}$ m. The bond length is actually $1.395 \times 10^{-10}$ m. The agreement isn't bad for a crude model!

## Exercise group 3.1

(1) 49.9988% at 20 °C, 49.9985% at $-30$ °C

(2) (a) The probability of a particular molecule being at altitude $h$ is given by $P(h) \propto \exp(-E_p/k_B T) = \exp(-mgh/k_B T)$. If we have a large number of molecules, we get $N(h) = N_{\text{total}} P(h) \propto \exp(-mgh/k_B T)$. This proportionality still holds if we consider a particular volume and divide by this volume, and $p \propto N/V$, so $p/p_0 = \exp(-mgh/k_B T)/e^0 = \exp(-mgh/k_B T)$.

    (b) Substitute $k_B = R/L$ and $m = M/L$ into the barometric formula.

    (c)

| $h/\text{m}$ | 920 | 1900 | 3660 |
|---|---|---|---|
| $p_{O_2}/\text{atm}$ | 0.18 | 0.16 | 0.13 |

    (d) 0.980%, vs. a sea-level abundance of 1.06%.

(3) (a) $P(E_2) = \dfrac{4}{\exp[\epsilon/k_B T] + 4}$. (b) In terms of the reduced temperature, we have $P(E_2) = \dfrac{4}{\exp(1/\theta) + 4}$. The percentage denatured would be $100 P(E_2)$. The denaturation curve is shown in Figure K.2. (c) We can't give an exact number for this temperature, but any answer near $\theta = 0.2$ or 0.3 would be reasonable. (d) 20%.

## Exercise group 3.2

(1) $\nu_0 = 8.6456 \times 10^{13}$ Hz

(2) 9

(3) (a) All four. (b) 0.0096.

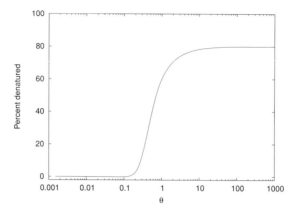

Figure K.2 Denaturation curve for a four-residue peptide with hydrophobic end residues on a lattice.

## Exercise group 3.3

(1) $1.6 \times 10^{-4}$ mol L

(2) (a) $7.19 \times 10^{-5}$ mol. (b) $1.1 \times 10^5$ L mol$^{-1}$cm$^{-1}$.

(3) (a) Suppose that the graph is convex, i.e. that it deviates downward from the line of best fit obtained at low concentrations. Then the apparent molar absorption coefficients would decrease with increasing concentration. The reverse would be true for a concave graph. These coefficients would not be of any real use. The correct procedure then would be to treat the $A$ vs. $c$ curve as an empirical relationship and to interpolate on this curve to estimate concentrations from absorbance measurements. (b) (i) The four concentrations are $3.0 \times 10^{-4}$, $5.95 \times 10^{-4}$, $8.92 \times 10^{-4}$ and $1.19 \times 10^{-3}$. (ii) 454 L mol$^{-1}$cm$^{-1}$.

(4) (a) [Chl $a$] = 0.006 05 g L$^{-1}$, [Chl $b$] = 0.001 82 g L$^{-1}$; 3.02 mg Chl $a$, 0.908 mg Chl $b$; 29.1 mg Chl $a$/g, 8.73 mg Chl $b$/g

   (b)

|  | $\varepsilon_\lambda$/L mol$^{-1}$cm$^{-1}$ | |
|---|---|---|
|  | 643 nm | 660 nm |
| Chlorophyll $a$ | $1.46 \times 10^4$ | $9.11 \times 10^4$ |
| Chlorophyll $b$ | $5.22 \times 10^4$ | $4.08 \times 10^3$ |

## Exercise group 3.4

(1) (a) $\epsilon$-ATP is the donor, and FITC is the acceptor. (b) 3.8 nm.

(2) FRET is most sensitive to distance when $R \sim R_0$. The only dye in this set that satisfies this condition is DPH.

(3) (a) Protein without lipid: 23 Å; protein with lipid: 42 Å. (b) The two helices involved in this experiment are close together (probably packed against each other) in the absence of lipid. Binding to a lipid induces a conformational change that moves these two helices apart.

## Review exercise group 3.5

(1) (a) The dipole moment of $O_2$ is identically zero. This frequency could have been measured by observing vibrational bands in the UV spectrum. Alternatively, it could have been done by Raman spectroscopy, a scattering technique not discussed in this book. (b) Somewhere around 1000–2000 K. (There's no point trying to be too precise here since the question isn't very precise.)

(2) (a) The energy level spacing between levels **1** and **2** is $2.84 \times 10^{-19}$ J. (b) $T = -8926$ K. The temperature calculated this way is negative. We are used to thinking of the Kelvin scale as starting

at zero. Note that a negative temperature is actually *hotter* than any normal temperature since, to return to equilibrium with surroundings of any given positive temperature, we would have to return the system to a Boltzmann distribution in which there were more molecules in the ground state than in the excited state, which means that the system would have to release energy. This is, to say the least, an unintuitive property of our usual absolute temperature scale.

(3) Fluorescence intensities depend on a number of details of the environment, such as the presence of quenchers. Moreover, the fluorescence intensity depends on the intensity of the exciting radiation. In order to reproducibly get the same intensities, we have to use the same spectrometer, and identically prepared solutions.

## Exercise group 5.1

(1) 2.25 u

## Exercise group 5.2

(1) $-911$ J
(2) $1.3 \times 10^5$ J. The work is positive, which indicates that work is being done *on* the system.
(3) (a) $-29$ kJ. The negative sign tells us that work is done *by* the system, i.e. that the system expends energy in the form of work. (b) The mass is raised by 1.3 km.
(4) (a) $-222$ J. (b) $-151$ J. (c) $-181$ J.
(5) (a) $w = -nRT \ln \left( \dfrac{V_f - nb}{V_i - nb} \right) - n^2 a \left( \dfrac{1}{V_f} - \dfrac{1}{V_i} \right)$. (b) $-2.713$ MJ. (c) $-2.740$ MJ. Whether the extra work was worth the effort will very much depend on how accurately we needed the answer. For most purposes, the ideal gas result would be accurate enough.
(6) $0.2$ J mol$^{-1}$. The work is tiny, especially when we consider the enormous change in pressure. For most purposes, ignoring pressure–volume terms for solids should therefore be safe.

## Exercise group 5.3

(1) According to the First Law of Thermodynamics, energy cannot be created, only transformed. The electric energy generated in the induction coil came from the magnetic fields produced by the power lines and thus ultimately from the energy generated by the power company. He was therefore stealing energy from the electric company.

## Exercise group 5.4

(1) 4.3 h
(2) 639 °C
(3) (a) 147 J. (b) The error in using a constant heat capacity is negligible.
(4) (a) 4.2 s. (b) 19 °C.

## Exercise group 5.5

(1) $U_m = -L\epsilon \exp(\epsilon/k_B T)(\exp(\epsilon/k_B T) + 4)^{-1}$. $C_{V,m}/R = \dfrac{4\epsilon^2}{k_B^2 T^2} \exp(\epsilon/k_B T)\left(\exp(\epsilon/k_B T)\right)$
$+ 4)^{-2} = \dfrac{4}{\theta^2} \exp(1/\theta)(\exp(1/\theta) + 4)^2$. The result is shown in Figure K.3.

## Exercise group 5.6

(1) 1.33 mol of liquid water and 0.67 mol of steam at 100 °C
(2) 106 g

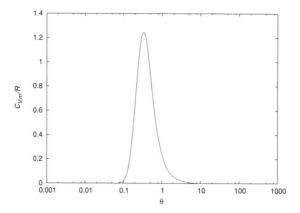

Figure K.3 Heat capacity of a four-residue peptide with hydrophobic end residues on a lattice.

(3) $c_p = 0.491 \, \mathrm{J\,K^{-1}g^{-1}}$ and $T_f = 1240 \, K$ (I got this number by careful measurement from the graph. Any answer within 10 degrees or so of this number would be reasonable.)

(4) (a) 694 kJ. (b) 14 °C.

## Exercise group 5.7

(1) 296.8 K
(2) 49 kJ mol$^{-1}$
(3) (a) $-206.61$ kJ mol$^{-1}$. (b) 59 g.
(4) (a) 0.002 10 mol. (b) 0.0126 mol. (c) 11.5 breaths min$^{-1}$. (d) $-585$ J.

## Exercise group 5.8

(1) $-57.53$ kJ mol$^{-1}$
(2) $-220$ kJ mol$^{-1}$

## Exercise group 5.9

(1) (a) $-1299.58$ kJ mol$^{-1}$. (b) $-1295.86$ kJ mol$^{-1}$.
(2) The balanced reaction is $CH_2O_{(s)} + O_{2(g)} \rightarrow CO_{2(g)} + H_2O_{(l)}$. $\Delta_r \nu_{gas} = 0$ so $\Delta H \approx \Delta U$.
(3) 134.46 kJ mol$^{-1}$ of heat produced
(4) (a) $-108.56$ kJ mol$^{-1}$. (b) $-19.01$ kJ.
(5) 2.4 g
(6) (a) 10.37 kJ K$^{-1}$. (b) $-5673$ kJ mol$^{-1}$. (c) $\Delta_f H^\circ(C_{12}H_{22}O_{11}, s) = -2194$ kJ mol$^{-1}$, a value in reasonable agreement, for a single experiment, with that given in the appendix.
(7) (a) $-203$ kJ mol. (b) 1.7 mol L$^{-1}$. (c) 174.5 kJ mol$^{-1}$.

## Exercise group 5.10

(1) $-2032$ kJ mol$^{-1}$
(2) 142 kJ mol$^{-1}$
(3) $-5634$ J mol$^{-1}$
(4) The enthalpy of reaction is $-2801.61$ kJ mol$^{-1}$ at 25 °C and $-2798.24$ kJ/mol at 37 °C. The difference (about 0.1%) is negligible for most purposes.
(5) $-144.1$ kJ mol$^{-1}$

## Exercise group 5.11

(1) $m_{\text{protein}} = 19\,\text{g}$, with a metabolic energy equivalent of 86 kcal or approximately 362 kJ
(2) (a) 32 days. (b) 30 days.
(3) (a) 2.96 Mcal. (b) 11 kcal.
(4) 15 kJ
(5) All other things being equal, the fat rat gains weight because he doesn't generate as much heat.
(6) 0.21 g
(7) (a) 172 kJ. (b) 4.3 g.
(8) 35 h
(9) 177 kcal

## Review exercise group 5.12

(1) In the following, words in parentheses are optional (a) molar (internal) energy (change) of combustion. (b) constant-volume heat capacity. (c) specific enthalpy (change) of reaction.
(2) (a) Mass of sample and temperature change. (b) To obtain the calorimeter's heat capacity. (c) The specific (or molar) internal energy change.
(3) Heat isn't a state function, so it doesn't make sense to talk of "the change in heat." To put it another way, bodies don't contain heat so the heat evolved during a process can't be thought of as the change of a property.
(4) $0.499\,\text{J}\,\text{K}^{-1}\text{g}^{-1}$
(5) (a) 351 kJ. (b) The dependence of the heat capacity on $T$ can't be ignored over a temperature range of 1400 K so the answer given above won't be very accurate. We need to know how $C_{p,m}$ depends on $T$. Then we could calculate the heat by

$$q = n \int_{293}^{1693\,\text{K}} C_{p,m}\, dT.$$

(6) $\Delta T = 33.5\,\text{K}$. This is not so large a temperature change as to make this process very dangerous (the final temperature will be well below the burn temperature for human skin), but it will clearly be necessary to stir the solution as we go to avoid creating hot spots due to slow diffusion of the heat away from the site of dissolution.
(7) (a) $-793.1\,\text{kJ}\,\text{mol}^{-1}$. (b) The heat capacities of the reactants and products were assumed to be independent of temperature. However, over the large temperature range studied (over 900 K), this is not a particularly good approximation. (c) $-793.1\,\text{kJ}\,\text{mol}^{-1}$.
(8) $36\,°\text{C}$
(9) $-13\,\text{kJ}$
(10) $26\,°\text{C}$
(11) $-108.95\,\text{kJ}\,\text{mol}^{-1}$
(12) $\Delta_f H°(\text{NO}_3^-, \text{aq}) = -207\,\text{kJ}\,\text{mol}^{-1}$; $\Delta_f H°(\text{NaNO}_3, \text{s}) = -467\,\text{kJ}\,\text{mol}^{-1}$
(13) $15\,300\,\text{t}\,\text{h}^{-1}$
(14) $50.9\,\text{kJ}\,\text{mol}^{-1}$

## Exercise group 6.1

(1) $\Delta S = 0$
(2) $\Delta S = \dfrac{q_{\text{rev}}}{T}$
(3) $20\,\text{J}\,\text{K}^{-1}\text{mol}^{-1}$
(4) $3.0\,\text{J}\,\text{K}^{-1}$
(5) (a) $35.7\,\text{kJ}\,\text{mol}^{-1}$. (b) $87.5\,\text{J}\,\text{K}^{-1}\text{mol}^{-1}$.

## Exercise group 6.2

(1) $\Delta S_{system,m} = -109.0\,\mathrm{J\,K^{-1}mol^{-1}}$, $\Delta S_{surroundings,m} = 138.8\,\mathrm{J\,K^{-1}mol^{-1}}$. The entropy change of the Universe is therefore positive so the process is thermodynamically allowed.

(2) $\Delta S_{glass} = 116.92\,\mathrm{J\,K^{-1}}$ and $\Delta S_{room} = -122.80\,\mathrm{J\,K^{-1}}$, therefore $\Delta S_{universe} = -5.88\,\mathrm{J\,K^{-1}} < 0$, indicating a process which *cannot* occur.

(3) $\Delta S < 0$, $\Delta U = 0$

(4) (a) $6.1\,^{\circ}$C. (b) $0.29\,\mathrm{J\,K^{-1}}$. The Second Law says that that entropy change for a thermodynamically allowed adiabatic process should be positive. Since temperature equilibration is a spontaneous process, this result is consistent with the Second Law.

(5) The reaction has a certain entropy change $\Delta_r S$. The reaction is exothermic so it will heat the contents of the adiabatic container, resulting in an entropy increase $\Delta_{heat} S$. Since the reaction is spontaneous, the Second Law requires that $\Delta_{heat} S + \Delta_r S > 0$, i.e. that $\Delta_r S > -\Delta_{heat} S$. Note that $\Delta_{heat} S > 0$ so that we cannot determine the sign of $\Delta_r S$ from this argument.

## Exercise group 6.3

(1) In order to determine whether or not a process is thermodynamically allowed, we must compute the change in entropy of the Universe, not just of the system. Most mixing processes are non-ideal; the particles interact so heat can be generated in the mixing process. If $q_{mix} > 0$ (i.e. heat flows into the system under isothermal conditions), then the surroundings must lose entropy to supply the heat needed. If this loss is greater than the gain in entropy of the system on mixing, the net change in entropy of the Universe is negative and the process in not thermodynamically allowed.

(2) The entropy of mixing is maximized when $X_1 = X_2 = \frac{1}{2}$.

## Exercise group 6.4

(1) (a) $k_B \ln 4$. (b) We see here a very clear illustration of entropy as a measure of our ignorance. The number of microstates is purely a function of our lack of knowledge of anything but the total amount which our friend has in his pocket. (c) If we knew that one of the coins was a 5¢ piece, then two of the above possibilities would be eliminated and the entropy would decrease (to $k_B \ln 2$) as our knowledge of the system increased.

(2) The number of molecules increases during the reaction. This represents an increase in the number of microstates so the entropy of the system increases.

(3) The reaction uses up a gas and forms a crystalline solid. The product has fewer microstates than the reactants so the entropy decreases.

(4) The entropy would be the same in both cases. In the first case, we would need to figure out how many ways we can put $N/10$ molecules on $N$ sites. In the second case, we could tell ourselves that we are choosing the $N/10$ unoccupied sites. Either way, the number of configurations is exactly the same.

(5) (a) 520. (b) We would calculate a larger number of conformations for the unfolded state.

## Exercise group 6.5

(1) $200.09\,\mathrm{J\,K^{-1}mol^{-1}}$

(2) $279.1\,\mathrm{J\,K^{-1}mol^{-1}}$

(3) $\Delta_{heating} S = 34.9\,\mathrm{J\,K^{-1}mol^{-1}}$, $\Delta_{fus} S = 9.68\,\mathrm{J\,K^{-1}mol^{-1}}$

(4) (a) The reaction is $2H^+_{(aq)} + CO_3^{2-}_{(aq)} \rightarrow H_2O_{(l)} + CO_{2(g)}$. Ions tend to organize polar solvent molecules, so the reactant side is more organized (has fewer microstates) than might first appear. The product side includes a gas, which has a lot of microstates. We would therefore expect the entropy to increase in this reaction. (b) $333.7 \, J \, K^{-1} mol^{-1}$.

(5) $14.897 \, 53 \, J \, K^{-1} mol^{-1}$

## Exercise group 6.6

(1) (a) 2 MW. (b) 500 K.
(2) 29 kW
(3) (a) 19 MW. (b) $3376 \, kJ \, kg^{-1}$. (c) $5.6 \, kg \, s^{-1}$. (c) $0.66 \, kg \, s^{-1}$.
(4) (a) $1.70 \times 10^9 \, m^3$. (b) 286 MW. (c) $6.69 \times 10^8 \, m^3 \, day^{-1}$. (d) Suppose that we can only get 30% efficiency from the electrical generator instead of the roughly 60% efficiency you would calculate from the operating temperatures. We would have to use roughly twice as much fuel to generate the same amount of electricity. However, in a cogeneration plant, the extra heat generated by burning the additional fuel replaces heat that would otherwise have had to have been generated in a furnace or boiler. If the cogeneration plant and furnace or boiler use the same fuel, then the efficiency of the electrical generator makes very little difference in terms of input costs. (It might make a difference in terms of operating profits depending on the price of electricity.) By contrast, in a conventional power plant, any extra fuel that has to be burned due to low efficiency is just wasted.

## Exercise group 6.7

(1) 255 kJ
(2) 19 W

## Review exercise group 6.8

(1) $p$, $S$ and $U$
(2) $210.21 \, J \, K^{-1} mol^{-1}$
(3) The Second Law says that the entropy of the Universe always increases. The argument given only considers the entropy change of the substance being frozen and not of its surroundings. A substance can be frozen if the sum of these two contributions to the entropy change of the Universe is positive.
(4) Entropy of vaporization is the entropy change in the process liquid $\rightarrow$ gas. Thus, $\Delta_{vap} S = S(gas) - S(liquid)$. A high entropy of vaporization could therefore be the result of one of two things: (1) the entropy of the gas could be unusually large, or (2) the entropy of the liquid could be unusually small. I can't think of any reason why (1) would be true for water. However, (2) is quite clearly true for water; water forms strong hydrogen bonds which result in a substantial degree of organization, and thus a low entropy.

   Based on this argument, you would expect that other compounds that form strong hydrogen bonds (alcohols, carboxylic acids, ... ) would also have large entropies of vaporization. This turns out to be true.
(5) (a) The reaction is

$$CH_3CH_2OH_{(l)} + 3O_{2(g)} \rightarrow 2CO_{2(g)} + 3H_2O_{(l)}.$$

There is a decrease in the number of moles of gas. Moreover, ethanol can form only one hydrogen bond while three water molecules can be involved in multiple hydrogen bonding interactions. Both of these factors indicate an increase in the organization of the products relative to the reactants and so a decrease in entropy.

(b) $-138.71\,\mathrm{J\,K^{-1}mol^{-1}}$

(c) We would have to adjust the entropies of the reagents and products to the new temperature. This is done by adding the entropy at $25\,^\circ$C to the change in entropy associated with heating each substance to $500\,^\circ$C. For water and ethanol, we would need to calculate the change in entropy on warming the liquid to the boiling point, the entropy of vaporization at the boiling point and the change in entropy on warming the vapor to the final temperature. For oxygen and carbon dioxide, we would only need to calculate the entropy change on warming these gases from $25\,^\circ$C to $500\,^\circ$C. Accordingly, we would need the following data:
   - the entropies at $25\,^\circ$C (from the table)
   - the boiling temperatures of water and of ethanol
   - the heats of vaporization of water and of ethanol at their respective boiling points
   - the specific heat capacities of ethanol and water in both the liquid and vapor phases, of oxygen and of carbon dioxide.

## Exercise group 7.1

(1) If $q > 0$, then $\Delta S \geq q/T > 0$.

(2) If $q < 0$, $\Delta S$ can be either positive or negative and still satisfy the Clausius inequality.

(3) (a) $-2.24\,\mathrm{kJ\,mol^{-1}}$. (b) No process can produce more work than an equivalent reversible process. The claim that the machine produces $3.5\,\mathrm{kJ\,mol^{-1}}$ is in error. Your friend shouldn't invest.

## Exercise group 7.2

(1) $\Delta G = -nRT \ln (V_2/V_1)$

## Exercise group 7.3

(1) $\Delta G = \Delta A + \Delta(pV)$ in general. $\Delta G = \Delta A - w_p$ at constant $p$.

## Exercise group 7.4

(1) $\Delta_r G^\circ = -1325.5\,\mathrm{kJ\,mol^{-1}} < 0$ therefore the reaction is thermodynamically allowed.

(2) (a) $118.89\,\mathrm{J\,K^{-1}mol^{-1}}$. (b) $188.84\,\mathrm{J\,K^{-1}mol^{-1}}$.

(3) $-92.1\,\mathrm{J\,K^{-1}mol^{-1}}$

(4) $-150.720\,\mathrm{kJ\,mol^{-1}}$

## Exercise group 7.5

(1) For this reaction,

$$Q = \frac{a_P}{a_A\, a_{H_2O}}.$$

Note that $a_{H_2O} = X_{H_2O}$. Adding an inert salt to the solution decreases the activity of water, which increases $Q$. Since

$$\Delta_r G_m = \Delta_r G_m^\circ + RT \ln Q$$

increased $Q$ can only make the free energy change more positive, and thus cannot make the reaction thermodynamically allowed.

(2) $\Delta_r G_m = -111.24\,\mathrm{kJ\,mol^{-1}} < 0$ so the reaction is thermodynamically allowed.

(3) The free energy change for the reaction $Pb^{2+}_{(aq)} + SO^{2-}_{4(aq)} \rightarrow PbSO_{4(s)}$ under the stated conditions is $-18.93\,\mathrm{kJ\,mol^{-1}}$, which is negative, so a precipitate will form.

(4) (a) $-4.15\,\text{MJ}\,\text{kg}^{-1}$. (b) $23\,\text{kJ}$.
(5) $116\,\text{MJ}$

## Exercise group 7.6

(1) $\Delta_r G_m = -2859.2\,\text{kJ}\,\text{mol}^{-1}$, so the maximum work is $2859.2\,\text{kJ}\,\text{mol}^{-1}$. In Example 7.3, we calculated that at $25\,^\circ\text{C}$, $\Delta_r G_m = -2859.28\,\text{kJ}\,\text{mol}^{-1}$. The two values are extremely close.
(2) (a) For the reaction $B_2H_{6(g)} + 3Cl_{2(g)} \rightarrow 2BCl_{3(l)} + 3H_{2(g)}$, $\Delta_r G = -880\,\text{kJ}\,\text{mol}^{-1}$. Since this is negative, the reaction is thermodynamically allowed. (b) At the new temperature, $\Delta_r G = -881\,\text{kJ}\,\text{mol}^{-1}$, which is still negative, so the reaction is still thermodynamically allowed.
(3) $w_{max} = 54\,\text{J}$; $h_{max} = 5.5\,\text{m}$

## Review exercise group 7.7

(1) The activity of the solvent is its mole fraction. Except for very concentrated solutions, the mole fraction of the solvent is usually very close to unity.
(2) $\Delta G$ gives the maximum **non-$pV$** work.
(3) Solids and liquids are incompressible. Therefore, $\Delta H = \Delta U + \Delta(pV) \approx \Delta U$. It follows then that $\Delta G = \Delta H + \Delta(TS) \approx \Delta U + \Delta(TS) = \Delta A$.
(4) (a) $213\,\text{kJ}\,\text{mol}^{-1}$. (b) $193\,\text{kJ}\,\text{mol}^{-1}$.
(5) $7.0\,\text{km}$
(6) (a) $-491.5\,\text{kJ}\,\text{mol}^{-1}$. (b) This quantity is related to the heat capacity: $H_m^\circ(298.15\,\text{K}) - H_m^\circ(0\,\text{K}) = \int_0^{298.15\,\text{K}} C_{p,m}\,dT$ (assuming that there are no phase transitions between 0 and 298.15 K).
(7) The Gibbs free energy change under the stated conditions is $-20.98\,\text{kJ}\,\text{mol}^{-1}$. Since the free energy change is negative, the reaction would be thermodynamically allowed. There is therefore no thermodynamic reason why this would not work.
(8) $\Delta_r S_m = \Delta_r S_m^\circ - R\ln Q$
(9) $34.4\,\text{kJ}\,\text{mol}^{-1}$
(10) (a) $-401\,\text{kJ}\,\text{mol}^{-1}$. (b) The sodium/sulfur battery has the larger energy storage capacity, by a factor of more than two. (c) Solids have unit activity, except under very unusual circumstances. Accordingly, $\Delta_r G = \Delta_r G^\circ$ as long as we have any reactants left. If any of the reactants or products were, for instance, solutes, then the work produced per mole consumed would change as the battery discharges. We would somehow have to integrate the work over these changing conditions.

## Exercise group 8.1

(1) $9.97 \times 10^{-15}$
(2) Using standard methods, we find a solubility of $483\,\text{mol}\,\text{L}^{-1}$. This absurdly large number means that nickel chloride is extremely soluble in water, although clearly the level of theory applied thus far in this course is not adequate to determine exactly how soluble it is.
(3) $6.6$
(4) $9.1$
(5) (a) $-314.4\,\text{kJ}\,\text{mol}^{-1}$. (b) $-314.3\,\text{kJ}\,\text{mol}^{-1}$. The result is virtually identical. (c) $1.0 \times 10^{-37}$ Pa. (d) The atmospheric gases at that altitude are not in chemical equilibrium. In particular, photochemical processes are important.

## Exercise group 8.2

(1) 0.604

## Exercise group 8.3

(1) (a)  We get the maximum amount of product when a reaction reaches equilibrium. In this case, $\Delta_r G_m^\circ$ is relatively large and positive. This means that the equilibrium constant

$$K = \exp(-\Delta_r G_m^\circ / RT)$$

will be very small. Since

$$K = \frac{(a_B)(a_C)}{a_A}.$$

we wouldn't expect to get much of the product C unless the activity of the reactant A were extremely large.

(b)  (A) consumes the desired product C, so it won't be much help.

(B) has a negative standard free energy change, and it consumes the undesired product B, so it would increase the yield (technically), but not by enough to make the synthesis of C practical, given the size of its $\Delta_r G_m^\circ$.

(C) has a negative free energy change and generates the reactant A. Because we're making a reactant, stoichiometry is going to limit the effectiveness of this reaction in increasing the yield.

We conclude that none of these reactions will help us very much.

(2) (a) $1.5 \times 10^{-5}$. (b) $2.5 \times 10^5$.

(3) $26.87 \, \text{kJ mol}^{-1}$

## Exercise group 8.4

(1) $91 \, \text{mol L}^{-1}$. This is an unreasonably large concentration, indicating that ATP hydrolysis provides enough free energy to reach any reasonable concentration of glucose in the cell.

(2) (a) 52.49 kJ. (b) $-8.56$ kJ. (c) 7.

## Exercise group 8.5

(1) Dissolving an ionic salt in water is generally an endothermic reaction.

(2) $9.3 \times 10^{-13}$ bar

(3) (a) 0.0259 bar. (b) 0.0816 bar. (c) 0.119 bar. Benzene should be most easily removed.

(4) (a) 97 °C. (b) My calculation did not take into account the temperature variation of $\Delta H^\circ$.

(5) Let's first consider an exothermic reaction. Le Chatelier's principle predicts that the equilibrium constant of an exothermic reaction decreases as the temperature increases. (The reactive system attempts to "soak up" the extra thermal energy by favoring the direction which absorbs heat.) Take $T_2 > T_1$. For an exothermic reaction, $\Delta_r H_m^\circ < 0$ so the right-hand side of the equation is negative. Thus, $K_2$ (at the higher temperature $T_2$) is lower than $K_1$, which is precisely as predicted by Le Chatelier's principle. Conversely, for an endothermic reaction, the right-hand side of the equation is positive implying that $K_2 > K_1$, which agrees with Le Chatelier's principle since in this case heat is absorbed by forming more products.

(6) (a) $-27.28 \, \text{kJ mol}^{-1}$. (b) $-37.96 \, \text{kJ mol}^{-1}$. (c) $126.9 \, \text{J K}^{-1}\text{mol}^{-1}$.

(7) (a) $\Delta_r G$ for the decomposition reaction is positive under normal atmospheric conditions so the decomposition is not thermodynamically allowed. This means that calcium carbonate is stable. (b) 518 °C.
(8) (a) 3.76. (b) The solubility goes up as the temperature goes down, so we need a smaller pressure to get a given amount of $CO_2$ into solution.
(9) 33.96 kJ mol$^{-1}$
(10) (a) $1.41 \times 10^{-4}$ mol L$^{-1}$. (b) $-350.2$ kJ mol$^{-1}$. (c) $-165.7$ kJ mol$^{-1}$. (d) $-143.5$ kJ mol$^{-1}$.

## Exercise group 8.6

(1) 6.8 °C
(2) −4.9 °C

## Review exercise group 8.7

(1) The process of dissolving gas molecules in water must generally be exothermic.
(2) If the reaction is endothermic, increasing the temperature should favor products. Conversely, we should decrease the temperature if the reaction is exothermic.
    Since the reaction converts two moles of gas to one, Le Chatelier's principle suggests that increasing the pressure on the system should favor the products.
(3) The vapor pressure of lead is $3.21 \times 10^{-22}$ mg m$^{-3}$ so yes, it's safe.
(4) 3.62
(5) The boiling point at this pressure is 59 °C so it should still be possible to observe liquid water on the surface of the planet.
(6) −628 kJ mol$^{-1}$
(7) The pH is 10.1 at 25 °C and 9.3 at 60 °C.
(8) 107 °C
(9) −1149.3 kJ mol$^{-1}$
(10) (a) 83.1 kJ mol$^{-1}$. (b) $1.4 \times 10^3$.
(11) [tetramer] = 21 $\mu$mol L$^{-1}$ and [dimer] = 8 $\mu$mol L$^{-1}$
(12) 196 g
(13) (a) D-glucose $+ 2P_i + 2\,NAD_{ox} + 2\,ADP \rightarrow 2$ pyruvate $+ 2\,NAD_{red} + 2\,H^+ + 2\,ATP + 2\,H_2O$, $\Delta_r G^{o\prime} = -73.1$ kJ mol$^{-1}$. (b) $\left.\frac{a_{ATP}}{a_{ADP}}\right|_{max} = 2.11 \times 10^6$.
(14) $7.2 \times 10^{-4}$ g
(15) $\Delta S^\circ = 22$ J K$^{-1}$mol$^{-1} > 0$. We would expect a positive $\Delta S^\circ$ for the breaking of a hydrogen bond, so this is consistent with the hypothesis.
(16) (a) 16 atm. (b) 36.50 kJ mol$^{-1}$. (c) 5.8 kJ. (d) −79 J. (e) Heat is clearly much more important. (Compare the answers to questions 16c and 16d, noting that the latter is in joules while the former is expressed in kilojoules.) We should therefore concentrate on heat if we want to know the $\Delta U$ of popcorn. Some data which would prove useful are the heat capacity of unpopped popcorn and the change in enthalpy associated with cooking the starch and protein. If we had these, we could calculate the heat required to bring the whole, unbroken kernel to 200 °C, the heat needed to vaporize the moisture at this temperature (calculated above), and the heat used to cook the solids. The sum of these quantities should be reasonably close to $\Delta U$.
(17) (a) 18.3 kJ mol$^{-1}$. (b) −10.8 kJ mol$^{-1}$. (c) $5.2 \times 10^{-4}$ mol L$^{-1}$bar$^{-1}$. (d) $2.6 \times 10^{-3}$ mol L$^{-1}$. (e) 308 cm$^3$. (f) 308 cm$^3$ is almost a third of a liter. Large bubbles will almost certainly be formed almost everywhere in the bloodstream. This could indeed cause serious problems.

## Exercise group 9.1

(1) 0.9890

## Exercise group 9.2

(1) 2.28
(2) $K_{sp} = 4.6 \times 10^{-7}$
(3) (a) In Equation (8.11) we see the ratio of equilibrium constants $K_1/K_2$. According to the Debye–Hückel theory, the activity coefficient of an ion only depends on the ionic strength and temperature. In these experiments, the ionic strength is constant, and activity coefficients depend only weakly on temperature, so the activity coefficients would roughly cancel in this ratio. (b) $13.0\,\mathrm{kJ\,mol^{-1}}$. (c) $K_a = 3.29 \times 10^{-6}$. (d) $31.3\,\mathrm{kJ\,mol^{-1}}$. (e) $30.0\,\mathrm{kJ\,mol^{-1}}$.

## Exercise group 9.3

(1) (a) 2.88. (b) $1.37 \times 10^{-3}\,\mathrm{mol\,L^{-1}}$.
(2) In pure water, $s = 8.9 \times 10^{-9}\,\mathrm{mol\,L^{-1}}$. In the sodium nitrate solution, $s = 9.9 \times 10^{-9}\,\mathrm{mol\,L^{-1}}$.
(3) (a) $-299.7\,\mathrm{kJ\,mol^{-1}}$. (b) $[H^+] = [F^-] = 0.0087\,\mathrm{mol\,L^{-1}}$. (c) 2.11.

## Review exercise group 9.4

(1) (a) $8.3 \times 10^{-5}\,\mathrm{mol\,L^{-1}}$. (b) 9.88.
(2) (a) $[P] = 1.01 \times 10^{-6}\,\mathrm{mol\,L^{-1}}$ and $[P_2] = 5.1 \times 10^{-7}\,\mathrm{mol\,L^{-1}}$. (b) $[P] = 8.3 \times 10^{-7}\,\mathrm{mol\,L^{-1}}$ and $[P_2] = 6.0 \times 10^{-7}\,\mathrm{mol\,L^{-1}}$.

## Exercise group 10.1

(1) $Ca_{(s)} + 2CO_{2(aq)} \rightarrow Ca^{2+}_{(aq)} + C_2O^{2-}_{4(aq)}$
(2) $2MnO^-_{4(aq)} + 5SO_{2(aq)} + H^+_{(aq)} + 2H_2O_{(l)} \rightarrow 2Mn^{2+}_{(aq)} + 5HSO^-_{4(aq)}$
(3) $2Bi(OH)_{3(s)} + 3SnO^{2-}_{2(aq)} \rightarrow 2Bi_{(s)} + 3SnO^{2-}_{3(aq)} + 3H_2O_{(l)}$
(4) $C_6H_{12}O_{6(aq)} + C_3H_3O^-_{3(aq)} + H_2O_{aq} \rightarrow C_6H_{11}O^-_{7(aq)} + C_3H_5O^-_{3(aq)} + H^+_{(aq)}$

## Exercise group 10.2

(1) $-1.17\,\mathrm{V}$
(2) $K = 0.0153$. This value of $K$ implies that the reaction is reactant favored, i.e. that there will be more $Hg_2^{2+}$ than $Hg^{2+}$ at equilibrium.
(3) (a) $3C_2H_5OH_{(aq)} + Cr_2O^{2-}_{7(aq)} + 8H^+_{(aq)} \rightarrow 3C_2H_4O_{(aq)} + 2Cr^{3+}_{(aq)} + 7H_2O_{(l)}$. (b) $0.2169\,\mathrm{V}$. (c) $-206\,\mathrm{kJ\,mol^{-1}}$.
(4) $0.45\,\mathrm{V}$
(5) $1.3 \times 10^{-7}\,\mathrm{mol\,L^{-1}}$
(6) 0.966

## Exercise group 10.3

(1) (a) $376\,\mathrm{kJ\,mol^{-1}}$. (b) $703\,\mathrm{kJ\,mol^{-1}}$.
(2) (a) $1.1446\,\mathrm{V}$. (b) $1.055\,\mathrm{V}$.
(3) (a) $86.5\,\mathrm{MJ\,m^{-3}}$. (b) $1.095\,72\,\mathrm{V}$.

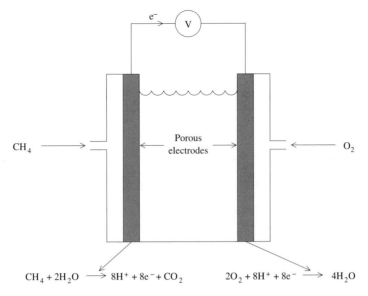

CH$_4$ + 2H$_2$O $\longrightarrow$ 8H$^+$ + 8e$^-$ + CO$_2$ $\qquad$ 2O$_2$ + 8H$^+$ + 8e$^-$ $\longrightarrow$ 4H$_2$O

Figure K.4 Methane/oxygen fuel cell.

### Review exercise group 10.4

(1) $3CN_{(aq)}^- + 2MnO_{4(aq)}^- + H_2O_{(l)} \rightarrow 3OCN_{(aq)}^- + 2MnO_{2(s)} + 2OH_{(aq)}^-$
(2) 0.455 28 V
(3) (a) $K = 6.29 \times 10^{-58}$. This is a tiny equilibrium constant, so at equilibrium (ignoring other reactions), we expect only a tiny amount of product to form. (b) This is an example of Le Chatelier's principle: the reaction of gold(III) ions with Cl$^-$ removes one of the products of the first reaction, which should favor the formation of more products.
(4) See Figure K.4 for the diagram. $E^\circ = 1.0597$ V.
(5) (a) 0.6 V. (b) $-307$ kJ mol$^{-1}$.

### Exercise group 11.1

(1) $1.3 \times 10^{-5}$ mol L$^{-1}$s$^{-1}$
(2) $1.0 \times 10^{-7}$ mol L$^{-1}$s$^{-1}$

### Exercise group 11.2

(1) (a) Certainly not, too many reactants. (b) Possibly. (c) Certainly not, too many bonds to make/break in one step.

### Exercise group 11.3

(1) (a) $v = kab$, $da/dt = -kab$. (b) $v = ka^2$, $da/dt = -2ka^2$. (c) $v = k_1a - k_{-1}bc$, $da/dt = -k_1a + k_{-1}bc$.

(2)

$$\frac{da}{dt} = \frac{db}{dt} = -k_1 ab + k_{-1}c$$

$$\frac{dc}{dt} = k_1 ab - k_{-1}c - 2k_2 c^2$$

$$\frac{dp}{dt} = k_2 c^2$$

## Exercise group 11.4

(1) $7.18 \times 10^{-25}\,\mathrm{bar^{-1}s^{-1}}$

## Review exercise group 11.5

(1) (a) $5.1 \times 10^{-17}\,\mathrm{m^3 mol^{-1}s^{-1}}$. (b) This is a third-order gas-phase elementary reaction. The role of the argon in this reaction is to carry away the excess energy released during bond formation. This situation (a so-called third body required to carry away excess energy) is one of the few where we commonly encounter third-order elementary processes. (c) First order with respect to argon, and second order with respect to I. (d) $7.3 \times 10^{-6}\,\mathrm{mol\,m^{-3}s^{-1}}$. (e) 4.2 ms.

(2) (a) $400\,\mathrm{L\,mol^{-1}}$. (b) $d[D]/dt = k_+[P]^2 - k_-[D]$, $d[P]/dt = -2k_+[P]^2 + 2k_-[D]$. (c) $d[D]/dt = -2 \times 10^4\,\mathrm{mol\,L^{-1}s^{-1}}$, so the reaction consumes the product, i.e. it proceeds in the reverse direction.

## Exercise group 12.1

(1) 1.6

(2) $v = k[O_3][NO]$ with $k = 3.6 \times 10^6\,\mathrm{L\,mol^{-1}s^{-1}}$

(3) $v = k[H^+][CH_3COCH_3]$ with $k = 3.86 \times 10^{-3}\,\mathrm{L\,mol^{-1}s^{-1}}$

(4) (a) $v = k[\text{ketone}][H^+]$. (b) $0.035\,\mathrm{L\,mol^{-1}s^{-1}}$.

## Exercise group 12.2

(1) The order of the reaction is 1. The rate constant is $1.0 \times 10^{-5}\,\mathrm{s^{-1}}$.

(2) The van't Hoff plot is shown in Figure K.5. If we fit the entire data set (solid line in the figure), we get a very poor fit, suggestive of curvature. We would then be tempted to reject the hypothesis that the data fit a simple rate law. However, the first three points (dashed line) fit a simple rate law accurately. Do we then assume that the last point is an experimental glitch? That view would certainly be defensible. In an ideal world, I would go back and measure that last point again, and maybe try to get some data beyond the range of concentrations already studied. This example shows the perils of having small data sets; we don't have enough points to either accept or reject that last point with any confidence. If things go really well, then four points is sometimes enough to extract kinetic parameters. Then there are cases like this one...

Anyway, if you believe that the first three points sit on a line and that the last point is bad, you would report an order of reaction of 0.82.

(3) (a) The plots of $\ln v$ vs. $\ln p_{HBr}$ and of $\ln v$ vs. $\ln p_{NO_2}$ are both linear. The slopes are $a = b \approx 1$. (b) $k = 0.0238\,\mathrm{min^{-1}}$.

## Review exercise group 12.3

(1) $\frac{1}{2}$

(2) (a) $-0.52\,\mathrm{mol\,L^{-1}min^{-1}}$. (b) $1.56\,\mathrm{mol\,L^{-1}min^{-1}}$.

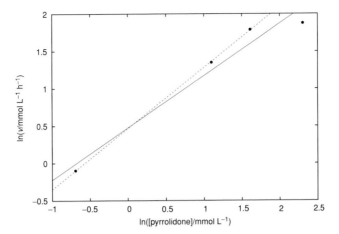

Figure K.5 van't Hoff plot for the oxidation of pyrrolidone by TPPFe and Bu$^t$OOH. The solid line is a fit of the entire data set. The dashed line is a fit of the first three points only.

(3) Rate law: $v = ka^{1/3}$. (Note that it's not necessary to round the exponent to $1/3$, but that it is necessary to (a) use each of experiments 1, 2 and 3 at least once to get the order with respect to A (i.e. make two determinations of the order) and (b) note that the rate is basically constant in experiments 1, 4 and 5, indicating that the order with respect to B is zero. ) Overall order: $1/3$. Rate constant (obtained by averaging estimates from all five experiments): $1.51 \times 10^4 \, \text{mol}^{2/3} L^{-2/3} s^{-1}$.

## Exercise group 13.1

(1) (a) 3.43 min. (b) 22.8 min.
(2) (a) A plot of $\ln [\text{In}^+]$ vs. $t$ is linear which confirms the first-order rate law. (b) $k = 1.01 \times 10^{-3} \, \text{s}^{-1}$. (c) 0.15 g.
(3) 27 d

## Exercise group 13.2

(1) 0.37 µg
(2) $2.0 \times 10^{15}$ y
(3) $4.51 \times 10^9$ y
(4) (a) 80 y. (b) 2010. (c) A graph of $\ln p$ vs. $t$ doesn't look linear, so no.
(5) 7.4–9.6 Gy
(6) If we plot $\ln \Delta T$ vs. $t$, we get the graph shown in Figure K.6. The graph is clearly curved, so we conclude that cooking roast beef does *not* obey Newton's law.

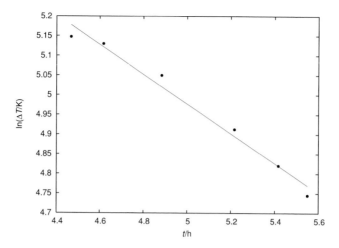

Figure K.6 Roast beef temperature vs. time data.

## Exercise group 13.3

(1) First order with $k = 3.4 \times 10^{-2}$ min$^{-1}$ (rate constant directly computed from the slope)
(2) 0.038 mol L$^{-1}$
(3) (a) The process (collide and stick) is simple enough to be elementary. The reaction would follow second-order kinetics if it were elementary. (b) A plot of $1/p_B$ vs. $t$ is linear, indicating that this is a second-order reaction. The rate constant is $2.279 \times 10^{-5}$ torr$^{-1}$min$^{-1}$ (directly computed from the slope).
(4) (a) Second order, $k = 1.50$ L mol$^{-1}$min$^{-1}$. (b) 465 days, far too long to be practical.
(5) Both the first- and second-order plots are curved. The data support neither a first- nor a second-order model.

## Exercise group 13.4

(1) $kt = \displaystyle\int_0^p \frac{\mathrm{d}p'}{(a_0 - 2p')(b_0 - 3p')}$
(2) (a)

$$t = \frac{1}{2k\,([H_2]_0 - [NO]_0)}\left[\frac{1}{[H_2]_0 - [NO]_0}\ln\left(\frac{[H_2]_0[NO]}{[NO]_0\,([NO] + [H_2]_0 - [NO]_0)}\right)\right.$$
$$\left. +\frac{1}{[NO]} - \frac{1}{[NO]_0}\right]$$

(b) 8671 s

## Exercise group 13.5

(1) (a) 9.2 ns. (b) 5.6 ns. (c) 26.1 Å.

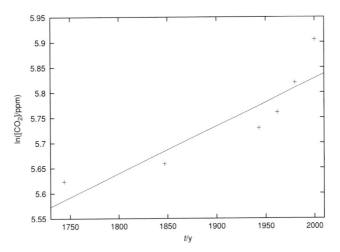

Figure K.7 Atmospheric $CO_2$ concentrations from question 2 of Exercise group 13.6.

## Review exercise group 13.6

(1) No. The half-life of a second-order reaction depends on the initial concentration of reactant, so at minimum we need to know this concentration to make sense of this number.

(2) The first-order plot is shown in Figure K.7. The data clearly do *not* obey a first-order rate law, and in fact increase much faster than exponentially.

(3) (a)

$$\frac{d(\Delta V)}{dt} = -\frac{\Delta V}{RC}$$

$$\therefore \frac{d(\Delta V)}{\Delta V} = -\frac{dt}{RC}$$

$$\therefore \int_{\Delta V_0}^{\Delta V} \frac{dx}{x} = -\frac{1}{RC} \int_0^t dt'$$

$$\therefore \ln x \big|_{\Delta V_0}^{\Delta V} = -\frac{1}{RC} t' \big|_0^t$$

$$\therefore -\frac{t}{RC} = \ln(\Delta V) - \ln(\Delta V_0) = \ln\left(\frac{\Delta V}{\Delta V_0}\right)$$

$$\therefore \frac{\Delta V}{\Delta V_0} = \exp(-t/RC)$$

$$\therefore V - V_{eq} = \Delta V_0 \exp(-t/RC)$$

$$\therefore V = V_{eq} + \Delta V_0 \exp(-t/RC)$$

(b) 691 µs

(4) There are three possible answers, depending on the rate equation you choose to work from:
**Using the [NO] rate law:**

$$t = \frac{-1}{2k} \int_{[NO]_0}^{[NO]} \frac{dx}{x^2 \left\{ [Cl_2]_0 - \frac{1}{2}([NO]_0 - x) \right\}}$$

**Using the [Cl$_2$] rate law:**

$$t = \frac{-1}{k} \int_{[Cl_2]_0}^{[Cl_2]} \frac{dx}{x \left\{ [NO]_0 - 2([Cl_2]_0 - x) \right\}^2}$$

**Using the [NOCl] rate law:**

$$t = \frac{1}{2k} \int_0^{[NOCl]} \frac{dx}{([NO]_0 - x)^2 \left( [Cl_2]_0 - \frac{1}{2}x \right)}$$

(5) $\dfrac{1}{a_0 - \frac{1}{2}b} = \dfrac{1}{a_0} + kt$

(6) (a) $\frac{2}{3}$. (b) $t_2 = m_0^{1/3} \left( \sqrt[3]{2} - 1 \right) / k$.

(7) The second-order plot is linear, which supports the hypothesis that DHF decays by recombination.

(8) (a) $v = k[\text{Si}][O_2]$

   (b)  (i) Given a large excess of oxygen, the oxygen concentration will be roughly constant during the experiment. Thus, $v = (k[O_2])[\text{Si}] = k'[\text{Si}]$ with $k' = k[O_2]$. A graph of $k'$ vs. [$O_2$] should therefore have a slope of $k$ and an intercept of 0.

      (ii) $k = 8.85 \times 10^{10} \, \text{L mol}^{-1}\text{s}^{-1}$. The intercept is $5083 \, \text{s}^{-1}$. This is smaller than any of the rate constants in the original data set, and there is some scatter in the values, so this value is in reasonable agreement with the expected intercept of zero.

      (iii) $22.8 \, \mu\text{s}$

## Exercise group 14.1

(1) (a) $2O_{3(g)} \rightarrow 3O_{2(g)}$. (b) $v = \left( \dfrac{k_1^2 k_3^2 k_4}{4k_{-1}^2} [N_2O_5]^2 [O_3]^2 \right)^{1/3}$. (c) Yes. Order w.r.t. $N_2O_5 =$ order w.r.t. ozone $= \frac{2}{3}$. (d) Yes.

(2) (a) $A \rightarrow P$. (b) $v = k_2 b \approx \dfrac{k_1 k_2 a^2}{k_{-1} + k_2}$. (c) 2.

(3) (a) $W + 2Y \rightarrow 2Z$. Reactants: W, Y. Product: Z. Intermediate: X. (b) A radical. (c) $v = \dfrac{1}{2} \dfrac{dz}{dt} = \dfrac{1}{2} k_2 y \sqrt{k_1 w / k_{-1}}$. (d) The steady-state approximation is applicable when either $k_{-1}w$ or $k_2$ is large compared to $k_1$. The steady-state approximation value of $x$ satisfies the quadratic equation

$$2k_{-1}wx^2 + k_2 yx - 2k_1 w^2 = 0$$

Taking the positive root and computing the reaction rate as above, we find

$$v = k_2 y \left( \frac{-k_2 y + \sqrt{k_2^2 y^2 + 16 k_1 k_{-1} w^3}}{4 k_{-1} w} \right)$$

(4) (a) **Mechanism 1:** Cl, $Cl_3$

    **Mechanism 2:** Cl, COCl

  (b) **Mechanism 1:**

    **Initiation:** $Cl_{2(g)} \rightarrow 2Cl_{(g)}$

    **Chain propagation:**

$$Cl_{(g)} + Cl_{2(g)} \rightarrow Cl_{3(g)}$$
$$Cl_{3(g)} + CO_{(g)} \rightarrow COCl_{2(g)} + Cl_{(g)}$$

    **Termination:** $2Cl_{(g)} \rightarrow Cl_{2(g)}$

    **None of the above:** $Cl_{3(g)} \rightarrow Cl_{(g)} + Cl_{2(g)}$

    **Mechanism 2:**

    **Initiation:** $Cl_{2(g)} \rightarrow 2Cl_{(g)}$

    **Chain propagation:**

$$Cl_{(g)} + CO_{(g)} \rightarrow COCl_{(g)}$$
$$COCl_{(g)} + Cl_{2(g)} \rightarrow COCl_{2(g)} + Cl_{(g)}$$

    **Termination:** $2Cl_{(g)} \rightarrow Cl_{2(g)}$

    **None of the above:** $COCl_{(g)} \rightarrow Cl_{(g)} + CO_{(g)}$

  (c) **Mechanism 1:** $v = \dfrac{k_2 k_3 [CO]}{k_{-2} + k_3 [CO]} \sqrt{\dfrac{k_1 [Cl_2]^3}{k_{-1}}}$

    **Mechanism 2:** $v = \dfrac{k_2 k_3 [CO]}{k_{-2} + k_3 [Cl_2]} \sqrt{\dfrac{k_1 [Cl_2]^3}{k_{-1}}}$

    The two mechanisms can be distinguished because the rates behave differently at high and low pressures of the two reactants. For instance, mechanism 1 predicts that the order with respect to CO will be zero at high pressure of CO, while mechanism 2 predicts first-order behavior with respect to CO under any experimental conditions.

(5) There are two ways to go about answering this question, each of which gives a slightly different result. If you assume that the first step is in quasi-equilibrium, we get the experimental rate law with $k = \frac{k_1 k_2}{k_{-1}}$. If you make this assumption, you would conclude that the first step is fast (in both directions) while the second step is slow.

    We can also apply the steady-state approximation to the $NOBr_2$ concentration. We get the rate law $v = \frac{k_1 k_2 [NO]^2 [Br_2]}{k_{-1} + k_2 [NO]}$, which reduces to the experimental rate law if $k_2 [NO] \ll k_{-1}$. The correspondence between the experimental rate constant $k$ and the rate constants appearing in the mechanism are exactly as above, i.e. $k = \frac{k_1 k_2}{k_{-1}}$. We get this result by assuming that the step controlled by $k_2$ is slow, along with the steady-state approximation which implies that the removal of the intermediate is fast. This means that the dissociation of $NOBr_2$ into NO and $Br_2$ is the fastest process in this reaction, the other two elementary reactions being relatively slow.

(6) If we apply the steady-state approximation to the intermediate, we get a rate law

$$v = \frac{k_1 k_2 a^2 c}{k_{-1} + k_2 c}$$

If $k_{-1} \gg k_2 c$, this reduces to $v \approx \frac{k_1 k_2}{k_{-1}} a^2 c$, which is a third-order rate law. This implies that the degradation of the dimer B back to monomers is a fast process.

## Exercise group 14.2

(1) (a) A $\rightarrow$ 3C. (b) $K = 4.4 \times 10^{-3} \, \text{mol}^2 \, \text{L}^{-2}$.

## Review exercise group 14.3

(1) (a) $v = \dfrac{k_1 k_2}{k_{-1} + k_2}$ [indole][$HSO_5^-$]. (b) $k = \dfrac{k_1 k_2}{k_{-1} + k_2}$. (c) $7.5 \times 10^4$ s or 21 h.

(2) (a) $2PAA \rightarrow 2AH + O_2$. (b) $v = \frac{k_1 k_2}{k_{-1}}$[PAA]$^2$[$H^+$]. (c) Second order. (d) Increasing the pH decreases the proton concentration, which should decrease the rate of reaction.

## Exercise group 15.1

(1) From an Eadie–Hofstee plot, we can calculate $v_{max} = 0.215\ \mu mol\,L^{-1}s^{-1}$ and $K_M = 159\ \mu mol\,L^{-1}$. However, the fit to the data is very poor, suggesting that the simple Michaelis–Menten mechanism may not be the correct model in this case.

(2) (a) $K_M = 5.5\ mmol/L$, $k_{-2} = 12\ s^{-1}$. (b) $k_1 > 2.2 \times 10^3\ L\,mol^{-1}s^{-1}$.

(3) $K_M = 21\ \mu mol\,L^{-1}$, specific activity $= 1.8\ nmol\,s^{-1}g^{-1}$

(4) (a) There is some scatter, but the Eadie–Hofstee plot is linear. This transport process therefore displays Michaelis–Menten kinetics. (b) $K_M = 31.8\ \mu mol/L$, $v_{max} = 0.903\ \mu mol\,min^{-1}g^{-1}$.

## Exercise group 15.2

(1) (a) The $v_{max}$ values are nearly constant and a plot of $K_M$ vs. $i_0$ is linear, so the data are compatible with competitive inhibition. (b) $v_{max} = 10.5\ \mu mol\,L^{-1}s^{-1}$, $K_S = 3.98\ mmol\,L^{-1}$, $K_I = 2.31\ mmol\,L^{-1}$.

(2) Competitive with $v_{max} = 745\ \mu mol\,L^{-1}s^{-1}$, $K_S = 83\ \mu mol\,L^{-1}$, $K_I = 18\ mmol\,L^{-1}$.

(3) I find the following for the Eadie-Hofstee plots at each of the BATA concentrations:

| [BATA]/$\mu mol\,L^{-1}$ | slope/$\mu mol\,L^{-1}$ | intercept/$\mu mol\,L^{-1}s^{-1}$ |
|---|---|---|
| 0   | −8.44   | 0.347 |
| 1.9 | −19.72  | 0.332 |
| 3.8 | −30.22  | 0.320 |
| 7.6 | −47.90  | 0.281 |

Both the intercept and slope are changing, so this is not an instance of competitive inhibition.

(4) If you create Eadie–Hofstee plots for each inhibitor concentration, you should get the following data:

| [E4P]/$\mu mol\,L^{-1}$ | $v_{max}$/$\mu mol\,g^{-1}min^{-1}$ | $K_M$/$\mu mol\,L^{-1}$ |
|---|---|---|
| 0   | 677 | 748 |
| 1.5 | 691 | 1376 |
| 3   | 501 | 954 |
| 4.5 | 521 | 2195 |
| 6   | 467 | 4284 |

The $v_{max}$ isn't constant, so that rules out competitive inhibition. (Despite the significant scatter, there is clearly an overall trend of $v_{max}$ decreasing with increasing [E4P].) If we had uncompetitive inhibition, $K_M$ would decrease with increasing [E4P]. However, the trend runs the other way, give or take some experimental scatter. Thus, it can't be uncompetitive inhibition either. Some other type of inhibition not studied in this course is operating here.

Incidentally, if you actually draw the Eadie–Hofstee plot (which you should), you'll see that the data are pretty messy. It's often difficult to get great data in some of these experiments. I should also note that there are one or two suspicious points that we might think about throwing out from the data set. This wouldn't affect our overall conclusion.

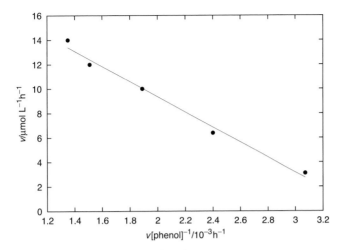

Figure K.8 Eadie-Hofstee plot for problem 1a from Exercise group 15.3.

## Review exercise group 15.3

(1) (a) See the Eadie–Hofstee plot in Figure K.8. There is some scatter, but no obvious curvature, so it seems that this biological remediation process does obey Michaelis–Menten kinetics.
(b) $K_M = 6.20 \, \text{mmol} \, L^{-1}$, $v_{max} = 21.8 \, \mu\text{mol} \, L^{-1} h^{-1}$.

(2) (a) We have the following conservation relations:

$$[\text{pol-}\alpha] + c_p = \alpha_0, \tag{K.1}$$

$$[\text{RF-C}] + c_r = r_0, \tag{K.2}$$

$$[\text{DNA}] + c_p + c_r = d_0, \tag{K.3}$$

where $\alpha_0$, $r_0$ and $d_0$ are, respectively, the total concentrations of pol-$\alpha$, RF-C and DNA. We are
told that

$$r_0 \gg d_0 \gg \alpha_0.$$

Since $c_r \leq d_0$ and $d_0 \ll r_0$, Equation (K.2) implies

$$[\text{RF-C}] \approx r_0.$$

Similarly, since $c_p \leq \alpha_0$ and $\alpha \ll d_0$, Equation (K.3) implies

$$[\text{DNA}] \approx d_0 - c_r.$$

Apply the SSA to $c_p$ and $c_r$. After some work, you should obtain

$$v = \frac{v_{max} d_0}{d_0 + K_S \left(1 + r_0/K_I\right)}$$

where $v_{max} = k_{-2}\alpha_0$, $K_S = (k_{-1} + k_{-2})/k_1$ and $K_I = k_{-3}/k_3$. (b) For [RF-C] $= 0$, $v_{max} =$
467 fmol min$^{-1}$ and $K_M = 0.105 \, \text{g} \, L^{-1}$. For [RF-C] $= 0.2$ mu, $v_{max} = 463$ fmol min$^{-1}$ and

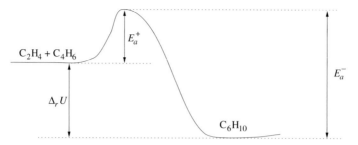

Figure K.9 Energy profile for the Diels–Alder reaction of ethene with 1,3-butadiene.

$K_M = 0.213 \, \text{g} \, \text{L}^{-1}$. Thus, $v_{max}$ is essentially constant, while $K_M$ increases with the inhibitor concentration, which is consistent with competitive inhibition. We calculate $K_S = 0.108 \, \text{g} \, \text{L}^{-1}$, $K_I = 0.205 \, \text{mu}$ and $v_{max} = 2.33 \, \text{nmol} \, \text{min}^{-1} \text{u}^{-1}$ (expressed as a specific activity).

(3) Use the enzyme conservation relation $e_0 = e + c_1 + c_2$, and apply the SSA to both intermediates. After some algebra, you get the Michaelis–Menten rate law with $v_{max} = \frac{k_3 e_0}{\alpha + 1}$ and $K_M = \frac{k_{-1}\alpha + k_3}{k_1(\alpha + 1)}$, where $\alpha = \frac{k_{-2} + k_3}{k_2}$.

(4) $v = \dfrac{e_0 \, [k_1 k_{-2} s - k_{-1} k_2 (s_0 - s)]}{k_1 s + k_{-1} + k_{-2} + k_2(s_0 - s)}$

## Exercise group 16.1

(1) $k_d = 8.0 \times 10^5 \, \text{s}^{-1}$, $k_a = 4.5 \times 10^{10} \, \text{L} \, \text{mol}^{-1} \text{s}^{-1}$

(2) (a) $\tau = (4k_a s_{eq} + k_d)^{-1}$. (b) The equilibrium constant decreases by a factor of 0.37.

(3) (a) Combine the equilibrium relation and the definition of $p_t$ to obtain $p_t = p_{eq} + 2k_a p_{eq}^2 / k_d$. Then isolate $p_{eq}$ from the equation for the relaxation time (from problem 2) and substitute into the equation for $p_t$. After rearranging, you get $\tau^{-2} = 8k_a k_d p_t + k_d^2$. If we plot $\tau^{-2}$ vs. $p_t$, the slope is $8k_a k_d$ and the intercept is $k_d^2$. (b) $k_a = 8.57 \times 10^8 \, \text{L} \, \text{mol}^{-1} \text{s}^{-1}$, $k_d = 1.95 \times 10^6 \, \text{s}^{-1}$. (c) $-15.1 \, \text{kJ} \, \text{mol}^{-1}$.

## Exercise group 17.1

(1) $E_a = 48.6 \, \text{kJ} \, \text{mol}^{-1}$, $A = 8.33 \times 10^4 \, \text{L} \, \text{mol}^{-1} \text{s}^{-1}$

(2) $E_a = 159 \, \text{kJ} \, \text{mol}^{-1}$, $A = 1.5 \times 10^{23} \, \text{L} \, \text{mol}^{-1} \text{s}^{-1}$. In this case, the Arrhenius plot shows an unusual amount of scatter, probably because of experimental errors. (Working in hot acid raises all kinds of problems.)

(3) (a) $E_a = 38.2 \, \text{kJ} \, \text{mol}^{-1}$, $A = 849 \, \text{min}^{-1}$. (b) $4.27 \times 10^{-5} \, \text{min}^{-1}$.

(4) (a) A plot of $\ln v$ vs. $1/T$ is linear, so the data are consistent with the Arrhenius equation. (b) $E_a = 24.5 \, \text{kJ} \, \text{mol}^{-1}$. (c) $38 \, ^\circ\text{C}$.

## Exercise group 17.2

(1) (a) See Figure K.9. (b) $271 \, \text{kJ} \, \text{mol}^{-1}$. (c) $\Delta^\ddagger H^+ = 110 \, \text{kJ} \, \text{mol}^{-1}$, $\Delta^\ddagger H^- = 269 \, \text{kJ} \, \text{mol}^{-1}$.

(2) If you calculate the activation parameters from the activation energy and pre-exponential factor, you get $\Delta^\ddagger H = 52 \, \text{kJ} \, \text{mol}$, $\Delta^\ddagger S = -136 \, \text{J} \, \text{K}^{-1} \text{mol}^{-1}$. If you use an Eyring plot, you get about the same enthalpy, but the entropy is closer to $-137 \, \text{J} \, \text{K}^{-1} \text{mol}^{-1}$.

(3) $A = 1.7 \times 10^{13} \, \text{s}^{-1}$. The condition $\Delta^\ddagger S = 0$ separates reactions in which the transition state has fewer microstates than the reactants (negative $\Delta^\ddagger S$, $A < 1.7 \times 10^{13} \, \text{s}^{-1}$, corresponding to

bond formation) from reactions in which the transition state has more microstates than the reactants (positive $\Delta^\ddagger S$, $A > 1.7 \times 10^{13}\,s^{-1}$, corresponding to bonds loosening on the way to the transition state). From Equation (17.16), we see that if $\Delta^\ddagger S_m = 0$, $A$ is $k_B T e / h$, i.e. it is just a small multiple of $k^\ddagger$ (Equation 17.8). Thus in this case, for reactants of sufficient energy, formation of the products is entirely associated with molecular vibrations, unlike the normal case in which statistical effects associated with the relative number of microstates of reactants and transition state (i.e. the activation entropy) modulate the pre-exponential factor.

(4) $\Delta^\ddagger S = -129\,J\,K^{-1}mol^{-1}$. This very negative value of $\Delta^\ddagger S$ tells us that the transition state is more organized than the reactants, i.e. that a complex is formed on the way to the transition state.

(5) (a)

| Substituent | $E_a/kJ\,mol^{-1}$ | $A/s^{-1}$ |
|---|---|---|
| −H | 145.44 | $2.63 \times 10^{12}$ |
| −$CH_3$ | 145.42 | $3.41 \times 10^{12}$ |
| −$OCH_3$ | 145.89 | $5.09 \times 10^{12}$ |
| −Cl | 146.22 | $3.13 \times 10^{12}$ |
| −$NO_2$ | 145.45 | $1.93 \times 10^{12}$ |

(b)

| Substituent | $\Delta^\ddagger H/kJ\,mol^{-1}$ | $\Delta^\ddagger S/J\,K^{-1}mol^{-1}$ |
|---|---|---|
| −H | 141.50 | −19.30 |
| −$CH_3$ | 141.48 | −17.14 |
| −$OCH_3$ | 141.95 | -13.82 |
| −Cl | 142.29 | −17.84 |
| −$NO_2$ | 141.52 | −21.86 |

There is very little variation in the enthalpy of activation. This means that the amount of energy required to reach the transition state is nearly independent of the substituents. On the other hand, there is significant variation in the entropy of activation. Since $\Delta^\ddagger S = S_{TS} - S_{reactant}$, this can either be due to a significant variation in the "looseness" of the transition state, or to variation in the number at microstates in the reactant. The former seems unlikely. The triangular transition state is probably quite similar in all cases. (The approximate constancy of the activation energy suggests this.) However, it may be that the isocyanide group is either more or less mobile in the reactant depending on substituent effects.

(6) There are at least three ways you can do this problem. Because there was significant scatter in the measurements, you will get slightly different values, depending on how you go about it. If you get the activation energy and pre-exponential factor using the raw data, then convert the exponential factor to $bar^{-1}s^{-1}$, and then compute the activation parameters, you get $\Delta^\ddagger H = 10.6\,kJ\,mol^{-1}$, $\Delta^\ddagger S = -101\,J\,K^{-1}mol^{-1}$ and $\Delta^\ddagger G = 40.6\,kJ\,mol^{-1}$. If you convert the rate constants to $bar^{-1}s^{-1}$ first, then get the activation energy and pre-exponential factor, and then calculate the activation parameters, you get $\Delta^\ddagger H = 7.7\,kJ\,mol^{-1}$, $\Delta^\ddagger S = -110\,J\,K^{-1}mol^{-1}$ and $\Delta^\ddagger G = 40.6\,kJ\,mol^{-1}$. If you use an Eyring plot, you get $\Delta^\ddagger H = 9.9\,kJ\,mol^{-1}$, $\Delta^\ddagger S = -103\,J\,K^{-1}mol^{-1}$ and $\Delta^\ddagger G = 40.6\,kJ\,mol^{-1}$. Whichever calculation you do, the value of $\Delta^\ddagger S$ is large and negative, which suggests that the transition state is a tightly bound complex of the two reactants.

## Exercise group 17.3

(1) The plot of $\ln k_\varrho$ vs. $\sqrt{I}$ is not linear, so Brønsted–Bjerrum theory is not adequate in this case. Brønsted–Bjerrum theory is based on Debye–Hückel theory, which assumes that charged particles

in solution organize in a certain way. The arrangement may, however, be substantially different for large polymeric particles than it is for smaller solutes.

(2) In Brønsted–Bjerrum theory, non-ideal effects enter through the equilibrium constant for accession to the transition state, $K^\ddagger$. For a first-order elementary reaction, $A \rightleftharpoons TS \rightarrow B$, we have

$$K^\ddagger = \frac{[TS]}{[A]} \frac{\gamma_{TS}}{\gamma_A}.$$

Debye–Hückel theory gives a first approximation to the activity coefficients. In this theory, the activity coefficients only depend on the charge. Since the charge of the transition state has to be the same as the charge of the reactant A, the two activity coefficients that appear in $K^\ddagger$ would cancel. As a side-note, in more sophisticated theories, the activity coefficients also depend on the size of a solute (Section 9.4). However, for a unimolecular reaction, where moving to the transition state only involves an internal rearrangement of some sort, the change in volume of the solute will be negligible. We would again conclude that $\gamma_{TS} \approx \gamma_A$, so that ionic strength should have very little (if any) effect on the rate constant.

(3) Consider an elementary reaction

$$A + B \underset{k_-}{\overset{k_+}{\rightleftharpoons}} C + D.$$

According to the Brønsted–Bjerrum theory,

$$k_+ = \frac{k^\ddagger K_+^\ddagger}{c^\circ} \frac{\gamma_A \gamma_B}{\gamma_{TS}}$$

$$\text{and } k_- = \frac{k^\ddagger K_-^\ddagger}{c^\circ} \frac{\gamma_C \gamma_D}{\gamma_{TS}}.$$

In these equations, $k^\ddagger$ is the frequency for motion through the transition state, $K_+^\ddagger$ is the equilibrium constant between the transition state and reactants, and $K_-^\ddagger$ is the equilibrium constant between the products and transition state. Note that $k^\ddagger$ is a universal temperature-dependent constant in transition-state theory (Equation 17.8), and not a quantity that depends on the specific reaction being studied. From our study of the relationship between kinetics and equilibrium, we know that the phenomenological equilibrium constant $K_{app}$ is related to the rate constants by

$$K_{app} = k_+/k_- = \frac{K_+^\ddagger}{K_-^\ddagger} \frac{\gamma_A \gamma_B}{\gamma_C \gamma_D}.$$

Now $K_+^\ddagger$ is related to the standard free energy change for accession to the transition state by

$$K_+^\ddagger = \exp(-\Delta_{(+)}^\ddagger G_m/RT)$$

$$\text{with } \Delta_{(+)}^\ddagger G_m = \Delta_f G_m^\circ(TS) - \left[\Delta_f G_m^\circ(A) + \Delta_f G_m^\circ(B)\right],$$

$$\text{and, similarly, } K_-^\ddagger = \exp(-\Delta_{(-)}^\ddagger G_m/RT)$$

$$\text{with } \Delta_{(-)}^\ddagger G_m = \Delta_f G_m^\circ(TS) - \left[\Delta_f G_m^\circ(C) + \Delta_f G_m^\circ(D)\right].$$

Putting these equations together, we get

$$K_{app} = \exp(\Delta_r G_m^\circ/RT) \frac{\gamma_A \gamma_B}{\gamma_C \gamma_D} = K \frac{\gamma_A \gamma_B}{\gamma_C \gamma_D},$$

$$\text{where } \Delta_r G_m^\circ = \Delta_f G_m^\circ(C) + \Delta_f G_m^\circ(D) - \left[\Delta_f G_m^\circ(A) + \Delta_f G_m^\circ(B)\right],$$

$$\text{or } K = K_{app} \frac{\gamma_A \gamma_B}{\gamma_C \gamma_D}.$$

This is exactly what we expect based on thermodynamics. $K_{app} = [C][D]/[A][B]$, so the thermodynamic equilibrium constant $K$ works out to be exactly the ratio of activities that we would have written down based on our study of chemical equilibrium. Moreover, the properties of the transition state have completely dropped out of the equation.

## Review exercise group 17.4

(1) (a) The two reactants have non-zero charges of the same sign. (b) $\Delta_r H > 0$.

(2) $372.75 \text{ kJ mol}^{-1}$

(3) (a) $v = k \dfrac{[OCl^-][I^-]}{[OH^-]}$ with $k = 60 \text{ s}^{-1}$. (b) The ionic strength changes substantially between experiments 3, 4 and 5, and yet the rate constant varies little. The experiment must involve some other salt used to keep a constant ionic strength, which really should be mentioned when the data are presented.

(4) (a) $v = k[H^+][HCOOH]$ with $k = 1.9 \times 10^{-4} \text{ kg mol}^{-1}\text{s}^{-1}$. (b) $E_a = 153 \text{ kJ mol}^{-1}$ and $A = 1.7 \times 10^{13} \text{ kg mol}^{-1}\text{s}^{-1}$.

(5) (a) Using the EA for the first step, we find $v = \frac{k_1 k_2}{k_{-1}} r$. (b) $E_{a(1)} + E_{a(2)} < E_{a(-1)}$ or $\Delta E_1 = E_{a(1)} - E_{a(-1)} < -E_{a(2)}$. In words, more energy must be released by reaction 1 than reaction 2 requires for activation. (This becomes clearer if you draw a reaction profile.)

(6) Since B is uncharged ($Z_B = 0$), the theory predicts that the rate constant will not depend on the ionic strength, i.e. that $k = 2.9 \times 10^3 \text{ L mol}^{-1}\text{s}^{-1}$ at any ionic strength. Brønsted–Bjerrum theory doesn't give us any guidance as to the value of the rate constant in reactions involving at least one neutral molecule, but that doesn't mean that the rate constant doesn't depend on the ionic strength. In fact, it is likely that ionic strength will have a measurable effect on the rate constant.

(7) (a) $v = \dfrac{k_1 k_2 [CH_3CH_2O][He]}{k_{-1}[He] + k_2}$. (b) If the pressure of helium is sufficiently low, we can ignore the term $k_{-1}[He]$ in the denominator of the rate expression. The rate becomes

$$v \approx \frac{k_1 k_2 [CH_3CH_2O][He]}{k_2} = k_1[He][CH_3CH_2O].$$

Since the helium concentration is constant ($d[He]dt = 0$ in this mechanism), the observed rate at low helium pressure would be $v \approx k_1'[CH_3CH_2O]$ where $k_1' = k_1[He]$. (c) $\Delta^\ddagger S = -32 \text{ J K}^{-1}\text{mol}^{-1}$.

(8) (a) $E_a = 74.5 \text{ kJ mol}^{-1}$, $A = 1.74 \times 10^{10} \text{ L mol}^{-1}\text{s}^{-1}$. (b) $\Delta^\ddagger S = -92.2 \text{ J K}^{-1}\text{mol}^{-1}$, a value consistent with a transition state with a bridging hydrogen atom.

(9) (a) $E_a = 128 \text{ kJ mol}^{-1}$, $A = 3.73 \times 10^{15} \text{ s}^{-1}$. (b) $\Delta^\ddagger S = 43 \text{ J K}^{-1}\text{mol}^{-1}$. This is a fairly large, positive entropy of activation which is likely associated with a significant loosening of the metal-ligand bond. Thus, early ligand loss seems most likely.

(10) Here are the calculated rate constants:

| $T/K$ | 300 | 350 | 400 | 450 | 500 |
|---|---|---|---|---|---|
| $k/\text{s}^{-1}$ | 0.35 | $1.2 \times 10^2$ | $1.0 \times 10^4$ | $3.3 \times 10^5$ | $5.3 \times 10^6$ |

The Arrhenius plot is perfectly linear (Figure K.10). From the slope and intercept of this graph, we calculate $E_a = 103 \text{ kJ mol}^{-1}$ and $A = 3.1 \times 10^{17} \text{ s}^{-1}$. On the other hand, using the enthalpy and entropy of activation, we get $E_a = 102 \text{ kJ mol}^{-1}$ and $A = 2.6 \times 10^{17} \text{ s}^{-1}$. These values are very similar to those we got from the graph. The small differences are attributable to a combination of factors: rounding in the intermediate steps of the calculation, and the somewhat arbitrary choice of 300 K for the transition-state theory calculations. The conclusion we can

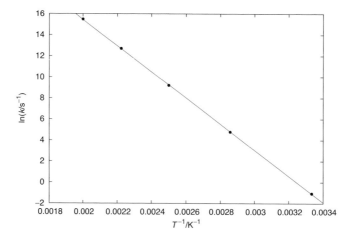

Figure K.10 Arrhenius plot for question 10.

reach is that the dependence of the pre-exponential factor and activation energy on $T$ predicted by transition-state theory are unimportant over temperature ranges of a few hundred kelvins.

## Exercise group 18.1

(1) $3.46 \times 10^{-10} \, m^2 \, s^{-1}$. This number is a little bit outside the experimental error of the measurement, but the agreement isn't bad considering how crude Stokes–Einstein theory is.

## Exercise group 18.2

(1) $2.2 \times 10^{10} \, L \, mol^{-1} s^{-1}$
(2) $1.76 \times 10^{11} \, L \, mol^{-1} s^{-1}$

## Review exercise group 18.3

(1) (a) If we assume that the radius of an iodine atom is half the radius of an iodine molecule, we get $D_I = 3.0 \times 10^{-9} \, m^2 \, s^{-1}$. (b) $R_{AB} = 2.1 \times 10^{-10} \, m$. This is similar to the bond length, suggesting that the reaction is diffusion-limited. The discrepancy could be due to any of a number of factors, the most obvious one being the limited precision of the rate constant.

(2) (a)   Substitute the Arrhenius-like equation for viscosity into the Stokes–Einstein equation to get

$$D_A = \frac{k_B T}{6\pi R_A A_\eta} \exp(-E_\eta / RT)$$

and similarly for $D_B$. Now applying Equation (18.9), we get, after a little simplification,

$$k_d = \frac{2RT}{A_\eta} \left( \frac{1}{R_A} + \frac{1}{R_B} \right) R_{AB} \exp(-E_\eta / RT).$$

Taking a natural logarithm and differentiating with respect to $T$, we get

$$\frac{d \ln k_d}{dT} = \frac{1}{T} + \frac{E_\eta}{RT^2}.$$

Applying the same operation to the Arrhenius equation, we get

$$\frac{d \ln k}{dT} = \frac{E_a}{RT^2}.$$

If we set these two results equal to each other, we get $E_a = E_\eta + RT$.

(b)  15.7 kJ mol$^{-1}$
(c)  18.2 kJ mol$^{-1}$

# Index